Cases in Laboratory Genetics and Genomics (LGG) Practice

Cases in Laboratory Genetics and Genomics (LGG) Practice

Xia Li
Genetics/Genomics Division, Sonora Quest Laboratories
and Department of Pathology, University of Arizona College of Medicine,
Phoenix, AZ, United States

ELSEVIER
ACADEMIC PRESS
An imprint of Elsevier

Academic Press is an imprint of Elsevier
125 London Wall, London EC2Y 5AS, United Kingdom
525 B Street, Suite 1650, San Diego, CA 92101, United States
50 Hampshire Street, 5th Floor, Cambridge, MA 02139, United States
The Boulevard, Langford Lane, Kidlington, Oxford OX5 1GB, United Kingdom

Notices

Knowledge and best practice in this field are constantly changing. As new research and experience broaden our understanding, changes in research methods, professional practices, or medical treatment may become necessary.

Practitioners and researchers must always rely on their own experience and knowledge in evaluating and using any information, methods, compounds, or experiments described herein. In using such information or methods they should be mindful of their own safety and the safety of others, including parties for whom they have a professional responsibility.

To the fullest extent of the law, neither the Publisher nor the authors, contributors, or editors, assume any liability for any injury and/or damage to persons or property as a matter of products liability, negligence or otherwise, or from any use or operation of any methods, products, instructions, or ideas contained in the material herein.

ISBN 978-0-323-99622-8

For information on all Academic Press publications
visit our website at https://www.elsevier.com/books-and-journals

Publisher: Mica H. Haley
Acquisitions Editor: Peter B. Linsley
Editorial Project Manager: Maria Elaine D. Desamero
Production Project Manager: Swapna Srinivasan
Cover Designer: Vicky Esser Pearson

Typeset by STRAIVE, India

Dedication

With great honor, this book is dedicated to our patients and the physicians who take care of them.

To my parents, who raised me to be independent; to my husband Jun Chen, who has supported my career from the day we met and who has moved all across the world with me; to my son Charles Chen, my daughter Carrie Chen, and my PhD mentor Professor Zhengcheng Geng, who gave me the inspiration and courage to write this book.

Contents

Contributors

Hanyin Cheng Hematologics, Inc., Seattle, WA, United States

Xia Li Sonora Quest Laboratories, Phoenix, AZ, United States

Guang Liu Sonora Quest Laboratories, Phoenix, AZ, United States

Mikako Warren Children's Hospital Los Angeles, Los Angeles, CA, United States

Dongbin Xu Hematologics, Inc., Seattle, WA, United States

Bo Yuan Department of Molecular and Human Genetics, Baylor College of Medicine, Houston, TX, United States

Foreword

In recent decades, a new paradigm has emerged for clinical laboratory testing, wherein classical cytogenetics and molecular genetics laboratories provide complementary services in both constitutional and cancer genomics. The need to train laboratory professionals in both areas was recognized by the American Board of Medical Genetics and Genomics (ABMGG), which now requires advanced training in both disciplines to be licensed in the Laboratory Genetics and Genomics subspecialty, replacing the individual board requirements and examinations in clinical cytogenetics and clinical molecular genetics. It is anticipated that similar training and licensure will ultimately be required for clinical laboratory staff members who are licensed at the state level. *Cases in Laboratory Genetics and Genomics (LGG) Practice* provides an excellent resource for students, laboratory trainees, practice providers, and genetic counselors who wish to become familiar with genomic testing in the laboratory, and who will use the case examples and figures in each chapter as a reference in their own practice of clinical genetics.

Cases in Laboratory Genetics and Genomics (LGG) Practice is written by Dr. Xia Li, who is board certified by the ABMGG in both clinical cytogenetics and molecular genetics. Dr. Li has directed laboratories in both the academic and private sectors and leveraged her vast clinical laboratory experience to compile the cases presented in the book. This volume contains 27 chapters and more than 100 clinical case examples that highlight the complementary cytogenetic and molecular genetic assays used as an aid in the diagnosis, prognosis, and therapy for patients with genomic disorders and cancer. Chapters 1 through 9 describe the background and approaches that are employed for patients with multiple congenital anomalies and developmental delay, sex chromosome abnormalities, imprinting disorders, and overgrowth syndromes, as well as contiguous gene deletion syndromes. Chapter 10 is focused on thrombosis and Chapter 11 introduces the topic of pharmacogenomics, a relatively new area for clinical testing. Chapters 12 through 27 encompass a variety of hematologic neoplasms, common solid tumors typically seen in the adult population, such as lung and colon cancer, as well as a chapter on pediatric solid tumors.

Each chapter begins with an overview of the topic, followed by at least one representative case example. For each case, a summary of the presenting features of the patient, a list of the tests that were performed, the results and integrated interpretation, and recommendations for future testing are provided. The most useful components of the case examples are the figures, which include representative karyotypes, fluorescence in situ hybridization images using a variety of DNA probes, chromosomal microarrays, and next-generation sequencing profiles. The figures will serve as an excellent teaching and reference set for laboratory staff as well as laboratory directors who are not familiar with the entities described in the book. *Cases in Laboratory Genetics and Genomics (LGG) Practice* will be a valuable study guide for pathologists specializing in molecular pathology, medical geneticists, genetic counselors, and laboratory geneticists.

Jaclyn A. Biegel

Preface

Genetic testing has entered a new era, in which personalized, predictive, and preventive medicine has become routine practice in the healthcare industry. As a result, the demand for qualified professionals in cytogenetics and molecular genetics has grown rapidly in clinical diagnostics laboratories to meet this market shift.

When I was a trainee of the American Board of Medical Genetics and Genomics (ABMGG) at Children's Hospital of Philadelphia, my mentor Dr. Nancy Spinner asked the fellows to present a case every Friday. I had the opportunity to review karyotype, fluorescence in situ hybridization (FISH), and microarray results for many cases and to assess if these test results were concordant in making the correct diagnosis for the patient. In the last 15 years, I have been the director of several different laboratories, all of which provide full services with cytogenetics and/or molecular genetics. This approach of reviewing multiple complementary tests for a patient has helped tremendously in making accurate diagnoses in my daily practice and has provided valuable information to clinicians for their patient care.

The primary audiences for this book are trainees or fellows in the Laboratory Genetics and Genomics (LGG) program organized through ABMGG, junior directors starting their first job, or directors managing both cytogenetics and molecular genetics labs. The book illustrates cases of real patients who we often see in our daily practice. It will help audiences to understand genetic abnormalities from an integrative viewpoint. Many laboratories may only perform some of these tests. Trainees and fellows may not have an opportunity to see cases with multiple genetic approaches. This book includes rare cases such as molar pregnancies, consanguinity, pharmacogenomics, and pediatric solid tumors, and many of these integrate chromosome analysis, FISH, microarray, Polymerase chain reaction (PCR), and next-generation sequencing (NGS) for both constitutional and oncology cases.

Finally, I sincerely hope that this book will be valuable for lab professionals in daily case analysis at the clinical lab and in preparing for the ABMGG board and American Society for Clinical Pathology (ASCP) technologist examinations. Ultimately, this book will help clinicians to understand the test results well so that they can make better decisions in caring for or treating their patients.

Xia Li

GENETICS/GENOMICS DIVISION, SONORA QUEST LABORATORIES AND DEPARTMENT OF PATHOLOGY, UNIVERSITY OF ARIZONA COLLEGE OF MEDICINE, PHOENIX, AZ, UNITED STATES

Acknowledgments

It has truly been a privilege and an honor to write this book. My husband, Jun Chen, always told me during this process that "this is more than just a book; it is the dedication of your career and passion for your patients." This makes me believe that all my time and efforts have been worth it.

I would like to thank:

- Dr. Gary Beauchamp, who accepted me as a postdoctoral fellow in 1998 so that I could start my life and career in the United States. He also supported the time I spent on cytogenetics training to begin my career in the clinical diagnostics field.
- Drs. Nancy Spinner, Ian Krantz, and Jaclyn Biegel, for accepting me as a trainee in the cytogenetics lab at CHOP. After finishing training, I became a lab director at AmeriPath Northeast.
- Dr. Xiao-Xiang Zhang, for being my long-term peer mentor since I started my training in 2008, and for guiding me in my career development over the last 15 years.
- everyone in the Genetics/Genomics Division at Sonora Quest Laboratories for delivering high-quality test results, making me proud that we have done our best for the patients.
- Drs. Zunyan Dai, Aurelia Meloni-Ehrig, Lisa Smith, and Liwen Lai, as well as Roba Eltalmas and Tracy Richard, for valuable inputs on the interpretations of the cases.
- Dr. Gordana Raca, for providing CMA images in Chapter 27.

A special thanks goes to Janet Orton for being the best partner one could ask for in managing our lab and for proofreading the manuscript.

Finally, I thank my son, Charles Chen, for editing the images and proofreading the entire manuscript, and Sonya Engle for her support in writing this book.

PART I

Inborn diseases

1

Multiple congenital anomalies and developmental delay

Xia Li and Guang Liu
SONORA QUEST LABORATORIES, PHOENIX, AZ, UNITED STATES

Background

Multiple congenital anomalies (MCAs) refer to two or more unrelated major structural malformations that cannot be explained by a known syndrome or sequence. Approximately 75% of babies have congenital anomalies as isolated incidents, and the remaining 25% have more than one major anomaly (https://www.cdc.gov/ncbddd/birthdefects/surveillancemanual/facilitators-guide/module-5/mod5-4.html). An estimated 295,000 newborns worldwide die within 28 days of birth every year due to congenital anomalies. Beyond mortalities, congenital anomalies are also significant contributors to infant and childhood chronic illnesses and disabilities. Although congenital anomalies may be the result of one or more genetic, infectious, nutritional, or environmental factors, it is often difficult to identify the exact causes. Among these factors, approximately 25% might be genetic [1].

Developmental delay is often present in patients with MCAs. Genetic factors are responsible for up to 40% of developmental delay cases such as global developmental delay/intellectual disability (GDD/ID). According to European guidelines on this type of disease, genetic testing is becoming a standardized diagnostic practice [2,3]. To investigate if the causes are indeed genetic, physicians usually order several genetic tests depending on the clinical presentations of the patients. The most common genetic tests used for this purpose include karyotyping, fluorescence in situ hybridization (FISH), chromosome microarray (CMA), single-gene sequencing, or next-generation sequencing (NGS). Each test can answer certain genetic questions, but not all of them. In the clinical cases illustrated below, some of the genetic testing mentioned above will be described in detail.

Cases in Laboratory Genetics and Genomics (LGG) Practice. https://doi.org/10.1016/B978-0-323-99622-8.00010-1

Case 1.1 Multiple congenital anomalies caused by an unbalanced translocation

Clinical indication

A 1-month-old baby girl presented with hypotonia and feeding difficulties and was admitted for further workup and swallow study to assess for micro-aspiration. Physical examination showed that the baby had a prominent forehead with frontal bossing, low-set ears, and a broad nasal bridge concerning possible genetic syndrome. An MRI of the brain demonstrated immature myelination. Differential diagnosis includes genetic syndromes (Trisomy, Prader-Willi, and Achondroplasia), inborn errors of metabolism, congenital hypothyroidism, neuromuscular disorder, or cerebral palsy.

Test ordered

- Chromosome analysis: Routine blood
- FISH: Prader-Willi syndrome
- Chromosome microarray (CMA)

Laboratory test performed

Chromosome analysis, also known as karyotyping, was performed in this case initially. This is the process by which photographs of chromosomes are taken to determine the chromosome complement of an individual. Thus, the number of chromosomes and any abnormalities can be identified. "Karyotype" also refers to the complete set of chromosomes in an individual. To visualize the chromosomes, the G-banding (Giemsa banding) technique is used in cytogenetics laboratories to produce a visible karyotype by staining condensed chromosomes. The metaphase chromosomes are treated with trypsin to partially digest the chromosomes. Then they are stained with Giemsa stain. Heterochromatic regions (relatively gene-poor) stain darkly in G-banding. In contrast, less condensed chromatin (Euchromatin, relatively gene-rich) stains lightly. From the pattern of bands, the chromosome abnormalities including numerical and structural aberrations can be identified by experienced cytogenetics technologists.

Prader-Willi syndrome FISH Test was performed subsequently. A FISH probe specific to the q11.2 region on chromosome 15 (*SNRPN*, supplied by Abbott Molecular, Inc.) is utilized. Ten metaphase and 50 interphase cells are analyzed for detection of the deletion. The analysis is also able to rule out a duplication for this locus, which is associated with autism in some individuals.

Next, the CMA test was conducted on this patient to verify the copy number changes associated with chromosome rearrangement. CMA is the assay designed to detect multiple recurring submicroscopic chromosomal aberrations. This technology can show genomic imbalances in up to 20% of patients with global developmental delay, intellectual disability, and multiple congenital anomalies. This assay will detect copy number variations (CNVs) only for genomic sequences represented in the array. CNVs identified as known variants in the general population and reported in the publicly available database of genomic variants will not be reported (www.dgv.tcag.ca). This assay will not detect balanced rearrangements (e.g., translocations and inversions).

Test results

Chromosome analysis on metaphases examined showed an unbalanced translocation involving the short arms of chromosomes 3 and 7, resulting in partial deletion of 7p and partial duplication of 3p (Fig. 1.1.1). These genetic alterations could be associated with congenital anomalies and/or developmental delays.

FISH for Prader-Willi syndrome: A fluorescent molecular probe (Abbott Molecular, Inc.), which is localized to the *SNRPN* gene region (15q11-q13) was hybridized for metaphase and interphase preparations. The fluorescence signal pattern was normal in the 10 metaphase cells and 50 interphase nuclei examined in this study. There was no evidence of a deletion in the Prader-Willi region on chromosome 15 using the *SNRPN* probe (Fig. 1.1.2).

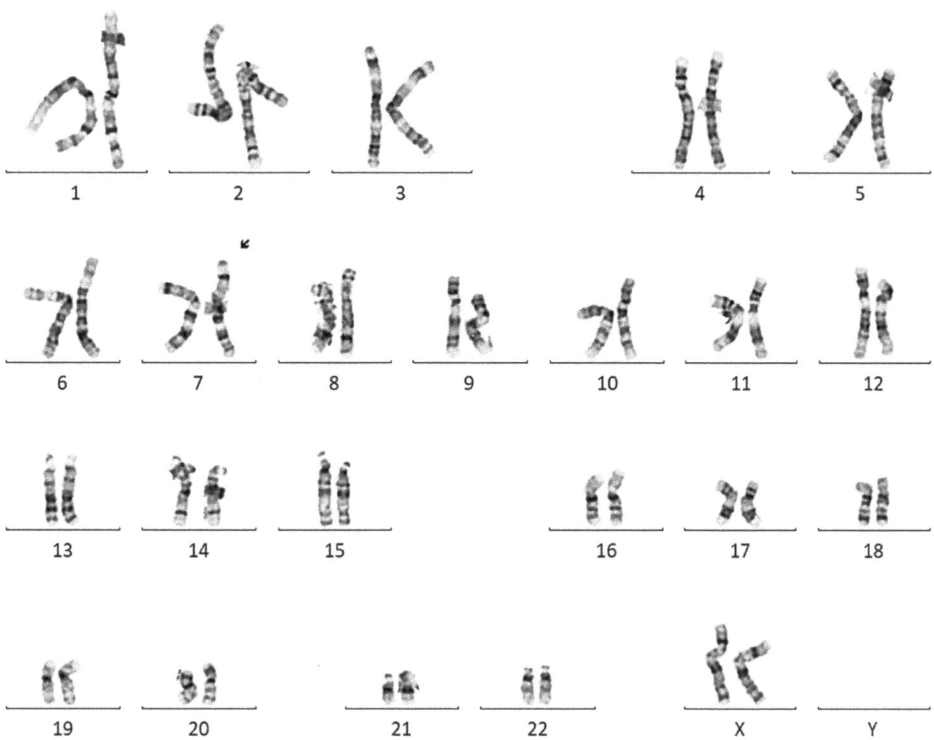

FIG. 1.1.1 The karyotype of the patient showed an unbalanced translocation between chromosomes 3 and 7. ISCN: 46, XX,der(7)t(3;7)(p26;p22)

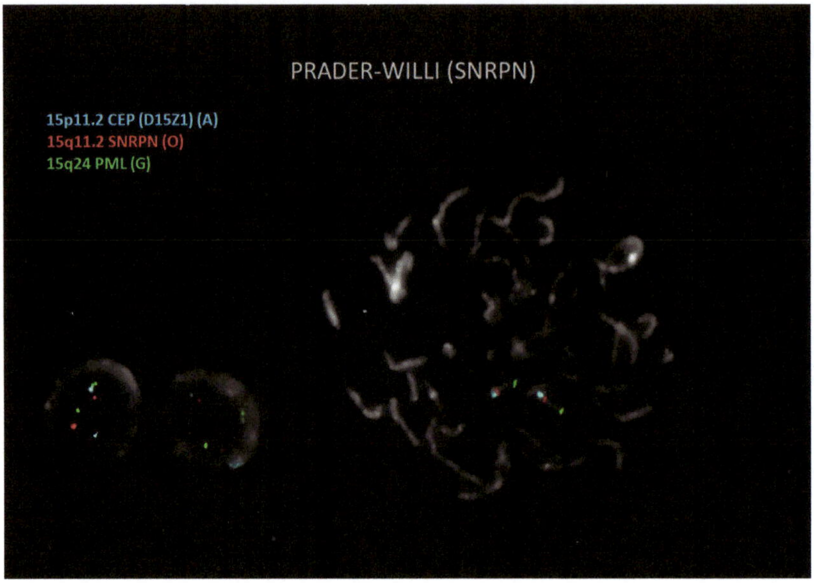

FIG. 1.1.2 FISH with probe *SNRPN* was performed on metaphase and interphase cells. All cells showed normal signal patterns. *PML* probe (15q24) is shown in *green*, *SNRPN* probe (15q11.2) is shown in *orange*, and *CEP15*(D15Z1) is shown in *aqua*. ISCN: ish 15q11-q13(SNRPNx2) [10].nuc ish(SNRPNx2)[50].

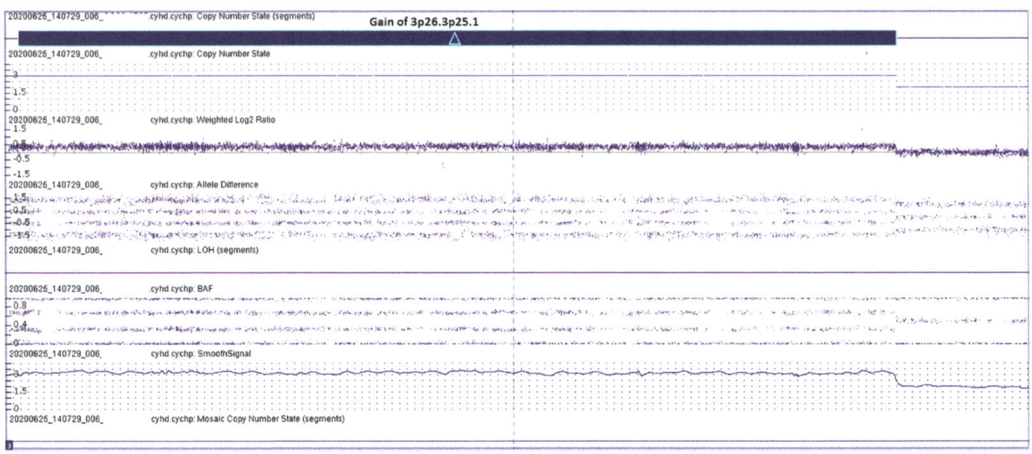

FIG. 1.1.3 The copy number state, Log_2R ratio, and B allele difference for 3q26.3 duplication. ISCN: arr[GRCh37] 3p26.3p25.1(232,786_14,264,680)x3

CMA analysis revealed a copy number gain of 14.0 Mb of DNA from chromosome 3 at band 3p26.3p25.1, encompassing 63 OMIM genes. CMA also revealed a copy number loss of 2.2 Mb of DNA from chromosome 7 at band 7p22.3, encompassing 18 OMIM genes. Both copy number variations are pathogenic with clinical significance (Figs. 1.1.3 and 1.1.4).

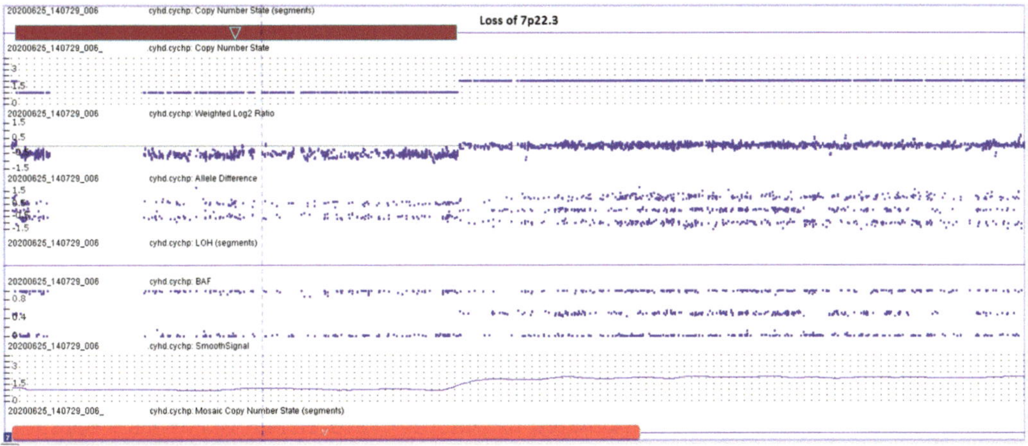

FIG. 1.1.4 The copy number state, Log_2R ratio, and B allele difference for 7q22 deletion. ISCN: arr[GRCh37] 7p22.3 (65,577_2,278,273)x1

Results with interpretations

Chromosome 3p duplication is a chromosome abnormality that occurs when there is an extra copy of genetic material on the short arm (p) of chromosome 3. The severity of the condition and the signs and symptoms depend on the size and location of the duplication, as well as which genes are involved. Features that often occur in people with chromosome 3p duplication include developmental delay, intellectual disability, behavioral problems, and distinctive facial features. Chromosome 3p duplication can be de novo or inherited from a parent with a balanced translocation. Treatment is based on the signs and symptoms present in each person (https://rarediseases.info.nih.gov/diseases/5343/chromosome-3p-duplication).

The symptoms and physical findings associated with Partial Monosomy 7p may vary in range and severity from case to case. Partial Monosomy 7p is also commonly characterized by premature closure of one or more fibrous joints (cranial sutures) between bones in the skull (craniosynostosis), potentially resulting in deformity of the skull and an unusually shaped head. The degree and severity of craniosynostosis may be variable, depending on the specific cranial sutures involved. Partial Monosomy 7p may be associated with various types of craniosynostosis, such as trigonocephaly or turricephaly. In some affected individuals, Partial Monosomy 7p may be associated with additional craniofacial abnormalities. Such features may include palpebral fissures, epicanthal folds, ptosis, dysplastic ears, a sunken nasal bridge, and/or other abnormalities. In some instances, Partial Monosomy 7p may also be characterized by musculoskeletal abnormalities. Reported features have included camptodactyly, unusually short hands, abnormalities of the thumbs, a deformity in which the top part of the foot is elevated, and the heel turned outward ("clubfoot" (i.e., talipes calcaneovalgus)), limited range of movement of certain joints, and/or other findings. Some affected individuals may also have various congenital heart defects. In addition, some affected individuals may have varying degrees of mental retardation and delays in the acquisition of skills requiring the coordination of mental and motor activities (psychomotor delays) (https://rarediseases.org/rare-diseases/chromosome-7-partial-monosomy-7p).

FISH testing using the *SNRPN* probe was negative and excluded the patient from Prader-Willi syndrome. The CMA results are concordant with the findings of the patient's chromosome analysis. The clinical presentations in this patient are well explained by Partial Monosomy 7p and 3p duplication identified by CMA and karyotype.

Future testing and recommendations

Parental chromosome analysis is indicated to rule out if either parent carries a balanced chromosome rearrangement that led to the unbalanced chromosome complement in the proband. Family members who are close relatives should be tested for a balanced translocation as well. If one of the parents carries the translocation, any future pregnancy should be tested prenatally. Genetic counseling is recommended.

Case 1.2 Recombinant chromosome 8 syndrome

Clinical indication

A 7-day-old baby girl was in the hospital's NICU with multiple congenital anomalies, atrial septal defect (ASD), congenital dislocation of the right knee, apnea, and seizure. Physical examination showed that the baby had deeply set eyes, a broad upturned nose, micrognathia, and a long slim body with a narrow chest, shoulders, and pelvis. Clinical suspicion was trisomy 8 with mosaicism.

Test ordered

- Chromosome analysis: Routine blood
- Chromosome microarray analysis (CMA)

Laboratory test performed

Initially, only chromosome analysis was ordered for this baby. Chromosome analysis identified an abnormal female chromosome complement with a derivative chromosome 8 resulting from a deletion of part of the short arm of chromosome 8 and a duplication of part of the long arm of chromosome 8. We called the physician and recommended CMA analysis of this baby for further characterization of this rearrangement and the specific gene regions involved. Then CMA was ordered and performed for this baby. The detailed chromosome analysis and CMA methods are described in this chapter, Case 1.1.

Test results

Chromosome analysis revealed an abnormal female chromosome complement in all cells examined with a derivative chromosome 8 generated by a deletion from 8pter to 8p23.1 and a duplication from 8q22.1 to 8qter (Fig. 1.2.1).

CMA analysis revealed a copy number loss of 6.8 Mb of DNA from chromosome 8 at band 8p23.3p23.1, encompassing 16 OMIM genes. CMA also revealed a copy number gain of 47.9 Mb of DNA from chromosome 8 at band 8q22.1q24.3, encompassing 189 OMIM genes. Both copy number variations are pathogenic with clinical significance (Fig. 1.2.2).

Results with interpretations

The results from karyotyping and CMA were concordant and consistent with a diagnosis of Recombinant 8 syndrome. The features of chromosome 8p deletion and 8q duplication may vary greatly in range and severity dependent upon the extent of the deletion and duplication. The symptoms and findings in patients with a chromosome 8q duplication

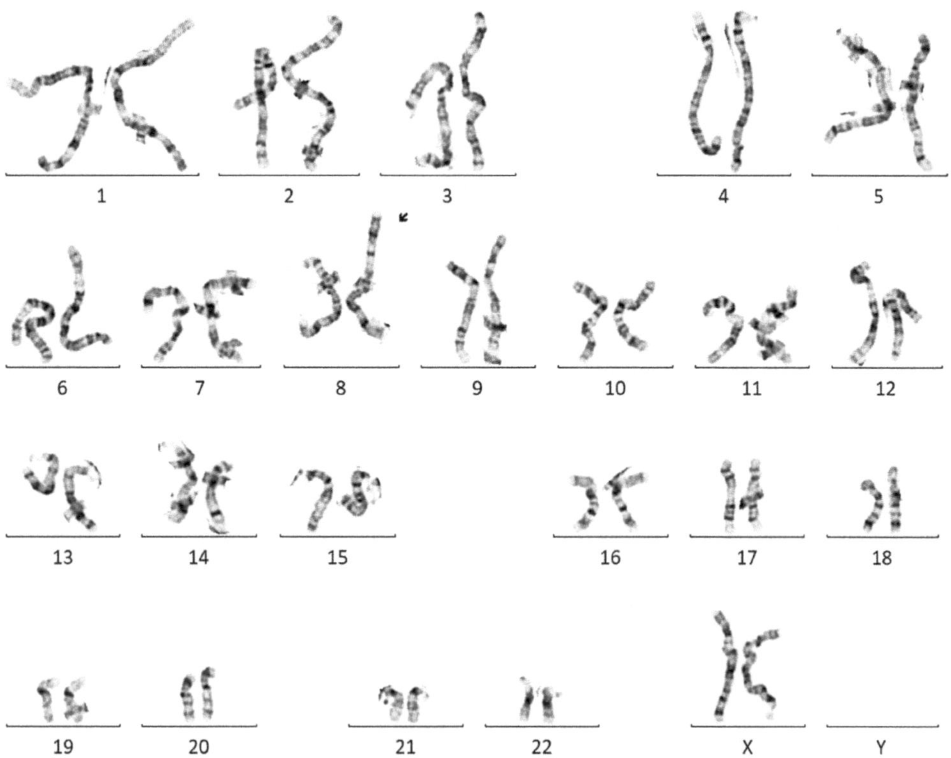

FIG. 1.2.1 The karyotype of the patient revealed a derivative chromosome 8 resulting from a deletion of 8pter to 8p23.1 and a duplication of 8q22.1 to 8qter. ISCN: 46,XX,der(8)(qter->q22.1::p23.1->qter)

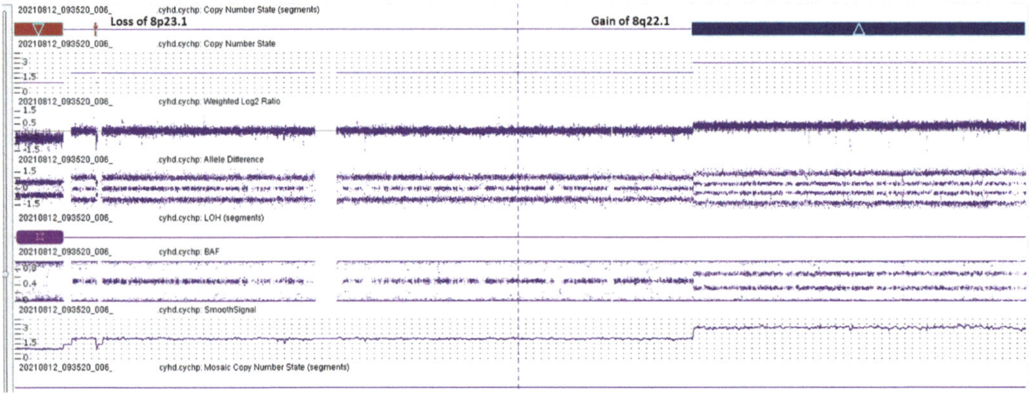

FIG. 1.2.2 The copy number state, Log2R ratio, and B allele difference for 8p23 deletion and 8q22.1 duplication. ISCN: arr[hg19] 8p23.1(11,652,034-11,898,980)x1, arr[hg19] 8q22.1q24.3(98,399,098-146,295,771)x3

may include developmental delay, seizure, intellectual disability, congenital heart defects, skeletal abnormalities, and distinctive facial features, which are consistent with this patient's clinical manifestation and overlap with trisomy 8 mosaicism [4] (https://rarediseases.org/rare-diseases/chromosome-8-monosomy-8p/; https://rarediseases.info.nih.gov/diseases/5359/mosaic-trisomy-8). This was the reason why the clinical suspicion of trisomy 8 mosaicism for this patient was indicated. The common features of chromosome 8p deletion include growth deficiency, mental retardation, malformations of the skull and facial region, and cardiac abnormalities. Some of these features were observed from the patient.

The pathogenic loss and gain seen on chromosome 8 are consistent with Recombinant Chromosome 8 syndrome (OMIM #179613). Alternative titles include Rec(8) Syndrome and San Luis Valley Syndrome. Recombinant chromosome 8 syndrome is characterized by duplication of 8q22.1-qter and deletion of 8pter-p23.1. Rec(8) is derived from the recombination of a parental pericentric inversion of chromosome 8. It is a chromosomal disorder found among individuals of Hispanic descent with ancestry from the San Luis Valley of southern Colorado and northern New Mexico. Affected individuals typically have impaired intellectual development, congenital cardiac defects, seizures, a characteristic facial appearance with hypertelorism, thin upper lip, anteverted nares, wide face, abnormal hair whorl, and other manifestations [5] (https://omim.org/entry/179613). This patient had congenital heart defects and seizures. A thorough clinical assessment of this patient and genetic counseling are advised.

Future testing and recommendations

Since the results from chromosome analysis and CMA may indicate a duplication/deletion of the recombinant chromosome from one of the parents carrying a pericentric inversion, parental chromosome analysis is highly recommended. An inversion carrier is generally phenotypically normal because it is a balanced rearrangement. However, an inversion carrier is at risk of producing abnormal gametes that may lead to unbalanced offspring. When an inversion is present, a loop needs to form to allow alignment and pairing of homologous segments of the normal and inverted chromosomes in meiosis I. When recombination occurs within the loop, it can lead to the production of unbalanced gametes; gametes with balanced chromosome complements (either normal or possessing the inversion) and gametes with unbalanced complements are formed. Therefore, a pericentric inversion can lead to the production of unbalanced gametes with both duplication and deletion of chromosome segments as observed in this patient [6].

Case 1.3 Multiple congenital anomalies caused by an unbalanced translocation and a deletion

Clinical indication

A 3-day-old baby girl presented with Intrauterine growth restriction (IUGR), preterm of 36 completed weeks of gestation, bacterial sepsis, transient tachypnea, hypotonia, and feeding difficulties. A physical exam showed microcephaly, a small jaw, wide-set eyes, and suspicion of an atrial septal defect (ASD). A follow-up echocardiogram confirmed ASD.

Test ordered

- Chromosome analysis: Routine blood
- Chromosome microarray analysis (CMA)

Laboratory test performed

This newborn baby was in the NICU. Initially, only chromosome analysis was ordered. Chromosome analysis identified a translocation between the long arms of chromosomes 2 and 13, and an interstitial deletion within the long arm of the same chromosome 13. We communicated with the physician and recommended chromosome microarray analysis on this baby for further characterization of these rearrangements and the specific gene regions involved. Then, CMA was ordered and performed on this newborn. The detailed chromosome analysis and CMA methods are described in this chapter, Case 1.1.

Test results

Chromosome analysis revealed an abnormal female chromosome complement in all cells examined with an unbalanced reciprocal translocation involving the long arms of chromosomes 2q31 and 13q14, and an interstitial deletion within the long arm of the same chromosome 13 between 13q12 and 13q14 (Fig. 1.3.1).

Chromosomal Microarray Analysis (CMA) revealed a copy number loss of 5.3 Mb of DNA from chromosome 2 at band 2q31.1q31.2, encompassing 33 OMIM genes

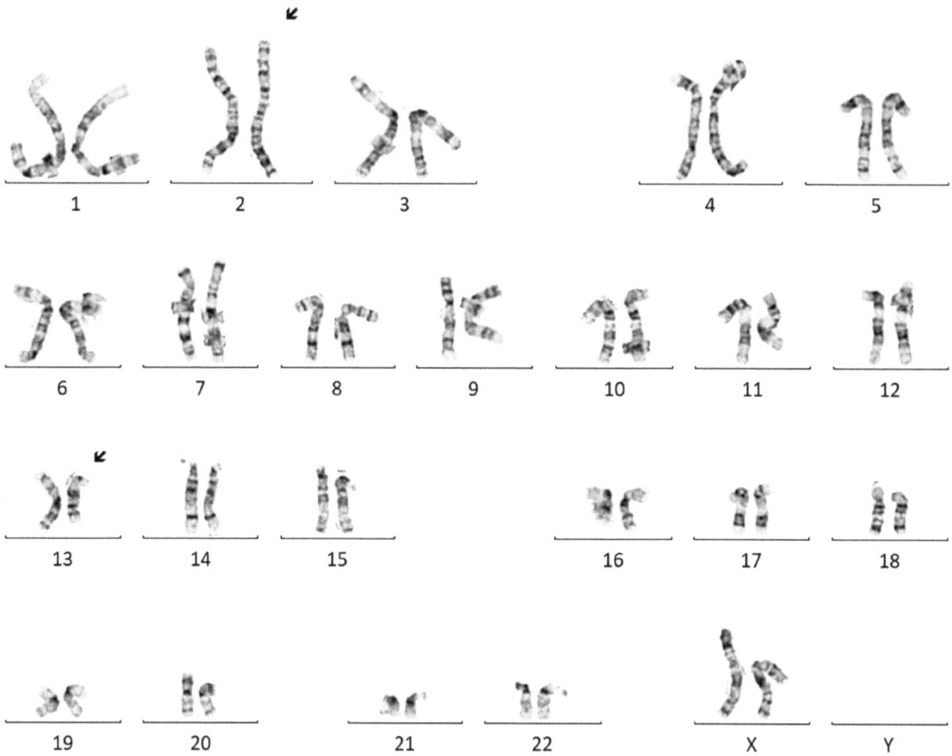

FIG. 1.3.1 The karyotype of the patient revealed an unbalanced reciprocal translocation between 2q31 and 13q14, and an interstitial deletion from the same chromosome 13 between 13q12 and 13q14. ISCN: 46,XX,der(2)t(2;13)(q31;q14),der(13)del(13)(q12q14)t(2;13)(q31;q14)

(Fig. 1.3.2). CMA also detected another copy number loss of 20.5 Mb of DNA from chromosome 13 at band 13q13.2q14.3, encompassing 92 OMIM genes. Both two deletions were pathogenic with clinical significance (Fig. 1.3.3).

Results with interpretations

The clinical phenotypes of the deletions greatly vary depending on the size and the location of the deleted region. The deletion of 5.3 Mb of DNA from chromosome 2 at band 2q31.1q31.2 was identified by CMA but not by chromosome analysis due to the size of the deletion and a translocation involving the same region. However, the reciprocal translocation was only detected by chromosome analysis, not by CMA. The chromosome 2q31 deletion syndrome may consist of limb anomalies and internal organ anomalies such as heart defects and ocular anomalies. Hemizygosity for *HOXD13* and *EVX2* genes were thought to cause the observed skeletal defects [7,8]. The deletion of 20.5 Mb of DNA from

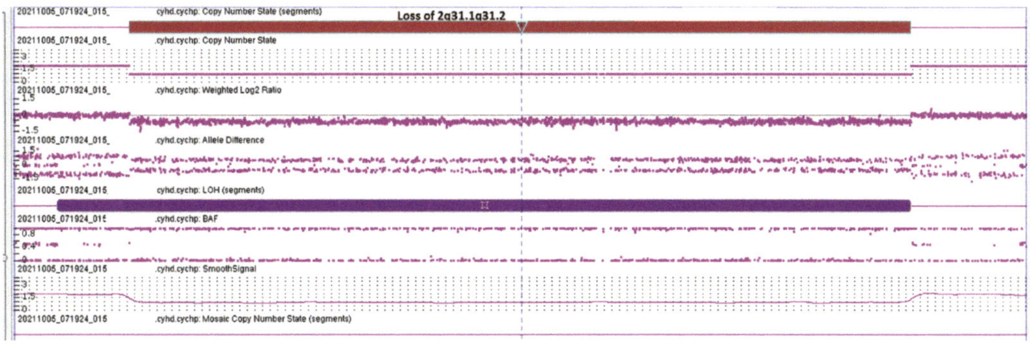

FIG. 1.3.2 The copy number state, Log2R ratio, and B allele difference for 2q deletion. ISCN: arr[hg19] 2q31.1q31.2 (173,795,601-179,110,381)x1

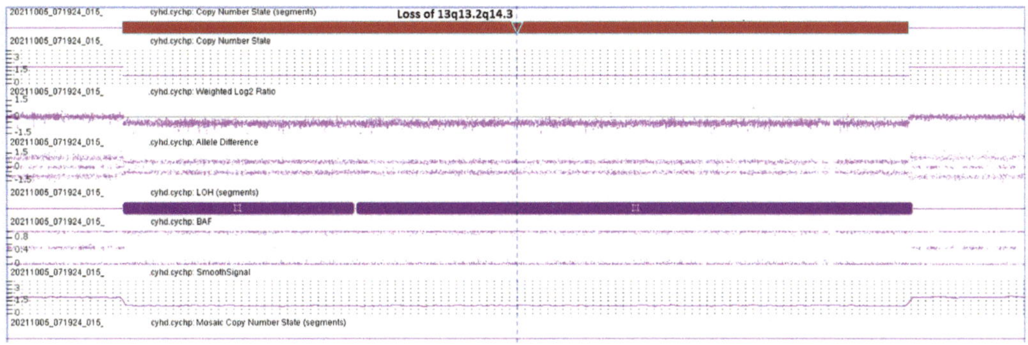

FIG. 1.3.3 The copy number state, Log2R ratio, and B allele difference for 13q deletion. ISCN: arr[hg19] 13q13.2q14.3 (34,702,245-55,163,462)x1

chromosome 13 at band 13q13.2q14.3 was detected by both chromosome analysis and CMA. The main clinical features of interstitial deletions of the long arm of chromosome 13 (13q) may include developmental delay, intellectual disability, behavioral problems, distinctive facial features, and increased susceptibility to tumors. A heterozygous germline mutation on one allele and a somatic mutation on the other allele of the *RB1* gene are associated with somatic mutation or autosomal dominant retinoblastoma (RB). Retinoblastoma is an embryonic malignant neoplasm of retinal origin. It almost always presents in early childhood and is often bilateral [9] (https://www.omim.org/entry/613884). Spontaneous regression occurs in some cases [10].

Future testing and recommendations

Since chromosome analysis and CMA identified an unbalanced translocation, and the deletions of chromosomes 2q and 13q, parental chromosome analysis is indicated to help clarify the origin of these abnormalities and assess the risk in future pregnancies. The deletion of chromosome 13q13.2q14.3 overlaps the *RB1* tumor suppressor gene. Germline heterozygous mutations and deletions in *RB1* cause a hereditary predisposition to retinoblastoma and other tumors such as pinealomas and other extraocular primary neoplasms. *RB1*-associated tumors arise when a second somatic *RB1* mutation leads to two nonfunctional *RB1* genes. A complete clinical examination and serial eye exams by an ophthalmologist are recommended [11].

Case 1.4 Diamond-Blackfan anemia

Clinical indication

A 5-month-old female presented with failure to gain weight and congenital malformations. Her hemoglobin was 7.1, better after the blood transfusion. She had myelodysplasia. The morphologic appearance of the marrow was very typical of Diamond-Blackfan anemia (DBA).

Test ordered

- Chromosome analysis: Routine blood
- FISH: MDS panel
- NGS bone marrow failure (BMF) and dyskeratosis congenita (DKC) panels
- Chromosome microarray (CMA)

Laboratory test performed

Methods for karyotyping, FISH, and microarray were described previously in Case 1.1. NGS BMF and DKC panels were performed according to [12,13].

Test results

Chromosome analysis on the bone marrow exhibited a normal female karyotype (Fig. 1.4.1).

FISH using probes for CEN8, 5q/-5, 7q/-7, and 20q was performed. The results were negative for all probes examined (Fig. 1.4.2).

NGS results for BMF and DKC panels were all negative. No mutations was identified.

Microarray analysis showed a deletion of 745 Kb from 3q29, including *RPL35A* (Fig. 1.4.3).

Results with interpretations

DBA is a disorder that primarily affects the bone marrow. In DBA patients, the bone marrow malfunctions and fails to make enough red blood cells, which carry oxygen to the body's tissues [14]. DBA is characterized by a profound isolated normochromic and usually macrocytic anemia with normal leukocytes and platelets, 50% have congenital

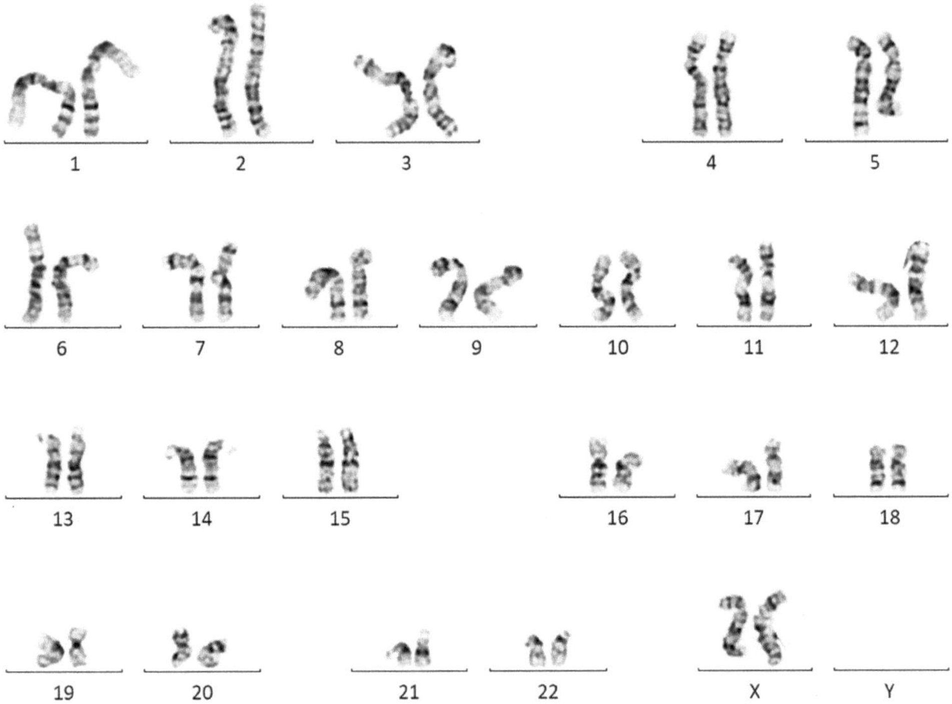

FIG. 1.4.1 The karyotype of the bone marrow showed a normal female karyotype. ISCN: 46,XX[20]

malformations, and 30% have growth retardation. The physical abnormalities include microcephaly, hypertelorism, ptosis, broad flat bridge of the nose, and micrognathia. Some have cleft lips or palate, webbed necks, absent thumbs, etc. The hematologic complications occur in 90% of affected individuals during the first year of life (median age of onset: 2 months). Eventually, 40% of affected individuals are corticosteroid-dependent, 40% are transfusion-dependent, and 20% go into remission. The phenotypic spectrum ranges from a mild form to a severe form of fetal anemia resulting in nonimmune hydrops fetalis. DBA is associated with an increased risk for Acute Myelogenous Leukemia (AML), Myelodysplastic Neoplasms (MDS), and solid tumors including osteogenic sarcoma [15,16]. The genetic testing for this patient showed normal results from karyotype, FISH MDS panel, NGS BMF, and DKC panels. However, microarray results were abnormal and revealed a 745 Kb deletion of chromosome region on 3q29 including *RPL35A*. This gene is associated with DBA. There are more than 13 genes that are associated with DBA. *RPL35A* mutations or deletions account for 3% of DBA patients [17]. Most of the DBA-associated gene mutations are inherited in an autosomal dominant manner. *GATA1* mutations are inherited in an X-linked manner. About 40%–45% of individuals with autosomal dominant DBA have inherited the mutation from a parent; ~55%–60% have a de novo mutation.

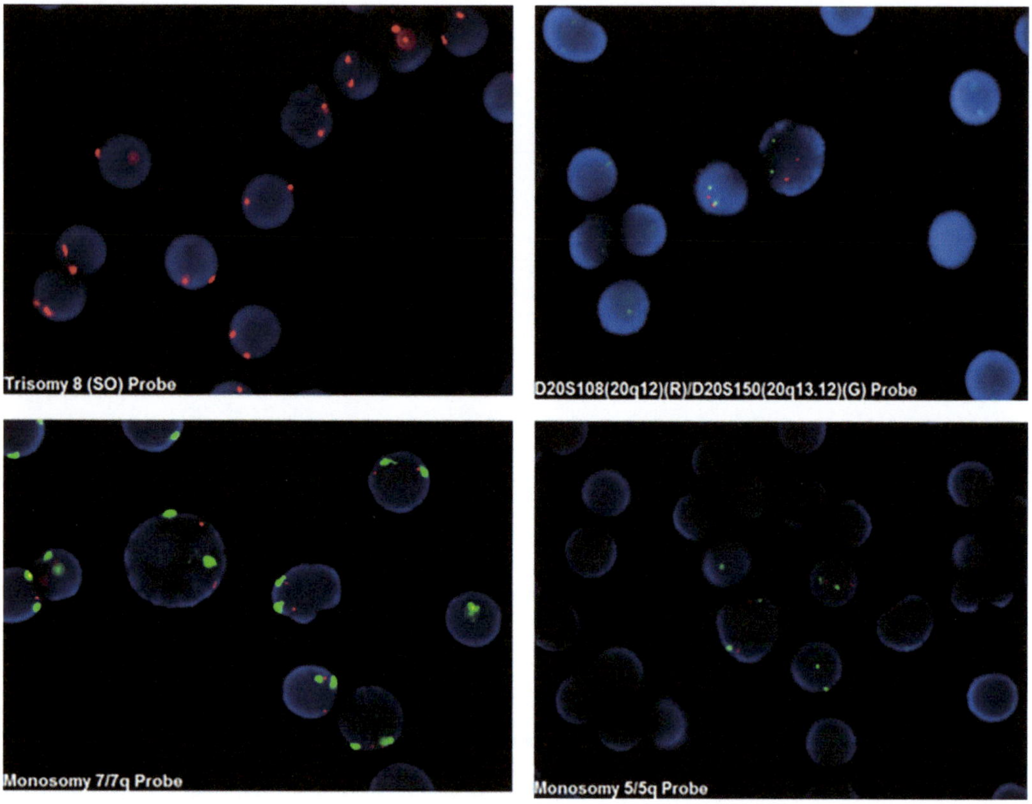

FIG. 1.4.2 FISH for the MDS panel was negative for trisomy 8, 5q/-5, 7q/-7, and 20q in all 200 cells examined. ISCN: nuc ish(D5S721,EGR1)x2[200/200],(CEP7,D7S486)x2[200/200],(D8Z2,D20S108)x2[200/200]

Future testing and recommendations

Family members who are at risk of inheriting this deletion of *RPL35A* should be tested by microarray. Genetic counseling is recommended.

Summary of key learning points

- Terminal deletion and duplication as identified by microarray usually indicate an unbalanced translocation between the two chromosomes involved.
- A duplication/deletion of a region on both ends of a single chromosome generally indicates a recombinant form of an inversion inherited from a parent.
- Identification of cryptical or subtle imbalance is critical for a phenotypically abnormal individual associated with a translocation. CMA is superior in identifying cryptic or submicroscopic copy number variations.
- Chromosome analysis is valuable in the detection and delineation of chromosomal structural rearrangements.

FIG. 1.4.3 Microarray showed a deletion of 745 Kb on 3q29 including the *RPL35A* gene. ISCN: arr[hg19] 3q29(197,093,758-197,838,262)x1

- Parental chromosome analysis is essential to rule out chromosome abnormalities from one of the parents, which is important for future risk assessment.
- People with DBA have an increased risk of developing AML, MDS, and other cancers. Half of them have multiple congenital anomalies.
- Mutations or deletions in more than 13 genes are associated with DBA. Most of these are inherited in an autosomal dominant manner, except for *GATA1*, which is inherited in an X-linked manner.

References

[1] R.E. Stevenson, Human Malformations and Related Anomalies, second ed., Oxford Monographs on Medical Genetics, Oxford University Press, Oxford; New York, 2006. xiv, 1495 p.

[2] D. Miclea, et al., Genetic testing in patients with global developmental delay/intellectual disabilities. A review, Clujul Med. 88 (3) (2015) 288–292.

[3] D. Miclea, et al., Diagnostic usefulness of MLPA techniques for recurrent copy number variants detection in global developmental delay/intellectual disability, Int. J. Gen. Med. 14 (2021) 4511–4515.

[4] M.C. Bonaglia, et al., A 2.3 Mb duplication of chromosome 8q24.3 associated with severe mental retardation and epilepsy detected by standard karyotype, Eur. J. Hum. Genet. 13 (5) (2005) 586–591.

[5] E. Sujansky, et al., Natural history of the recombinant (8) syndrome, Am. J. Med. Genet. 47 (4) (1993) 512–525.

[6] R.L. Nussbaum, R.R. McInnes, H.F. Willard, Thompson & Thompson Genetics in Medicine, eighth ed., Elsevier, Philadelphia, PA, 2016. xi, 546 pages.

[7] B. Dimitrov, et al., 2q31.1 microdeletion syndrome: redefining the associated clinical phenotype, J. Med. Genet. 48 (2) (2011) 98–104.

[8] S. Puvabanditsin, et al., 2q31.1 microdeletion syndrome: case report and literature review, Clin. Case Rep. 3 (6) (2015) 357–360.

[9] N. Rapini, et al., De novo 13q13.3-21.31 deletion involving RB1 gene in a patient with hemangioendothelioma of the liver, Ital. J. Pediatr. 40 (2014) 5.

[10] B.L. Gallie, et al., Retinoma: spontaneous regression of retinoblastoma or benign manifestation of the mutation? Br. J. Cancer 45 (4) (1982) 513–521.

[11] A.H. Skalet, et al., Screening children at risk for retinoblastoma: consensus report from the American association of ophthalmic oncologists and pathologists, Ophthalmology 125 (3) (2018) 453–458.

[12] L. Dyer, X. Li, J. Denton, B. Jones, E. Liston, D. Hilton, A. Mathur, K. Zhang, C.A. Valencia, Gene dosage defects in primary immunodeficiencies and related disorders: a pilot study, J. Transl. Genet. Genom. 1 (2017) 23–27.

[13] V. Gadoury-Levesque, et al., Frequency and spectrum of disease-causing variants in 1892 patients with suspected genetic HLH disorders, Blood Adv. 4 (12) (2020) 2578–2594.

[14] S. Ball, Diamond Blackfan anemia, Hematology Am. Soc. Hematol. Educ. Program 2011 (2011) 487–491.

[15] C. Sieff, Diamond-Blackfan anemia, in: M.P. Adam, et al. (Eds.), GeneReviews((R)), 1993. Seattle, WA.

[16] J. Boultwood, A. Pellagatti, J.S. Wainscoat, Haploinsufficiency of ribosomal proteins and p53 activation in anemia: diamond-Blackfan anemia and the 5q-syndrome, Adv. Biol. Regul. 52 (1) (2012) 196–203.

[17] I. Boria, et al., The ribosomal basis of diamond-Blackfan anemia: mutation and database update, Hum. Mutat. 31 (12) (2010) 1269–1279.

2

Molar pregnancy

Xia Li

SONORA QUEST LABORATORIES, PHOENIX, AZ, UNITED STATES

Background

A molar pregnancy is an abnormal form of pregnancy characterized by the abnormal growth of trophoblasts, in which a nonviable fertilized egg implants in the uterus and will fail to come to term. Therefore, it is called a "hydatidiform mole" [1,2]. Hydatidiform moles are a relatively common complication of pregnancy, making up 1 in 1500 pregnancies [3].

There are two types of hydatidiform moles: complete moles or partial moles. A complete mole is formed by a single or two sperm combining with an egg without a nucleus (DNA). The single sperm can reduplicate, forming a diploid with a "complete" 46-chromosome set. Among complete moles, 85% are 46,XX and 15% are 46,XY. In contrast, a partial mole occurs when a normal egg is fertilized by one or two sperm, which then reduplicates itself, yielding the genotypes 69,XXY, 69,XXX, 69,XYY (triploid), or 92, XXXY (tetraploid). Studies showed that 85% of triploid partial moles contain one haploid maternal set of chromosomes and two haploid paternal sets of chromosomes, which is known as diandric triploidy (paternally derived). The remaining 15% of triploids have two haploid maternal chromosomes and one haploid paternal chromosome. These are called digynic triploidy (maternally derived), which are not associated with molar changes [3].

Complete hydatidiform moles have a 2%–4% risk of developing into choriocarcinoma in Western countries and 10%–15% in Eastern countries, and a 15% risk of becoming an invasive mole. Incomplete moles can become invasive (<5% risk) but are not associated with choriocarcinoma. Complete moles account for 50% of all cases of choriocarcinoma [3].

The current clinical practice for the detection of hydatidiform moles includes morphological analysis with p57 immunostaining followed by ploidy testing with FISH and/or karyotype to identify complete mole versus partial mole. Single-nucleotide polymorphism (SNP) microarray can also detect complete mole and partial mole, in which some of the chromosomes will show loss of heterozygote if they are of monospermic origin [4,5]. The cases illustrated below are examples of using karyotypes to identify molar pregnancy.

Cases in Laboratory Genetics and Genomics (LGG) Practice. https://doi.org/10.1016/B978-0-323-99622-8.00014-9

Case 2.1 Complete mole

Clinical indication

A product of conception (POC) was received from a 35-year-old pregnant woman for a possible molar pregnancy by ultrasound and human chorionic gonadotropin (hCG) level.

Test ordered

- Chromosome analysis on POC

Laboratory test performed

Chromosome analysis on POC was performed (detailed method for karyotyping described in Chapter 1).

Test results

There were two balanced translocations between the long arms of chromosomes 2 and 13 with the same breakpoints (Fig. 2.1.1). Subsequently, parental chromosome analysis

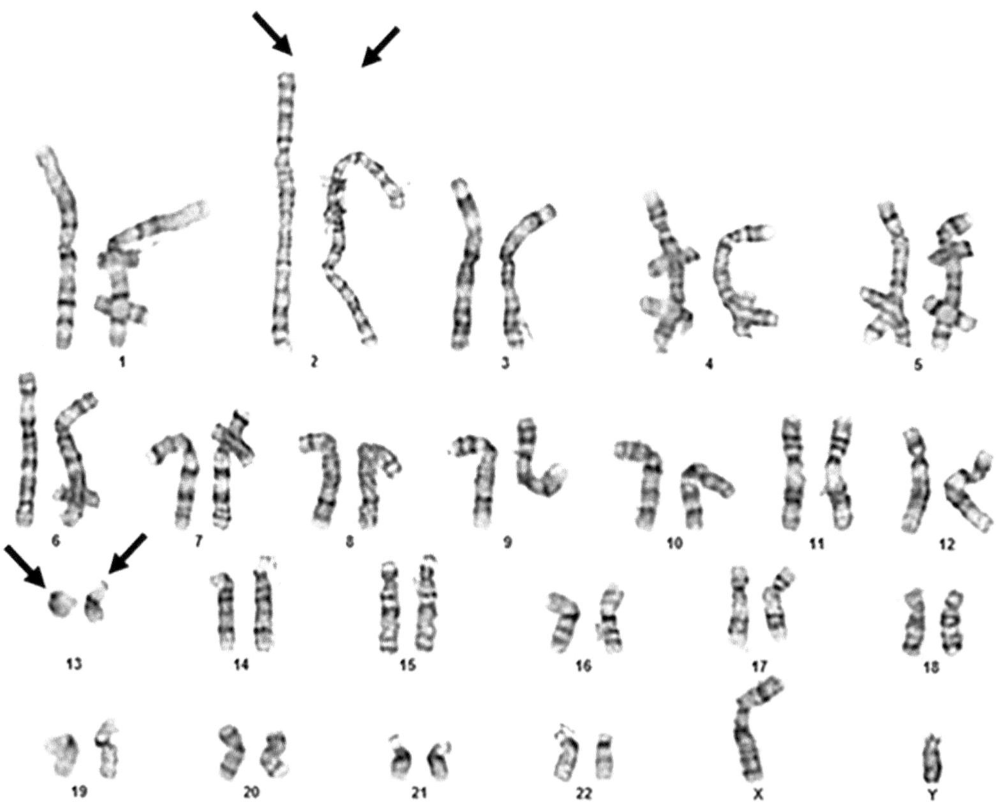

FIG. 2.1.1 Karyotype from the POC showed a balanced translocation between two copies of chromosomes 2 and 13. ISCN: 46,XY,t(2;13)(q36:q12.1)x2

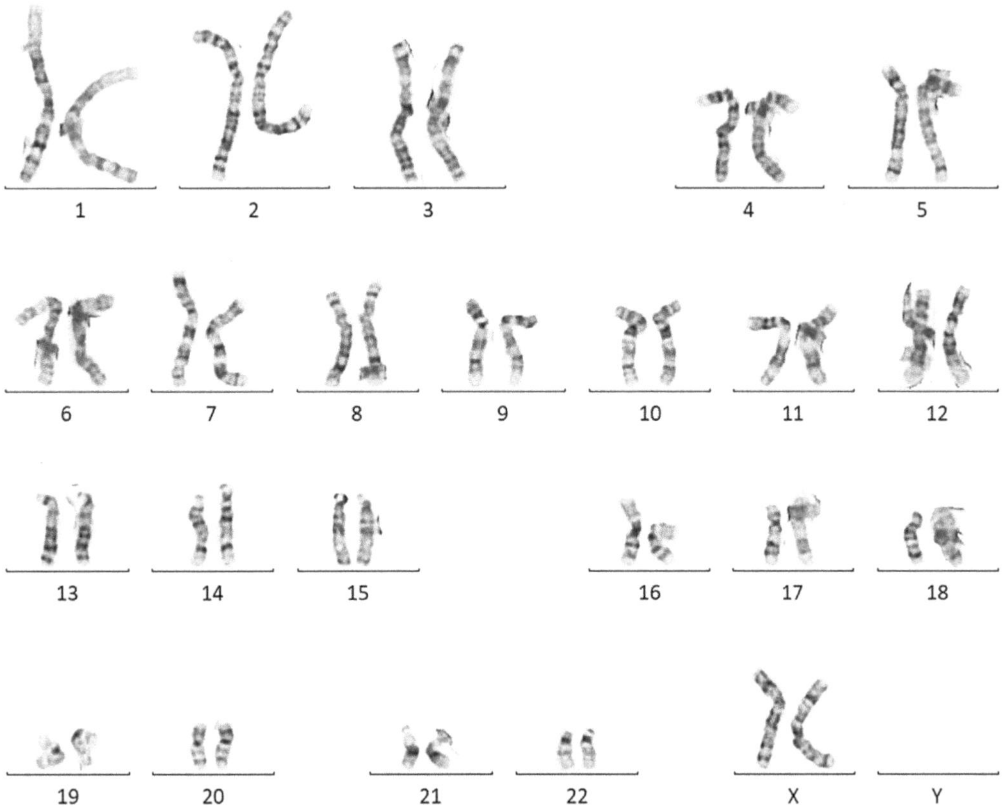

FIG. 2.1.2 Karyotype from the mother showed a normal chromosome complement 46,XX. ISCN: 46,XX

was recommended and performed. The mother's karyotype showed a normal chromosome complement 46,XX (Fig. 2.1.2). Father's karyotype revealed a balanced translocation between chromosomes 2 and 13 (Fig. 2.1.3).

Results with interpretations

The original POC showed a balanced translocation on two copies of chromosomes 2 and 13. The breakpoints were also the same between these two translocations. It is very unlikely that parents are consanguineous with the same breakpoints from the same translocations. Therefore, parental chromosome analysis was recommended. As a result, the mother showed a normal chromosome complement (46,XX), and the father revealed a balanced translocation [46,XY,t(2;13)]. It was concluded that the two derivative chromosomes were both from the father. The POC consisted of two sperms plus one empty egg, which was consistent with the diagnosis of a complete mole. These results were also concordant with the findings from the pathophysiology analysis.

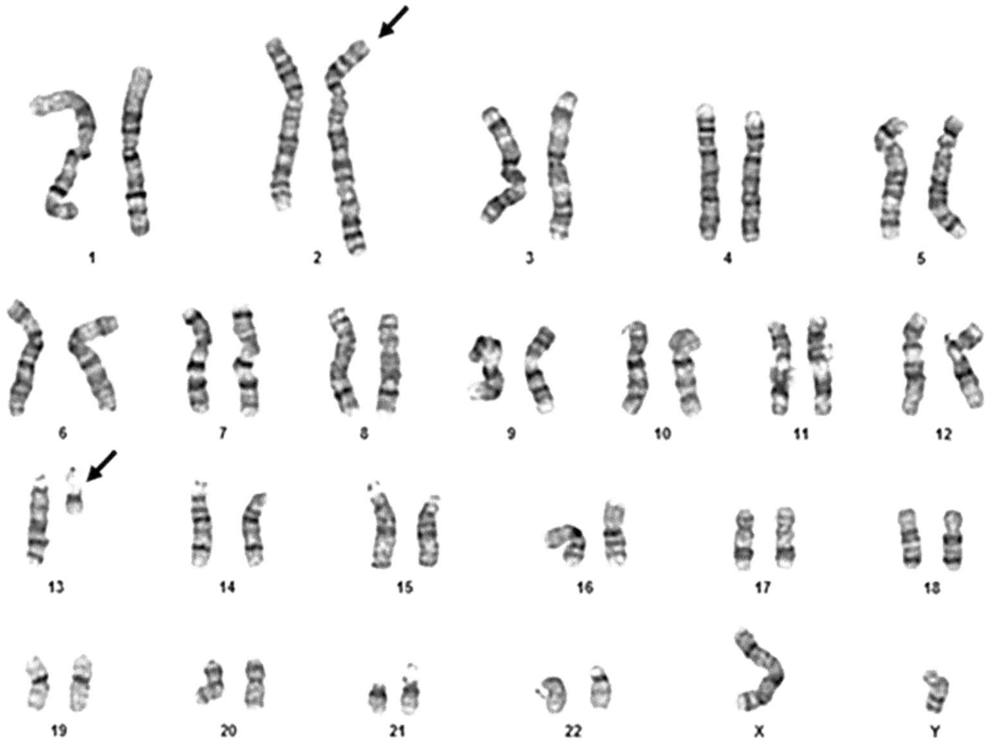

FIG. 2.1.3 Karyotype from the father showed a balanced translocation between chromosomes 2 and 13. ISCN: 46,XY,t(2;13)(q36:q12.1)

Future testing and recommendation

One study showed that the overall recurrence risk for molar pregnancy was 1.8% (1 in 55). Of 2578 complete moles, the subsequent pregnancy was affected by a hydatidiform mole in 27 (1.9%) cases, or a 20-fold increase compared with the background risk [6]. Prenatal genetic testing for future pregnancy and genetic counseling are recommended.

Case 2.2 Partial mole

Clinical indication

A 25-year-old female experienced vaginal bleeding and missed spontaneous abortion. Ultrasound examination showed embryonic demise. Genetic testing was recommended to rule out a partial mole.

Test ordered

- Chromosome analysis on POC
- FISH: Prenatal aneuploidy screen

Laboratory test performed

Chromosome analysis on POC was performed (see method for karyotyping described in Chapter 1).

Test results

Chromosome analysis revealed three copies of each chromosome (triploidy) from all metaphases examined. These results suggested a diagnosis of a possible partial mole (Fig. 2.2.1).

FISH for prenatal aneuploidy screen was performed using FISH probes (Abbott Molecular, Inc.; 150 nuclei examined per probe set) targeting specific sequences from chromosomes X, Y, 13, 18, and 21. The signal patterns were abnormal with three copies of chromosomes 13, 18, and 21, two copies of the X chromosome, and one copy of the Y chromosome. This result was consistent with a triploidy in a male individual (Fig. 2.2.2).

Results with interpretations

All cells analyzed showed a 69,XXY chromosome complement, which is called a triploid. FISH results were concordant with the finding from the karyotype. Triploidy is a common

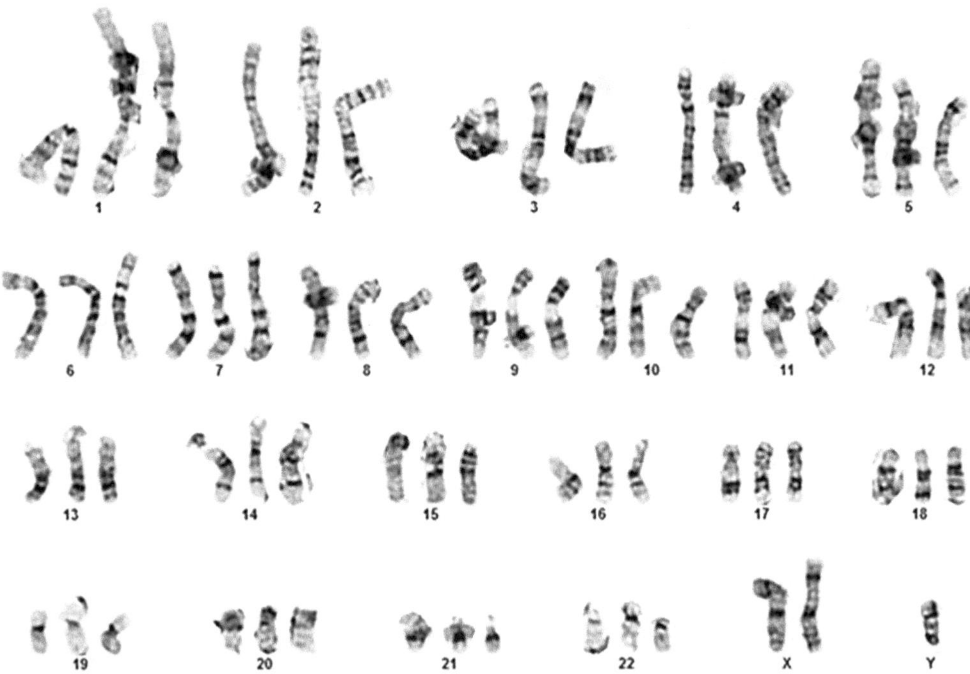

FIG. 2.2.1 The karyotype from the POC showed a triploid chromosome complement with 3 copies of each chromosome. ISCN: 69,XXY

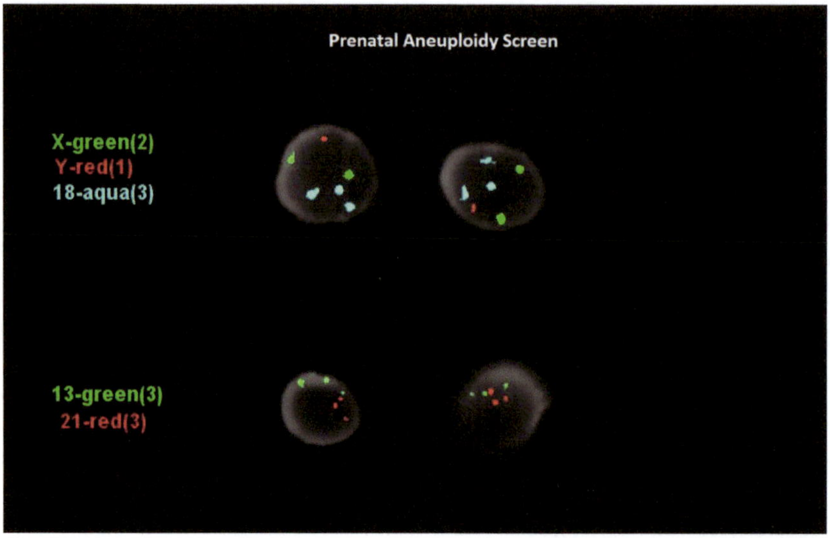

FIG. 2.2.2 FISH for prenatal aneuploidy panel showed 3 copies of chromosomes 13, 18, 21, and sex chromosome (XXY). ISCN: nuc ish(DXZ1x2,DYZ3x1,D18Z1x3)[150/150],(RB1,D21S341)x3[150/150]

occurrence in human gestation and is present in 2%–3% of pregnancies. It often culminates in early spontaneous abortion, but occasionally results in the birth of an abnormal fetus or newborn. Triploidy may be the result of either digyny (extra haploid set from the mother) or diandry (extra haploid set from the father). Digynic triploidy predominates in fetuses, and diandry accounts for about 50%–60% of early spontaneous abortions [7].

Future testing and recommendations

Genetic counseling is recommended. PCR analysis during diagnosis on DNA samples from both maternal and POC with microsatellite markers is indicated to confirm whether it is diandry or digyny. The association with molar changes can be addressed by morphology analysis in the pathology department [5,8].

Summary of key learning points

- The concepts of the complete mole and partial moles are described.
- Complete mole is associated with molar changes. For partial moles, only diandric triploidy is associated with molar changes.
- Technologies used to identify molar pregnancy include karyotyping, FISH, and chromosome microarray. Correlation with morphological analysis by pathologists is essential.

References

[1] M.J. Seckl, N.J. Sebire, R.S. Berkowitz, Gestational trophoblastic disease, Lancet 376 (9742) (2010) 717–729.

[2] R.A. Fisher, G.J. Maher, Genetics of gestational trophoblastic disease, Best Pract. Res. Clin. Obstet. Gynaecol. 74 (2021) 29–41.

[3] L. Duffy, et al., The diagnosis of choriocarcinoma in molar pregnancies: a revised approach in clinical testing, J. Clin. Med. Res. 7 (12) (2015) 961–966.

[4] H. Usui, et al., Genome-wide single nucleotide polymorphism array analysis unveils the origin of heterozygous androgenetic complete moles, Sci. Rep. 9 (1) (2019) 12542.

[5] Y.P. Wong, et al., Diagnostic utility of p57 immunohistochemistry and DNA ploidy analysis by fluorescence in situ hybridization in hydatidiform moles, Malays. J. Pathol. 43 (3) (2021) 341–351.

[6] N.J. Sebire, et al., Risk of recurrent hydatidiform mole and subsequent pregnancy outcome following complete or partial hydatidiform molar pregnancy, BJOG 110 (1) (2003) 22–26.

[7] D.E. McFadden, W.P. Robinson, Phenotype of triploid embryos, J. Med. Genet. 43 (7) (2006) 609–612.

[8] M. Kolarski, et al., Genetic counseling and prenatal diagnosis of triploidy during the second trimester of pregnancy, Med. Arch. 71 (2) (2017) 144–147.

3

Sex chromosomal abnormalities

Guang Liu and Xia Li
SONORA QUEST LABORATORIES, PHOENIX, AZ, UNITED STATES

Background

Sex chromosome abnormalities (SCAs) are among the most common of all human genetic diseases, with an overall incidence of approximately 1 in 400 live births [1–3]. SCAs can be either numerical (aneuploidies) or structural and can be present in all cells or mosaic forms. The most common sex chromosome aneuploidies are 45,X (Turner syndrome), 47,XXY (Klinefelter syndrome), 47,XYY, and 47,XXX [3]. Sex chromosome structural abnormalities are also common, which include deletions, duplications, translocations, rings, inversions, isochromosomes, etc. Structural abnormalities of the X chromosome in males are generally associated with more severe phenotypic consequences than in females. This is mainly due to skewed X-chromosome inactivation in females with X-chromosome abnormalities [4]. Sex chromosome mosaicism is not unusual. For example, about half of the patients with Turner syndrome have monosomy X (45,X). The remaining 50% of patients with Turner syndrome have a mosaic chromosomal component (45,X with mosaicism) [5]. The phenotype of individuals having mosaicism for sex chromosomes is variable and usually related to the proportion and distribution of the two cell lines in different tissues. Also, the percentage of abnormal cells in the peripheral blood may not accurately reflect the distribution of abnormal cells in other tissues. Various clinical indications raise the possibility of a sex chromosome abnormality and thus the need for chromosomal, FISH, microarray, or molecular studies. These indications include ambiguous genitalia, short stature, primary or secondary amenorrhea, delay in the onset of puberty, and infertility [3]. In this chapter, a few cases with sex chromosome abnormalities are illustrations of how patients can benefit from genetic testing.

Case 3.1 Female with 45,X/46,XY mosaicism

Clinical indication

A 15-year-old girl was referred to the endocrine clinic for investigation of her primary amenorrhea, short stature, and delay in the onset of puberty. Physical examination revealed her height of 134 cm, broad chest with widely spaced nipples, and absent breast development. Pelvic examination showed normal external female genitalia. Blood tests

Cases in Laboratory Genetics and Genomics (LGG) Practice. https://doi.org/10.1016/B978-0-323-99622-8.00022-8

revealed a normal growth hormone (GH) level and an elevated follicle-stimulating hormone (FSH) level. The uterus and ovaries were not visualized by pelvic images.

Test ordered

- Chromosome analysis: Routine blood
- FISH: *SRY*/*CEPX*/*DYZ1*
- Chromosome microarray analysis (CMA)

Laboratory test performed

Chromosome analysis, FISH, and CMA were performed on peripheral blood (PB) (detailed methods are described in Chapter 1).

Test results

Chromosome analysis revealed two cell lines. The first cell line is represented by 16 cells, characterized by only one copy of the X chromosome (monosomy X) (Fig. 3.1.1).

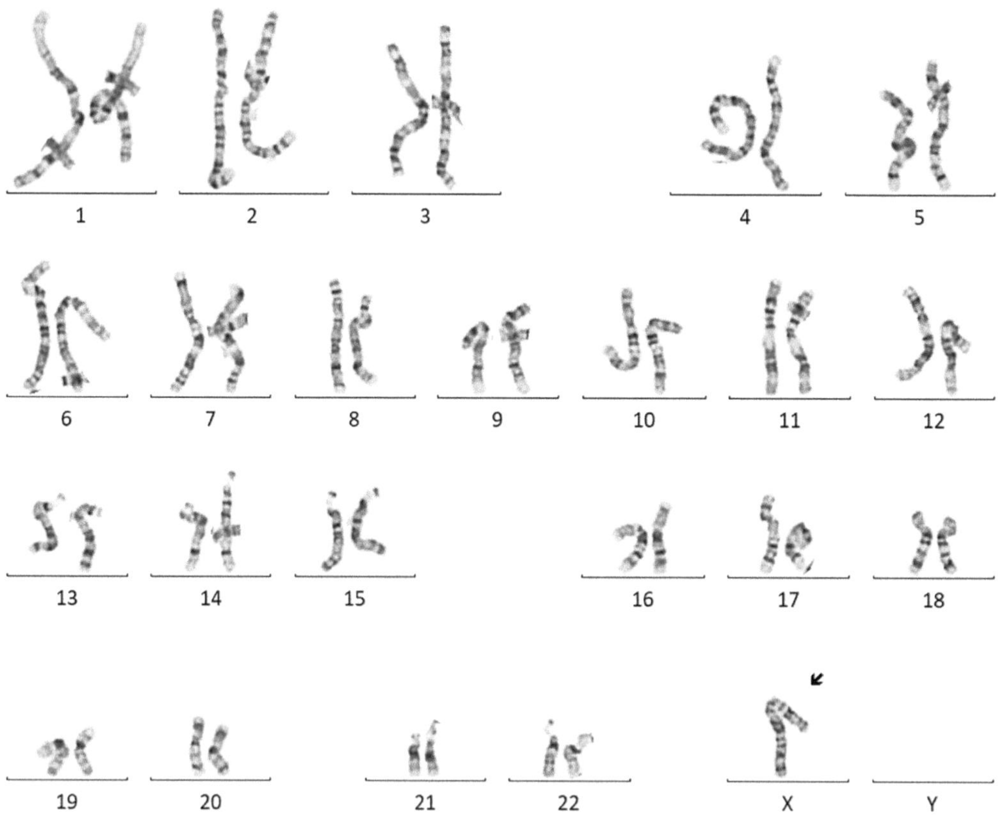

FIG. 3.1.1 The karyotype from the PB showed a monosomy X. ISCN: mos 45,X[16]/46,XY[4]

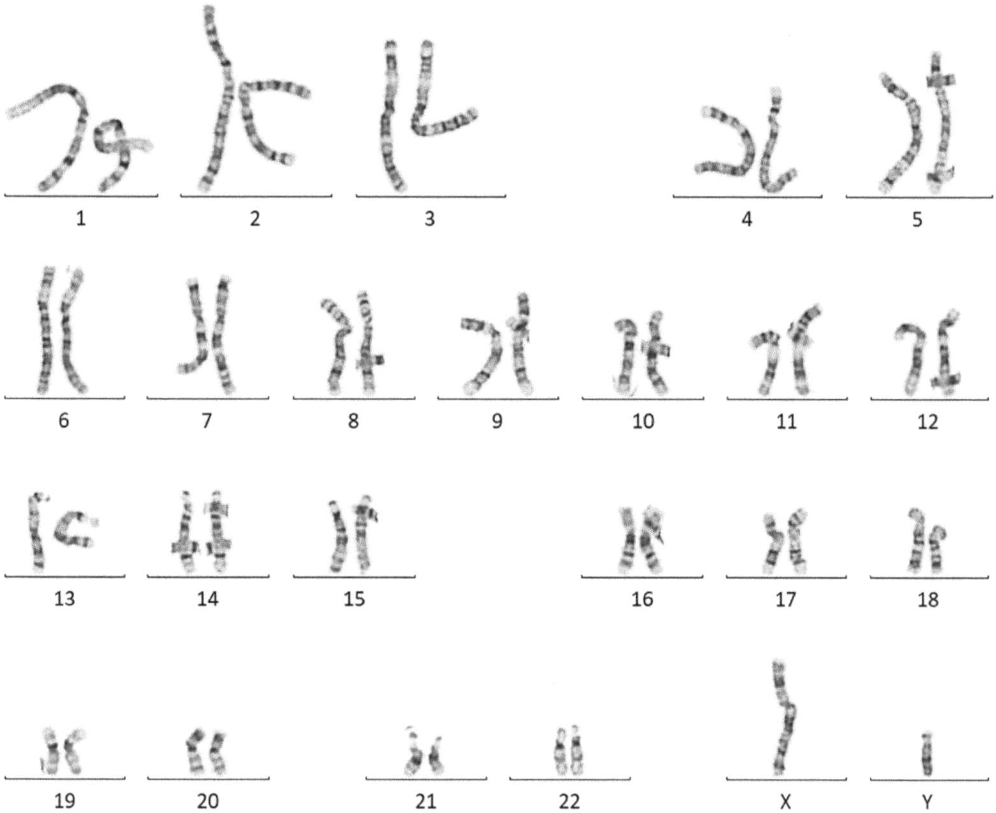

FIG. 3.1.2 The karyotype from the PB showed a normal male karyotype: 46,XY.

The second cell line is represented by four cells, characterized by the presence of one copy of the X chromosome and one copy of the Y chromosome (46,XY) (Fig. 3.1.2).

Fluorescent molecular probes, which are localized to the centromere of the X chromosome (*CEPX*; *DXZ1*), the *SRY* gene region on the Y chromosome at Yp11.3, and the sat III Y at Yq12 (*DYZ1*) were hybridized to metaphase and interphase preparations. The fluorescent signal pattern also revealed two cell lines: (1). 8/10 metaphase and 89/100 interphase cells showed only one copy of the centromere of the X chromosome, consistent with a 45,X karyotype (Figs. 3.1.3); (2) 2/10 metaphase and 11/100 interphase cells showed one copy of centromere of the X chromosome, one copy of *SRY*, and one copy of Yq12, consistent with 46,XY karyotype (Fig. 3.1.4).

Chromosomal microarray analysis (CMA) results were mosaic for cells that have only one X chromosome (monosomy X) (Figs. 3.1.5) and cells with XY chromosomes (Fig. 3.1.6). The Y chromosome appears to be in about 20% of the cells, consistent with the concurrent chromosome and FISH analysis, which revealed mosaicism for two cell lines (Fig. 3.1.6). The whole genome view also revealed the mosaic status of the sex chromosome (Fig. 3.1.7).

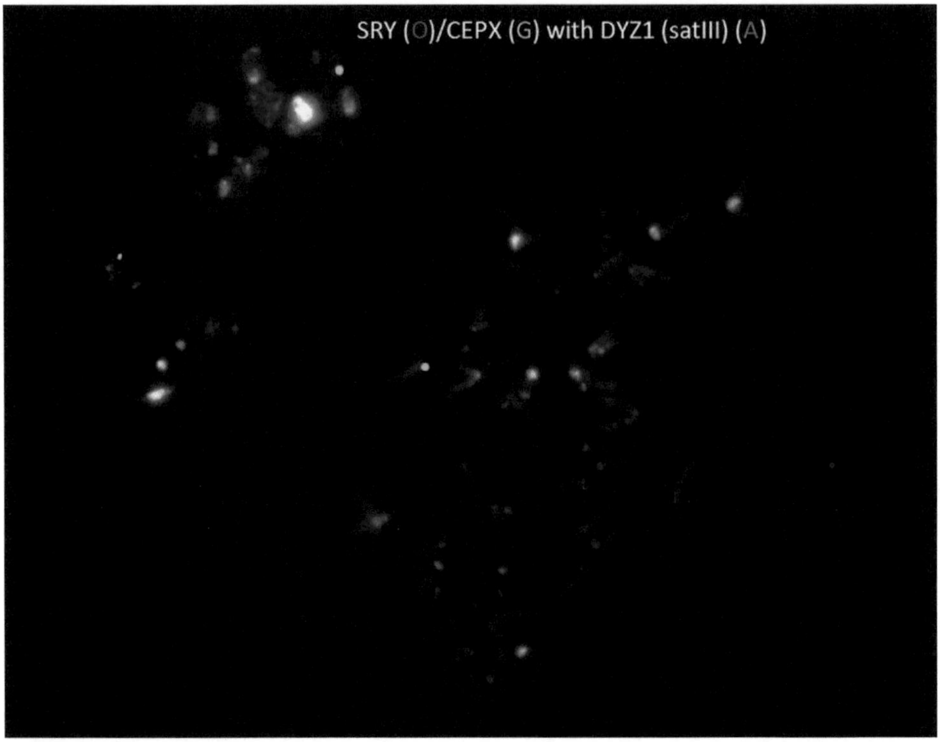

FIG. 3.1.3 FISH with probes localized to *CEPX* (DXZ1), the *SRY* gene, and *DYZ1* (sat III) were hybridized to metaphase and interphase cells. The cells showed abnormal signal patterns with monosomy X. *CEPX* probe (Xp11.1-q11.1) was shown in *green*, *SRY* probe (Yp11.3) was shown in *orange*, and *DYZ1* was shown in *aqua*. ISCN: ish Xp11.1-q11.1 (DXZ1x1),Yp11.3(SRYx0),Yq12(satIII)x0 [8].nuc ish(DXZ1x1,SRYx0,Yq12x0)[89]

Results with interpretations

The results of all three tests (chromosome analysis, FISH, and CMA) were concordant and consistent with the clinical diagnosis of mosaic Turner syndrome. The most common features of Turner syndrome include short stature and ovarian dysgenesis, which are consistent with this patient's clinical findings. Other features associated with this syndrome may include extra folds of skin on the neck (webbed neck), low hairline at the back of the neck, lymphedema of the hands and feet, skeletal abnormalities, kidney problems, and heart defects. Most patients have normal intelligence, but some have developmental delays, learning disabilities, and behavioral problems. The phenotype of individuals having mosaicism is variable and usually related to the proportion and distribution of the two cell lines in different tissues. Also, the percentage of abnormal cells in the peripheral blood may not accurately reflect the distribution of abnormal cells in other tissues. The 45,X/46,XY mosaicism

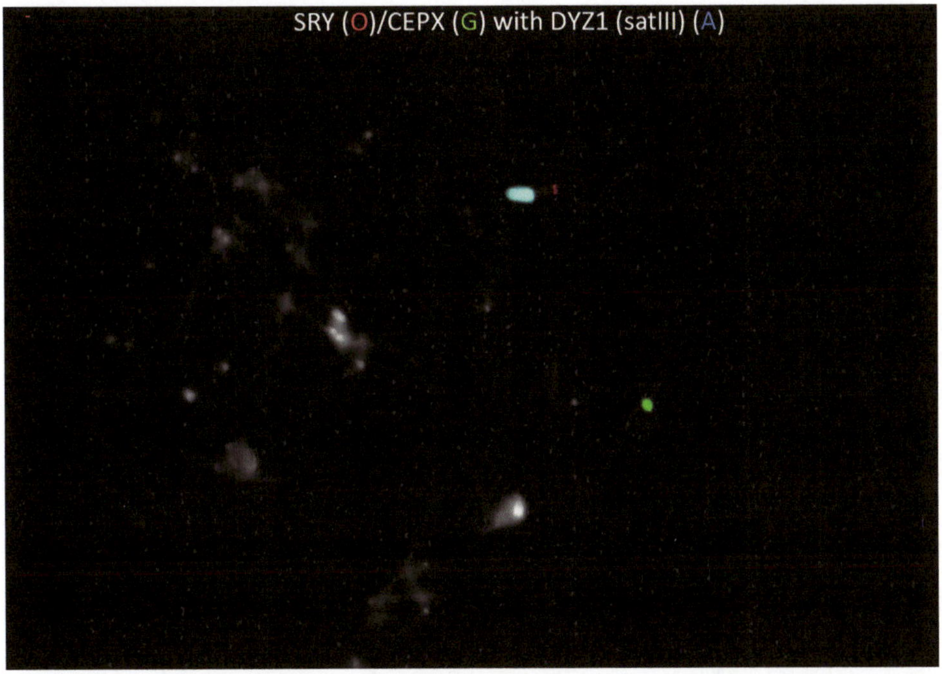

FIG. 3.1.4 FISH with probes localized to *CEPX* (*DXZ1*), the *SRY* gene, and *DYZ1* (sat III) were hybridized to metaphase and interphase cells. The cell showed normal male signal patterns. *CEPX* probe (Xp11.1-q11.1) was shown in *green*, *SRY* probe (Yp11.3) was shown in *orange*, and *DYZ1* was shown in *aqua*. ISCN: ish Xp11.1-q11.1(DXZ1x1),Yp11.3 (SRYx1),Yq12(satIIIx1) [2].nuc ish(DXZ1x1,SRYx1,Yq12x1) [11]

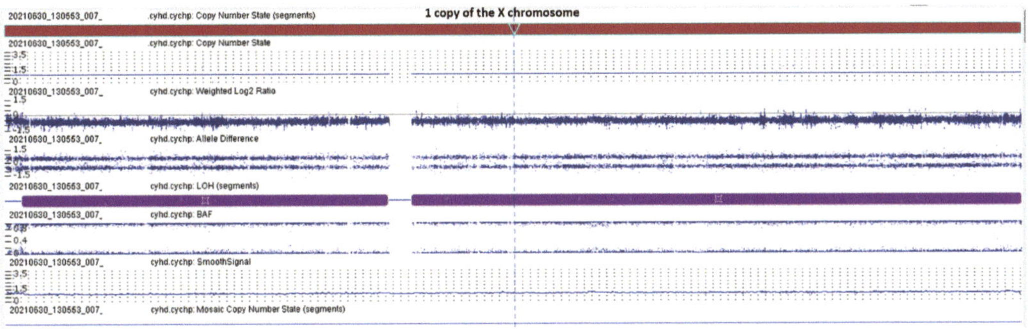

FIG. 3.1.5 The copy number state, Log2R ratio, and B allele difference for the X chromosome. ISCN: arr[GRCh37] Xp22.33q28(168,567-155,233,731)x1

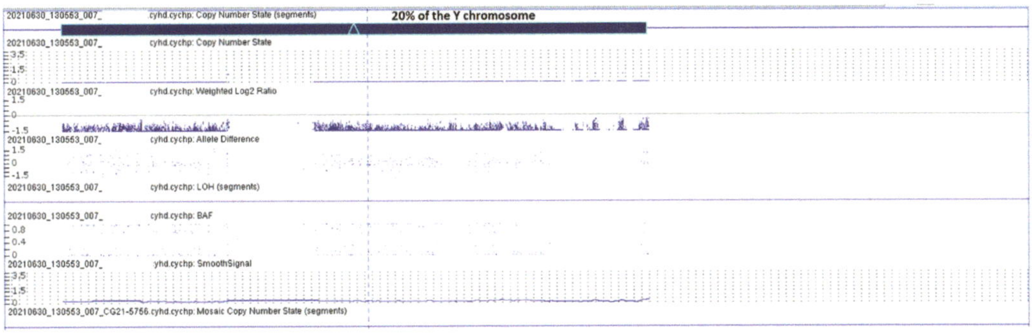

FIG. 3.1.6 The copy number state, Log2R ratio, and B allele difference for the Y chromosome. ISCN: arr[GRCh37] Yp11.31q11.23(2,650,141-28,721,110)x[0.2]

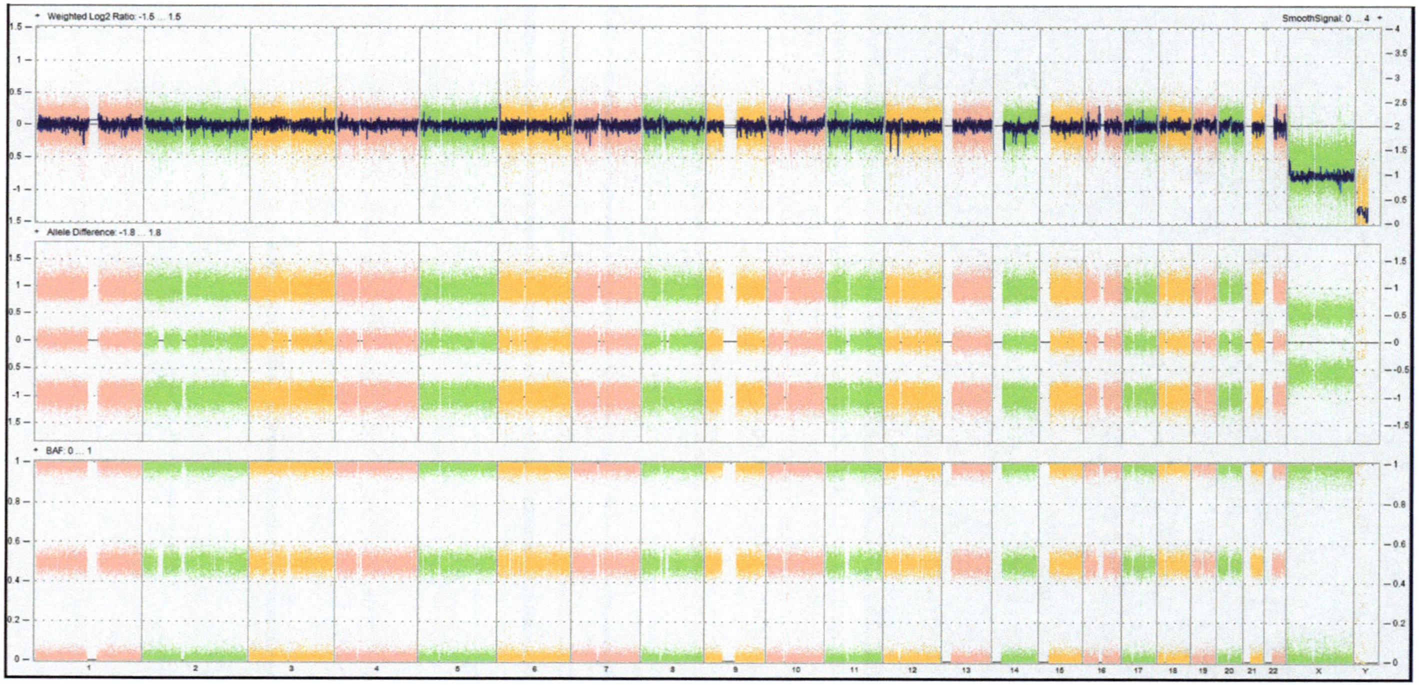

FIG. 3.1.7 The whole genome view showed the copy number state for all chromosomes.

accounts for 10%–12% of cases of Turner syndrome. Patients with 45,X/46,XY may have a highly variable phenotype, ranging from nearly normal males to ambiguous genitalia to females with features of Turner syndrome. In the case of female genitalia, this individual is at increased risk of gonadoblastoma due to the presence of a Y chromosome [6].

Future testing and recommendations

The risk of gonadoblastoma is increased in phenotypic females with 45,X/46,XY mosaicism. Therefore, early investigation and individual management including prophylactic gonadectomy should be considered [7]. A thorough clinical assessment of this patient and genetic counseling are recommended.

Case 3.2 Sex reversal

Clinical indication

A 5-day-old baby girl, with intrauterine growth restriction, was born full-term. Upon physical examination, the baby had a cyst of oral soft tissue, a single umbilical artery, and female external genitalia (normal-appearing clitoris, vaginal opening, and labia majora and minora). The gonads were not palpable. A pelvic image did not show a uterus, fallopian tubes, or gonads.

Test ordered

- Chromosome analysis: Routine blood
- FISH: *SRY*/*CEPX*/*DYZ1*
- Chromosome microarray analysis (CMA)

Laboratory test performed

Chromosome analysis, FISH, and CMA were performed on peripheral blood (PB) (detailed methods are described in Chapter 1).

Test results

Chromosome analysis on metaphases examined revealed an abnormal male chromosome complement in all cells examined with a duplication within the short arm of the X chromosome from band p22.33 to p21.1 (Fig. 3.2.1).

FISH for the *SRY* gene was performed. Fluorescent molecular probes (Abbott Molecular, Inc.), which are localized to the centromere of the X chromosome, the *SRY* gene region on the Y chromosome (at Yp11.3), and the sat III Y (at Yq12; internal control) were hybridized to metaphase and interphase preparations. The fluorescence signal pattern showed an *SRY*-positive XY (male) sex complement in the 10 metaphases and 50 interphase cells examined (Figs. 3.2.2 and 3.2.3).

Chromosomal microarray analysis (CMA) revealed a male (XY) chromosome complement with a copy number gain of 31.3Mb of DNA from chromosome X at band Xp22.33p21.1, encompassing 106 OMIM genes, including *NR0B1* (*DAX1*) of clinical significance (Figs. 3.2.4). The whole genome view also revealed the duplication of chromosome Xp (Fig. 3.2.5).

Results with interpretations

The results of chromosome analysis, FISH, and CMA are concordant. Chromosomal analysis revealed an abnormal male karyotype with a duplication in the short arm of the X

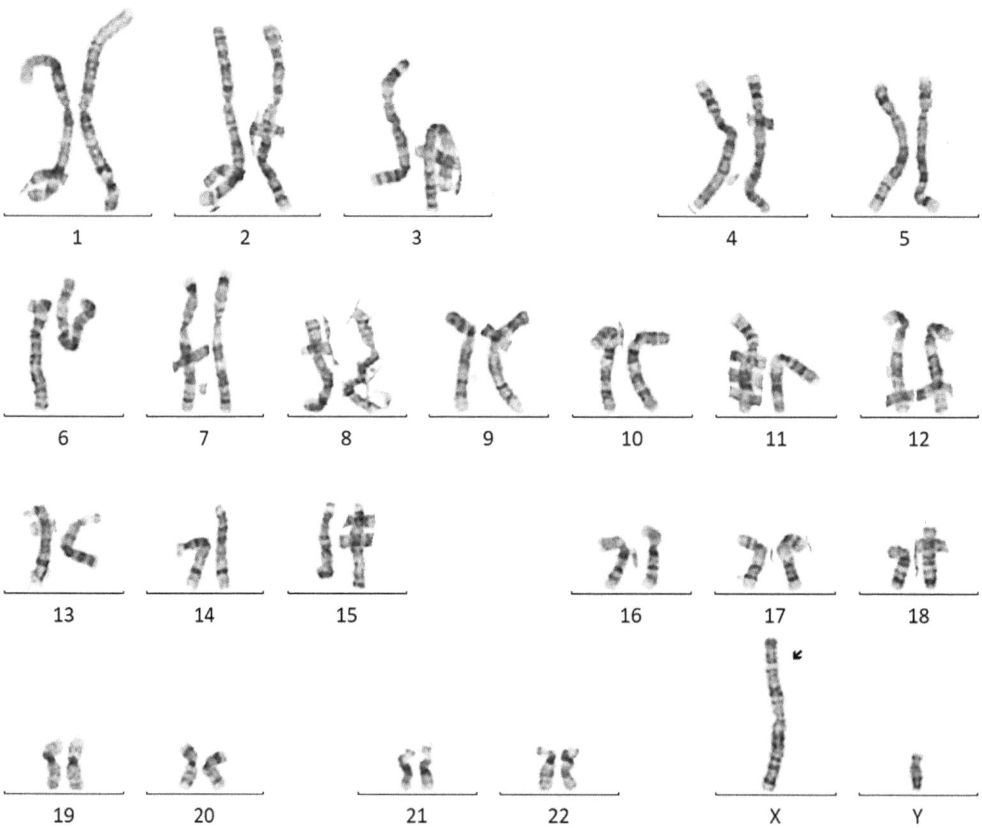

FIG. 3.2.1 The karyotype of the patient showed a duplication within the short arm of the X chromosome from band p22.33 to p21.1. ISCN: 46,Y,dup(X)(p22.33p21.1)

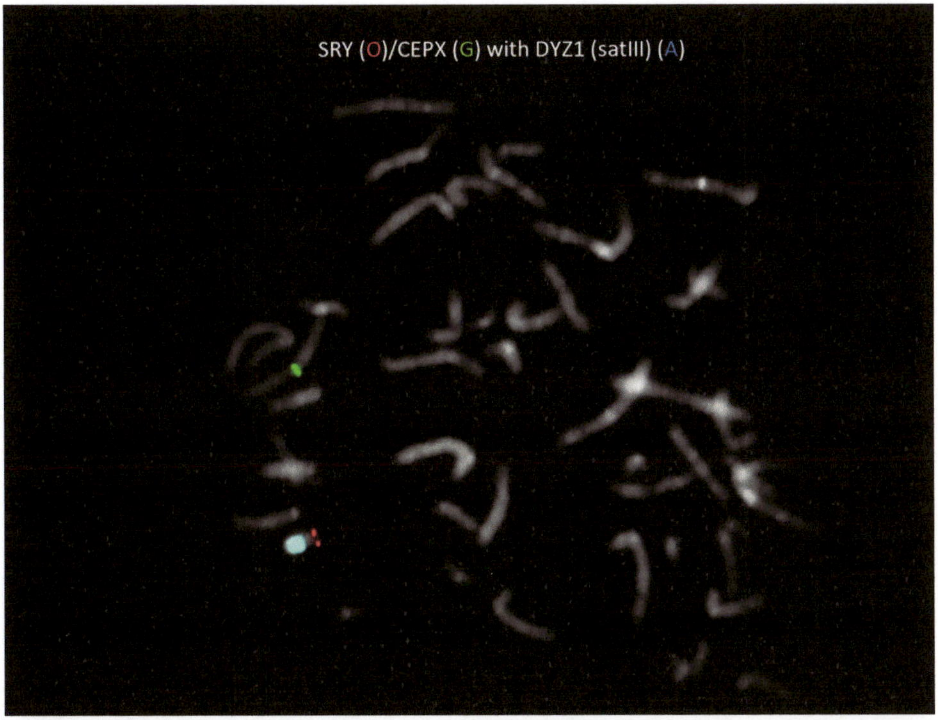

FIG. 3.2.2 FISH *SRY* gene and control probe was performed on metaphase cells. All cells showed normal male signal patterns. *SRY* probe (Yp11.3) was shown in *red*, X centromere probe (Xp11.1-q11.1) was shown in *green*, and Y heterochromatin (Yq12) was shown in *aqua*. ISCN: ish Yp11.3(SRYx1),Xp11.1-q11.1(DXZ1x1),Yq12(sat III x 1).nuc ish(SRYx1,DXZ1x1,Yq12x1)

chromosome from band p22.33 to p21.1. FISH showed a male signal pattern with the *SRY* gene present. CMA confirmed cytogenetic findings and more accurately indicated a duplication of 31.3 Mb of DNA on Xp with 106 OMIM genes, including the *NR0B1* (*DAX1*) gene.

XY individuals with a duplication of part of the short arm of the X chromosome including *NR0B1* and an intact *SRY* gene might show male-to-female sex reversal [8], which is consistent with the patient's clinical findings. Duplications of this size usually also result in developmental delay, mental impairment, and congenital anomalies. The patient had intrauterine growth restriction and some congenital abnormalities. Phenotypic females with a Y chromosome are at risk of gonadoblastoma [9]. *NR0B1* is responsible for a sex-reversal syndrome in humans, referred to as dosage-sensitive sex reversal, in which XY individuals carrying duplications of Xp21, part of the short arm of the X chromosome, develop as females [10,11].

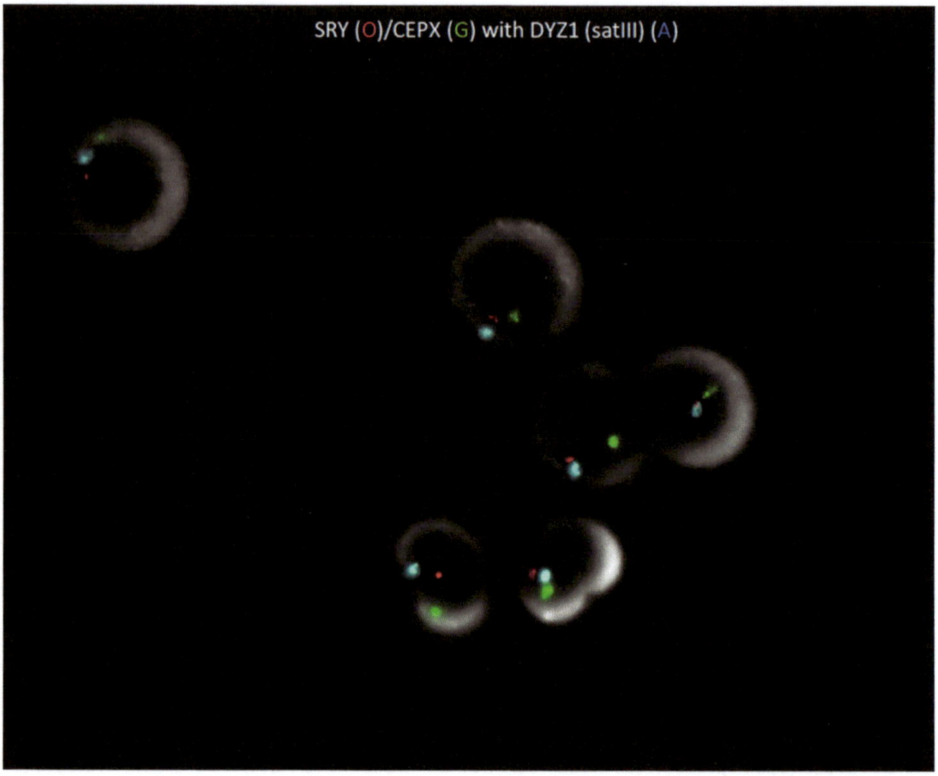

FIG. 3.2.3 FISH *SRY* gene and control probe was performed on interphase cells. All cells showed normal male signal patterns. *SRY* probe (Yp11.3) was shown in *red*, X centromere probe (Xp11.1-q11.1) was shown in *green*, and Y heterochromatin (Yq12) was shown in *aqua*.

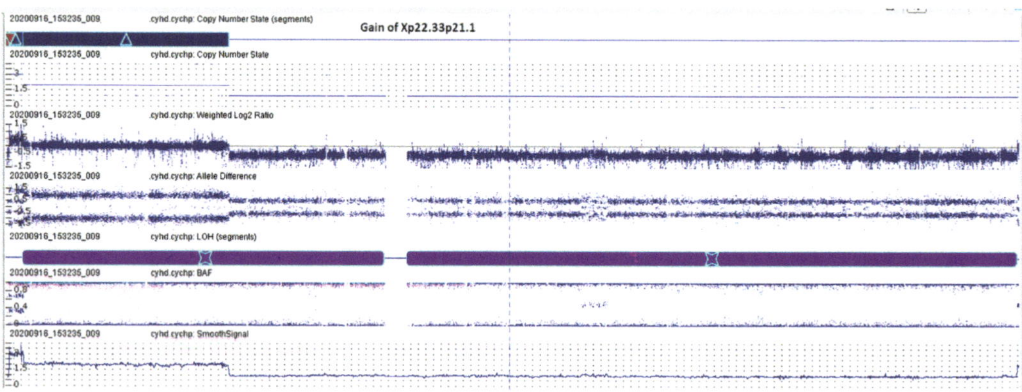

FIG. 3.2.4 The copy number state, Log2R ratio, and B allele difference for Xp duplication. ISCN: arr[GRCh37] Xp22.33p21.1(2,760,762-34,086,722)x2

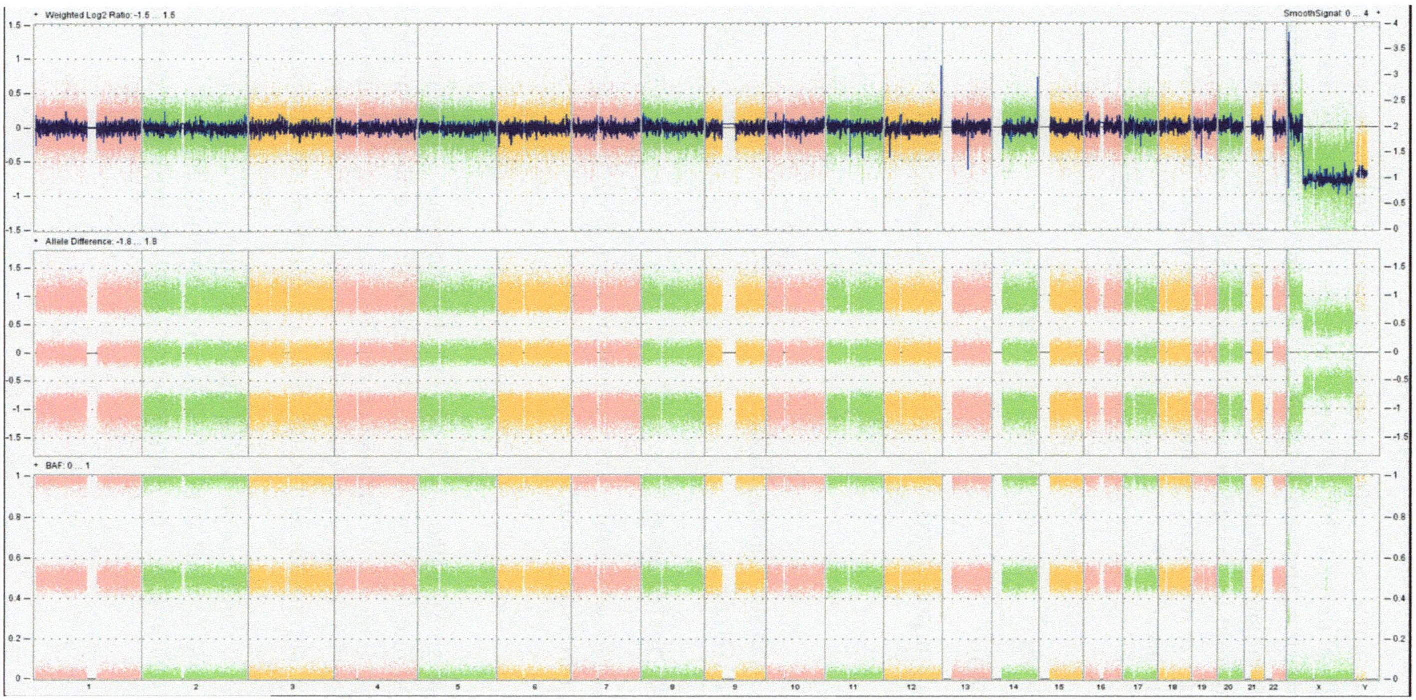

FIG. 3.2.5 The whole-genome view showed Xp duplication.

Future testing and recommendations

Genetic counseling and a thorough clinical assessment of this patient are recommended. Parental (especially maternal) chromosome analysis is indicated to determine the potential origin of this abnormality. Female carriers of Xp duplications are generally healthy and fertile due to preferentially inactivating the duplicated chromosome and thereby protecting the individual from increased gene expression. If the mother is a carrier of the Xp duplication, the risk of recurrence is increased, and prenatal diagnosis should be offered for future pregnancies.

Case 3.3 Variant turner syndrome

Clinical indication

A 6-month-old infant presented with indeterminate sex. Peripheral blood was sent for genetic testing.

Test ordered

- Chromosome analysis: Routine blood
- FISH: *SRY*/*CEPX*/*DYZ1*

Laboratory test performed

Chromosome analysis and FISH were performed on peripheral blood (PB) (detailed methods are described in Chapter 1).

Test results

Chromosome analysis revealed an abnormal chromosome complement with two cell lines. The first cell line, seen in 24 cells, showed only one X chromosome resulting in monosomy X [45,X] (Fig. 3.3.1). The second cell line, seen in 26 cells, showed one normal X chromosome plus an isodicentric chromosome consisting of possibly a portion of the short arm of the Y chromosome and the long arm of the Y chromosome [46,X,idic(Y)?(p11.31)] (Fig. 3.3.2).

Fluorescent molecular probes localized to the X chromosome centromere (*DXZ1*), Yp11.3 (*SRY*), and Yq12 (*DYZ1*) were hybridized to 10 metaphase and 50 interphase preparations. The fluorescence signal pattern showed three cell lines (Fig. 3.3.3).

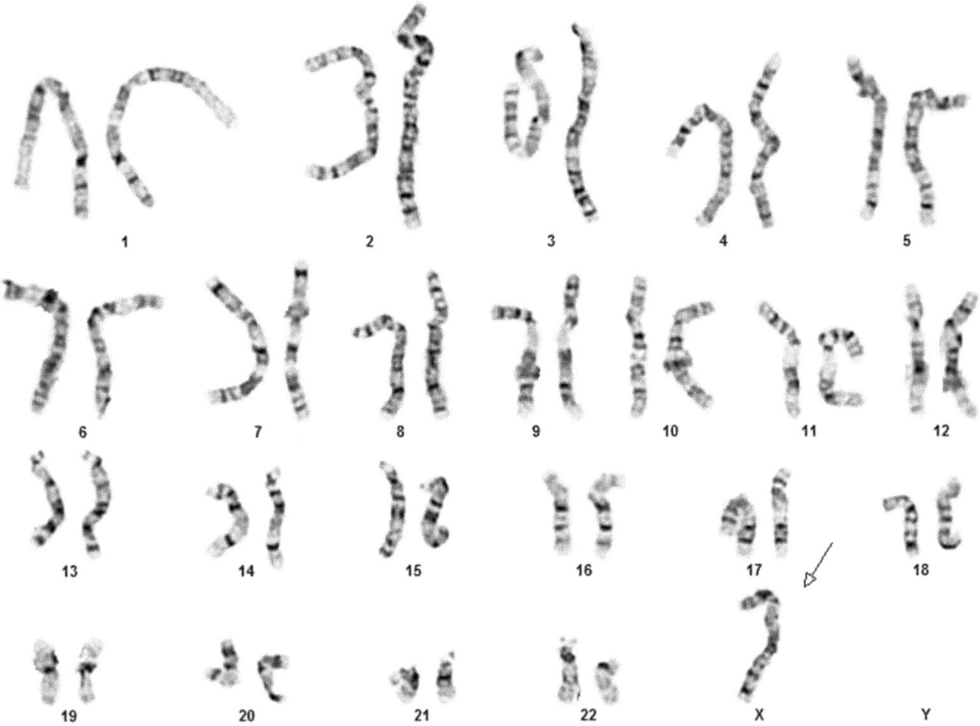

FIG. 3.3.1 The karyotype showed monosomy X (45,X). ISCN: mos 45,X[24]/46,X,idic(Y)?(p11.31)[26]

The first abnormal cell line showed one copy of the X centromere without the presence of the *SRY* gene region or Yq12 in 5 metaphases and 30 interphase cells. This suggests a cell line with one X chromosome and missing a second sex chromosome.

The normal cell line exhibited one copy of the X chromosome centromere, one copy of the *SRY*, and one copy of Yq12 in 5 metaphase and 20 interphase cells.

The second abnormal cell line showed one copy of the X chromosome centromere, one copy of the *SRY* gene region, and two copies of Yq12 in 5 metaphases and 20 interphase cells (see metaphase image in Fig. 3.3.4).

Results with interpretations

Phenotypes with sex chromosome mosaicism vary from male to female or individuals with ambiguous genitalia depending on the location of the chromosome breakpoints

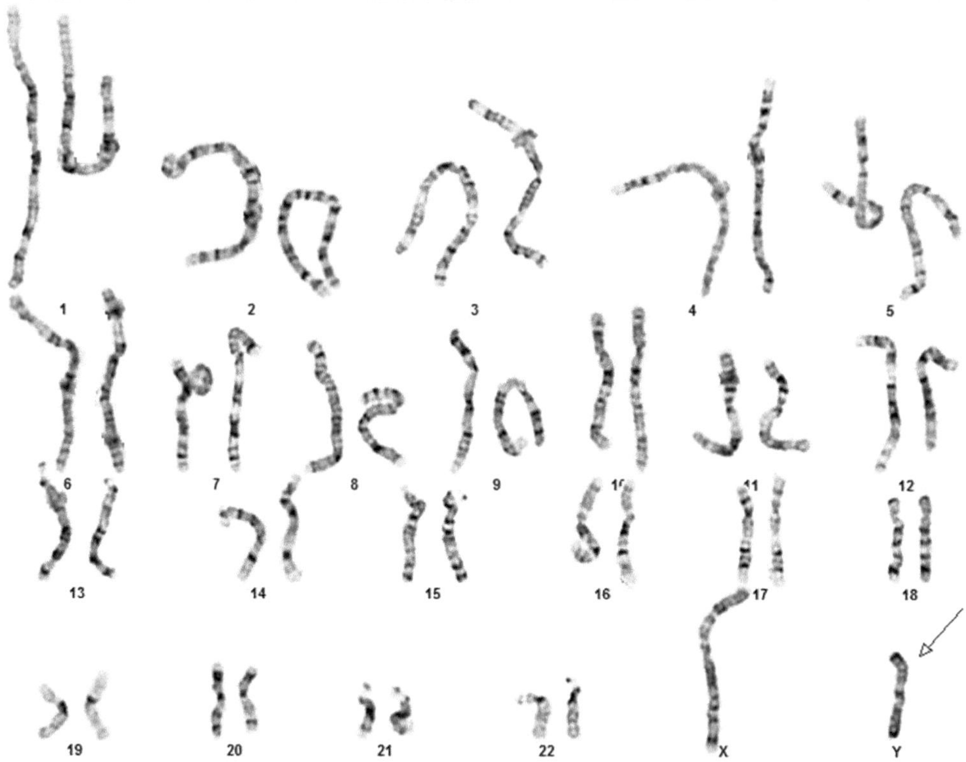

FIG. 3.3.2 The karyotype showed an isodicentric Y chromosome for the long arm [46,X,idic(Y)?(p11.31)].

as well as on the proportion of each cell line. The presence of the Y chromosome material in females greatly increases the risk of gonadoblastoma. Please note that the proportion of each cell line in the peripheral blood may not accurately reflect the distribution of these two cell lines in other tissues.

Results from FISH were mostly concordant with those of the karyotype. The first abnormal cell line corresponded to the 45,X; the second abnormal cell line was concordant with the isodicentric Yp11.3. The cell line with one *SRY*, one X centromere, and one Yq12 was the normal cell line that was not seen in the karyotype, likely due to a small percentage of the cells.

Future testing and recommendations

Molecular studies and microarray analysis may be helpful in further characterization of these findings. Increased surveillance, including periodic testis/ovary ultrasound and biopsy if clinically indicated, should be considered in individuals with abnormal genitalia. Clinical correlation and genetic counseling are highly recommended.

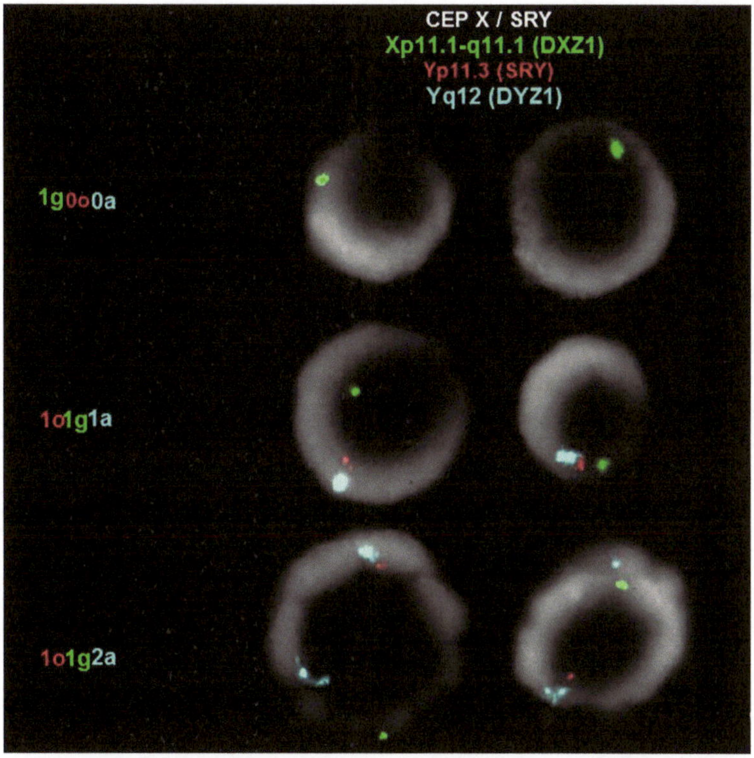

FIG. 3.3.3 FISH showed three cell lines with different signal patterns. ISCN: ish Yp11.3(SRYx0),Xp11.1-q11.1(DXZ1x1), Yq12(sat IIIx0) [5] nuc ish(SRYx0,DXZ1x1,Yq12x0)[30]/ ish Yp11.3(SRYx1),Xp11.1-q11.1(DXZ1x1),Yq12(sat IIIx1) [5] nuc ish(SRYx1,DXZ1x1,Yq12x1)[20] ish Yp11.3(SRYx1),Xp11.1-q11.1(DXZ1x1),Yq12(sat IIIx2) [5] nuc ish(SRYx1,DXZ1x1, Yq12x2)[20]

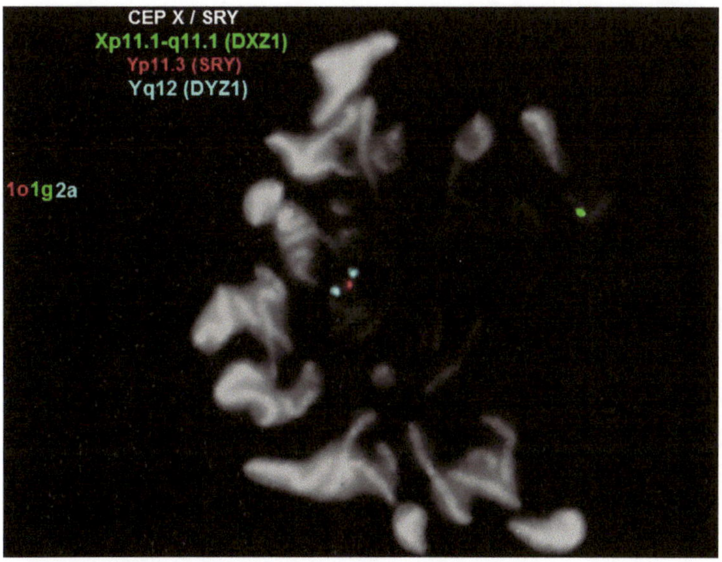

FIG. 3.3.4 FISH showed metaphase for the second abnormal cell line (1o1g2a).

Case 3.4 Indeterminate sex with an abnormal Y chromosome

Clinical indication

A 23-year-old female had an abnormal prenatal screen by NIPT, concerning trisomy 18. Amniotic fluid was sent for genetic testing.

Test ordered

- Chromosome analysis: Routine blood
- FISH: Prenatal aneuploidy screen
- Chromosome microarray (CMA)

Test results

Chromosome analysis revealed a normal male chromosome complement with 46,XY in 15 colonies examined (Fig. 3.4.1).

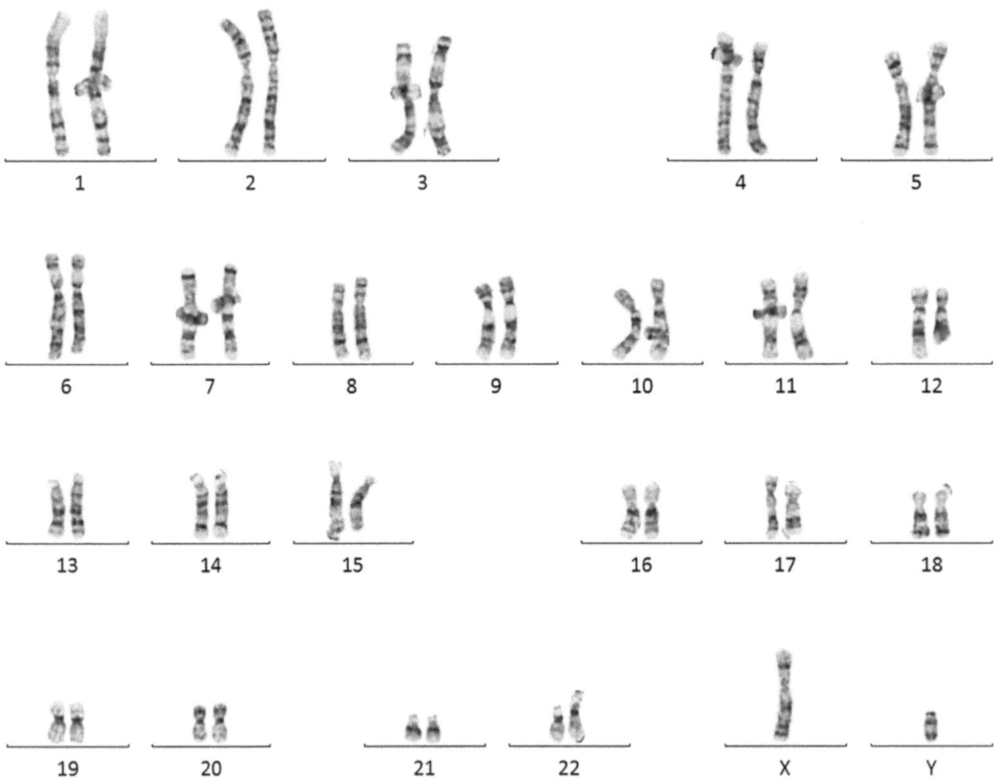

FIG. 3.4.1 The karyotype showed 46,XY from the amniotic fluid sample. ISCN: 46,XY

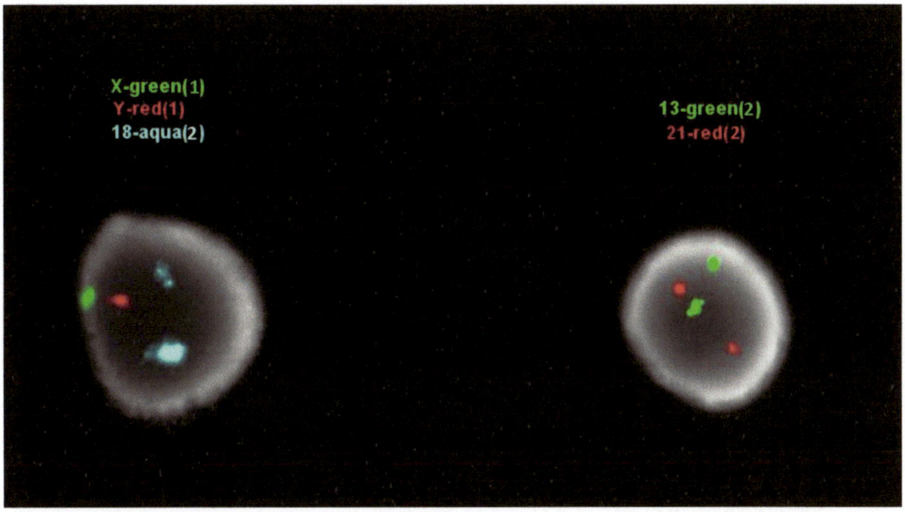

FIG. 3.4.2 FISH prenatal aneuploidy screening showed normal results for chromosomes 13, 18, 21 in a male fetus. ISCN: nuc ish(DXZ1x1,DYZ3x1,D18Z1x2)[100],(RB1,D21S341)x2[100]

FISH with prenatal aneuploidy screening showed no trisomy for chromosomes 13, 18, and 21, and suggested a male fetus (Fig. 3.4.2).

CMA detected an approximately 16 Mb mosaic gain of Ypter-q11.221 and an approximately 43 Mb mosaic loss of Yq11.221-qter from this fetus (Fig. 3.4.3A for the X chromosome and Fig. 3.4.3B for the Y chromosome).

The predicted karyotype for the abnormal cell line was 46,X,psu idic(Y)(q11.221). After we reexamined the G-banded amniotic fluid slides, the abnormal cell line with two cells (with likely appreciation of the abnormal Y chromosome) was observed in addition to the normal cell line (46,XY) (Fig. 3.4.4). Supplemental FISH analysis with probes detecting the X centromere (*DXZ1*) and sex-determining region of the Y chromosome (*SRY*) exhibited two positive *SRY* probe signals at the psu idic(Y)(q11.221) chromosome and one X chromosome in 45/50 (90%) of metaphase cells from cultured amniotic fluid. The remaining four metaphase cells had a single *SRY*, and one cell did not have the *SRY* signal (Fig. 3.4.5). Reanalysis of the FISH prenatal aneuploidy screen shows abnormal results with one copy of the X chromosome and two copies of the Y chromosome (Fig. 3.4.6).

Results with interpretation

The original karyotype and FISH from amniotic fluid did not detect the abnormal cell line due to poor morphology of the G-banding. Chromosome microarray identified mosaic gain of the pseudo dicentric Y chromosome from Ypter to Yq11.221 and mosaic loss of Y chromosome from Yq11.221 to Yqter. To confirm the CMA finding, FISH on metaphases from amniotic fluid showed two copies of *SRY* signals and one copy of the

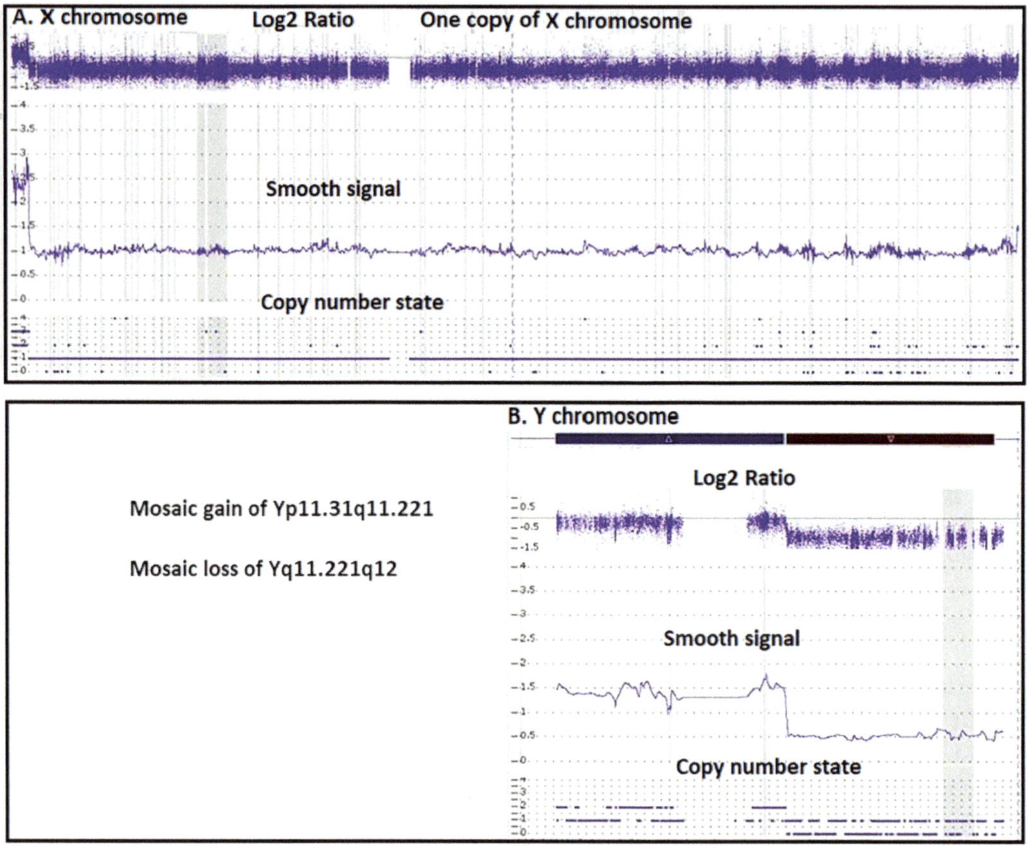

FIG. 3.4.3 CMA results showed one copy of the X chromosome (A), a mosaic gain of Yp11.31q11.221, and a mosaic loss of Yq11.221q12 (B). ISCN: arr[GRCh37] Yp11.31q11.221(1-16053069)x2[0.5], Yq11.221q12(16203970-59363566)x0[0.5]

centromere of the X chromosome, concordant with the results of CMA. Next, a reanalysis of G-banding metaphases showed 2 out of 50 (4%) colonies with a possible idic(Y)(q11.221), and FISH for prenatal aneuploidy screen also showed two centromere signals for the Y chromosome in 3 out of 100 (3%) the cells scored.

Future testing and recommendations

The outcome in the majority of prenatally identified cases with this chromosomal complement is a normal male phenotype at birth, albeit with a high risk of infertility. If ultrasound indicates male genitalia, a male phenotype is to be anticipated. On the other hand, if female genitalia are identified by ultrasound, the patient usually has gonadal dysgenesis. Please note that the determination of male fetal sex does not rule out the presence of ambiguous genitalia or the existence of ovotestes or other gonadal disorders. Additionally,

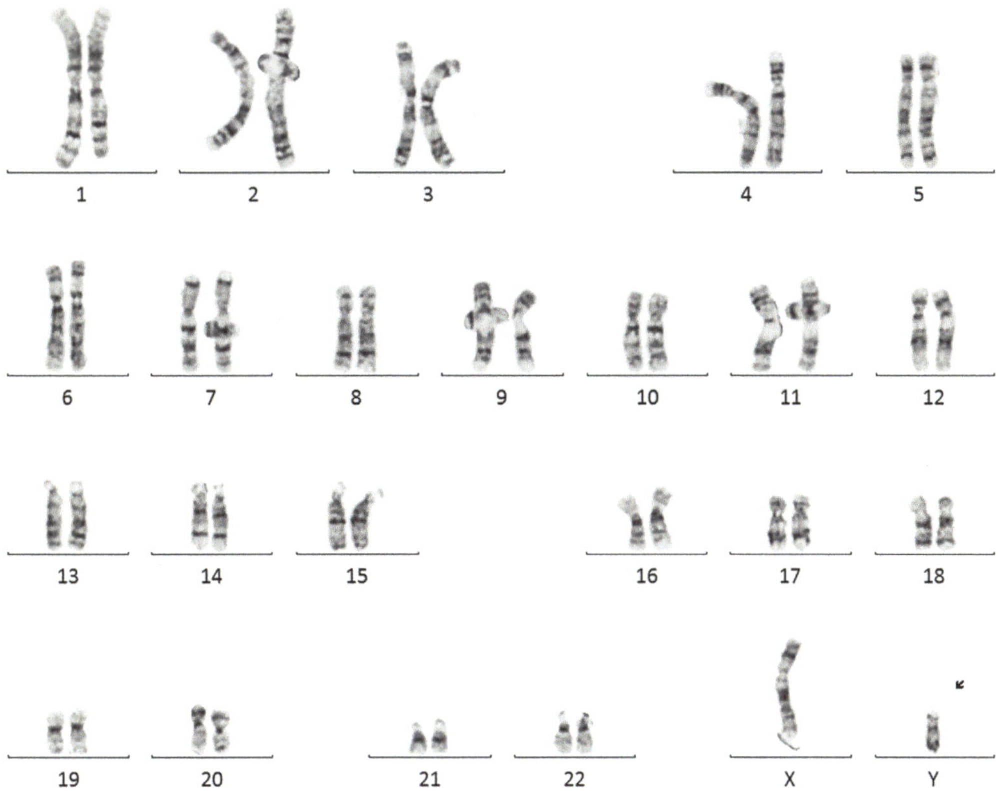

FIG. 3.4.4 The karyotype showed the abnormal Y chromosome in 2 out of 48 cells reexamined. ISCN: 46,X,psu idic(Y)(q11.221)[2]/46,XY[48]

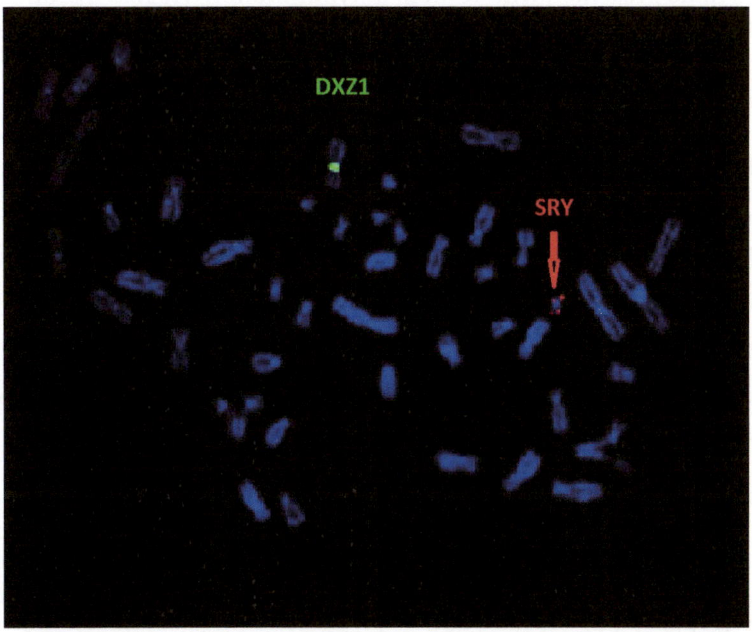

FIG. 3.4.5 FISH showed two copies of *SRY* and one copy of the X chromosome centromere from the metaphase examined.

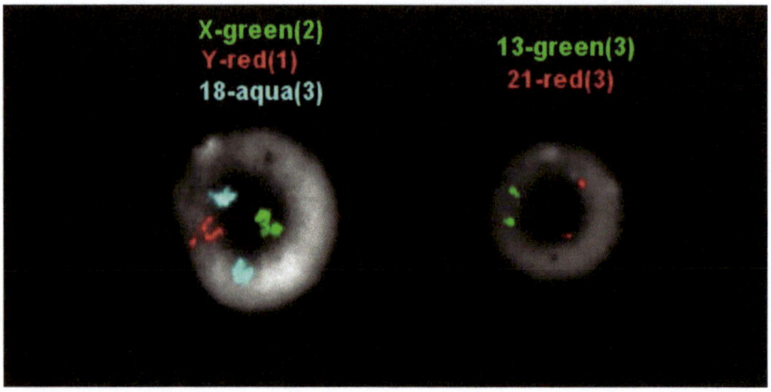

FIG. 3.4.6 Reanalysis of the FISH prenatal aneuploidy screening showed abnormal results with one copy of the X chromosome and two copies of the Y chromosome. ISCN: nuc ish(DXZ1x1,DYZ3x2,D18Z1x2)[3/100],(RB1,D21S341)x2 [100]

dysgenetic gonads in any individual with Y chromosome material are at risk for malignant transformation, initially as gonadoblastoma and subsequently as malignant germinoma [12,13]. Close follow-up with primary physicians and oncologists, parental chromosome analysis, and genetic counseling are recommended.

Case 3.5 Klinefelter syndrome (47,XXY)

Clinical indication

A 25-year-old male presented with testicular hypofunction. He had disruptive mood dysregulation disorder and ADHD. He was evaluated by a neurologist to determine his present neurocognitive functioning ability and to provide recommendations that will help with the course of care.

Test ordered

- Chromosome analysis: Routine blood
- FISH: Postnatal aneuploidy screening
- Chromosome microarray analysis (CMA)

Laboratory test performed

Chromosome analysis, FISH, and CMA were performed on peripheral blood (PB) (detailed methods described in Chapter 1).

Test results

Chromosome analysis revealed the presence of three sex chromosomes with two X chromosomes and one Y chromosome. This karyotype is found in males with Klinefelter syndrome (KS) (Fig. 3.5.1).

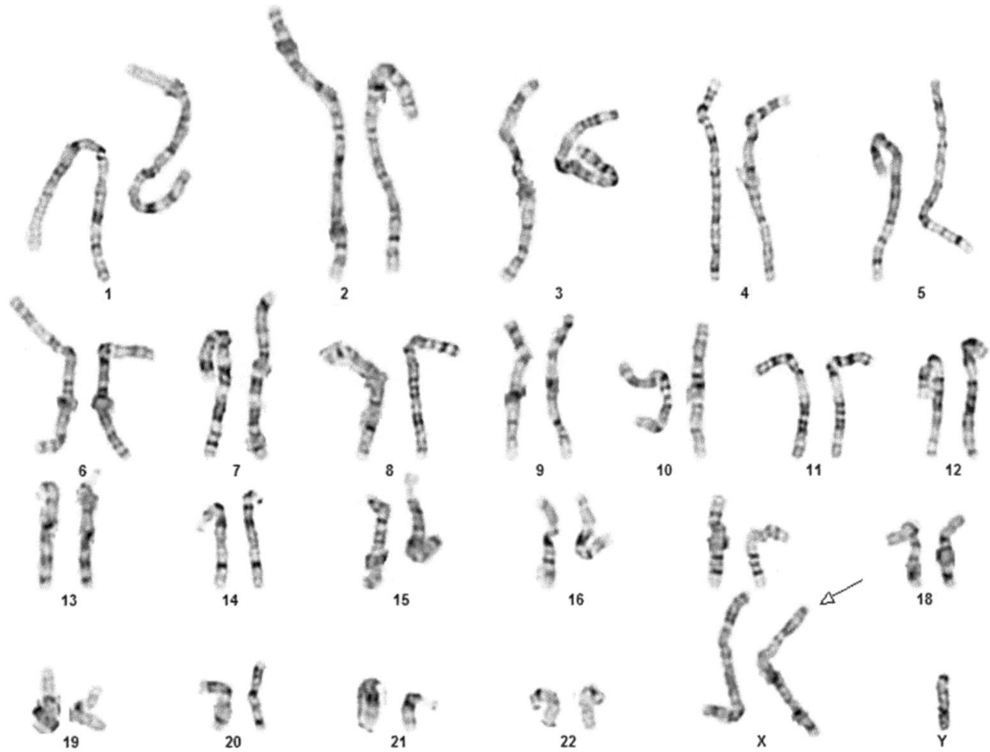

FIG. 3.5.1 Chromosome analysis revealed a gain of the X chromosome. ISCN: 47,XXY

Next, 100 cells were scored using FISH probes targeting specific sequences from chromosomes X, Y, 13, 18, and 21. Analysis revealed signals that were normal for chromosomes 13, 18, and 21. However, the signal pattern for X and Y indicated XXY, consistent with a diagnosis of KS (Fig. 3.5.2).

Chromosomal Microarray Analysis (CMA) revealed a copy number gain of the entire X chromosome, concordant with the findings from chromosome analysis and FISH (Fig. 3.5.3). In this case, the two copies of the X chromosomes are not identical (heterozygous SNP probes can be appreciated). For learning purposes, CMA results from a different patient illustrating two copies of the X chromosome that are completely identical with no heterozygous probes observed are shown (Fig. 3.5.4).

Results with interpretations

Patients with KS typically have small testes, cryptorchidism, hypospadias or micropenis, delayed puberty, gynecomastia, decreased body hair, and infertility. These patients are generally taller than their peers. Learning disabilities, delayed speech, and delayed language development are more common for individuals with this syndrome. Additionally,

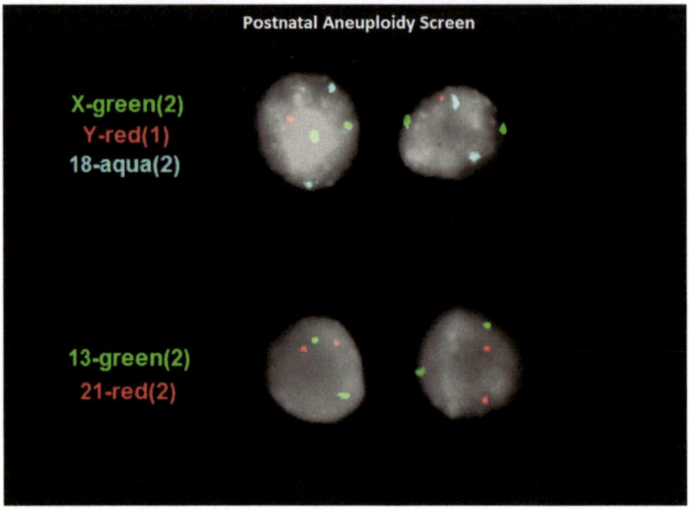

FIG. 3.5.2 FISH for postnatal aneuploidy screening showed a gain of the centromere of the X chromosome in all cells examined from a male patient. Chromosomes 13, 18, and 21 were normal with two copies. ISCN: nuc ish(DXZ1x2, DYZ3x1,D18Z1x2)[100],(RB1x2,D21S341x2)[100]

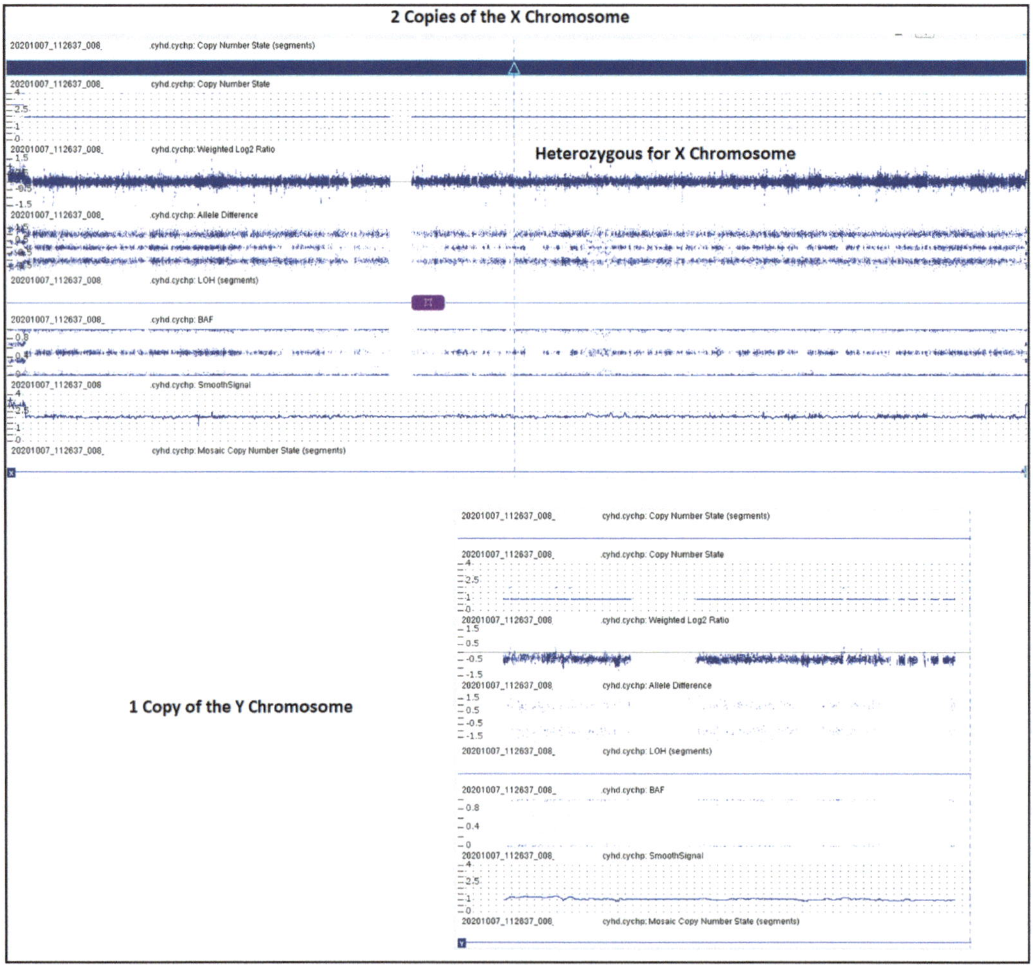

FIG. 3.5.3 Chromosome microarray showed two copies of the X chromosome and one copy of the Y chromosome from this patient. Note that the two X chromosomes are not identical for many of these SNP probes. ISCN: arr[hg19] (X)x2,(Y)x1

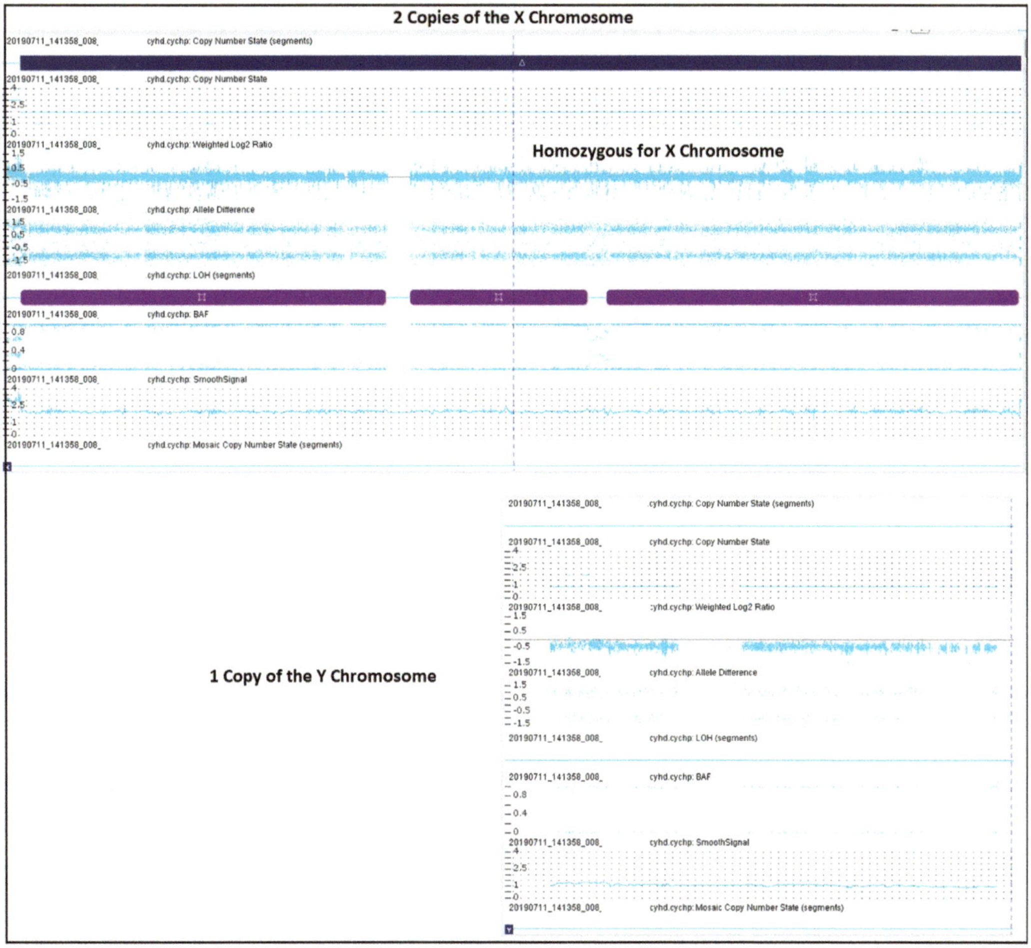

FIG. 3.5.4 Chromosome microarray showed two copies of the X chromosome and one copy of the Y chromosome from another patient. Note that the two X chromosomes are identical in all of the SNP probes.

adults with KS have an increased risk of developing breast cancer and systemic lupus erythematosus, which is a chronic inflammatory disease. Personality characteristics vary among affected individuals. It is common for individuals with this condition to be reserved, sensitive, and timid [1–4].

Future testing and recommendations

Both KS and mosaic Klinefelter syndrome (46,XY/47,XXY) are not inherited. They usually occur as a random event during the formation of reproductive cells (eggs and sperm). During cell division, an error called nondisjunction occurs resulting in a reproductive cell with an abnormal number of chromosomes. KS is diagnosed through karyotype and chromosome microarray on a blood sample. Genetic counseling is recommended.

Case 3.6 Klinefelter syndrome variant (48,XXYY syndrome)

Clinical indication

A 12-year-old male presented with moderate intellectual disabilities and an unspecified developmental disorder of speech and language.

Test ordered

- Chromosome analysis: Routine blood
- Chromosome microarray analysis (CMA)

Laboratory test performed

Chromosome microarray was performed on peripheral blood (PB) (detailed methods described in Chapter 1).

Test results

Chromosome analysis revealed 48,XXYY in all metaphase cells examined (Fig. 3.6.1). Chromosomal Microarray Analysis (CMA) revealed copy number gains of an entire X chromosome and an entire Y chromosome, consistent with a diagnosis of 48,XXYY syndrome, or KS variant (Figs. 3.6.2 and 3.6.3).

Results with interpretations

The 48,XXYY syndrome is estimated to affect 1 in 18,000–40,000 male newborns. Although patients with 48,XXYY syndrome have phenotypes that are very similar to the KS, sharing features such as infertility and hyper-gonadotropic hypogonadism [5], they have distinct phenotypes including mental retardation and psychiatric disorders [6,7]. The syndrome manifests later in life with abdominal adiposity, small testicles, delayed development, behavioral disorders, learning disabilities, delayed puberty, and skeletal deformities [8].

Future testing and recommendations

48,XXYY is not inherited. It usually occurs as a random event during the formation of reproductive cells (eggs and sperm). During cell division, an error called nondisjunction occurs resulting in a reproductive cell with an abnormal number of chromosomes. Most of the 48,XXYY syndrome has the extra sex chromosomes coming from a sperm cell. Genetic counseling is recommended to provide better disease management for the patient.

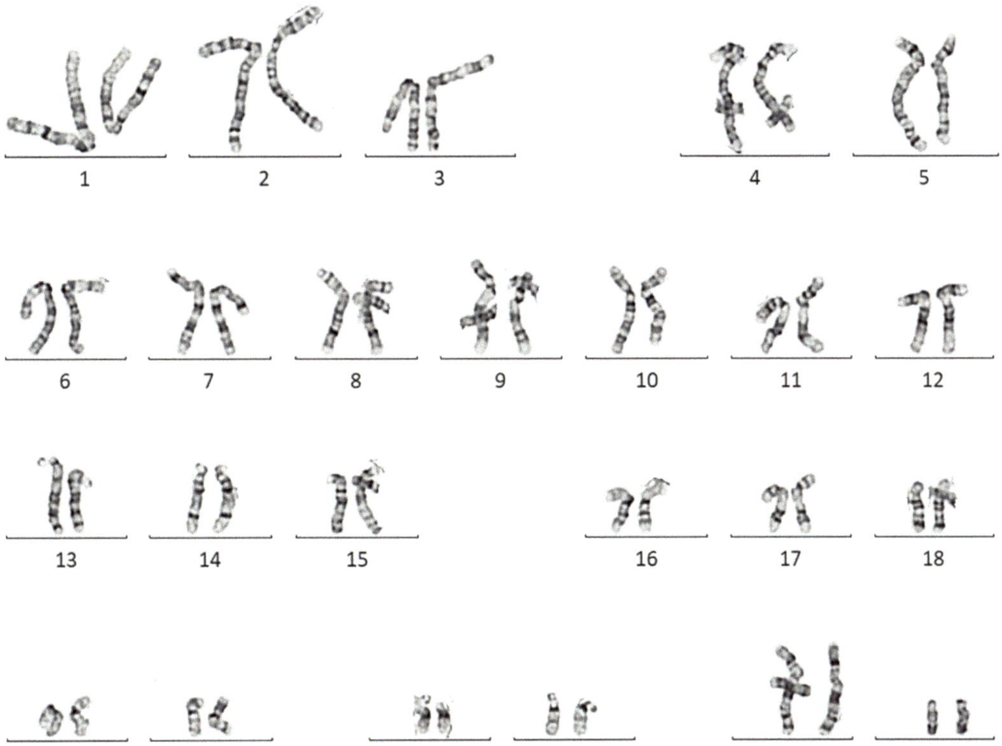

FIG. 3.6.1 Chromosome analysis revealed two copies of the X chromosome and two copies of the Y chromosome. ISCN: 48,XXYY

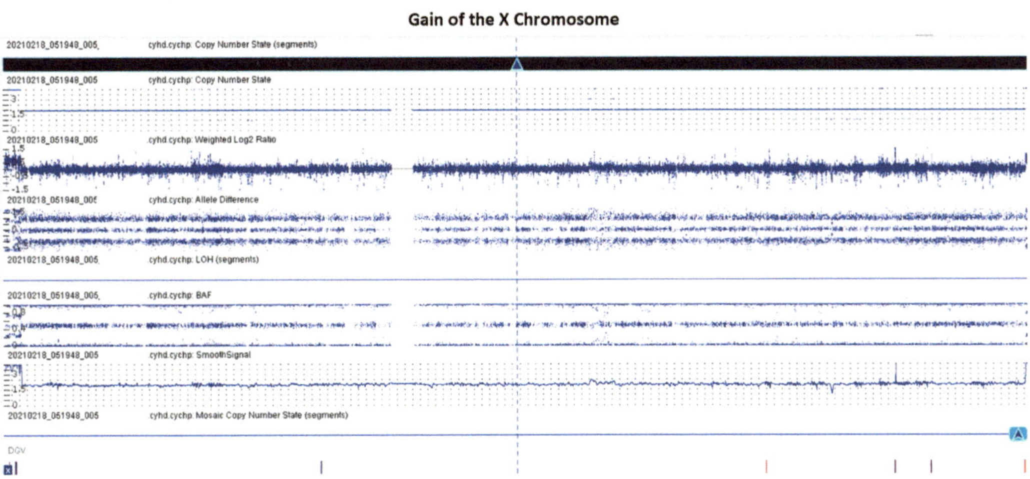

FIG. 3.6.2 Chromosome microarray showed two copies of the X chromosome. ISCN: arr[hg19] (X)x2,(Y)x2

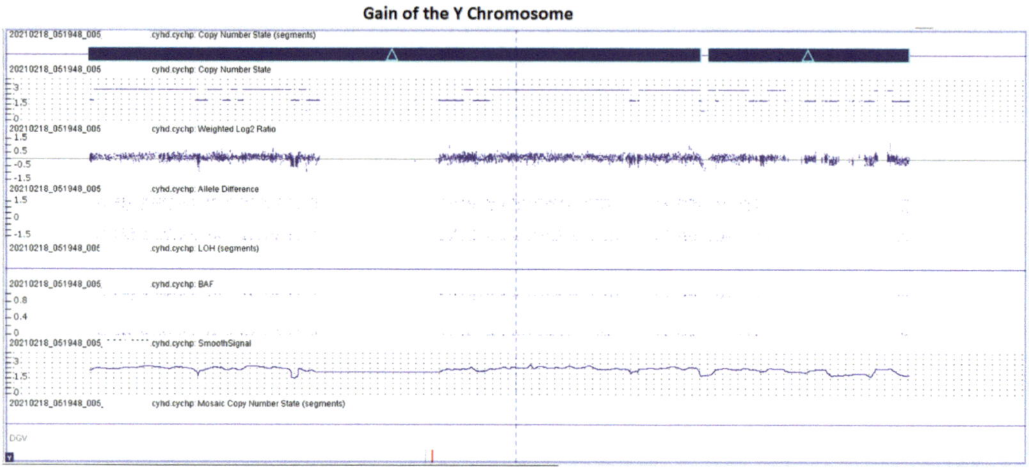

FIG. 3.6.3 Chromosome microarray showed two copies of the Y chromosome.

Summary of key learning points

1. Identification of the Y chromosome material in phenotypic females is important due to the increased risk of gonadoblastoma.
2. CMA provided more accurate information such as size, breakpoints, and more specific gene regions involved, which is helpful for genotype-phenotype correlation.
3. Chromosome analysis may miss submicroscopic genetic alterations, including microdeletions and microduplications. CMA can identify those that could have been missed by karyotyping.
4. Parental chromosome analysis is essential to determine the origin of the abnormality and assess the recurrence risk.

References

[1] J. Nielsen, Follow-up of 25 unselected children with sex chromosome abnormalities to age 12, Birth Defects Orig. Artic. Ser. 26 (4) (1990) 201–207.

[2] J. Nielsen, M. Wohlert, Sex chromosome abnormalities found among 34,910 newborn children: results from a 13-year incidence study in Arhus, *Denmark*, Birth Defects Orig. Artic. Ser. 26 (4) (1990) 209–223.

[3] R.L. Nussbaum, R.R. McInnes, H.F. Willard, Thompson & Thompson Genetics in Medicine, eighth ed., Elsevier, Philadelphia, 2016. xi, 546 pages.

[4] D.J. Wolff, et al., Laboratory guideline for Turner syndrome, Genet. Med. 12 (1) (2010) 52–55.

[5] N. Shankar Kikkeri, S. Nagalli, Turner Syndrome, *StatPearls*, Treasure Island, FL, 2022.

[6] C.H. Gravholt, et al., Clinical practice guidelines for the care of girls and women with Turner syndrome: proceedings from the 2016 Cincinnati international turner syndrome meeting, Eur. J. Endocrinol. 177 (3) (2017) G1–G70.

[7] F. Matsumoto, et al., Variation of gonadal dysgenesis and tumor risk in patients with 45,X/46,XY mosaicism, Urology 137 (2020) 157–160.

[8] A. Swain, et al., Dax1 antagonizes Sry action in mammalian sex determination, Nature 391 (6669) (1998) 761–767.

[9] M. Garcia-Acero, et al., Gene dosage of DAX-1, determining in sexual differentiation: duplication of DAX-1 in two sisters with gonadal dysgenesis, Mol. Biol. Rep. 46 (3) (2019) 2971–2978.

[10] M. Barbaro, et al., Multiplex ligation-dependent probe amplification analysis of the NR0B1(DAX1) locus enables explanation of phenotypic differences in patients with X-linked congenital adrenal hypoplasia, Horm. Res. Paediatr. 77 (2) (2012) 100–107.

[11] M. Barbaro, et al., Multigeneration inheritance through fertile XX carriers of an NR0B1 (DAX1) locus duplication in a kindred of females with isolated XY gonadal dysgenesis, Int. J. Endocrinol. 2012 (2012), 504904.

[12] L.A. Gole, et al., Gonadal mosaicism 45,X/46,X,psu dic(Y)(q11.2) resulting in a Turner phenotype with mixed gonadal dysgenesis, Singapore Med. J. 49 (4) (2008) 349–351.

[13] R.J.M. Gardner, D.J. Amor, Gardner and Sutherland's chromosome abnormalities and genetic counseling, in: Oxford Monographs on Medical Genetics, fifth ed., Oxford University Press, Oxford; New York, 2018. xii, 714 pages.

4

Consanguinity

Xia Li

SONORA QUEST LABORATORIES, PHOENIX, AZ, UNITED STATES

Background

Consanguinity is the kinship of two individuals characterized by the sharing of (a) common ancestor(s) [1,2]. It refers to marriage or a reproductive relationship between two closely related individuals. The degree of relatedness between two individuals defines the proportion of genes shared between them. Rare autosomal recessive diseases are more frequent in the offspring of these couples due to the high risk of homozygosity by descent [3]. Consanguinity is mostly seen in the Middle East, West Asia, and North Africa as well as immigrants from these communities now residing in North America, European, and Australia. In Arab countries, first-cousin marriages may reach 25%–30% of all marriages [4]. Due to the high incidence of congenital and genetic disorders in the offspring of consanguineous couples, it is very important to provide preconception and premarital consultation to these couples [5]. In many cases, the health providers are not aware of the consanguinity until the test results from the offspring are available. In genetic diagnostic laboratories, chromosome single-nucleotide polymorphism (SNP) microarray is the most frequently used assay to identify consanguinity because it can detect large regions of homozygosity (ROH) or absence of heterozygosity (AOH) in patients [6]. In this chapter, we will illustrate such cases with multiple ROH regions.

Case 4.1 Multiple congenital anomalies due to family history of consanguinity

Clinical indication

An 11-year-old boy presented multiple congenital anomalies including a localized visual field defect in the right eye, torticollis, flexion deformity, right finger joints, right congenital diaphragmatic hernia, cystic malformation of the lung, developmental delay, and learning difficulties. He has a family history of consanguinity.

Test ordered

- Chromosome analysis: Routine blood
- Chromosome microarray (CMA)

Cases in Laboratory Genetics and Genomics (LGG) Practice. https://doi.org/10.1016/B978-0-323-99622-8.00030-7

Laboratory test performed

Chromosome analysis and chromosome microarray test methods are described in Chapter 1.

Test results

Chromosome analysis showed a normal male chromosome complement, 46,XY (Fig. 4.1.1).

The chromosome SNP microarray did not detect copy number variations from this patient. However, 36 AOH regions were detected. The clinical significance of these regions was uncertain (Figs. 4.1.2 and 4.1.3). Family history and thorough clinical assessment are indicated to illustrate the significance of these findings.

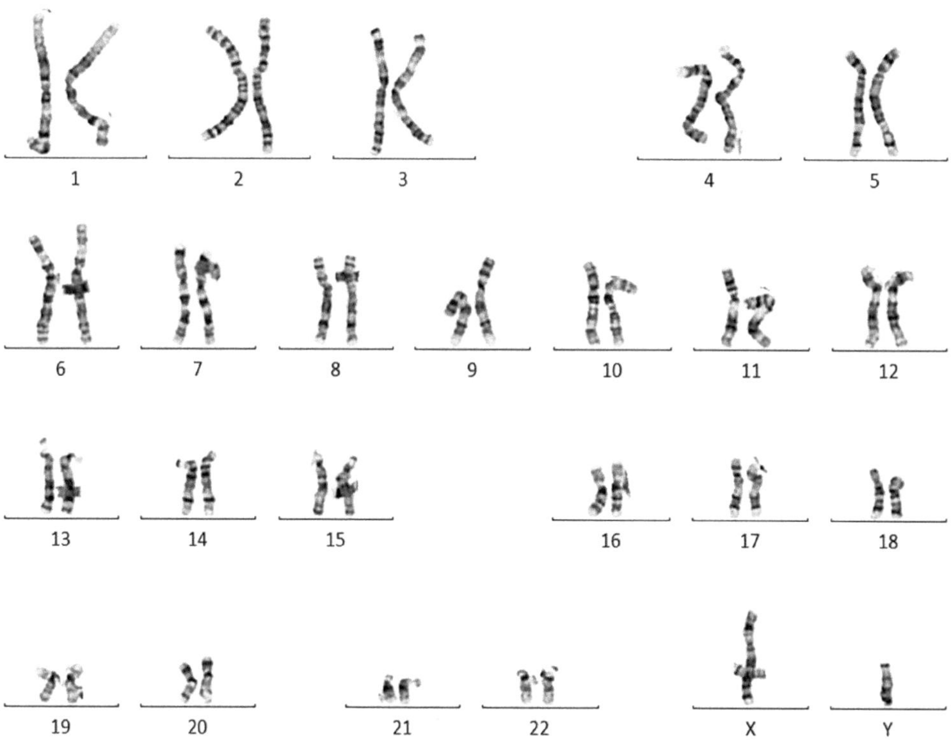

FIG. 4.1.1 The karyotype showed a normal male chromosome complement, 46,XY from this patient. ISCN: 46,XY

FIG. 4.1.2 AOH regions of chromosomes 1–10. All AOH regions were marked by *purple* bars. In the marked regions by *purple* bars, there was no heterozygosity shown. ISCN: arr[hg19] 1q32.1q32.3(202,203,757-213,840,987)x2 hmz; arr[hg19] 2p25.3p14(15,702-66,062,158)x2 hmz; arr[hg19] 2p12p11.2(76,157,542-89,129,064)x2 hmz; arr[hg19] 2q31.2q35(180,569,099-220,771,575)x2 hmz; arr[hg19] 3p14.3p14.2(56,797,105-62,393,949)x2 hmz; arr[hg19] 3q26.33q29(181,324,493-194,226,576)x2 hmz; arr[hg19] 4q21.1q32.3(76,603,925-166,723,333)x2 hmz; arr[hg19] 5p15.33p15.1(2,627,821-16,969,407)x2 hmz; arr[hg19] 5q14.3q21.3(87,073,971-109,353,601)x2 hmz; arr[hg19] 6q13q16.2(70,740,290-100,457,411)x2 hmz; arr[hg19] 6q22.33q25.3(129,797,390-156,640,656)x2 hmz; arr[hg19] 7p14.1p12.3(38,153,649-47,270,301)x2 hmz; arr[hg19] 7p12.3p11.1(48,666,956-58,019,983)x2 hmz; arr[hg19] 7q11.21q21.12(62,461,703-87,157,051)x2 hmz; arr[hg19] 8p23.1p12(8,107,313-34,893,701)x2 hmz; arr[hg19] 8q13.3q22.2(73,419,785-101,492,927)x2 hmz; arr[hg19] 8q22.3q24.23(103,761,836-136,666,924)x2 hmz; arr[hg19] 8q24.23q24.3(137,013,587-146,292,734)x2 hmz; arr[hg19] 9p24.3p21.3(2,123,638-20,482,992)x2 hmz; arr[hg19] 9p21.2p13.1(25,845,455-40,087,758)x2 hmz; arr[hg19] 9q21.11q31.2(71,013,799-110,850,958)x2 hmz; arr[hg19] 10q11.21q21.1(42,433,539-53,061,500)x2 hmz

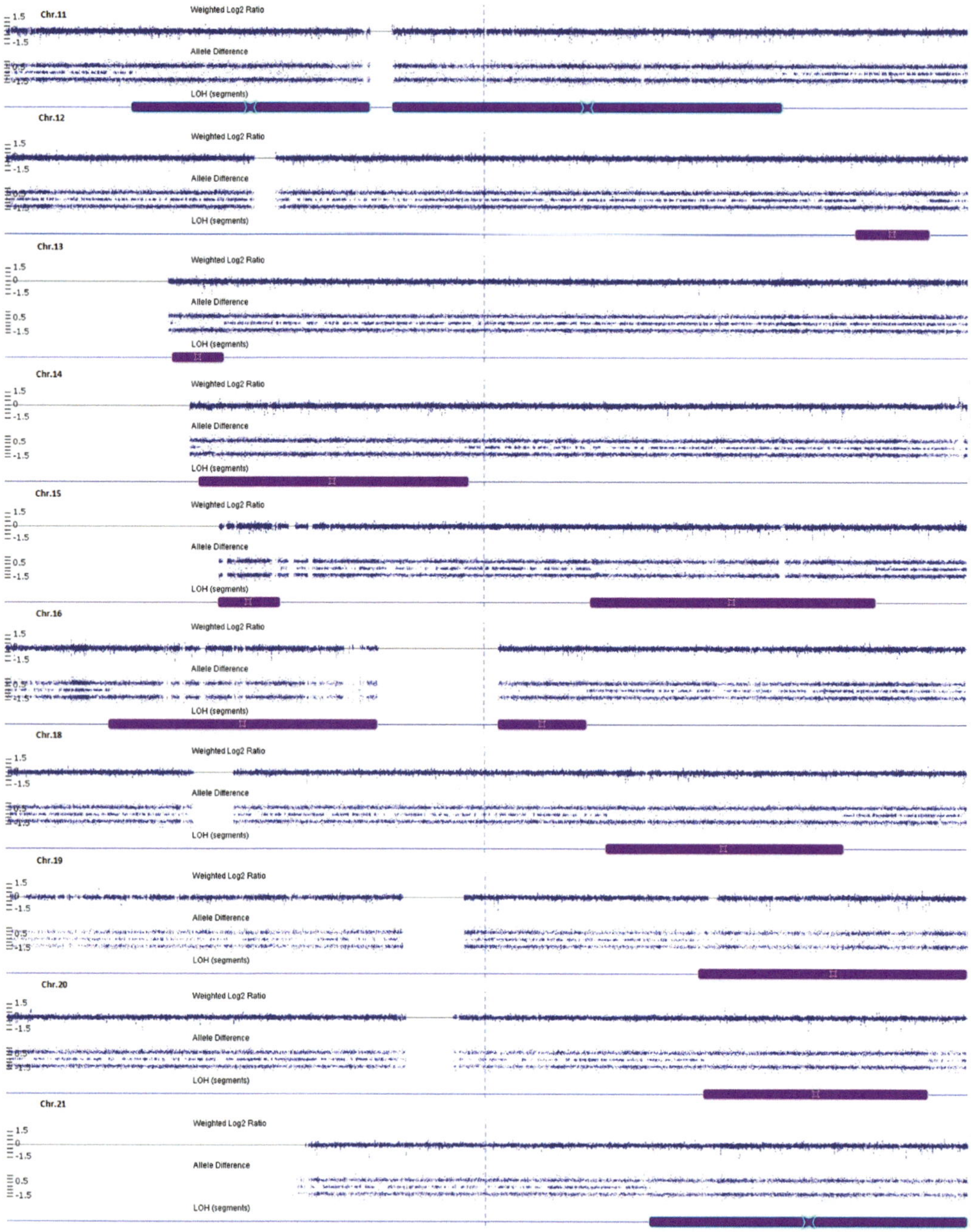

FIG. 4.1.3 AOH regions of chromosomes 11–21. All AOH regions were marked by *purple* bars. In the marked regions by *purple* bars, there was no heterozygosity shown. ISCN: arr[hg19] 11p15.1p11.12(17,742,205-51,563,636)x2 hmz; arr [hg19] 11q11q22.3(54,794,726-109,179,796)x2 hmz; arr[hg19] 12q24.23q24.32(118,190,004-128,413,009)x2 hmz; arr [hg19] 13q12.11q12.13(19,922,501-26,154,513)x2 hmz; arr[hg19] 14q11.2q22.1(21,628,945-51,967,896)x2 hmz; arr [hg19] 15q11.2q13.1(22,752,398-29,580,538)x2 hmz; arr[hg19] 15q22.2q26.1(62,574,003-92,746,498)x2 hmz; arr [hg19] 16p13.2p11.1(9,664,528-35,220,544)x2 hmz; arr[hg19] 16q11.2q12.2(46,464,488-54,798,311)x2 hmz; arr[hg19] 18q21.2q22.2(48,882,697-68,037,283)x2 hmz; arr[hg19] 19q13.2q13.43(42,676,915-59,097,752)x2 hmz; arr[hg19] 20q13.12q13.33(45,827,497-60,410,121)x2 hmz; arr[hg19] 21q22.11q22.3(32,278,815-48,084,820)x2 hmz

Results with interpretations

Chromosome analysis demonstrated a normal male karyotype. No copy number variations (deletions or duplications) were identified in this patient by SNP microarray analysis; however, the SNP microarray detected 36 regions of the AOH regions greater than 5 Mb. The combined total length of the homozygous segments was approximately 833.7 Mb and encompassed 28.9% of the genome. Although this result is not diagnostic of a specific condition, it raises the possibility of a recessive disorder within one of these regions. Additionally, these results could indicate a close familial relationship between this individual's parents. Based on the percentage of homozygous segments and family history, it was confirmed that these AOH regions are due to sibling mating. According to Kearney et al., a long-contiguous stretch of homozygosity (LCSH) describes an uninterrupted region of homozygous alleles with a genomic copy number state of 2. When LCSH is over 25%, the parental relationship is parent/child or full siblings. In this case, it was the latter [7,8].

Future testing and recommendations

These results should be correlated with the patient's family history for interpretation. If a specific autosomal recessive disorder is suspected in this patient, additional molecular testing may be indicated. Genetic counseling is recommended.

Case 4.2 Multiple developmental disorders due to consanguinity and Charcot-Marie-tooth disease type 1A

Clinical indication

A 15-year-old male presented with deficits in the frontal lobe and executive functions. He was diagnosed with a developmental disorder of motor function, speech delay, attention-deficit/hyperactivity disorder (ADHD), anxiety disorder, depression, and posttraumatic stress disorder (PTSD) at various points in his development.

Test ordered

- Chromosome microarray (CMA): Routine blood.

Laboratory test performed

The CMA test method was described in Chapter 1.

Test results

The SNP array analysis showed 20 regions of the AOH greater than 5 Mb. The combined total length of the homozygous segments was approximately 388.3 Mb and encompassed 13.5% of the genome (Fig. 4.2.1).

FIG. 4.2.1 AOH regions of chromosomes 1–18. All AOH regions were marked by *purple* bars. In the marked regions by *purple* bars, there was no heterozygosity shown. ISCN: arr[hg19] 1p36.13p35.2(19,153,428-32,071,674)x2 hmz; arr [hg19] 1q41q42.2(215,004,616-231,317,171)x2 hmz; arr[hg19] 2q31.3q32.2(181,327,569-189,827,546)x2 hmz; arr [hg19] 2q36.1q37.1(224,669,669-235,384,171)x2 hmz; arr[hg19] 3p23p14.1(31,687,524-65,359,644)x2 hmz; arr[hg19] 3q11.1q21.2(93,536,053-124,686,733)x2 hmz; arr[hg19] 4p15.1p14(31,316,889-37,790,259)x2 hmz; arr[hg19] 4q21.21q22.3(82,387,265-96,301,458)x2 hmz; arr[hg19] 4q32.3q34.3(164,662,212-181,153,885)x2 hmz; arr[hg19] 5q31.3q35.1(140,121,427-168,844,301)x2 hmz; arr[hg19] 6p21.1p11.1(40,754,148-58,741,497)x2 hmz; arr[hg19] 6q11.1q22.31(61,968,745-123,867,334)x2 hmz; arr[hg19] 6q25.1q27(152,476,647-170,908,114)x2 hmz; arr[hg19] 8p21.3p11.23(21,706,073-37,256,045)x2 hmz; arr[hg19] 8q13.2q21.13(69,346,828-82,257,060)x2 hmz; arr[hg19] 8q22.1q22.3(96,871,803-103,830,188)x2 hmz; arr[hg19] 12p12.1p11.1(23,982,558-34,835,837)x2 hmz; arr[hg19] 12q11q13.13(37,857,750-53,125,363)x2 hmz; arr[hg19] 15q11.2q13.3(22,752,398-31,566,867)x2 hmz; arr[hg19] 18q12.1q22.3(28,536,054-69,297,115)x2 hmz

Chromosomal microarray analysis (CMA) also revealed a copy number gain of 1.4 Mb of DNA from chromosome 17 at band 17p11.2 (Fig. 4.2.2). This duplication, which includes five OMIM genes, is overlapping with Charcot-Marie-Tooth disease type 1A (CMT1A). *PMP22* is one of these five OMIM genes in the duplicated region from this patient (Fig. 4.2.3).

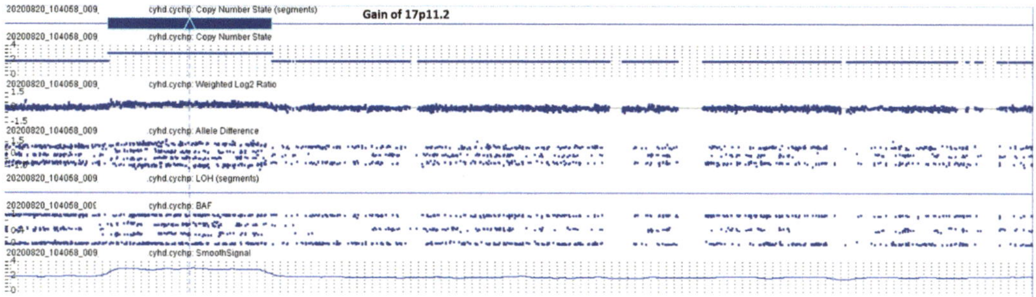

FIG. 4.2.2 Chromosomal microarray analysis also revealed a copy number gain of 1.4 Mb of DNA from chromosome 17 at band 17p11.2. This duplication includes *PMP22*, which is associated with CMT1A. ISCN: arr[hg19] 17p12(14,087,933-15,479,924)x3

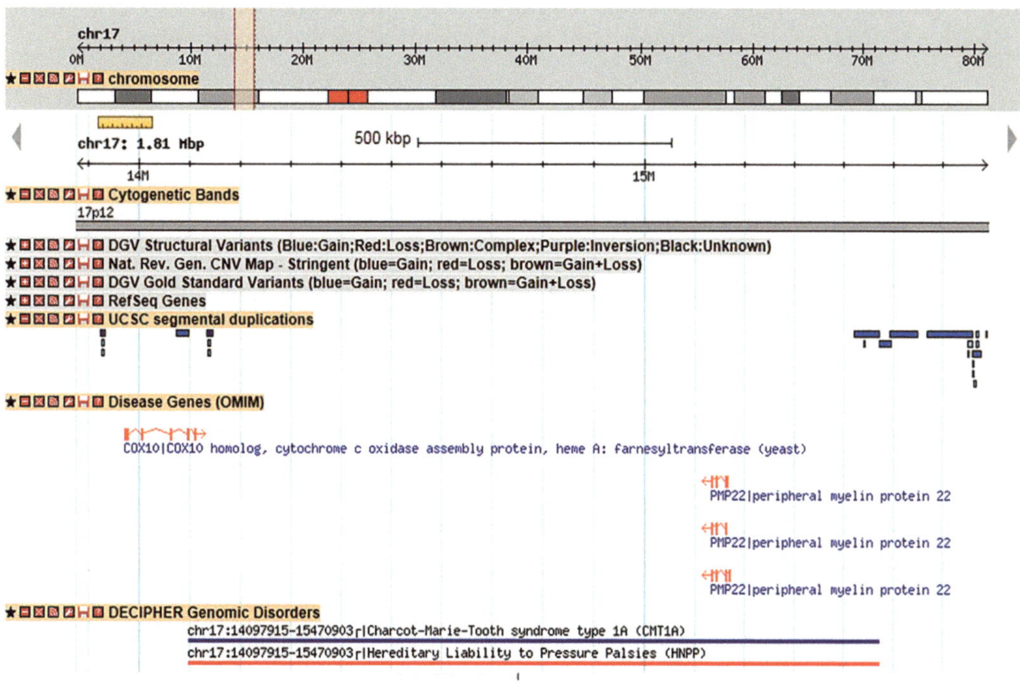

FIG. 4.2.3 A screenshot of the database of genomic variants showed that *PMP22* is one of the five OMIM genes in the duplicated region from this patient.

Results with interpretations

From the first finding, the chromosome microarray identified 20 regions of AOH. Based on the percentage of homozygosity, this result indicates a possibility of second-degree mating (half-siblings or uncle-niece) [8,9]. Although this result is not diagnostic of a specific condition, it raises the possibility of a recessive disorder within one of these regions. A thorough evaluation of the patient's clinical presentations and the genes in the AOH regions with clinical follow-up are indicated.

The second finding with a gain of 1.4 Mb of DNA from chromosome 17 at band 17p11.2 is consistent with a diagnosis of CMT1A. This is an inherited demyelinating peripheral neuropathy. Individuals affected with CMT1A usually have distal muscle weakness and atrophy, bilateral foot drop, and sensory loss. This disorder is slowly progressive, and the onset of symptoms often occurs in the second or third decade of life. This is an autosomal dominant disorder associated with distal muscle wasting and weakness. Most cases are due to an approximate 1.5 Mb duplication in 17p12 involving the *PMP22* gene locus [10–15].

Future testing and recommendations

For multiple AOH regions identified, the results should be correlated with the patient's family history for interpretation. If a specific autosomal recessive disorder is suspected in this patient, additional molecular testing may be indicated. For gain of 17p, follow-up with a neurologist for treatment and genetic counseling are recommended.

Summary of key learning points

- The concept of consanguinity is described.
- There are risks in the offspring of a consanguineous couple with autosomal recessive disorders.
- Further molecular testing is indicated if any specific disorders are suspected.

References

[1] M. Fareed, M. Afzal, Genetics of consanguinity and inbreeding in health and disease, Ann. Hum. Biol. 44 (2) (2017) 99–107.

[2] W.J. Schull, J.V. Neel, The effects of parental consanguinity and inbreeding in Hirado, Japan. V. Summary and interpretation, Am. J. Hum. Genet. 24 (4) (1972) 425–453.

[3] W.B. Coleman, G.J. Tsongalis, Essential Concepts in Molecular Pathology, second ed., Elsevier, San Diego, CA, 2019. pages cm.

[4] G.O. Tadmouri, et al., Consanguinity and reproductive health among Arabs, Reprod. Health 6 (2009) 17.

[5] H. Hamamy, Consanguineous marriages: preconception consultation in primary health care settings, J Community Genet. 3 (3) (2012) 185–192.

[6] B. Levy, R.D. Burnside, Are all chromosome microarrays the same? What clinicians need to know, Prenat. Diagn. 39 (3) (2019) 157–164.

[7] L. Grote, et al., Variable approaches to genetic counseling for microarray regions of homozygosity associated with parental relatedness, Am. J. Med. Genet. A 164A (1) (2014) 87–98.

[8] K.L. Sund, et al., Regions of homozygosity identified by SNP microarray analysis aid in the diagnosis of autosomal recessive disease and incidentally detect parental blood relationships, Genet. Med. 15 (1) (2013) 70–78.

[9] K.L. Sund, C.W. Rehder, Detection and reporting of homozygosity associated with consanguinity in the clinical laboratory, Hum. Hered. 77 (1–4) (2014) 217–224.

[10] L.G. Shaffer, et al., Diagnosis of CMT1A duplications and HNPP deletions by interphase FISH: implications for testing in the cytogenetics laboratory, Am. J. Med. Genet. 69 (3) (1997) 325–331.

[11] W. Marques Jr., et al., 17p duplicated Charcot-Marie-tooth 1A: characteristics of a new population, J. Neurol. 252 (8) (2005) 972–979.

[12] G. Kuhlenbaumer, et al., Clinical features and molecular genetics of hereditary peripheral neuropathies, J. Neurol. 249 (12) (2002) 1629–1650.

[13] K. Inoue, et al., The 1.4-Mb CMT1A duplication/HNPP deletion genomic region reveals unique genome architectural features and provides insights into the recent evolution of new genes, Genome Res. 11 (6) (2001) 1018–1033.

[14] J.H. Lee, B.O. Choi, Charcot-Marie-tooth disease: seventeen causative genes, J. Clin. Neurol. 2 (2) (2006) 92–106.

[15] C.J. Shaw, et al., Comparative genomic hybridization using a proximal 17p BAC/PAC array detects rearrangements responsible for four genomic disorders, J. Med. Genet. 41 (2) (2004) 113–119.

5

Uniparental disomy and imprinting disorders

Xia Li

SONORA QUEST LABORATORIES, PHOENIX, AZ, UNITED STATES

Background

Uniparental disomy (UPD) refers to the situation in which both chromosomes of a pair or the regions of chromosome in any individual have been inherited from a single parent [1]. There are two types of uniparental disomy: uniparental isodisomy and uniparental heterodisomy. Uniparental isodisomy refers to two chromosomes or regions of chromosomes that are identical (usually due to monosomy rescue by duplication); uniparental heterodisomy means that the two different chromosomes are inherited from one parent (usually due to trisomy rescue). The predicted phenotype of UPD can vary depending on the chromosome involved or whether the regions of UPD contain genetically imprinted genes. Humans inherit two complete sets of chromosomes, one from the mother and one from the father. Most autosomal genes are expressed from both maternal and paternal alleles. Imprinted genes show expression from only one allele of the gene pair, and their expression is determined by the parent during the production of the gametes. This phenomenon is called genomic imprinting. There are three types of phenotypic effects seen in UPD: (1) the phenotypes related to imprinted genes (only one of the two copies is normally turned on, which copy is active depending on the parent of origin); (2) the phenotypes related to unmasking autosomal recessive disorders; (3) the phenotypes related to aneuploidy that will produce mosaicism. Genomic imprinting is caused by the methylation of cytosine bases in the CpG dinucleotides in the DNA molecule. These CpG islands are key regulatory elements of genes. Almost all imprinted genes have a CpG-rich differentially methylated region (DMR), which usually relates to the silence of the allele expression [2].

There are many genetic diseases related to errors in the imprinting of specific genes and chromosomes. Prader-Willi and Angelman syndromes are classic examples of imprinting disorders. These syndromes are seen in UPD for chromosome 15. About 70% of Prader-Willi syndrome (PWS) cases are caused by a deletion on one chromosome 15 involving bands 15q11.2-q13, which can be detected using high-resolution chromosome analysis, FISH, or chromosomal microarray. There are 20%–30% of PWS due to maternal UPD (missing the paternal chromosome 15) and imprinting defects. A DNA

Cases in Laboratory Genetics and Genomics (LGG) Practice. https://doi.org/10.1016/B978-0-323-99622-8.00031-9

methylation analysis will confirm the underlying genetic mechanism. The remaining <1% of PWS is caused by an imprinting center defect, which requires DNA sequencing [3].

Approximately 70% of individuals with Angelman syndrome have a deletion of the mother's copy of 15q11.2-q13. In Angelman syndrome (AS), paternal UPD of chromosome 15 is rarer and is observed in approximately 5% of the cases (missing the maternal chromosome 15).

Angelman syndrome is caused by the absence of a maternally contributed gene known as *UBE3A* and can be the result of maternal deletion, maternal *UBE3A* mutation, paternal UPD, and abnormalities in the maternal imprinting center on chromosome 15q11-13 region [4,5]. DNA methylation is typically the first test ordered. It can identify 80% of individuals with AS. If DNA methylation analysis is normal, then sequencing of *UBE3A* is indicated (https://www.ncbi.nlm.nih.gov/books/NBK1144/).

There are also other imprinting disorders associated with UPD. Maternal UPD for chromosomes 2, 7, 14, and 15 and paternal UPD for chromosomes 6, 11, 15, and 20 are associated with growth and behavioral abnormalities. UPD of maternal chromosome 7 is associated with intrauterine growth restriction, a phenotype such as Russell-Silver syndrome [6]. A recent review summarized imprinting disorders including chromosome 15q11.2-q13.3 duplication, Silver-Russell syndrome, Beckwith-Weidemann syndrome, *GNAS* gene-related inactivation disorders (e.g., Albright hereditary osteodystrophy), uniparental chromosome 14 disomy, chromosome 6q24-related transient neonatal diabetes mellitus, parent of origin effects in 15q11.2 BP1-BP2 deletion (Burnside-Butler) syndrome, and 15q11-q13 single-gene imprinted disorders [7].

Many studies showed autosomal recessive disorders caused by UPD without imprinting genes involved. The disorders being reported include spinal muscular atrophy with paternal UPD for *ASAH1* gene [8], cystic fibrosis with paternal isodisomy of chromosome 7 [9], cartilage-hair hypoplasia with maternal UPD of chromosome 9 [10], α-thalassemia with maternal UPD for chromosome 16 [11], β-thalassemias with paternal UPD of 11p14.3-11p15 [12], and Bloom syndrome with maternal UPD for chromosome 15 [13].

In this chapter, a few case reports are shown to demonstrate the clinical utility of multiple diagnostic approaches used for identifying an imprinting disorder. The last case is to illustrate the identification of autosomal recessive disorder, Gaucher disease, due to paternal UPD.

Case 5.1 Prader-Willi syndrome

Clinical indication

A 28-year-old female had an abnormal pregnancy history. Previous chromosome analysis on her amniotic fluid had additional material of unknown origin on the long arm of chromosome 15 [add(15)(q25)].

Test ordered

- Chromosome analysis: Amniotic fluid
- AneuVysion FISH: Chromosomes 13, 18, 21, X, and Y
- Chromosome microarray (CMA)
- DNA methylation PCR: PWS/AS

Laboratory test performed

Chromosome analysis was performed (a detailed method for karyotyping was described in Chapter 1). FISH was performed using probes (Abbott Molecular, Inc.) targeting specific sequences from chromosomes X, Y, 13, 18, and 21, with 100 cells examined for each probe. This test will not detect numerical abnormalities of other chromosomes, structural abnormalities of any chromosome, or mosaicism. The microarray method was described previously (see Chapter 1). For DNA methylation PCR, bisulfite conversion of DNA was performed, then followed by DNA methylation PCR using primers designed to amplify the PWS/AS region. Gel electrophoresis was run to separate the bands, along with the positive and negative controls for PWS/AS so that the band from the fetus can be illustrated.

Test results

The patient had her first amniocentesis. The results showed trisomy 15 in 6 out of 15 (40%) colonies (Fig. 5.1.1). She was not convinced by the results and asked for chromosome analysis on the second amniocentesis and cordocentesis, which is percutaneous umbilical blood sampling (PUBS). The second amniocentesis results revealed trisomy 15 in 1 out of 15 (6.7%) colonies (Fig. 5.1.2) and 1 out of 20 (5%) cells in cordocentesis examined, respectively (Fig. 5.1.3). Finally, AneuVysion FISH showed a normal signal pattern for chromosomes 13, 18, and 21 in a female fetus (Fig. 5.1.4). Because all trisomy 15 instances could be associated with trisomy rescue resulting in uniparental disomy for chromosome 15, microarray and DNA methylation testing were performed. Microarray results showed a large region of heterodisomy and ~4 Mb of isodisomy with the absence of heterozygosity of chromosome 15, indicating a possible UPD involving the 15q11-q13 imprinting region associated with PWS/AS (Fig. 5.1.5). Genotyping data from microarray with markers close to the centromere and telomere indicated that the fetus had maternal UPD for chromosome 15. Arrows refer to the informative markers (Fig. 5.1.6). Finally, DNA methylation analysis showed PWS-positive results in the fetus (Fig. 5.1.7).

Results with interpretations

It is known that paternal deletion or maternal uniparental disomy leads to PWS (mental retardation, obesity, and dysmorphic facial features). *SNRP* is associated with

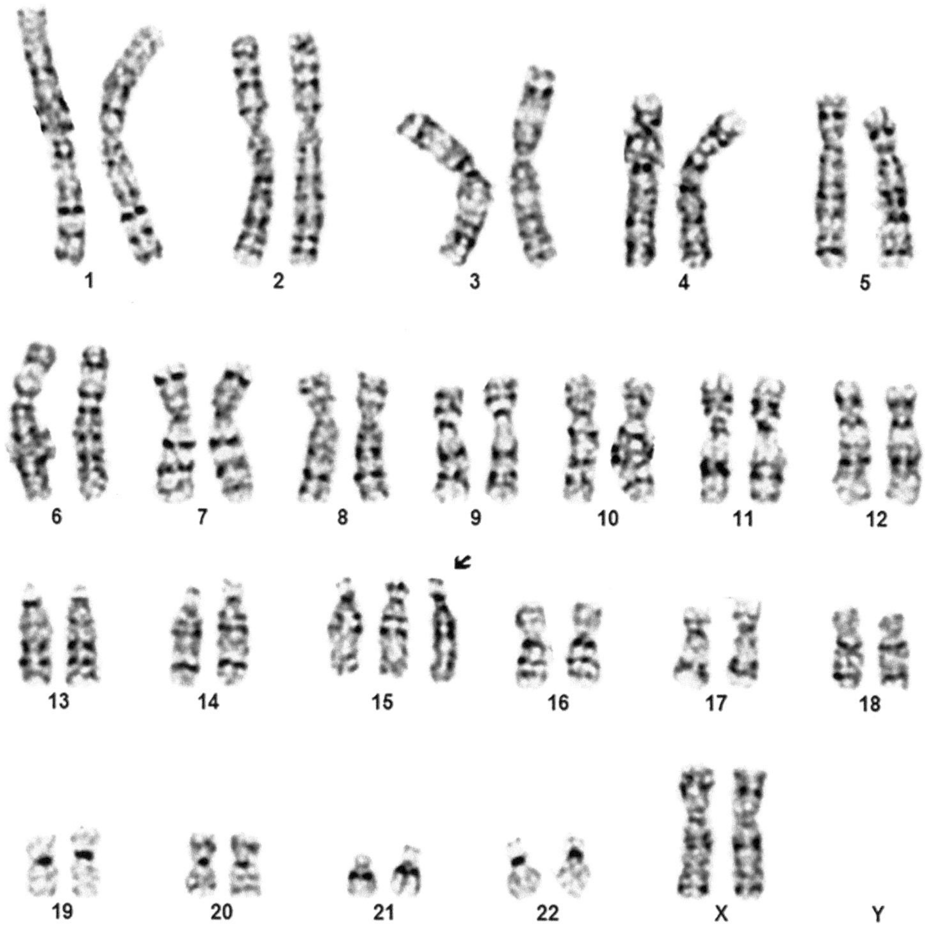

FIG. 5.1.1 Karyotype from first amniocentesis showed trisomy 15 in 6 out of 15 colonies. ISCN: 47,XX,+15 [6]/46,XX [9]

the imprinting center; Maternal deletion, paternal disomy, or deletion of *UBE3A* leads to AS (profound MR, no speech development, and uncontrolled laughter). Two mechanisms result in UPD: trisomy rescue and monosomy duplication. In this case, trisomy 15 was identified from karyotypes of amniocentesis and cordocentesis. Therefore, a UPD 15 has resulted from trisomy rescue. Microarray analysis detected a partial isodisomy on distal chromosome 15, indicating a possible UPD 15. Genotyping data from microarray showed that some informative markers close to the centromere and telomere are of maternal origin. Furthermore, methylation PCR confirmed that the diagnosis of this case is PWS. The conclusion was that postzygotic trisomy rescue resulted in the loss of paternal chromosome 15 and rendered PWS in the fetus.

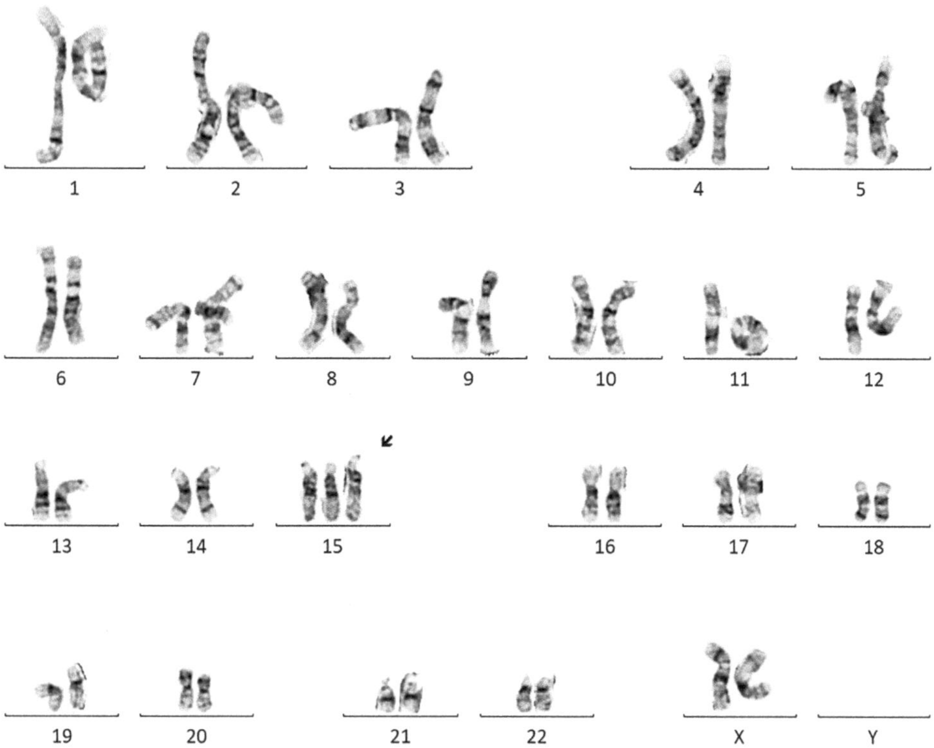

FIG. 5.1.2 Karyotype from second amniocentesis showed trisomy 15 in 1 out of 15 colonies. ISCN: 47,XX,+15 [1]/46, XX [14]

Future testing and recommendation

PWS is caused by an absence of expression of imprinted genes in the paternally derived PWS/AS region (i.e., 15q11.2-q13) of chromosome 15. Several genetic mechanisms are causing PWS: paternal deletion, maternal uniparental disomy 15, and rarely an imprinting defect. The risk to the sibs of an affected child of having PWS depends on the genetic mechanism that renders PWS, also known as missing paternal expression of genes in the 15q11.2-q13 region. The risk to sibs is typically <1% if the affected child has a deletion or uniparental disomy, up to 50% if the affected child has an imprinting defect, and up to 25% if a parental chromosome translocation is present. Prenatal testing is possible for future pregnancies at increased risk if the underlying genetic mechanism is known (https://www.ncbi.nlm.nih.gov/books/NBK1330/). Genetic counseling is recommended.

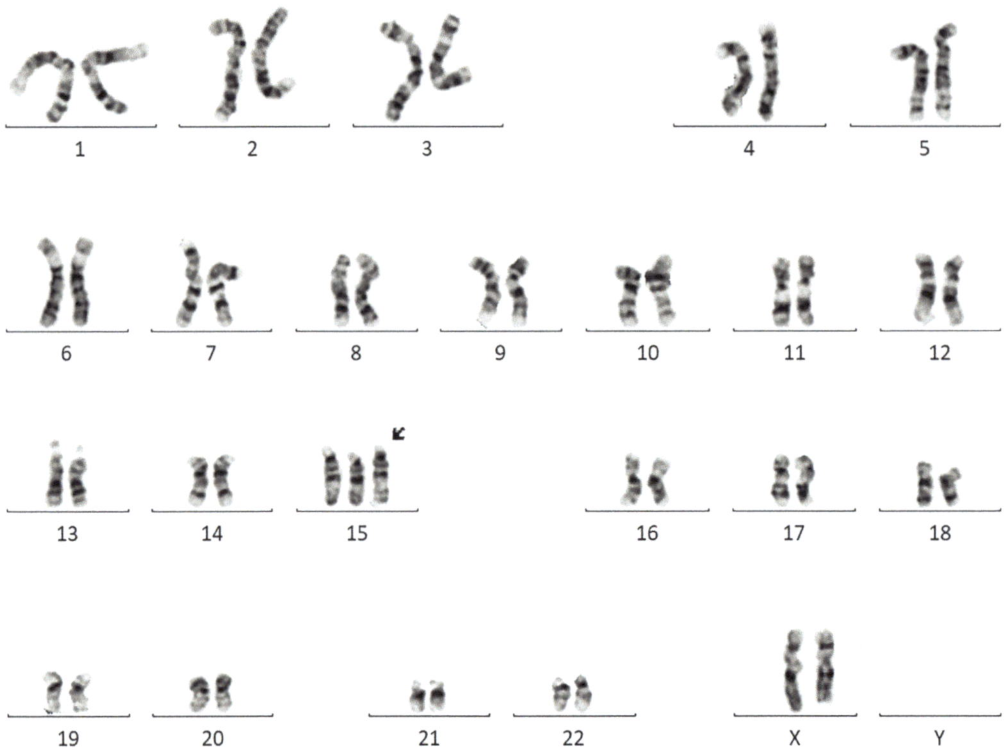

FIG. 5.1.3 Karyotype from cordocentesis showed trisomy 15 in 1 out of 20 cells. ISCN: 47,XX,+15 [1]/46,XX[19]

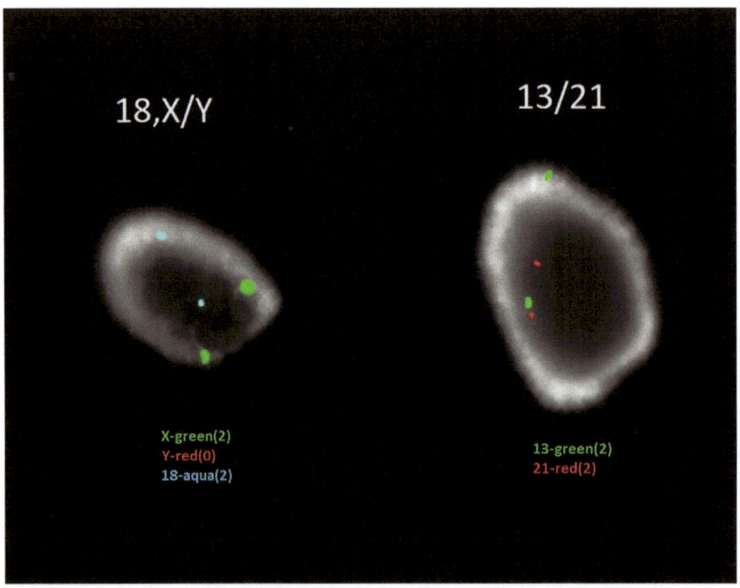

FIG. 5.1.4 AneuVysion FISH showed a normal signal pattern for chromosomes 13, 18, and 21 in a female fetus. ISCN: nuc ish(DXZ1,D18Z1)x2[100],(RB1,D21S341)x2[100]

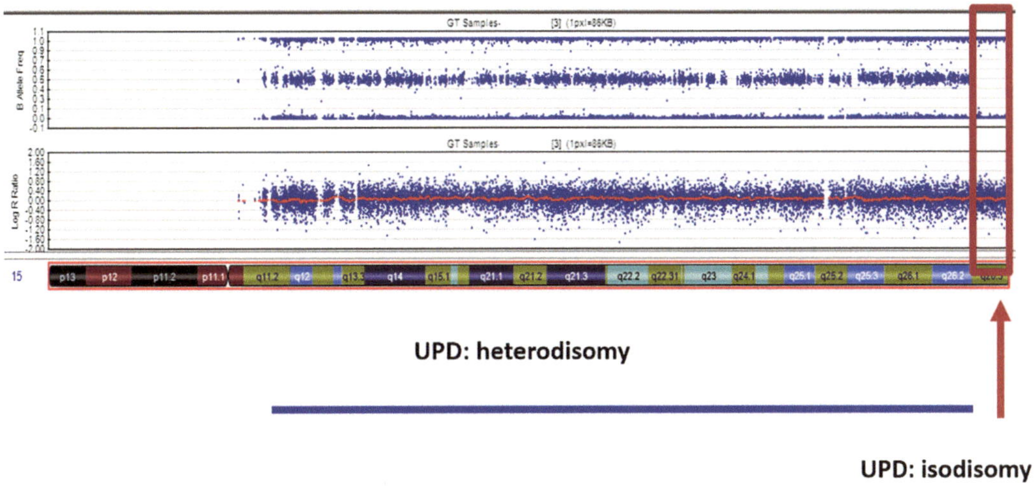

FIG. 5.1.5 Microarray results showed a large region of heterodisomy and ~4 Mb of isodisomy with the absence of heterozygosity of chromosome 15, indicating a possible UPD involving the 15q11-q13 imprinting region associated with PWS/AS. ISCN: arr[GRCh37]15q26.3(98503933_102429049)x2 hmz

Markers close to centromere

		Dad	Mom	Fetus
Probe Name	Chr	GType	GType	GType
rs7402159	15	BB	AB	AB
rs2843721	15	BB	BB	BB
rs9330233	15	AB	AA	AA
rs3561796	15	AB	AA	AA
rs8040193	15	AB	AA	AA
rs8034562	15	AB	BB	BB
rs1290025	15	AB	BB	BB
rs7359214	15	BB	BB	BB
rs1291552	15	AB	AA	AA
rs1210202	15	AB	AA	AA
rs1148638	15	BB	AA	AA
rs2289815	15	BB	AA	AA
rs765763	15	BB	AA	AA
rs1290207	15	AB	BB	BB
rs1289830	15	AB	BB	BB
rs7171787	15	AB	AB	AB
rs722410	15	AA	AB	AB
rs8029108	15	AB	AA	AA
rs8042342	15	BB	BB	BB
rs1579821	15	AB	AA	AA
rs4778298	15	BB	AB	AB
rs3751567	15	BB	BB	BB
rs2289816	15	AB	BB	BB
rs4134803	15	BB	AA	AA
rs2289817	15	AB	AA	AA
rs2289819	15	AB	BB	BB
rs4778471	15	BB	AB	AB
rs7174982	15	AB	AB	AB
rs6606813	15	AB	AA	AA

Markers close to telomere

		Dad	Mom	Fetus
Probe Name	Chr	GType	GType	GType
rs8037807	15	AB	BB	BB
rs8031873	15	AA	AA	AA
rs11633427	15	AA	NC	AA
rs3736298	15	BB	BB	BB
rs2456491	15	AB	AB	AA
rs375606	15	AB	AB	AA
rs405805	15	BB	BB	BB
rs352716	15	AB	AA	AA
rs8040299	15	BB	BB	BB
rs3934044	15	BB	BB	BB
rs3905942	15	AB	BB	BB
rs352743	15	AB	BB	BB
rs352744	15	AB	AA	AA
rs9328589	15	AB	AA	AA
rs12324219	15	BB	BB	BB
rs7182334	15	AB	AB	AA
rs1399026	15	BB	AB	BB
rs7182651	15	BB	AA	AA
rs2203348	15	AA	AA	AA
rs7165135	15	AB	AB	BB
rs7180058	15	BB	AB	AA
rs2203349	15	AB	AB	BB
rs6598500	15	AB	AB	AA
rs8036482	15	BB	BB	BB
rs1513981	15	BB	AB	BB
rs7177306	15	AB	BB	BB
rs10152940	15	AB	AB	BB
rs13329121	15	AA	AA	AA
rs8024127	15	AB	AB	BB

FIG. 5.1.6 Genotyping data from microarray with markers close to the centromere and telomere indicated that the fetus had maternal UPD for chromosome 15. *Arrows* refer to the informative genotypes from the mother.

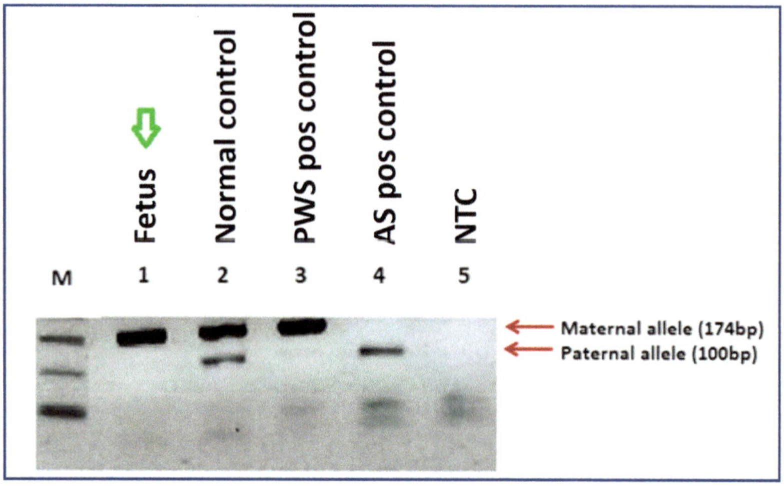

FIG. 5.1.7 Gel electrophoresis from DNA methylation analysis showed PWS positive results in the fetus as indicated in lane 1 by a *green* arrow. Lane 2: normal control, lane 3: PWS-positive control, lane 4: AS-positive control, lane 5: PCR negative control. M: DNA marker.

Case 5.2 Prader-Willi/Angelman syndrome

Clinical indication

A newborn baby boy presented with hypotonia and bilateral cryptorchidism, suspecting Prader-Willi syndrome.

Test ordered

- Chromosome analysis: Routine blood
- FISH: PWS/AS
- Chromosome microarray (CMA)

Laboratory test performed

Karyotyping, FISH, and microarray methods were described in Chapter 1.

Test results

Chromosome analysis was performed initially. Of the 20 metaphases examined, all exhibited an interstitial deletion of the long arm of chromosome 15 (Fig. 5.2.1).

Fluorescent molecular probes, which are localized to *D15Z1* (15p11.2; internal control), *SNRPN* (15q11-q13), and *PML* (15q24; internal control) were hybridized to metaphase preparations. The fluorescence signal pattern was abnormal in the 10 metaphase cells examined. There was a evidence for a deletion in the Prader-Willi region on chromosome 15 using the *SNRPN* probe (Fig. 5.2.2).

FIG. 5.2.1 The karyotype from peripheral blood showed an interstitial deletion of 15q11.2q13 in 20 cells examined. ISCN: 46,XY,del(15)(q11.2q13)

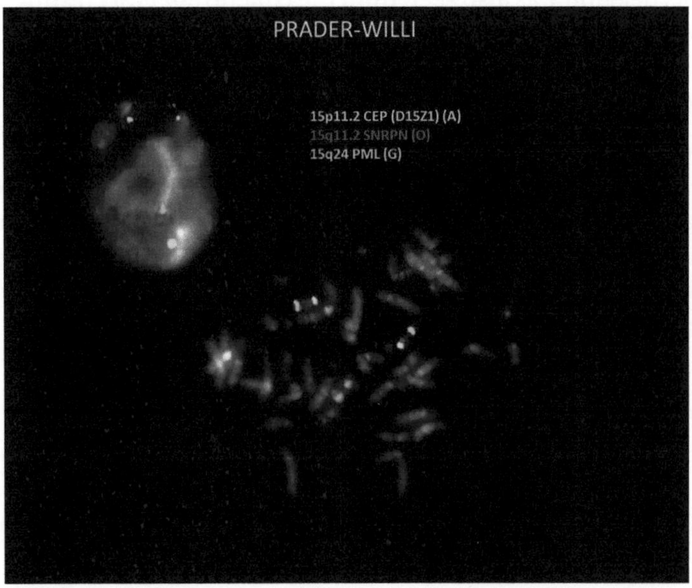

FIG. 5.2.2 FISH using probes for PWS/AS showed a deletion of *SNRPN*. ISCN: ish 15p11.2(D15Z1x2),15q11-13 (SNRPNx1),15q24(PMLx2)[10]

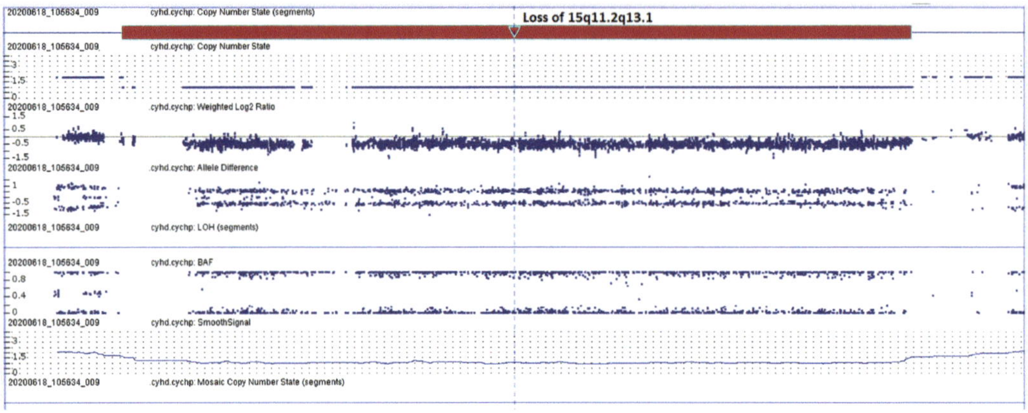

FIG. 5.2.3 Microarray results showed a 5.3 Mb deletion of chromosome 15. ISCN: arr[hg19] 15q11.2q13.1(23,213,406-28,561,671)x1

Chromosomal microarray analysis revealed a copy number loss of chromosome band 15q11.2q13.1 in the Prader-Willi/Angelman syndrome critical region of 5.3 Mb in size, spanning 20 OMIM genes. This finding is consistent with a clinical diagnosis of Prader-Willi or Angelman syndrome depending on the parental origin of the deleted chromosome 15. This finding confirms the chromosome and FISH results reported for this patient (Fig. 5.2.3).

Results with interpretations

Of the metaphases examined, all exhibited an interstitial deletion of the long arm of chromosome 15. This deletion was also seen from the concurrent FISH testing using probes for PWS/AS. Microarray analysis was performed and confirmed the deletion.

Future testing and recommendations

The great majority of deletion-associated cases have been de novo in origin. However, recurrence secondary to a parental rearrangement has been documented, and parental chromosome analyses are recommended. Potential parental origin is dependent upon clinical diagnosis (Prader-Willi or Angelman syndrome). A methylation study is indicated to differentiate whether the patient has PWS or AS. Clinical correlation and genetic counseling are highly recommended.

Case 5.3 Angelman syndrome

Clinical indication

A 22-month-old male presented with global developmental delay and dysmorphic craniofacial features.

Test ordered

- Chromosome microarray (CMA)
- DNA methylation PCR: PWS/AS

Laboratory test performed

CMA methods were described in Chapter 1. The DNA methylation method was described previously in Case 5.1.

Test results

CMA analysis identified UPD for the entire chromosome 15 (see Fig. 5.3.1). DNA Methylation PCR for PWS/AS has confirmed that this patient is positive for Angelman syndrome (Fig. 5.3.2).

Results with interpretations

Microarray identified UPD 15 from the patient. Based on the SNP analysis, the region of homozygosity revealed uniparental isodisomy. Methylation PCR for PWS/AS confirmed that this UPD was associated with Angelman syndrome (paternal UPD 15).

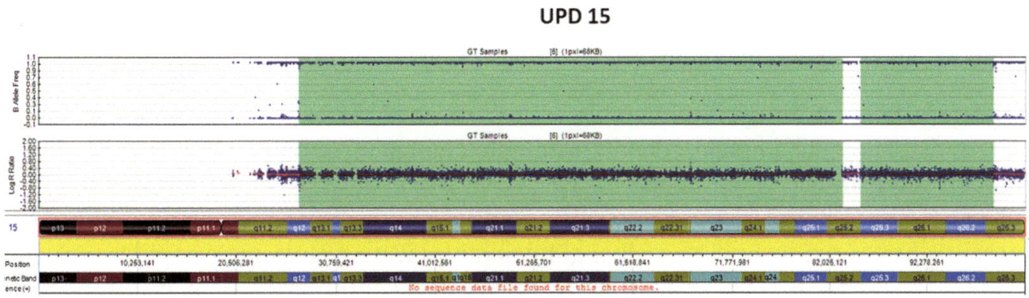

FIG. 5.3.1 Microarray results showed UPD for the entire chromosome 15 from the proband. ISCN: arr[hg19] 15q11.2q26.3 (18,427,103-100,338,915)x2 hmz

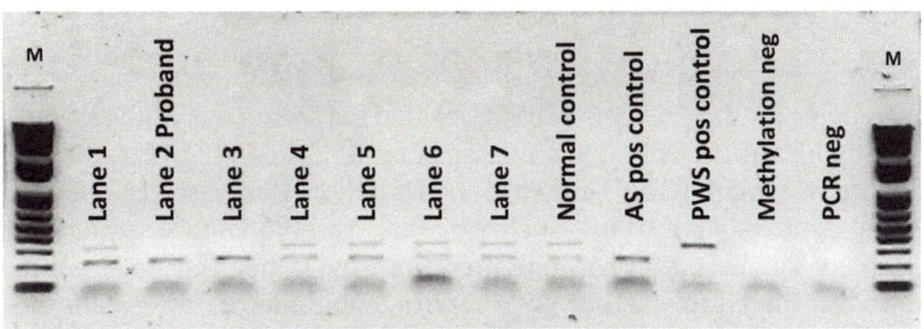

FIG. 5.3.2 Methylation PCR showed the band pattern positive for Angelman syndrome for the proband (lane 2).

Future testing and recommendations

The risk of recurrence of uniparental disomy is less than 1%. Abnormal changes (mutations) within *UBE3A* have been detected in 10%–20% of individuals with Angelman syndrome. Loss of function of this gene causes all the cardinal clinical features of Angelman syndrome. Most cases of AS result from typical large de novo deletions of 15q11-q13 and are expected to have a low (<1%) risk of recurrence. AS caused by paternal UPD is also expected to have a <1% risk [14]. Genetic counseling is recommended for assessing recurrent risk.

Case 5.4 Gaucher disease

Clinical indication

A 22-month-old girl presented with thrombocytopenia and hepatosplenomegaly. Her enzyme analysis showed reduced glucocerebrosidase. Based on these findings, Gaucher disease was suspected. Peripheral blood was sent for genetic testing.

Test ordered

- *GBA* gene custom Sanger sequencing
- Chromosome microarray (CMA)

Laboratory test performed

GBA gene custom Sanger sequencing was performed using the conventional Sanger sequencing method. To proceed with automated Sanger sequencing, the DNA sequence of interest was used as a template for a special type of PCR called chain-termination PCR. All ddNTPs were mixed in a single reaction, and each of the four dNTPs had a unique fluorescent label. In the second step, all oligonucleotides were run in single capillary gel electrophoresis within the sequencing machine. Next, a computer reads each band of the capillary gel, in order, using fluorescence to call the identity of each terminal ddNTP. The output, called a chromatogram, showed the fluorescent peak of each nucleotide along the length of the template DNA.

Test results

Sanger sequencing for the *GBA* gene was performed for 12 exons. A homozygous mutation in exon 11 was identified. This was a missense mutation c.1448T>C (p.L444P) and was reported as pathogenic by The Human Gene Mutation Database (HGMD). This result was consistent with a diagnosis of Gaucher disease type II. Parental sequencing analyses were performed. The mutation was inherited from the father, but not from the mother. Microarray was ordered to rule out UPD. As expected, microarray results showed UPD for the entire chromosome 1 (Fig. 5.4.1).

UPD chromosome 1

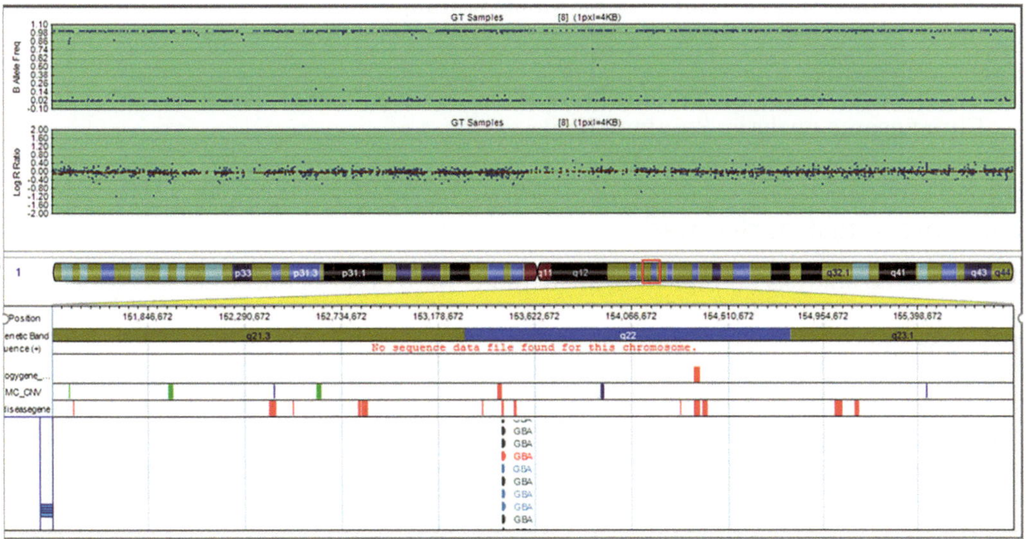

FIG. 5.4.1 Microarray results showed paternal UPD of chromosome 1 from the proband. The *GBA* gene is located on chromosome 1 at band q22. ISCN: arr[hg19]1p36.33q44 (37,714-247,195,072)x2 hmz

Results with interpretations

Sanger sequencing detected a homozygous mutation c.1448T>C (p.L444P) in the *GBA* gene from this patient. This is a pathogenic mutation associated with Gaucher disease type II [15,16]. Microarray confirmed that the patient has paternal UPD suggesting that he carried two mutation alleles from the father according to the results of the Sanger sequencing from both parents. These results confirmed the diagnosis of Type II Gaucher disease. This diagnosis significantly impacts the medical management of liver and bone issues for this patient. Microarray is a very useful tool for identifying copy number variation (CNV), loss of heterozygosity (LOH), and UPD, in many cases finding genes that are associated with autosomal recessive and imprinting disorders.

Future testing and recommendations

Genetic counseling is recommended for recurrent risk assessment.

Case 5.5 Uniparental disomy 7

Clinical indication

The patient was a 4-day-old newborn baby with late prematurity. The maternal medical history revealed polysubstance abuse. He had syphilis with a postnatal diagnosis of double outlet right ventricle with D-malposed great arteries, mild right ventricular outflow tract (RVOT) obstruction, and possible single coronary artery.

Test ordered

- Chromosome analysis: Routine blood
- Chromosome microarray (CMA)

Laboratory test performed

Chromosome analysis and microarray methods were described in Chapter 1.

Test results

Chromosome analysis revealed a normal male karyotype (Fig. 5.5.1). CMA analysis identified uniparental disomy UPD for the entire chromosome 7. The length of the homozygous segment was approximately 159 Mb, encompassing 5.5% of the genome (Fig. 5.5.2).

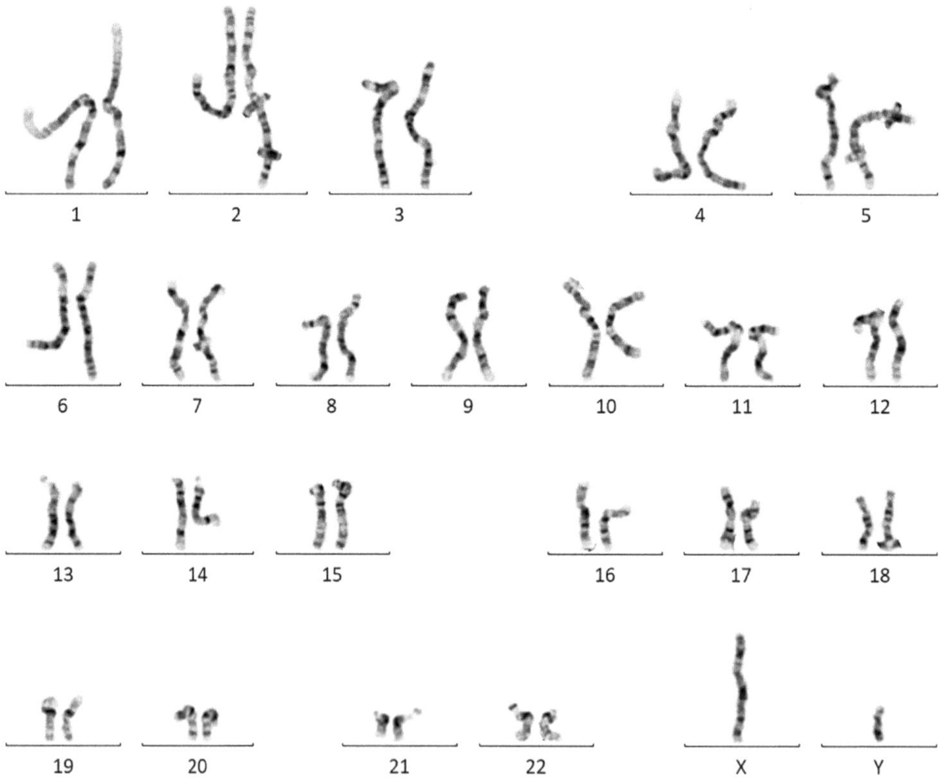

FIG. 5.5.1 Chromosome analysis revealed a normal male karyotype. ISCN: 46,XY

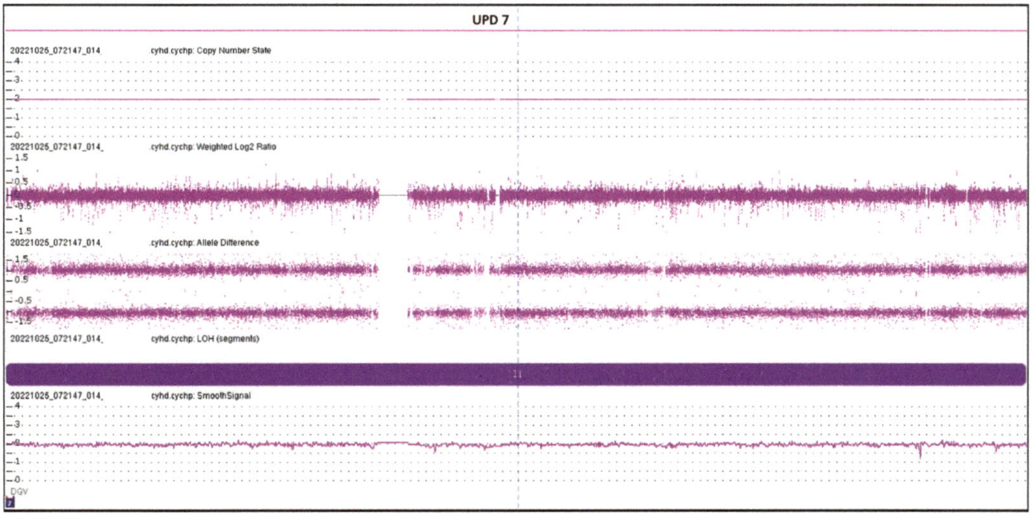

FIG. 5.5.2 Microarray results showed uniparental disomy for chromosome 7. ISCN: arr[hg19] 7p22.3q36.3 (44,167_159,119,220)x2 hmz

Results with interpretations

Microarray identified UPD7 from the patient. Silver-Russell syndrome (SRS) is an imprinting disorder that is associated with UPD7. Hypomethylation of the imprinted control region 1 (ICR1) at 11p15.5 causes SRS in 35%–50% of individuals, and maternal uniparental disomy (mUPD7) causes SRS in 7%–10% of individuals (https://www.ncbi.nlm.nih.gov/books/NBK1324/). Maternal UPD causes people to have two active copies of some imprinted genes and no active copies of others (https://medlineplus.gov/genetics/chromosome/7/#conditions). However, it is unclear whether this patient has matUPD7.

SRS is a clinically and genetically heterogeneous syndrome that is characterized by severe intrauterine and postnatal growth retardation and typical characteristic facial dysmorphisms. It has been associated with maternal uniparental disomy (UPD) for chromosome 7. Further characteristic findings are cardiac anomalies and genital dysmorphia. Asymmetry of body and limb anomalies is present in more than 50% of patients. According to the study from Karaca et al., one case of matUPD7 was recognized among 13 SRS patients. There were no significant differences between the clinical features of maternal UPD7 cases and other SRS cases except for congenital heart defects. Echocardiographic studies revealed ventricular septal defect (VSD), patent ductus arteriosus (PDA), and patent foramen ovale (PFO) in patients with matUPD7 [17].

Paternal uniparental disomy of chromosome 7 most likely does not have any phenotypic expression, except in cases of homozygosity for a recessive disease mutation for which only the father is a carrier (e.g., cystic fibrosis, congenital chloride diarrhea, and sensorineural hearing loss).

Future testing and recommendations

DNA methylation study is indicated to further characterize whether this patient has paternal or maternal UPD7. Genetic counseling is recommended.

Summary of key learning points

- Understand the concepts of uniparental disomy and imprinting disorders.
- Become familiar with how uniparental disomy affects the phenotype with or without imprinting genes (resulting in autosomal recessive diseases).
- Become familiar with the genetic etiology of Prader-Willi and Angelman syndromes.
- Understand technologies used to identify uniparental disomy and imprinting disorders including methylation analysis, karyotyping, FISH, CMA, and sequencing.
- Accurate diagnosis of Gaucher disease will help to improve the management of the patients.

References

[1] W.P. Robinson, Mechanisms leading to uniparental disomy and their clinical consequences, Bioessays 22 (5) (2000) 452–459.

[2] J.G. Falls, et al., Genomic imprinting: implications for human disease, Am. J. Pathol. 154 (3) (1999) 635–647.

[3] R.M. Kliegman, Nelson Textbook of Pediatrics, twenty-first ed., Elsevier, Philadelphia, MO, 2019. pages cm.

[4] S.J. Kim, et al., Unique and atypical deletions in Prader-Willi syndrome reveal distinct phenotypes, Eur. J. Hum. Genet. 20 (3) (2012) 283–290.

[5] T. Ohta, et al., Imprinting-mutation mechanisms in Prader-Willi syndrome, Am. J. Hum. Genet. 64 (2) (1999) 397–413.

[6] M.G. Butler, Genomic imprinting disorders in humans: a mini-review, J. Assist. Reprod. Genet. 26 (9–10) (2009) 477–486.

[7] M.G. Butler, Imprinting disorders in humans: a review, Curr. Opin. Pediatr. 32 (6) (2020) 719–729.

[8] B.G. Giraldez, et al., Uniparental disomy as a cause of spinal muscular atrophy and progressive myoclonic epilepsy: phenotypic homogeneity due to the homozygous c.125C>T mutation in ASAH1, Neuromuscul. Disord. 25 (3) (2015) 222–224.

[9] F. Fares, et al., Paternal isodisomy of chromosome 7 with cystic fibrosis and overgrowth, Am. J. Med. Genet. A 140 (16) (2006) 1785–1788.

[10] T. Sulisalo, et al., Uniparental disomy in cartilage-hair hypoplasia, Eur. J. Hum. Genet. 5 (1) (1997) 35–42.

[11] P.K. Au, et al., A fetus with Hb Bart's disease due to maternal uniparental Disomy for chromosome 16, Hemoglobin 40 (1) (2016) 66–69.

[12] C. Bento, et al., Beta Thalassemia major due to acquired uniparental disomy in a previously healthy adolescent, Haematologica 98 (1) (2013) e4–e6.

[13] T. Woodage, et al., Bloom syndrome and maternal uniparental disomy for chromosome 15, Am. J. Hum. Genet. 55 (1) (1994) 74–80.

[14] H.J. Stalker, C.A. Williams, Genetic counseling in Angelman syndrome: the challenges of multiple causes, Am. J. Med. Genet. 77 (1) (1998) 54–59.

[15] J.T. Brown, et al., Polymorphisms in the glucocerebrosidase gene and pseudogene urge caution in clinical analysis of Gaucher disease allele c.1448T>C (L444P), BMC Med. Genet. 7 (2006) 69.

[16] R. Tammachote, et al., A common and two novel GBA mutations in Thai patients with Gaucher disease, J. Hum. Genet. 58 (9) (2013) 594–599.

[17] E. Karaca, et al., First genetic screening for maternal uniparental disomy of chromosome 7 in Turkish silver-Russell syndrome patients, Iran. J. Pediatr. 22 (4) (2012) 445–451.

6

Pallister-Killian syndrome

Guang Liu and Xia Li
SONORA QUEST LABORATORIES, PHOENIX, AZ, UNITED STATES

Background

Pallister-Killian syndrome (PKS) is a rare genetic disorder caused by mosaic tetrasomy 12p due to an extra isochromosome 12p. It is not inherited and is the result of a random event. PKS is characterized by craniofacial dysmorphisms including a prominent forehead with sparse frontotemporal hair, significant hypotonia in infancy and early childhood, variable developmental delay and intellectual disability, seizures, skin pigmentation, diaphragmatic hernia, congenital heart defects, and other systemic abnormalities [1,2]. A hallmark of this syndrome is tissue-limited mosaicism. The extra isochromosome 12p is rarely found in peripheral blood lymphocytes but can be found in skin fibroblasts and other tissues such as buccal smears, chorionic villi, and amniocytes. Studies have shown that phytohemagglutinin (PHA) used in lymphocyte cultures promotes the growth of normal cells, which leads to the underrepresentation or disappearance of abnormal cells [3]. The percentage of cells containing the extra isochromosome 12p decreases in the blood with age. In lymphocytes, it has been found in fetal blood and newborn infants but has never been seen beyond childhood [4]. The prevalence of PKS has been estimated to be 1 in 20,000; however, PKS is likely underdiagnosed due to the difficulty of making a cytogenetic diagnosis in lymphocytes from a peripheral blood test and requiring sampling of other tissues. This abnormality can be detected by chromosome microarray analysis (CMA) using peripheral blood samples [5]. This is likely explained by the fact that CMA is performed on the total genomic DNA directly prepared from the blood and does not require cell culture with PHA [6].

Case 6.1 Pallister-Killian syndrome

Clinical indication

A 4-day-old baby boy in NICU presented with extremely weak muscle tone and feeding difficulties. Clinical examination showed that the baby had significant hypotonia, a broad forehead, decreased scalp hair density, wide-set eyes, epicanthal folds, low-set ears, flat nasal bridge, micrognathia, micropenis, and hyperpigmented skin on his back.

Cases in Laboratory Genetics and Genomics (LGG) Practice. https://doi.org/10.1016/B978-0-323-99622-8.00023-X

Test ordered

- Chromosome analysis: Routine blood
- Chromosome microarray (CMA)

Laboratory test performed

Chromosome analysis, FISH, and CMA were performed on peripheral blood (PB) (detailed methods were described in Chapter 1).

Test results

Cytogenetic analysis revealed two cell lines. Two of the 68 cells examined showed an abnormal male chromosome complement with an extra isochromosome consisting of the short arm of chromosome 12, resulting in a tetrasomy 12p (Fig. 6.1.1). The remaining 66 cells showed a normal male karyotype (46,XY) (Fig. 6.1.2).

CMA revealed a copy number gain of 34.6 Mb of DNA from chromosome 12 at band 12p13.33p11.1, encompassing 227 OMIM genes (Figs. 6.1.3). The whole genome view from CMA is shown in Fig. 6.1.4.

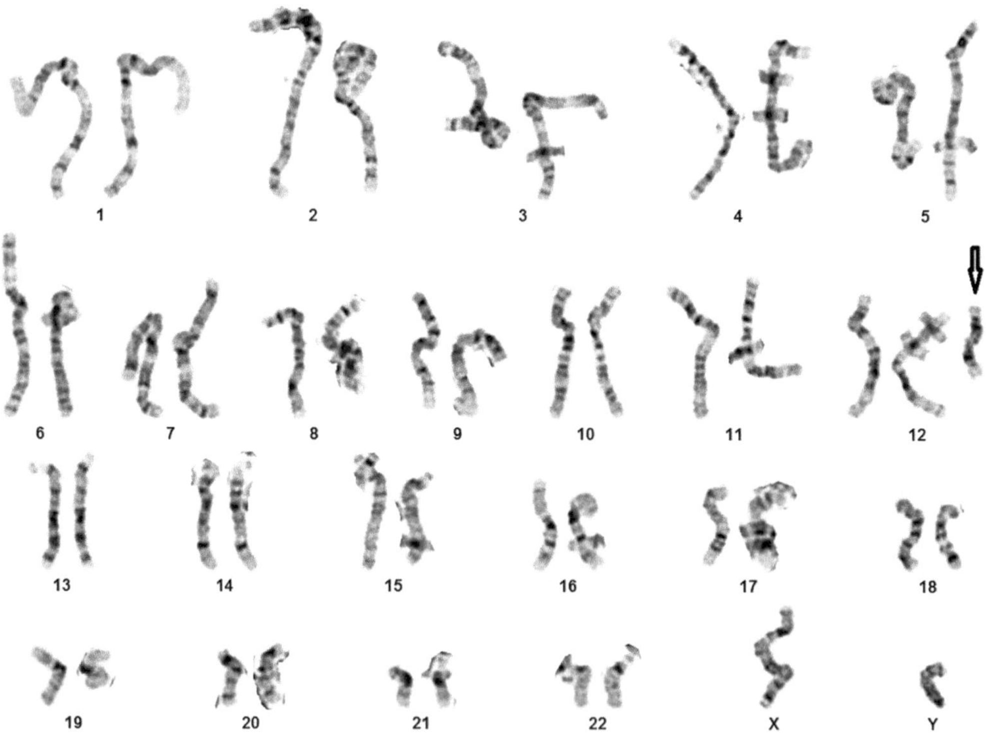

FIG. 6.1.1 The karyotype of the patient showed an extra isochromosome 12p in 2 out of 68 cells examined. ISCN: mos 47,XY,+i(12)(p10)[2]/46,XY[66]

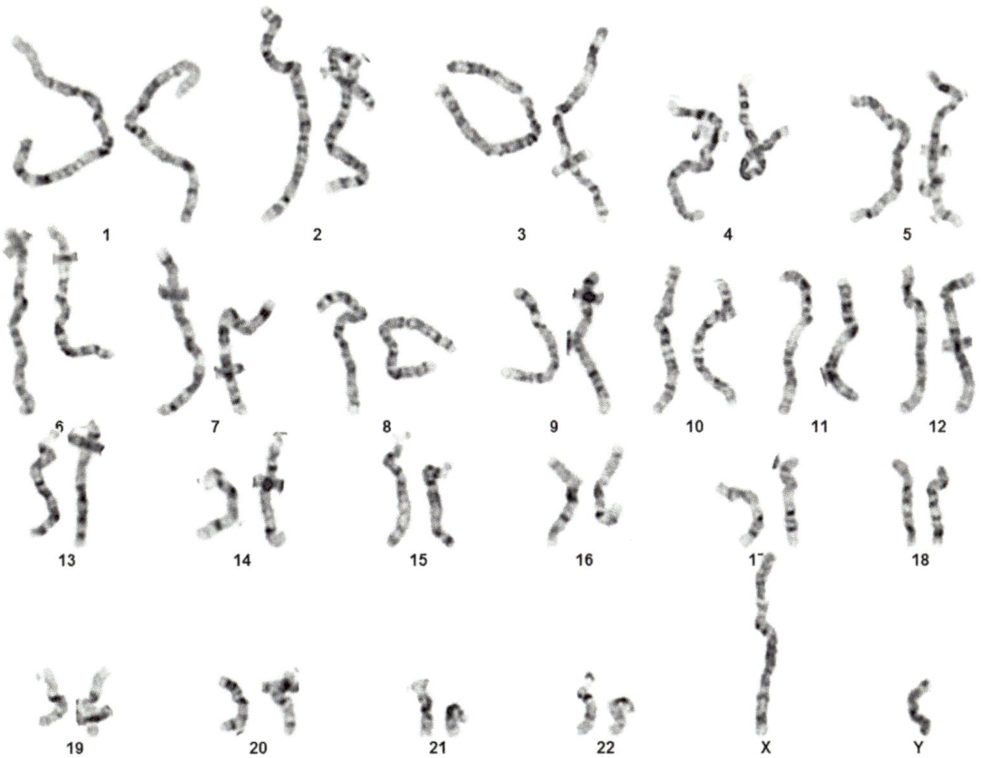

FIG. 6.1.2 The karyotype of the patient showed a normal male karyotype in 66 out of 68 cells examined.

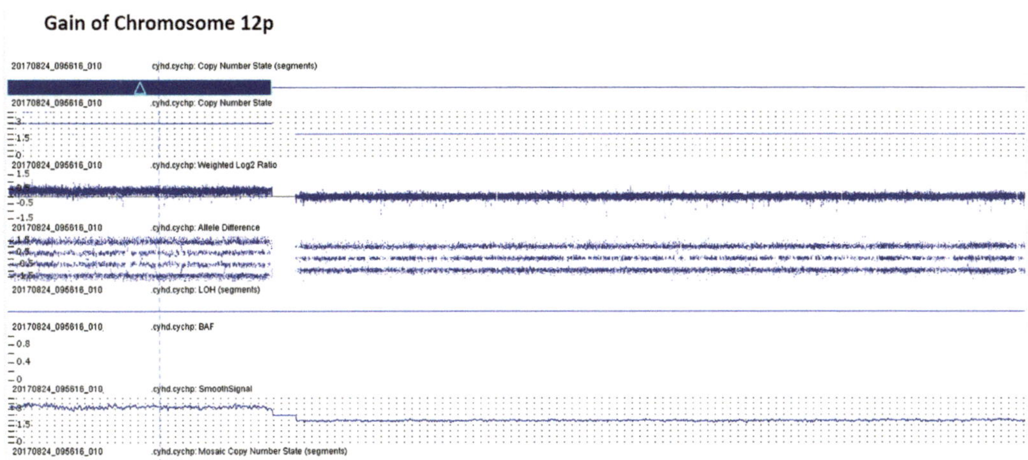

FIG. 6.1.3 The copy number state, Log_2R ratio, and B allele difference for a gain of 12p. ISCN: arr[hg19] 12p13.33p11.1 (173,786-34,835,837)x3

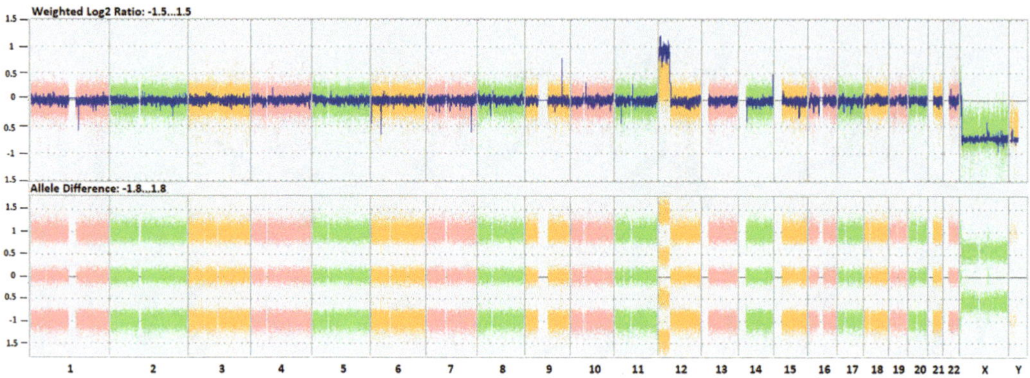

FIG. 6.1.4 The whole-genome view showed a gain of 12p. ISCN: arr[hg19] 12p13.33p11.1(173,786-34,835,837)x3

Results with interpretations

The results from chromosome analysis and CMA are concordant and are consistent with the diagnosis of PKS. Clinical features of PKS include hypotonia, intellectual disability, developmental delay, sparse hair, areas of unusual pigmentation, speech anomalies, and distinctive facial features. The patient had many signs and symptoms described in a typical PKS diagnosis. Chromosome analysis identified very low-level mosaicism of an extra isochromosome 12p in peripheral blood (2 out of 68 cells) due to the selective growth advantage over the karyotypically normal cell population. CMA showed mosaicism for tetrasomy 12p and confirmed the findings from chromosome analysis.

Future testing and recommendations

PKS is not inherited and occurs spontaneously in a child by chance. All cases reported to date have been sporadic. Therefore, no further testing is indicated for this patient. A high-resolution karyotype of both parents should be performed to look for other small chromosomal rearrangements such as cryptic pericentric inversions, which might play a causal role in duplications or isochromosome formation [7]. A thorough clinical assessment of this patient and genetic counseling are recommended.

Case 6.2 Pallister-Killian syndrome

Clinical indication

A newborn (baby boy) presented with a coarse face, hypotonia, and dysmorphic features. The clinical suspicion was genetic disorders.

Test ordered

- Chromosome microarray (CMA)
- Chromosome analysis: Routine blood

Laboratory test performed

Chromosome microarray and chromosome analysis (karyotyping) methods were described in Chapter 1.

Test results

Chromosome analysis was performed initially. Of the 53 metaphases analyzed, one exhibited an isochromosome 12p (Fig. 6.2.1). The remaining 52 cells showed a normal male chromosome complement (46,XY) (Fig. 6.2.2).

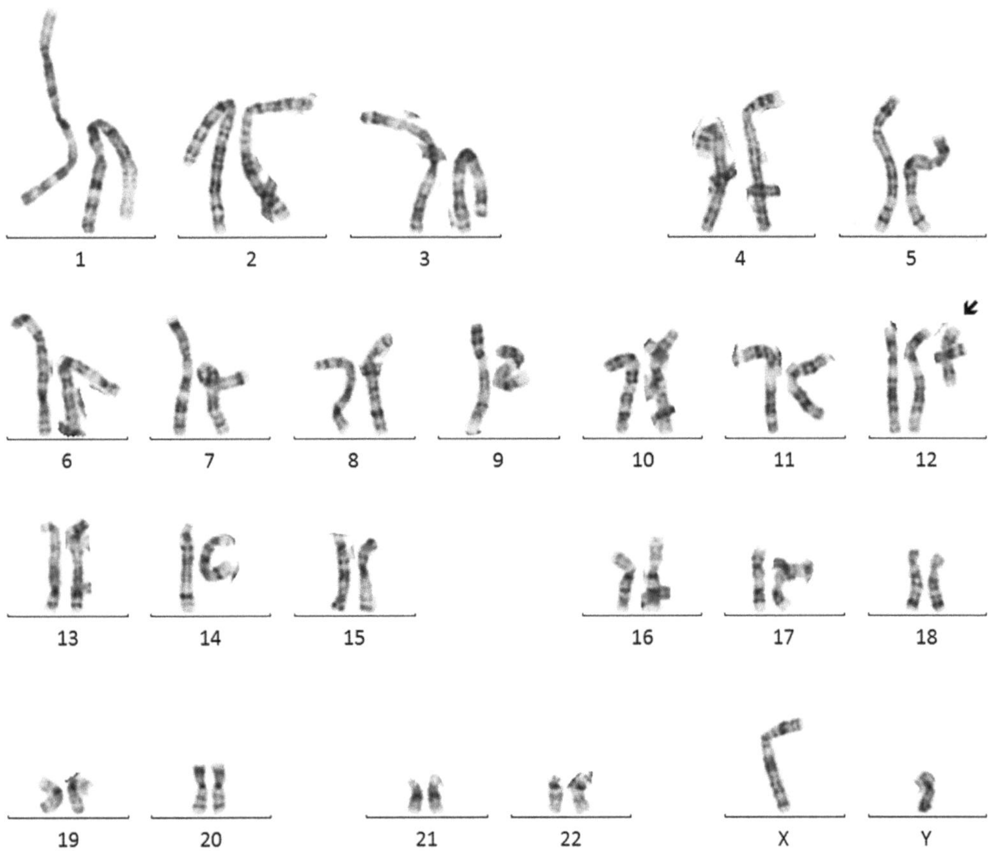

FIG. 6.2.1 The karyotype from the patient showed 1 out of 53 cells with an isochromosome 12 for the short arm, which resulted in a tetrasomy 12p. ISCN: 47,XY,+i(12)(p10)[1]/46,XY[52]

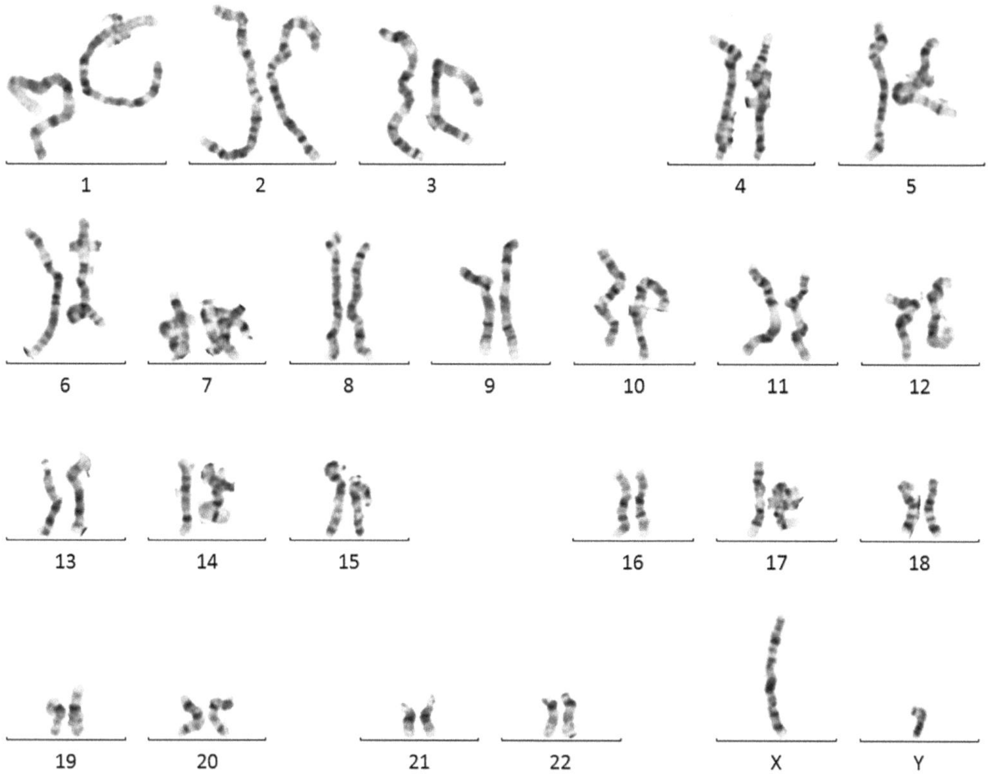

FIG. 6.2.2 The karyotype from the patient showed 52 out of 53 cells with a normal male karyotype.

CMA revealed a copy number gain of 37.7 Mb of DNA from chromosome 12 at band 12p13.33p11.1, encompassing 247 OMIM genes (Fig. 6.2.3). The whole-genomic view from CMA was shown in Fig. 6.2.4.

Results with interpretations

This gain of 37.7 Mb of DNA from chromosome 12 is consistent with the diagnosis of PKS, which is characterized by hypotonia, intellectual disability, developmental delay, sparse hair, areas of unusual pigmentation, speech anomalies, and distinctive facial features. Some affected individuals have been reported with multiple dysmorphic features including epicanthic folds, a flat face, a long philtrum, short palpebral fissures, an upturned nose, and small, low-set ears with an overfolded helix. Other commonly reported features include seizures, diaphragmatic hernias, congenital heart defects, and other systemic abnormalities.

Chromosome analysis exhibited an isochromosome 12p in 1/53 cells examined. Along with the microarray results, the diagnosis of mosaic PKS was confirmed. The percentage of

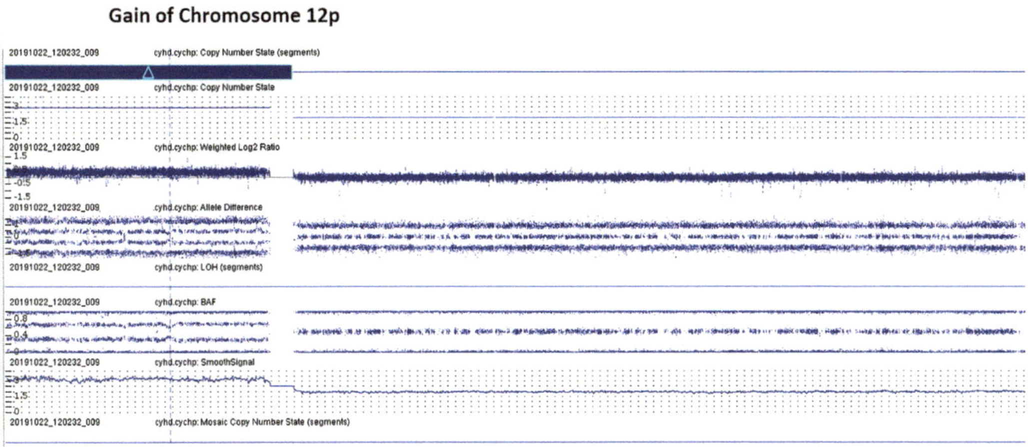

FIG. 6.2.3 Chromosome microarray results showed a copy number gain of 37.7 Mb of DNA from chromosome 12 at band 12p13.33p11.1. ISCN: arr[hg19] 12p13.33q11(173,786-37,869,301)x3

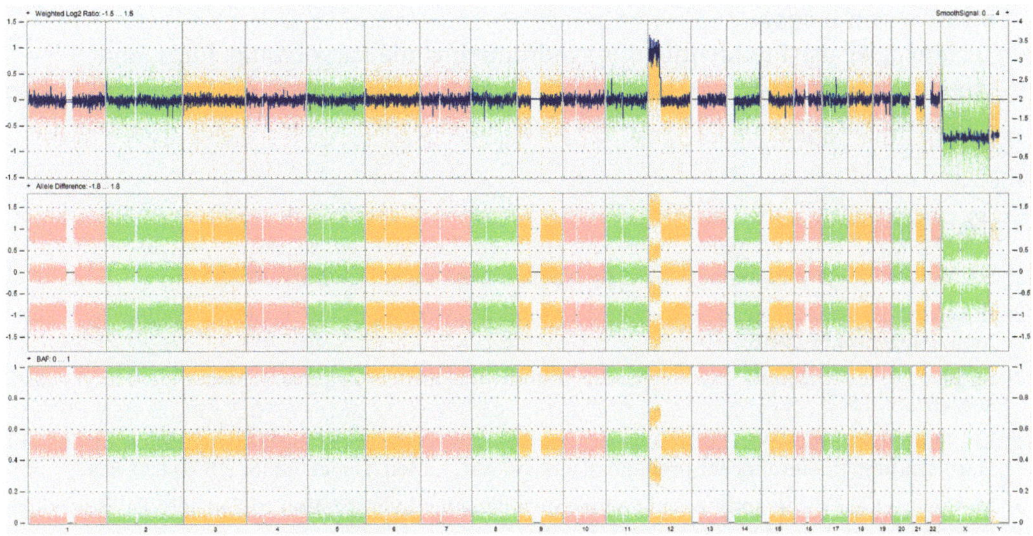

FIG. 6.2.4 The whole-genome view showed a gain of 12p. ISCN: arr[hg19] 12p13.33q11(173,786-37,869,301)x3

abnormal cells with i(12p) is usually higher in fibroblasts than it is in blood. Therefore, this is called mosaic syndrome.

PKS is frequently caused by a mosaic gain of an isochromosome 12p, which results in tetrasomy 12p. In rare cases, other more complex chromosomal changes involving chromosome 12 are responsible for this disorder; therefore, chromosome analysis on this patient is recommended along with CMA testing.

Future testing and recommendation

There is no specific therapy for individuals with Pallister-Killian mosaic syndrome. Affected children may benefit from early intervention programs and special education (https://rarediseases.info.nih.gov/diseases/8421/pallister-killian-mosaic-syndrome).

Genetic counseling is recommended and may be a benefit for affected individuals and other family members.

Summary of key learning points

- Understand the clinical presentation of PKS.
- The diagnostic testing results may be discordant because i(12p) is usually higher in fibroblast than it is in blood.
- CMA should be used as first-tier testing when PKS is considered as a differential diagnosis because CMA is more sensitive than chromosome analysis in identifying tetrasomic cells in peripheral blood.
- Given that the percentage of tetrasomy 12p cells in peripheral blood decreases as the individual with PKS ages, tests should be performed as early as possible to avoid invasive skin biopsy.
- Technologies used to identify this syndrome include karyotyping, FISH, and chromosome microarray. Correlation with clinical assessment of the patients is also indicated.

References

[1] S.L. Gersen, The Principles of Clinical Cytogenetics, Springer, New York, 2012.

[2] B. Karaman, et al., Pallister-Killian syndrome: clinical, cytogenetic and molecular findings in 15 cases, Mol. Cytogenet. 11 (2018) 45.

[3] M. Blyth, et al., Pallister-Killian syndrome: a study of 22 British patients, J. Med. Genet. 52 (7) (2015) 454–464.

[4] J.H. Priest, J.M. Rust, P.M. Fernhoff, Tissue specificity and stability of mosaicism in Pallister-Killian +i (12p) syndrome: relevance for prenatal diagnosis, Am. J. Med. Genet. 42 (6) (1992) 820–824.

[5] M.N. Lee, et al., Using array-based comparative genomic hybridization to diagnose Pallister-Killian syndrome, Ann. Lab. Med. 37 (1) (2017) 66–70.

[6] L.K. Conlin, et al., Utility of SNP arrays in detecting, quantifying, and determining meiotic origin of tetrasomy 12p in blood from individuals with Pallister-Killian syndrome, Am. J. Med. Genet. A 158A (12) (2012) 3046–3053.

[7] B. Doray, et al., Pallister-Killian syndrome: difficulties of prenatal diagnosis, Prenat. Diagn. 22 (6) (2002) 470–477.

7

Fragile X syndrome

Xia Li
SONORA QUEST LABORATORIES, PHOENIX, AZ, UNITED STATES

Background

FMR1 disorders include fragile X syndrome (FXS), fragile X-associated tremor/ataxia syndrome (FXTAS), and fragile X-associated primary ovarian insufficiency (FXPOI). FXS is one of the manifestations of *FMR1*-related disorders and the most common inherited cause of mental retardation [1]. It occurs predominantly in males and less frequently in females [2]. Affected patients may show autistic features, learning disabilities, or mental retardation. In addition, learning difficulties, attentional problems, anxiety, aggressive behavior, stereotypes, and mood disorders are also commonly seen in FXS. Increases in the number of CGG repeats in the fragile X gene (*FMR1*) result in perturbed expression and functionality of the FMR1 protein. The range of normal alleles is within 5–44 repeats. Intermediate alleles (gray zone) have 45–54 repeats, and premutation alleles have 55–200 repeats. A repeat size greater than 200 is associated with fragile X syndrome. Females with repeat sizes of 56–200 are considered premutation carriers as they are at risk of having offspring with the fragile X gene of more than 200 CGG repeats [3]. Most premutation carriers do not show fragile X syndrome features; however, some high-repeat carriers (>100 repeats) in both males and females have been described as having learning difficulties, behavioral problems, and even intellectual disabilities [2,4,5].

Older males with premutation may develop neurologic symptoms such as tremors and ataxia (abnormal gait) called fragile X tremor ataxia syndrome (FXTAS). FXTAS is a late-onset, progressive disorder affecting cognition and behavior. Approximately 20%–25% of women with a premutation experience FXPOI, which is characterized by infertility, decreased ovarian function, early menopause, or irregular cycles (Note: Women with a full mutation are not at risk for FXPOI and have no associated infertility) (https://fragilex.org/understanding-fragile-x/info-series/females-fragile-x/).

Full mutations of the fragile X gene (*FMR1*) are associated with fragile X syndrome in >99% of cases. Skewed X-inactivation and methylation of the normal and mutation alleles may affect the phenotypic presentation of female patients [6]. The diagnosis of *FMR1* disorder is established by using molecular techniques to identify trinucleotide repeat expansion in the 5′UTR of *FMR1* with abnormal gene methylation for alleles with >200 CGG repeats. The test consists of a polymerase chain reaction (PCR) of genomic DNA purified

Cases in Laboratory Genetics and Genomics (LGG) Practice. https://doi.org/10.1016/B978-0-323-99622-8.00027-7

from whole blood, followed by fragment sizing on a capillary electrophoresis platform and conversion of product size to the number of CGG repeats. In this chapter, two cases with FXS are illustrated to show how genetic testing identifies the CGG repeats.

Case 7.1 Fragile X syndrome in a male with a full mutation

Clinical indication

A 7-year-old boy presented with delays in talking, anxiety, and hyperactive behavior. He has large ears, a long face, a prominent jaw, and a prominent forehead. Fragile X syndrome was suspected.

Test ordered

- Fragile X testing of peripheral blood (AmplideX PCR/CE *FMR1*, a reflex to AmplideX mPCR *FMR1* if needed)

Laboratory test performed

AmplideX *FMR1* PCR was performed according to the following protocol. Genomic DNA was extracted via MagnaPure compact (Roche). Then DNA was assayed for CGG-repeat expansion of the *FMR1* locus by analysis of DNA fragments generated via PCR and capillary electrophoresis (ABI (Hitachi) 3130xl Genetic Analyzer) using the AmplideX *FMR1* PCR kit (Asuragen, Inc.). The length of the PCR products (as represented by peaks) was determined using GeneMapper software version 5.0 (ABI). The length/size of the amplicons (in base pairs) was converted to the number of CGG repeats by referencing them to those of the process control. Normal and mutation categories of the *FMR1* allele were determined according to the American College of Medical Genetics and Genomics guidelines. Samples with premutation (CGG repeat sizes between 55 and 200) and full mutation (CGG repeat sizes >200) were further tested using AmplideX *FMR1* mPCR to confirm these mutations and to assess the methylation status.

The AmplideX mPCR *FMR1* Kit was used as a reflex test when premutation or full mutation was identified in the patient. It employs an innovative PCR-only approach for the detection of methylation status in the *FMR1* gene. These reagents determine the extent of methylation of each allele in both male and female samples and thus eliminate the need to run tedious and time-consuming Southern blots.

Test results

AmplideX *FMR1* PCR showed >200 CGG, which is consistent with a diagnosis of full mutation for fragile X syndrome (Fig. 7.1.1). Next, AmplideX mPCR was performed, and the methylation status was 100% for this full mutation allele (Fig. 7.1.2).

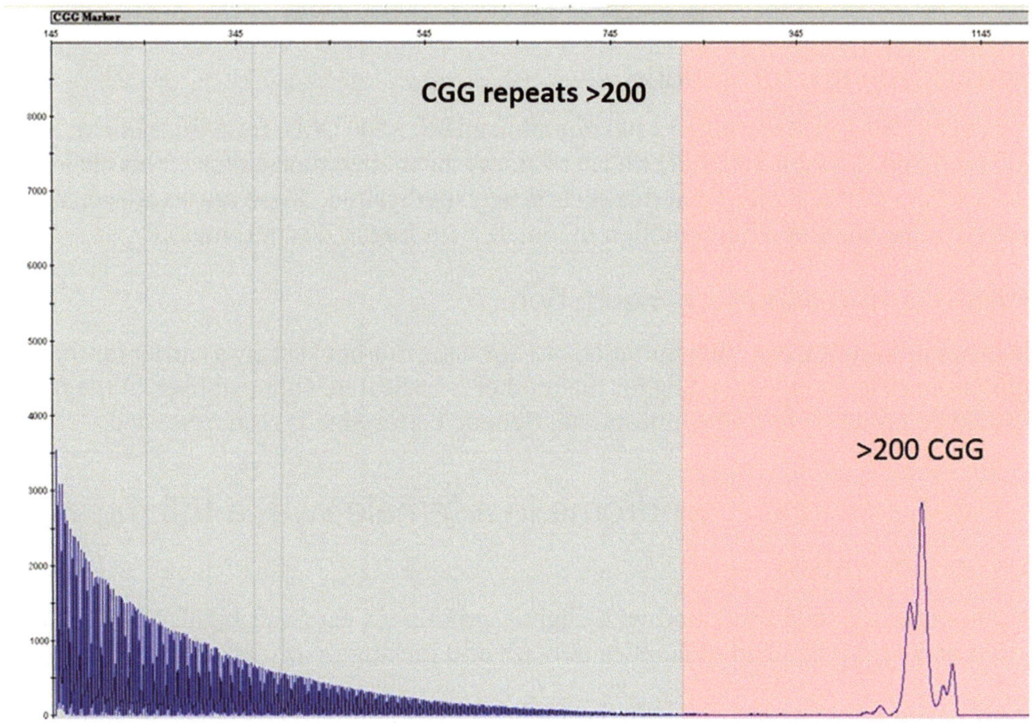

FIG. 7.1.1 AmplideX *FMR1* PCR results showed a full mutation (>200 CGG repeats) from this male patient. Nomenclature: c.-128_-126[(1050_1100)]

FIG. 7.1.2 AmplideX mPCR results showed a full mutation (>200 CGG repeats, fully methylated) from this male patient.

Results with interpretations

AmplideX *FMR1* PCR identified a full mutation allele (>200 CGG) from this patient. It was then reflexed to AmplideX *FMR1* mPCR to assess the methylation status of this allele. The results from mPCR showed that this allele is fully methylated. These results are consistent with the diagnosis of a full mutation in a male with fragile X syndrome.

Future testing and recommendations

In general, if the son has a full mutation of *FMR1*, the mother is likely a carrier for the premutation allele. Fragile X testing for the mother is indicated. Other siblings in the family should be tested if clinically appropriate. Genetic counseling is recommended.

Case 7.2 Fragile X syndrome in a female with a full mutation

Clinical indication

A 20-year-old female presented with slightly prominent ears and highly flexible finger joints and wrists. She had difficulties in math and picking up on "social cues."

Test ordered

- Fragile X testing on peripheral blood (AmplideX PCR/CE *FMR1*, a reflex to AmplideX mPCR *FMR1* if needed)

Laboratory test performed

AmplideX *FMR1* PCR and mPCR methods were described in Case 7.1.

Test results

Amplidex *FMR1* PCR showed two CGG repeat alleles: 30, >200 CGG, which is consistent with a diagnosis of full mutation in females for fragile X syndrome (Fig. 7.2.1). Next, AmplideX mPCR was performed, and the methylation status was 100% for this full mutation allele and 66% for the normal allele with 30 CGG repeats (Fig. 7.2.2).

Results with interpretations

AmplideX *FMR1* PCR identified a normal allele and a full mutation allele (30, >200 CGG) from this patient. Then it was reflexed to AmplideX *FMR1* mPCR to assess the methylation status of this allele. The results from mPCR showed that this full mutation allele is fully methylated, and the normal allele was 66% methylated. These results are consistent with the diagnosis of a full mutation in a female with fragile X syndrome. The phenotype manifestation in this female patient was milder than those males with full mutation allele due to skewed X-inactivation.

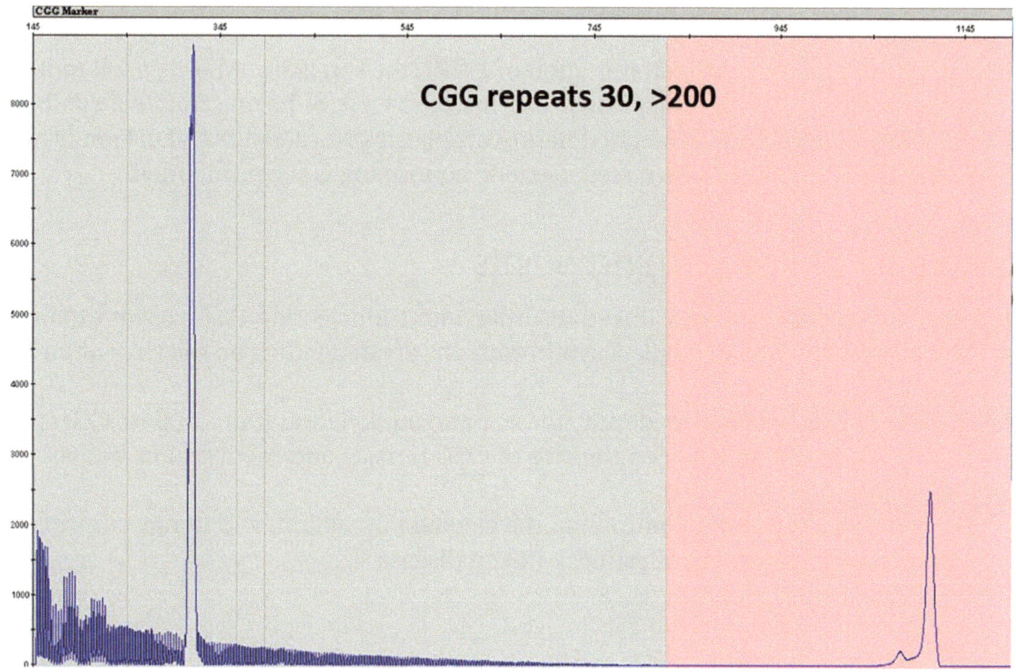

FIG. 7.2.1 AmplideX *FMR1* PCR results showed a full mutation (>200 CGG repeats) in addition to a normal allele (CGG repeats 30) from this female patient. Nomenclature: c.-128_-126[(600_800)]

FIG. 7.2.2 AmplideX mPCR results showed a full mutation (>200 CGG repeats, fully methylated) and a normal allele (30 CGG repeats, 66% methylated) from this female patient.

Future testing and recommendations

In general, if the mother is a full mutation of *FMR1*, the son likely inherits a full mutation from the mother. All of her offspring are at increased risk of having fragile X syndrome, FXTAS, FXPOI, or fragile X-associated neuropsychiatric disorders (FXAND). Family members who are at risk should be tested. Genetic counseling is recommended.

Summary of key learning points

- Fragile X syndrome is an X-linked disorder with trinucleotide CGG repeat expansion.
- Clinical presentation of fragile X syndrome can vary depending on the sizes of the CGG repeats.
- *FMR1* testing is designed to identify the size and methylation status of the CGG repeats.
- A clinical correlation between the size of CGG repeats and methylation status is indicated.
- Skewed X-inactivation is common in the general population, and it may contribute to heterozygous females manifesting X-linked disease.

References

[1] L. Boyle, W.E. Kaufmann, The behavioral phenotype of FMR1 mutations, Am. J. Med. Genet. C Semin. Med. Genet. 154C (4) (2010) 469–476.

[2] A. Stembalska, et al., Fragile X syndrome in females—a familial case report and review of the literature, Dev. Period Med. 20 (2) (2016) 99–104.

[3] J.E. Hunter, et al., *FMR1* disorders, in: M.P. Adam, et al. (Eds.), GeneReviews® [Internet], University of Washington, Seattle, Seattle, WA, 1993–2023. 20301558.

[4] W. Saldarriaga, et al., Mosaicism in Fragile X syndrome: a family case series, J. Intellect. Disabil. 26 (3) (2022) 800–807.

[5] W. Saldarriaga, et al., Fragile X syndrome, Colomb. Med. 45 (4) (2014) 190–198.

[6] D. Heine-Suner, et al., Fragile-X syndrome and skewed X-chromosome inactivation within a family: a female member with complete inactivation of the functional X chromosome, Am. J. Med. Genet. A 122A (2) (2003) 108–114.

8

Overgrowth syndrome

Bo Yuan[a] and Xia Li[b]

[a]DEPARTMENT OF MOLECULAR AND HUMAN GENETICS, BAYLOR COLLEGE OF MEDICINE, HOUSTON, TX, UNITED STATES [b]SONORA QUEST LABORATORIES, PHOENIX, AZ, UNITED STATES

Background

Overgrowth syndromes generally present with inherent health concerns and, in some instances, an increased risk of tumor predisposition that necessitates prompt diagnosis and appropriate referral [1]. The term overgrowth generalizes abnormally tall stature and is used to describe three types of phenotypes: (1) prenatal overgrowth, which refers to newborns who are large for gestational age (LGA) and is commonly seen in maternal diabetes and overgrowth syndromes such as Beckwith-Wiedemann syndrome (BWS); (2) postnatal overgrowth, which refers to individuals having an accelerated growth pattern starting in childhood or adolescence. This group is consistent with mainly endocrine abnormalities (such as thyroid, growth hormone, sex hormones, or glucocorticoid). Other etiologies include familial tall stature (constitutional tall stature), precocious puberty, obesity, Marfan syndrome (https://marfan.org/conditions/marfan-syndrome/), homocystinuria (https://rarediseases.info.nih.gov/diseases/10770/homocystinuria), Klinefelter syndrome, and 47,XYY syndrome [2]; (3). Segmental overgrowth is a phenotype of excessive growth that is confined to one or a few regions of the body, for example, macrocephaly or a single digit. The segmental or mosaic overgrowth often occurs with overactivation mutations of the PI3K/AKT/mTOR (phosphoinositide-3-kinase/protein kinase B/mammalian target of rapamycin) pathway [3]. Overgrowth syndromes can be associated with various causes such as hormone imbalance, life-threatening hypoglycemia (e.g., Beckwith-Wiedemann Syndrome, BWS), seizures (Sotos syndrome), developmental delay (Sotos syndrome, Weaver syndrome), and increased susceptibility to malignancy (Wilms' tumor, hepatoblastoma, etc.).

Somatic variations have been well described in cancer and become increasingly recognized in human congenital diseases, such as somatic overgrowth syndromes. The disease-causing variants often occur during a postzygotic mutational event, leading to mosaicism that is described by two (or more) genetically distinct populations of cells within one individual. The genetic basis of somatic overgrowth syndrome is heterogeneous. The most common and best-described somatic growth syndromes are caused by somatic variations

Cases in Laboratory Genetics and Genomics (LGG) Practice. https://doi.org/10.1016/B978-0-323-99622-8.00026-5

of the cellular pathways that promote growth, such as activating variations in the PI3K (phosphatidylinositol 3-kinase)/AKT/mTOR (mammalian target of rapamycin) pathway [4]. Other mechanisms, such as activating or amplification of *GPR101* leading to growth hormone hypersecretion [5], and loss-of-function of *PTEN*, have also been described. Phenotype variability has been observed for somatic overgrowth syndromes, due to the variability in functional alteration caused by different variations, the timing of variation during embryonic development, and the level of mosaicism. Identification of the genetic variation can inform the diagnosis of somatic overgrowth syndromes; however, the molecular diagnosis is usually challenging because of genetic and phenotypic heterogeneity, low-level mosaicism, and/or tissue-specific manifestation.

The approach to identifying patients with overgrowth syndrome should start with detailed family history and physical examination, followed by genetic testing including karyotyping, FISH, microarray, and/or NGS. These tests can be done through saliva, blood, or skin tissue according to clinical appropriateness. In this chapter, two cases are illustrated to demonstrate the technologies used for the diagnosis of overgrowth syndrome.

Case 8.1 Sotos syndrome

Clinical indication

A 3-year-old boy presented with chronic medical conditions including bronchopulmonary dysplasia, atrial septal defect, ventricular septal defect, history of cardiac pacemaker placement, nephrocalcinosis, cryptorchidism (S/P surgical correction), and tracheostomy. He had chronic lung disease and was ventilator dependent. Routine ultrasound was performed at birth to screen for any renal abnormalities, which showed a large mass on top of his right kidney. CT with contrast showed a right-sided suprarenal mass likely related to neuroblastoma. Genetic tests were performed. Chromosome analysis revealed an abnormal chromosome 5q in every cell analyzed suspecting Sotos syndrome. The pathology report was consistent with neuroblastoma, and FISH with *MYCN* was not amplified. The mass was completely resected with no recurrence, and chemotherapy was not recommended.

Tests ordered

- Chromosome analysis: Bone marrow
- FISH: *MYCN*
- Chromosome analysis: Peripheral blood (PHA-stimulated)

Laboratory test performed

Chromosome analysis and FISH methods were described in Chapter 1.

Test results

Chromosome analysis of the bone marrow revealed unknown material on the 5q at the telomere. Due to the low resolution of the banding from the bone marrow, it was unclear whether this was a deletion or additional material on 5q (Fig. 8.1.1).

FISH was performed on interphase nuclei using probes localized to the *MYCN* (2p24.3) and D2Z1 (2cen) gene regions. Two hundred nuclei were examined, and the results demonstrated a gain of chromosome 2 (*MYCN*) but did not meet the criteria for *MYCN* amplification in 180/200 (90%) of cells scored (Fig. 8.1.2). The patient had surgery to remove the adrenal mass.

High-resolution chromosome analysis on PHA-stimulated blood specimen was performed. The results revealed a terminal deletion of chromosome 5 at band q35.1 (Fig. 8.1.3).

Results with interpretations

FISH detected gains of three to four copies of chromosome 2. These results do not meet the criteria for *MYCN* amplification.

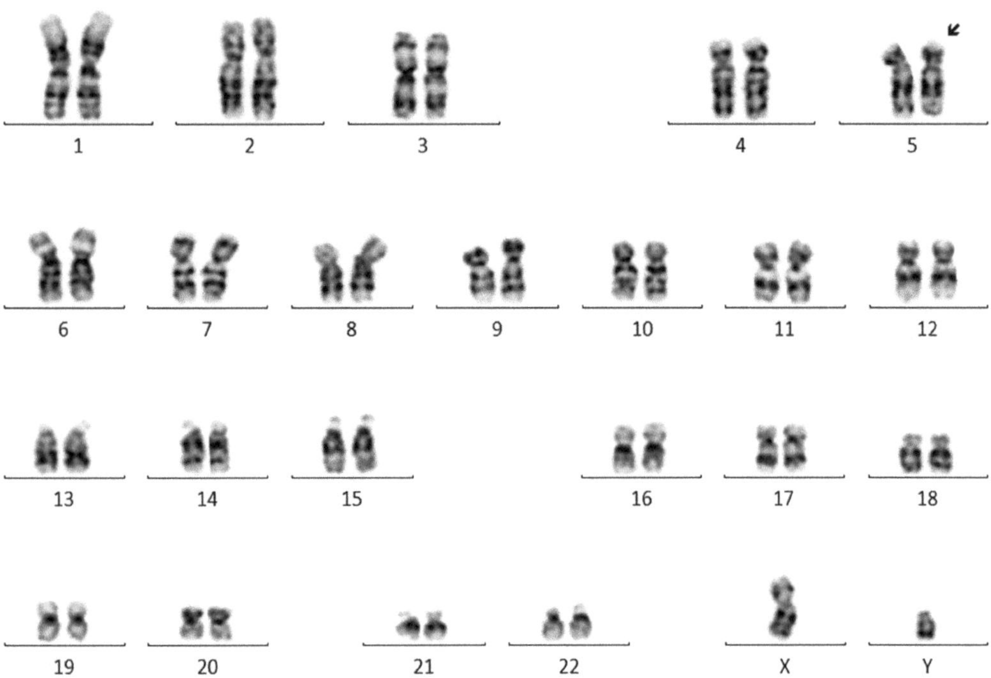

FIG. 8.1.1 Chromosome analysis on bone marrow revealed an abnormal banding on 5q. ISCN: 46,XY,add(5)(q34)?c[20]

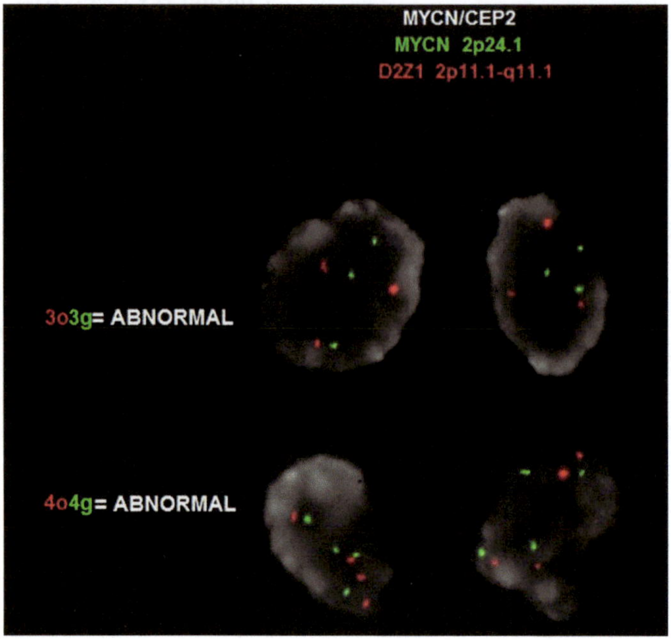

FIG. 8.1.2 FISH using the *MYCN* probe showed a gain of chromosome 2 with no evidence for *MYCN* amplification. ISCN: nuc ish(*MYCN*,D2Z1)x3-4[180/200]

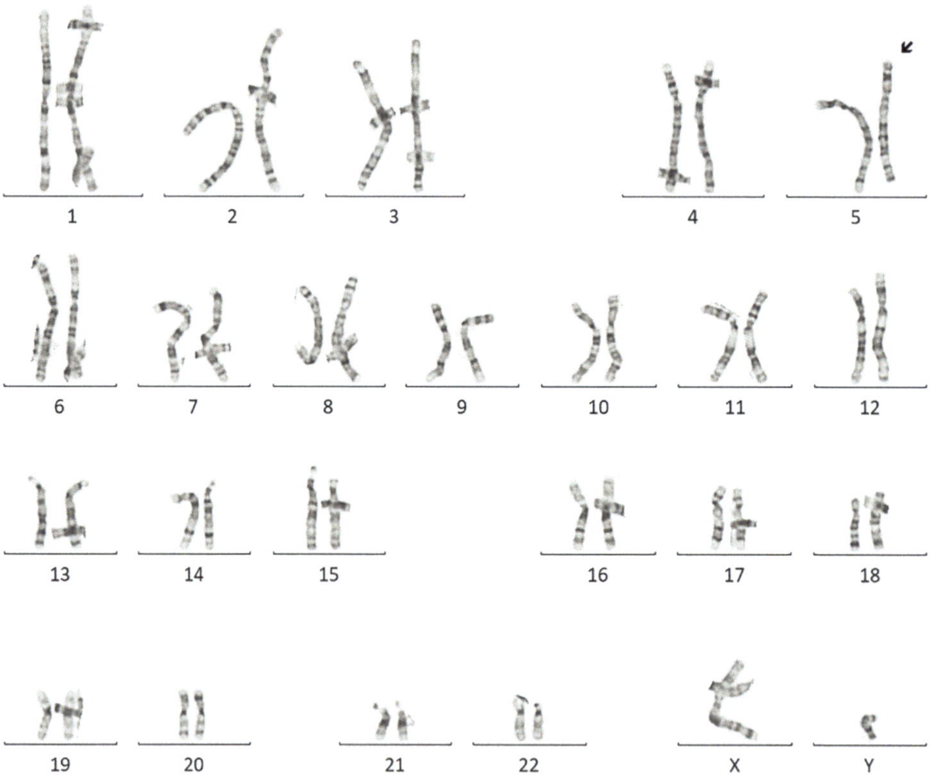

FIG. 8.1.3 High-resolution chromosome analysis on peripheral blood demonstrated a terminal deletion of chromosome 5 at q35.1. ISCN: 46,XY,del(5)(q35.1)

Neuroblastoma, the most common extracranial solid tumor of childhood, is thought to originate from undifferentiated neural crest cells. Amplification of the *MYC* family member, *MYCN*, is found in ~25% of cases and correlates with high-risk disease and poor prognosis. Currently, amplification of *MYCN* remains the best-characterized genetic marker of risk in neuroblastoma [6,7]. The nonamplification of *MYCN* is associated with a favorable prognosis [8].

For chromosome analysis of the bone marrow, all cells examined exhibited material of unknown origin on 5q34. Since this abnormality was seen in every cell examined, it is unclear whether this is an acquired or constitutional abnormality. A peripheral blood sample for constitutional chromosome analysis was recommended to rule out a constitutional finding.

Finally, a high-resolution chromosome analysis was performed. All cells analyzed showed a terminal deletion of the long arm of chromosome 5 at band q35.1. This is consistent with a diagnosis of Sotos syndrome.

Sotos syndrome is a disorder characterized by a distinctive facial appearance, overgrowth in childhood, and learning disabilities or delayed development of mental and movement abilities. Characteristic facial features include a long, narrow face, a high forehead, flushed (reddened) cheeks, and a small pointed chin. This is an autosomal dominant condition. *NSD1* located at 5q35.3 is the gene associated with Sotos syndrome. This gene is deleted from the patient, which explains the cause of the patient's phenotype with multiple health issues https://www.ncbi.nlm.nih.gov/books/NBK1479/.

Future testing and recommendations

Genetic counseling and parental chromosome studies are strongly recommended. Chromosome microarray analysis is indicated to examine the exact length of the deletion and to better delineate the genes associated with the patient's phenotype.

Case 8.2 Somatic overgrowth syndrome with *PIK3CA* mutation

Clinical indication

A 2-year-old female presented with overgrowth, hemihypertrophy, hemimegalencephaly, posterior plagiocephaly, macrosomia, capillary malformations, and hemangioma. Differential diagnoses include Beckwith-Wiedemann syndrome, megalencephaly-capillary malformation-polymicrogyria, Klippel-Trenaunay syndrome, and Sturge-Weber syndrome. Blood and skin samples were submitted for genetic testing.

Test ordered

- Macrocephaly and overgrowth syndrome NGS panel on the blood sample (send out)
- Beckwith-Wiedemann syndrome (BWS) methylation panel on the blood sample (send out)

- Whole-genome sequencing (WGS) on the blood sample (send out)
- Trio whole-exome sequencing (WES) on skin biopsy sample from the affected body region and blood sample of the proband, as well as blood samples of the unaffected parents

Laboratory test performed

NGS, BWS methylation, and WGS tests were performed in other labs. The methodology of WES has been described previously (PMID 24088041 and 25326635).

Test results

The macrocephaly and overgrowth syndrome panel, Beckwith-Wiedemann syndrome methylation panel, and the macrocephaly deletion/duplication panel were all negative for the submitted blood sample (data not shown).

WGS was then requested, and single-nucleotide variants (SNVs) and small insertion/deletion (indel), copy number variants (CNVs) and structural variants (SVs), and absence of heterozygosity (AOH) were analyzed. However, WGS was also negative for the individual's blood sample (data not shown).

A trio WES was subsequently performed and revealed a pathogenic variant, c.1357G>A (p.E453K), in the *PIK3CA* gene with a variant allele fraction (VAF) of 10% in this individual's skin biopsy, consistent with somatic origin. This variant was identified in the blood sample at 1.7% VAF and was not present in the parental samples (Fig. 8.2.1A). Sanger sequencing detected the mosaic variant in the skin biopsy, but not in the blood sample (Fig. 8.2.1B).

A targeted reanalysis of WGS data was performed focusing on the *PIK3CA* variant identified by WES, which remains negative (data not shown).

Results with interpretations

PIK3CA c.1356G>A (p.E453K) identified by WES is a recurrent pathogenic gain-of-function variant that has been observed in multiple individuals with overgrowth syndrome [9]. Additionally, this variant has also been reported in cancer samples (COSMIC [10]: COSM12584). Variants detected in *PIK3CA* as the cause of overgrowth syndromes include megalencephaly-capillary malformation-polymicrogyria syndrome (MCAP) [MIM:602501], CLOVE syndrome [MIM:612918], and Cowden syndrome 5 (CWS5) [MIM:615108]. These syndromes are all inherited in an autosomal dominant manner, according to OMIM [11].

Sequencing depth is critical for somatic variant detection. The limit of detection (LOD) for the somatic variant is correlated with sequencing depth. It has been shown in one study that, to reach a 95% sensitivity and above 95% positive predictive value (PPV), ~40× sequencing depth is needed to detect variants with 20% VAF with 95% sensitivity, ~94× for 10% VAF, ~294× for 5% VAF, and ~1085× are needed for 2% VAF [12]. The sequencing depth of WES is averaged around 80–120×; thus, it is sufficient to provide

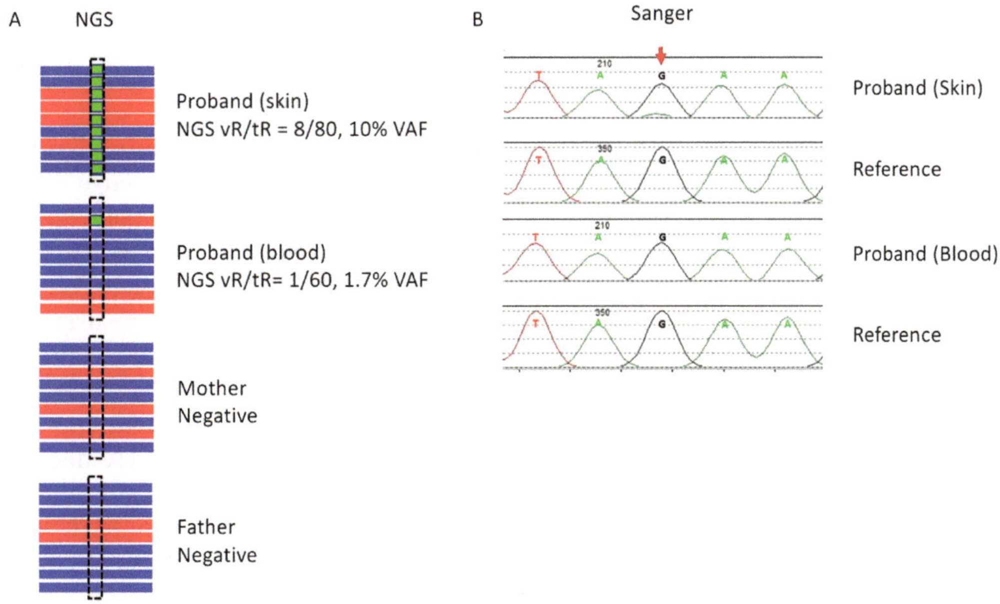

FIG. 8.2.1 *PIK3CA* c.1357G>A (p.E453K) variant was identified in the blood sample at 1.7% VAF by WES and was not present in the parental samples (A). Sanger sequencing detected the mosaic variant in the skin biopsy, but not in the blood sample (B).

accurate detection for somatic variants above 10% VAF. However, variants with VAF under 10% may not be reliably detected by WGS, because the sequencing depth of WGS is averaged around 30–50×.

Testing of the affected tissue where the disease manifests would more likely yield diagnostic variants. Blood is generally not considered as the high-yield specimen for mosaicism testing of somatic diseases. First, blood cells may not be the tissue manifesting the genetic disease, and therefore, disease-causing variants are less likely to be present in blood cells. One exception may be immunological or hematological disorders in that blood is the primary tissue manifesting the disease [13]. Additionally, blood cells undergo self-renewal during hematopoiesis and negative or positive selection for growth advantages, skewing VAF in blood compared with the manifesting tissue [14].

PIK3CA-related somatic overgrowth syndromes are caused by gain-of-function variants and are broadly characterized by cutaneous vascular malformations with segmental overgrowth involving multiple tissues or body regions [9]. Testing specimens of biopsied tissue affected by overgrowth is likely to have a higher yield compared with other unaffected tissues such as blood, saliva, and skin, which are thought of as "surrogate" tissues since they are easier to access [9]. The selection of genetic testing for somatic overgrowth syndrome also requires special consideration. Genetic testing using methods with enhanced sensitivity and specificity for somatic variant detection is preferred. For example, somatic overgrowth and vascular malformation gene sequencing panels targeting specific variants or genes that are involved in the PI3K/AKT/mTOR pathway [15,16] with high coverage (e.g., above 1000×) should be considered for *PIK3CA*-related diseases. Other methods, such as droplet digital PCR (ddPCR) [17,18], have been emerging to provide a highly sensitive molecular diagnosis for somatic overgrowth syndromes.

Although the contribution of SNV/indel to somatic overgrowth syndromes has been well studied and the current genetic testing resource is ample, the knowledge of CNV in association with somatic overgrowth syndromes is scarce. One prominent example may be the *GPR101*-associated overgrowth syndromes. Both duplication of *GPR101* of germline and somatic origins have been demonstrated to cause X-linked acrogigantism (XLAG) characterized by early onset excessive growth due to growth hormone dysregulation [5,19,20]. Somatic activating variants in *GPR101* have also been associated with pituitary adenomas; however, the association has been controversial [19–21]. The somatic duplications involving *GPR101* have been successfully identified by ddPCR and high-density chromosome microarray [5]; however, the resource for somatic CNV detection has not been widely available as a clinical test.

In conclusion, the finding of a somatic mosaic pathogenic variant c.1357G>A (p. E453K) in the *PIK3CA* gene is consistent with the phenotypes of this individual. Detection of low-level somatic variants requires high sequencing depth. The molecular diagnostic yield will be higher if the affected tissue is used for testing.

Future testing and recommendations

Genetic counseling and regular follow-up with physicians are recommended.

Summary of key learning points

- Somatic overgrowth syndromes are caused by postzygotic pathogenic variants.
- Genetic testing for somatic diseases requires special consideration of sample types and testing methodology.
- Chromosome analysis on bone marrow can reveal constitutional abnormalities, which require confirmation on peripheral blood.
- Neuroblastoma with no amplification of *MYCN* is associated with a favorable prognosis.
- Sotos syndrome is a disorder that is associated with overgrowth in childhood and many other learning disabilities.

References

[1] J. Manor, S.R. Lalani, Overgrowth syndromes-evaluation, diagnosis, and management, Front. Pediatr. 8 (2020) 574857.

[2] T. Urakami, Tall stature in children and adolescents, Minerva Pediatr. 72 (6) (2020) 472–483.

[3] K.M. Keppler-Noreuil, et al., Somatic overgrowth disorders of the PI3K/AKT/mTOR pathway & therapeutic strategies, Am. J. Med. Genet. C Semin. Med. Genet. 172 (4) (2016) 402–421.

[4] D.A. Fruman, et al., The PI3K pathway in human disease, Cell 170 (4) (2017) 605–635.

[5] A.F. Daly, et al., Somatic mosaicism underlies X-linked acrogigantism syndrome in sporadic male subjects, Endocr. Relat. Cancer 23 (4) (2016) 221–233.

[6] C. Huang, et al., Therapeutic potential of targeting MYCN: a case series report of neuroblastoma with MYCN amplification, Medicine (Baltimore) 99 (25) (2020) e20853.

[7] M. Huang, W.A. Weiss, Neuroblastoma and MYCN, Cold Spring Harb. Perspect. Med. 3 (10) (2013) a014415.

[8] M.L. Schmidt, et al., Favorable prognosis for patients 12 to 18 months of age with stage 4 nonamplified MYCN neuroblastoma: a Children's Cancer Group Study, J. Clin. Oncol. 23 (27) (2005) 6474–6480.

[9] G. Mirzaa, et al., PIK3CA-associated developmental disorders exhibit distinct classes of mutations with variable expression and tissue distribution, JCI Insight 1 (9) (2016).

[10] J.G. Tate, et al., COSMIC: the catalogue of somatic mutations in cancer, Nucleic Acids Res. 47 (D1) (2019) D941–D947.

[11] J.S. Amberger, et al., OMIM.org: leveraging knowledge across phenotype-gene relationships, Nucleic Acids Res. 47 (D1) (2019) D1038–D1043.

[12] H.T. Shin, et al., Prevalence and detection of low-allele-fraction variants in clinical cancer samples, Nat. Commun. 8 (1) (2017) 1377.

[13] J. Aluri, et al., Immunodeficiency and bone marrow failure with mosaic and germline TLR8 gain of function, Blood 137 (18) (2021) 2450–2462.

[14] I.M. Campbell, et al., Somatic mosaicism: implications for disease and transmission genetics, Trends Genet. 31 (7) (2015) 382–392.

[15] F. Chang, et al., Molecular diagnosis of mosaic overgrowth syndromes using a custom-designed next-generation sequencing panel, J. Mol. Diagn. 19 (4) (2017) 613–624.

[16] S.E. Sheppard, et al., Cerebrofacial vascular metameric syndrome is caused by somatic pathogenic variants in PIK3CA, Cold Spring Harb. Mol. Case Stud. 7 (6) (2021).

[17] K.S. Yeung, et al., Somatic PIK3CA mutations in seven patients with PIK3CA-related overgrowth spectrum, Am. J. Med. Genet. A 173 (4) (2017) 978–984.

[18] A.M. Piacitelli, et al., Characterization of a severe case of PIK3CA-related overgrowth at autopsy by droplet digital polymerase chain reaction and report of PIK3CA sequencing in 22 patients, Am. J. Med. Genet. A 176 (11) (2018) 2301–2308.

[19] D. Iacovazzo, et al., Germline or somatic GPR101 duplication leads to X-linked acrogigantism: a clinico-pathological and genetic study, Acta Neuropathol. Commun. 4 (1) (2016) 56.

[20] G. Trivellin, et al., Gigantism and acromegaly due to Xq26 microduplications and GPR101 mutation, N. Engl. J. Med. 371 (25) (2014) 2363–2374.

[21] P. Kamenicky, J. Bouligand, P. Chanson, Gigantism, acromegaly, and GPR101 mutations, N. Engl. J. Med. 372 (13) (2015) 1264.

9

Contiguous gene syndrome

Bo Yuan[a] and Xia Li[b]

[a]DEPARTMENT OF MOLECULAR AND HUMAN GENETICS, BAYLOR COLLEGE OF MEDICINE, HOUSTON, TX, UNITED STATES [b]SONORA QUEST LABORATORIES, PHOENIX, AZ, UNITED STATES

Background

Contiguous gene syndrome, also known as a genomic disorder, is caused by genomic rearrangements that change the dosage and/or organization of multiple genes in a contiguous chromosomal region [1,2]. These rearrangements, balanced or unbalanced, may disrupt the genes and chromosome regions, or alter the copy number of a gene. Unbalanced rearrangements that change the copy number of a gene are named copy number variants (CNVs). These events may be at a microscopic or submicroscopic level with sizes ranging from kilobases (kb) to megabases (Mb) [3]. CNVs of a genomic interval containing dosage-sensitive genes may result in genomic disorders, often sporadic and characterized by a wide spectrum of clinical phenotypes with extensive heterogeneity [4].

The mutagenic mechanisms of CNVs are variable. Nonallelic homologous recombination (NAHR), mediated by long homologous repeats such as low-copy repeats (LCR), has been shown to be the major molecular mechanism leading to disease-causing recurrent genomic rearrangements [3]. For example, an unequal crossover between two major LCR clusters flanking the Smith-Magenis Syndrome (SMS) critical region on human chromosome 17p11.2 in direct orientation, termed proximal and distal SMS-repeats (REPs), resulting in the ~3.6 Mb common recurrent SMS deletion is observed in ~70%–80% of individuals with SMS (MIM# 182290) as well as the reciprocal ~3.6 Mb common recurrent 17p11.2 duplication that is observed in ~60%–70% of individuals with Potocki-Lupski syndrome (MIM# 610883) [5–8]. While NAHR-mediated rearrangements can be recurrently observed in unrelated individuals, other nonrecurrent rearrangements may be observed in sporadic individuals with similar phenotypes when the rearrangement affects the same critical gene; however, other phenotypes, perhaps unrelated to the ones caused by the recurrent rearrangement, may also present if the nonrecurrent rearrangement extends to other disease-associated loci due to variable genomic contents being included. Replication-based mechanisms, such as fork stalling and template switching/microhomology-mediated break-induced replication (FoSTeS/MMBIR), have been proposed to generate nonrecurrent rearrangements independent of LCRs [9,10]. Nonrecurrent rearrangements have been shown to complicate the diagnosis of a well-established genomic disorder and even result in concomitant genomic disorders in a single rearrangement event, such as the *PMP22-RAI1* contiguous gene deletions and *PMP22-RAI1* contiguous

Cases in Laboratory Genetics and Genomics (LGG) Practice. https://doi.org/10.1016/B978-0-323-99622-8.00007-1

gene deletions that result in blended disease phenotypes involving both central and peripheral nervous system abnormalities [11,12].

Clinical diagnosis of genomic disorders is challenging especially for the cases caused by nonrecurrent genomic rearrangements. Molecular diagnosis is helpful to enable careful genotype-phenotype correlation and support clinical diagnosis and management.

This chapter will illustrate variable phenotypic and clinical outcomes of two cases with A20 haploinsufficiency (HA20) caused by two different contiguous gene deletions affecting different genomic content but all including the same critical gene *TNFAIP3*.

More than 60 patients with HA20 have been reported in literature. Most of these patients carry heterozygous pathogenic SNV/indels in the *TNFAIP3* gene. Reports of large contiguous gene deletions are rarely reported, and such reports also focused on neurodevelopmental problems and congenital malformation while immunological phenotypes were not carefully reviewed. Incomplete penetrance or variable expressivity of HA20 has been suspected due to the presence of *TNFAIP3* deletions in individuals who do not have HA20. Atypical phenotypes may also be observed for individuals molecularly diagnosed with HA20. CNVs affecting contiguous genes and multiple disease-associated loci may be included. Therefore, careful variant interpretation and clinical correlation are needed to fully evaluate the clinical significance of a CNV and guide medical management.

Finally, this chapter will also show a few cases with large contiguous gene deletions or duplications that affect patients. All cases cover multiple genes spanning several syndromes in the deleted or duplicated regions.

Case 9.1 Haploinsufficiency of A20 (HA20) with 3.4 Mb deletion

Clinical indication

Case 9.1 is a 16-year-old Caucasian female with a long history of intermittent fevers, oral ulcers, lymphoproliferation, recurrent infections, eczema, and celiac disease. Abnormal blood test results over time revealed continuously elevated erythrocyte sedimentation rates and C-reactive protein levels, mild neutropenia with hypogranular neutrophils, an inverted CD4+/CD8+ T cell ratio, elevated IgG, IgA, IgM, and IgE levels, and elevated serum cytokine levels including significantly elevated interferon gamma (IFN-γ) level [13].

Test ordered

- Chromosomal microarray (CMA) was performed at 6 years of age
- A trio exome sequencing was performed at 15 years of age.

Laboratory test performed

Chromosome microarray methods were described in Chapter 1.

Clinical exome sequencing was performed as previously described (PMID 24088041, 25326635).

Test results

In Case 9.1, a heterozygous pathogenic deletion from 6q23.2q23.3 of approximately 3.4 Mb was identified by CMA (Fig. 9.1.1) and was confirmed to be de novo by targeted fluorescence in situ hybridization (FISH) studies of both parents. Exome did not identify any variants that were associated with the patient's phenotype.

Results with interpretations

The deletion includes the *TNFAIP3* gene. Defects of the *TNFAIP3* gene are the cause of the autoinflammatory syndrome, familial, Behcet-like 1 (MIM# 191163, also known as A20 haploinsufficiency [HA20]) inherited in an autosomal dominant manner. HA20 is characterized by autoinflammation, recurrent fevers, ulcers, ocular inflammation, and arthritis, consistent with this patient's clinical phenotypes. The deletion also includes the *IFNGR1* gene. Monoallelic protein-truncating variants of *IFNGR1* cause autosomal dominant immunodeficiency 27B with mycobacteriosis (MIM# 615978) via a dominant-negative mechanism, and biallelic loss-of-function variants of *IFNGR1* result in autosomal recessive immunodeficiency 27A with mycobacteriosis (MIM# 209950). Heterozygous carriers of the autosomal recessive form of *IFNGR1* deficiency are asymptomatic; thus the heterozygous deletion of *IFNGR1* identified in Case 9.1 is unlikely to be related to her recurrent infection. Trio exome sequencing was performed to evaluate for additional genetic changes underlying her disease, but was unrevealing. Notably, no sequence variant of *IFNGR1* was identified by exome sequencing. Therefore, the identification of the heterozygous deletion involving *TNFAIP3* is consistent with the patient's clinical phenotypes and provides a molecular diagnosis to guide treatment.

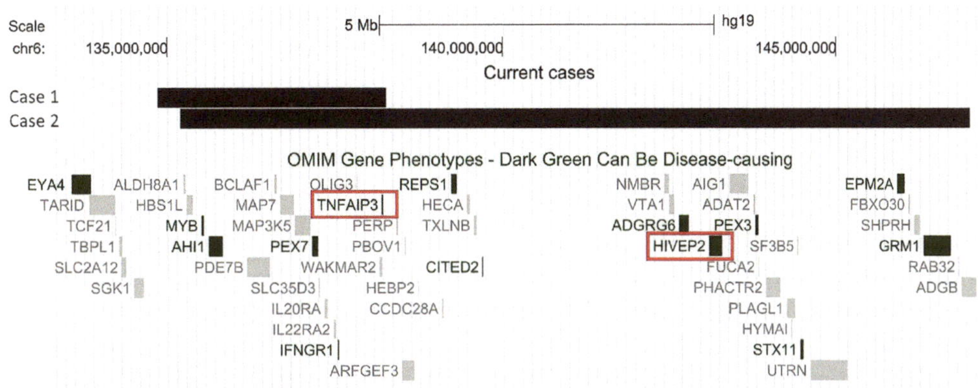

FIG. 9.1.1 Deletions were detected in Cases 9.1 and 9.2. *Black segments* indicate the chromosomal intervals of deletions in each case. OMIM genes are illustrated under the deletion segments. Disease-associated genes are in *green color*. Two genes associated with autosomal dominant diseases, *TNFAIP3* and *HIVEP2*, are highlighted in *red boxes*.

Future testing and recommendation

Case 9.1 was placed on antibacterial, antifungal, and antiviral agents at various times for infections, but she used these with poor adherence. From the ages of 6–13 years, she received acyclovir to control her oral ulcers, but it was unsuccessful. Upon her genetic diagnosis of HA20, anakinra was recommended but declined. She is now being maintained with TNF-alpha inhibition.

Case 9.2 Haploinsufficiency of A20 (HA20) with 11.7 Mb deletion

Clinical indication

Case 9.2 was an African American male presented with prenatal intrauterine growth restriction (IUGR), premature birth at 36 weeks of gestation, failure to thrive, developmental delay, intellectual disability, cryptorchidism, cytopenias, and recurrent oral ulcers. Immune thrombocytopenia (ITP) was first diagnosed at age 3 and ran a recurrent and refractory course.

Test ordered

- Chromosome microarray (CMA) was performed at 7 years of age.
- A proband-only exome sequencing was also performed.

Laboratory test performed

Chromosome microarray and exome sequencing methods are described in Chapters 1 and 8.

Test results

In Case 9.2, a heterozygous deletion of approximately 11.7 Mb in size from 6q23.3q24.3 was identified by CMA (Fig. 9.1.1). The mother had a normal karyotype, and the father's specimen was not available for analysis.

Results with interpretations

The deletion identified by CMA contains more than 45 genes, including *TNFAIP3*. This finding is consistent with the patient's immunological phenotypes including recurrent oral ulcers and thrombocytopenia. The deletion also includes *HIVEP2*, a gene located approximately 4.9 Mb distal to *TNFAIP3*. *HIVEP2* encodes a transcription factor critical to brain development. Defects in *HIVEP2* cause autosomal dominant mental retardation 43 (MRD43, MIM# 616977), consistent with the patient's neurodevelopmental disorders including developmental delay and intellectual disability. Exome sequencing of the proband did not reveal any additional genetic changes contributing to the complex phenotype.

Future testing and recommendation

Case 9.2 was managed for years with multimodal immunosuppression treatment, but the effects were mostly transient with limited efficacy. After the HA20 diagnosis was obtained, hematopoietic stem cell transplantation (HSCT) from a 10/10 HLA-matched, unrelated donor was performed due to the occurrence of life-threatening bleeding. The conditioning regimen consisted of a fully myeloablative regimen (busulfan, cyclophosphamide, fludarabine, and alemtuzumab). On the day of transplantation, he complained of chest pain and showed signs of acute kidney injury. After the stem cell infusion, he developed abdominal pain, loss of consciousness, and irrecoverable cardiopulmonary arrest. Postmortem evaluation showed changes consistent with noninfectious, cytokine-mediated multiorgan injury.

Case 9.3 Contiguous gene syndrome with duplication of 22q11.2q12.1

Clinical indication

A 17-year-old female presented with preauricular sinus and cyst, congenital malformations of the anterior segment of the eye, disorders of psychological development, and vesicoureteral reflux.

Test ordered

- Chromosome analysis: Routine blood
- FISH: DiGeorge syndrome 22q11.2 (*TUPLE1*)
- Chromosome microarray (CMA)

Laboratory test performed

Chromosome analysis, FISH, and CMA methods were described previously in Chapter 1.

Test results

Chromosome analysis identified two cell lines: one cell line with 46 cells exhibited a normal female karyotype (Fig. 9.3.1), and the second cell line with 4 cells revealed a marker chromosome along with otherwise a normal female karyotype (Fig. 9.3.2).

A FISH with DiGeorge/VCF probe localized to the *TUPLE1* gene region (22q11.2) was hybridized to metaphase and interphase preparations. The fluorescence signal pattern was abnormal in both the metaphase and interphase nuclei examined. From the metaphase analysis, 5 of 30 cells examined showed 2 normal chromosomes 22 and a

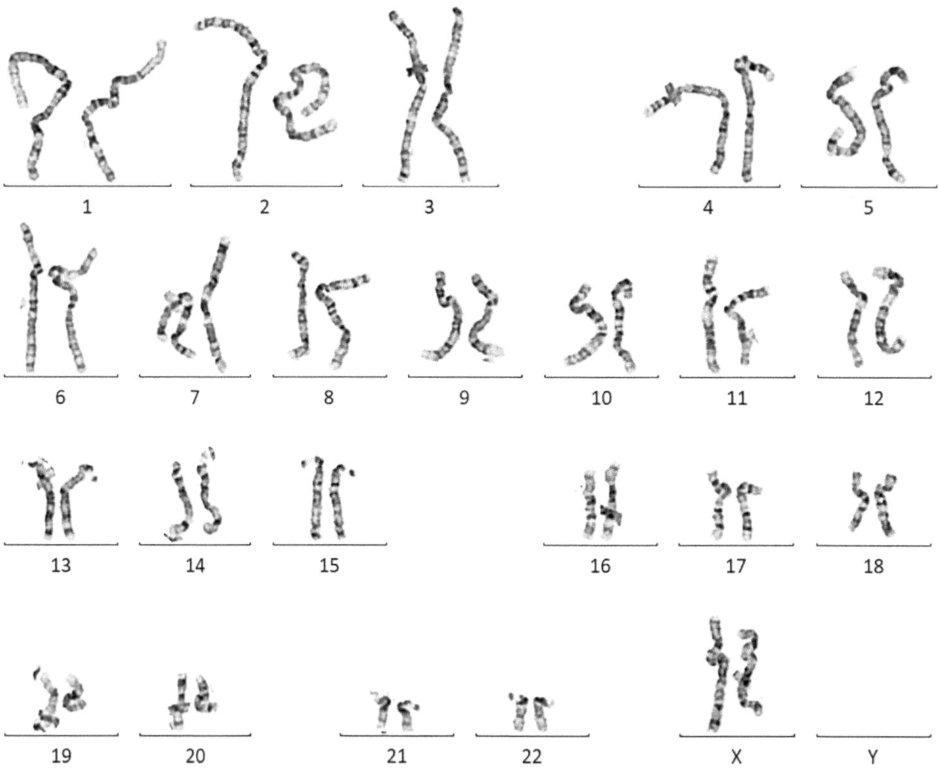

FIG. 9.3.1 Chromosome analysis revealed a normal female karyotype in 46 out of 50 cells examined. ISCN: 46,XX[46]

duplication on a marker chromosome. From the interphase analysis, 12 of 100 cells examined showed duplication appearing as 4 orange *TUPLE1-HIRA* signals and 2 green *ARSA* signals (Fig. 9.3.3 for abnormal cell line, and Fig. 9.3.4 for normal cell line in interphase and metaphase cells).

Chromosomal Microarray Analysis (CMA) revealed a copy number gain of 10.2 Mb of DNA from chromosome 22 at band 22q11.1q12.1, encompassing 110 OMIM genes of clinical significance. CMA also identified a second gain of 890 kb of DNA from the telomere of chromosome 22 at band 22q13.33, encompassing 27 OMIM genes of uncertain clinical significance (Fig. 9.3.5).

Results with interpretations

Chromosome analysis identified two cell lines. The abnormal cell line had a marker chromosome in addition to the normal female karyotype. FISH also revealed two cell lines. The abnormal cell line showed three to four copies of *TUPLE1* on the marker chromosome, concordant with the cytogenetic finding. CMA results also confirmed a gain of 10.2 Mb of DNA from 22q11.2q12.1 including *TUPLE1*.

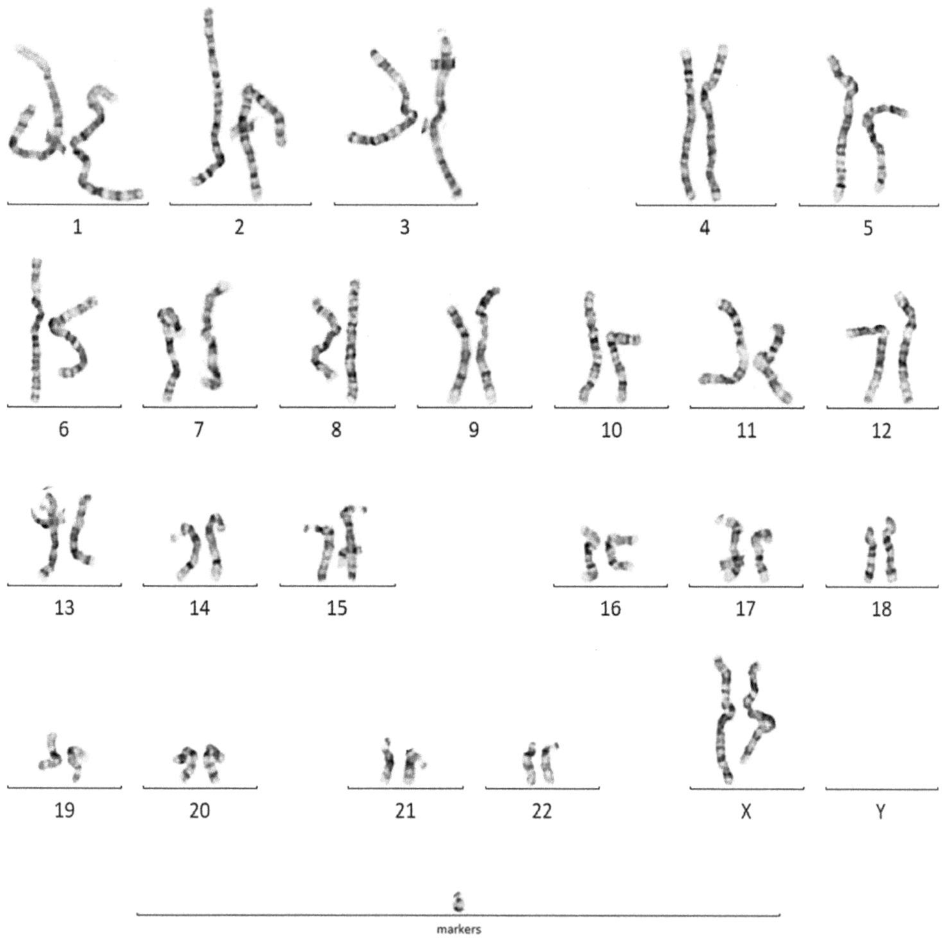

FIG. 9.3.2 Chromosome analysis revealed an abnormal female karyotype with a marker chromosome in 4 out of 50 cells examined. ISCN: 47,XX,+mar[4]

This large duplication overlaps multiple syndromes that include Cat eye syndrome Type I, 22q11.2 microduplication syndrome, and 22q11.2 distal duplication syndrome.

The clinical features of Cat eye syndrome (CES) are extremely variable. It is characterized by the combination of coloboma of the iris and anal atresia with fistula, down slanting palpebral fissures, preauricular tags and/or pits, frequent occurrence of heart and renal malformations, and normal or near-normal mental development. Most patients harbor a small supernumerary bi-satellited marker chromosome (sSMC) that results in partial trisomy of 22pter-22q11 (https://omim.org/entry/115470). In one-third of cases, this extra chromosome is present in a mosaic state. Although rare, other cytogenetic anomalies have been reported (https://www.orpha.net/consor/cgi-bin/OC_Exp.php?Lng=GB&Expert=195) [14].

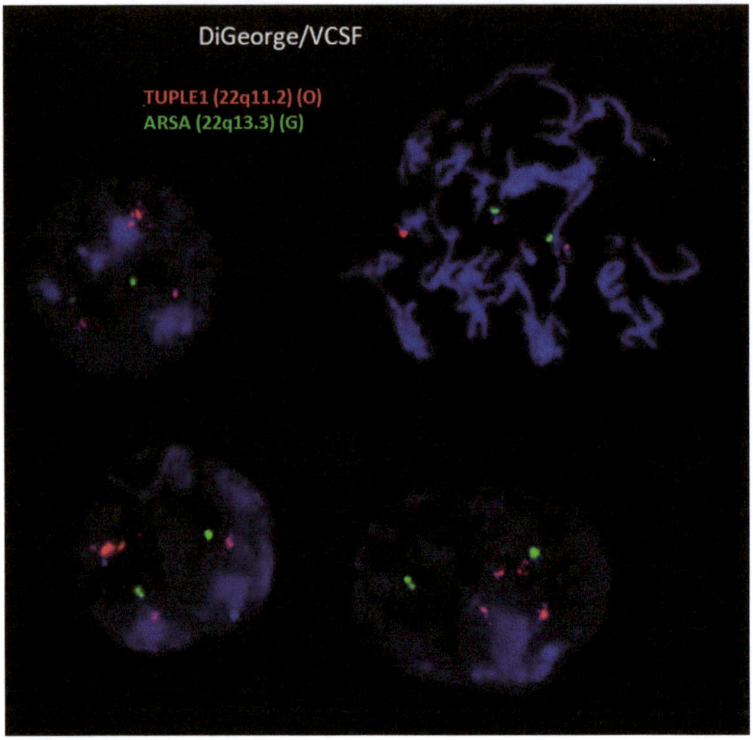

FIG. 9.3.3 FISH with a *TUPLE1* probe showed three to four copies in both metaphase and interphase cells. ISCN: ish 22q11.2(TUPLE1x3)[5/30].nuc ish(TUPLE1x3-4)[12/100]

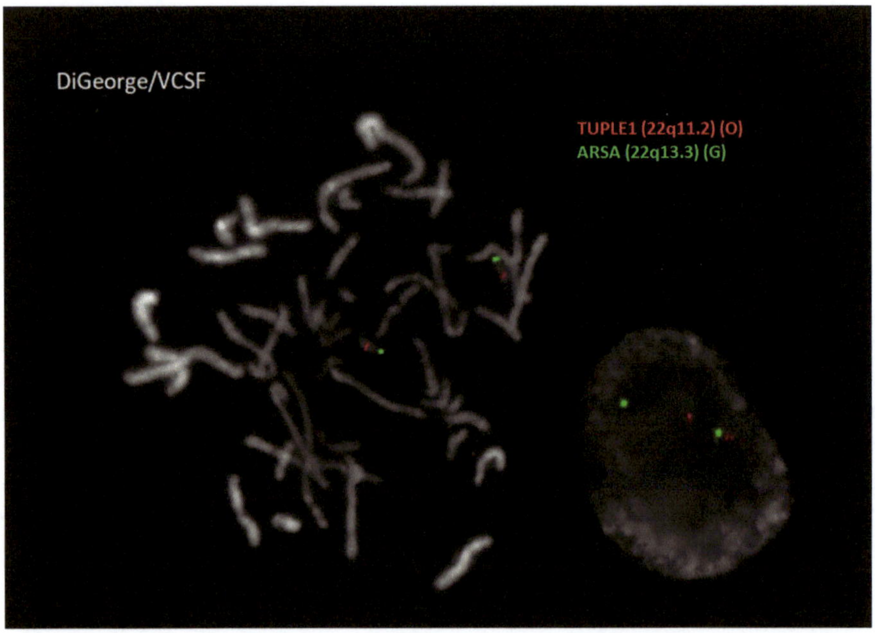

FIG. 9.3.4 FISH with a *TUPLE1* probe showed a normal signal pattern (two copies) in both metaphase and interphase cells.

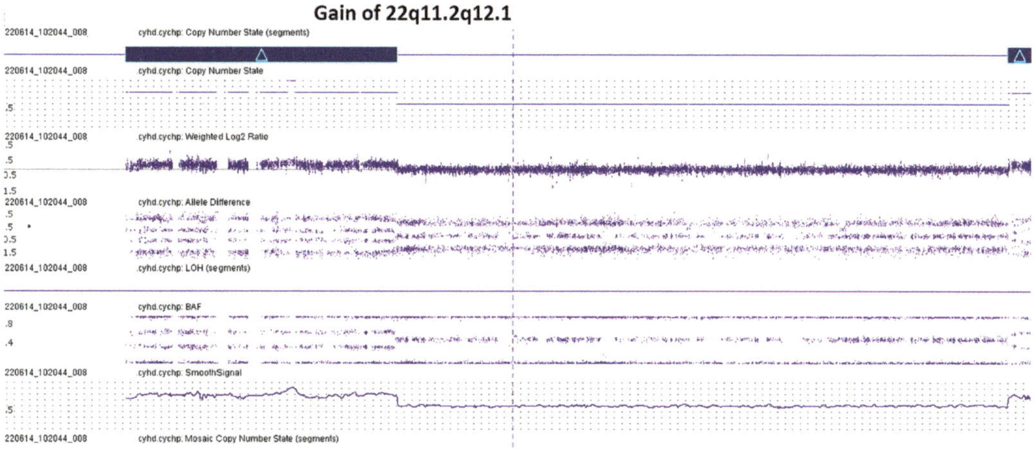

FIG. 9.3.5 CMA identified a gain of 10.2 Mb of DNA from chromosome 22 at band 22q11.1q12.1 of clinical significance, and a gain of 890 kb of DNA from 22q13.33 of uncertain clinical significance. ISCN: arr[hg19] 22q11.1q12.1(16,888,900-27,094,711)x3 arr[hg19] 22q13.33(50,307,514-51,197,838)x3

This duplication is most consistent with Cat eye syndrome; however, this gain is larger than the typical gain seen in CES and extends into the DG/VCFS region involving additional genes. Therefore, the phenotype for this patient may be variable depending on the genes involved.

The 22q11.2 microduplication syndrome phenotype appears to be relatively mild and highly variable. Findings range from no phenotype to learning disability, mental retardation, delayed psychomotor development, growth retardation, and hypotonia. The high frequency with which the 22q11.2 duplication is found in a normal parent of a proband suggests that many individuals can harbor a duplication of 22q11.2 with no discernible phenotypic effect (https://www.omim.org/entry/608363) [15].

Children with distal chromosome 22q11.2 duplication syndrome have features including developmental delay, language delay, dysmorphic facies, and epilepsy [16–18]. Duplications of this region have also been observed in mildly affected and unaffected family members, consistent with variable expressivity or incomplete penetrance of the phenotype. The various phenotypic expressions and incomplete penetrance observed for distal 22q11.2 duplications make it exceedingly difficult to ascribe pathogenicity to these duplications. Given the observed enrichment of the duplication in patient samples versus healthy controls, some authors proposed that it is likely that distal 22q11.2 duplications represent a susceptibility/risk locus for speech and mild developmental delay [19].

CMA also identified the second gain of 890 kb DNA in this case. The clinical presentation for cases with overlapping 22qter duplications includes mild to moderate mental retardation, microcephaly, and similar mild dysmorphic features. One report of two siblings with a larger duplication partially overlapping the same region as this patient with mostly similar genes involved presented with mild dysmorphic features, such as hypertelorism, divergent strabismus, high nasal bridge, and a bulbous nose tip. They both developed obesity and learning difficulties. One of the siblings had neuropsychiatric symptoms including maladaptive behavior, epilepsy, repetitive monotonous playing, and bipolar affective disorder [20]. However, this study is limited and larger in size involving additional genes.

The *SHANK3* gene is of particular interest, in which mutations and deletions are associated with an autism spectrum disorder. Mutations in this gene also cause schizophrenia type 15 and are a major causative factor in the neurological symptoms of 22q13.3 deletion syndrome, which is also known as Phelan-McDermid syndrome (PHMDs) (https://www.ncbi.nlm.nih.gov/gene/85358). PHMDS is an autosomal dominant disorder and can be caused by heterozygous contiguous gene deletion at chromosome 22q13 or by a mutation in the *SHANK3* gene. This syndrome is characterized by neonatal hypotonia, global developmental delay, absent to severely delayed speech, moderate to profound intellectual disability, normal to accelerated growth, autistic features, and additional variable findings (https://omim.org/entry/606232). *SHANK3* is also a candidate gene possibly related to neuropsychiatric problems; however, it is unclear whether duplication of the *SHANK3* gene is causative of disease [21].

There was no exact duplication reported that was the same as that of this patient; therefore, the clinical significance of this gain is currently unknown.

Future testing and recommendation

Chromosome analysis and FISH on this patient are indicated to rule out a structural rearrangement. A parental FISH analysis is recommended to discriminate between a familial variant and a de novo genomic alteration. A thorough clinical assessment of this patient and genetic counseling are recommended.

Case 9.4 Contiguous gene syndrome with duplication of 6q16.1q23.3

Clinical indication

A 16-year-old male had a left central tympanic membrane perforation, which was fixed through surgery. He also had developmental delays and distinctive features.

Test ordered

- Chromosome microarray on peripheral blood

Laboratory test performed

The chromosome microarray method was described previously in Chapter 1.

Test results

Chromosomal Microarray Analysis (CMA) revealed a copy number gain of 41.0 Mb of DNA from chromosome 6 at band 6q16.1q23.3, encompassing 145 OMIM genes of clinical significance, consistent with the diagnosis of Chromosome 6q duplication syndrome or partial trisomy 6q (Fig. 9.4.1). In addition, two regions of the absence of heterozygosity (AOH) were detected with uncertain clinical significance (Fig. 9.4.2).

Results with interpretations

CMA identified a gain of 41 Mb containing 145 OMIM genes. This duplication is pathogenic. Features that often occur in people with chromosome 6q duplication include developmental delay, intellectual disability, behavioral problems, and distinctive facial features. The severity of the condition and symptoms depend on the genes involved and the size of the duplication. In most cases, chromosome 6q duplication occurs de novo or is inherited from a parent with a chromosomal rearrangement. Rarely, it is inherited from a parent with the same duplication (https://rarediseases.info.nih.gov/diseases/5353/chromosome-6q-duplication) [22].

The clinical presentation of this duplication is not well documented in the literature. There is only one reported case with a similar duplication in a patient diagnosed with Prader-Willi-like syndrome [23]. Therefore, it is unclear whether this patient will be affected by the phenotype described in this study.

CMA analysis also showed two regions of the AOH greater than 5 Mb. The combined total length of the homozygous segments was approximately 14.8 Mb and encompassed 0.51% of the genome. The implications of such copy-neutral segments showing

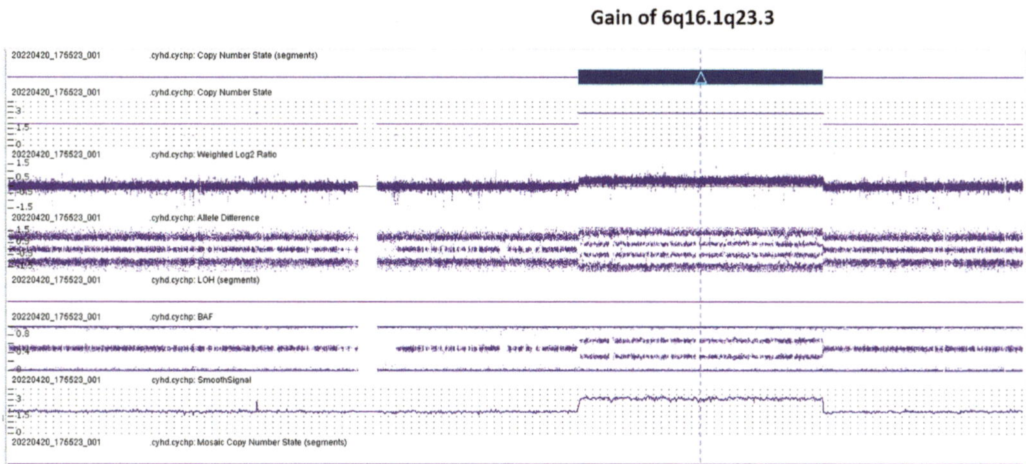

FIG. 9.4.1 CMA identified a pathogenic gain of 41.0 Mb of DNA from chromosome 6 at band 6q16.1q23.3, encompassing 145 OMIM genes. ISCN: arr[hg19] 6q16.1q23.3(95,837,354-136,915,573)x3

FIG. 9.4.2 CMA also detected two AOH regions that were larger than 5 Mb. ISCN: arr[hg19] 4q12q13.2(58,840,643-68,058,681)x2 hmz arr[hg19] 5q31.2q31.3(136,921,616-142,521,025)x2 hmz

homozygosity are unclear at present. In theory, if the degree of homozygosity is significant, parental relatedness is a likely cause. Although this result is not diagnostic of a specific condition, it raises the possibility of a recessive disorder within one of these regions.

Future testing and recommendation

Chromosome analysis is highly recommended to rule out a structural rearrangement. Parental chromosome analysis is indicated to determine whether these genetic imbalances are familial or de novo in origin. If a specific disorder is suspected in this patient, additional testing may be indicated. Genetic counseling is recommended.

Case 9.5 DiGeorge/Velo-cardio-facial (VCF) syndrome (22q11.2 deletion syndrome)

Clinical indication

A 2-day-old newborn baby boy presented with feeding problems, moderate right ventricular enlargement, and other congenital anomalies from the physical examination. He had moderate hypoglycemia on the basic metabolic panel (BMP). DiGeorge/VCF syndrome is suspected.

Test ordered

- Chromosome microarray (CMA)
- FISH: DiGeorge (*TUPLE1*)

Laboratory test performed

Chromosome microarray and FISH methods were described in Chapter 1.

Test results

Fluorescent molecular probes, which are localized to the *TUPLE1* gene region (22q11.2) and *ARSA* gene region (22q13; internal control), were hybridized to metaphase preparations. The fluorescence signal pattern was abnormal in the 10 metaphase cells examined. Analysis revealed a deletion of the *TUPLE1* gene region. This deletion is consistent with the diagnosis of DiGeorge/Velo-cardio-facial (VCF) syndromes (Fig. 9.5.1).

Chromosomal Microarray Analysis (CMA) revealed a copy number loss of 2.55 Mb of DNA from chromosome 22 at band 22q11.21. This deletion, which includes 43 OMIM genes, has been recognized as chromosome 22q11.2 deletion syndrome (DiGeorge/VCF syndrome, OMIM#188400, 192430) (Fig. 9.5.2).

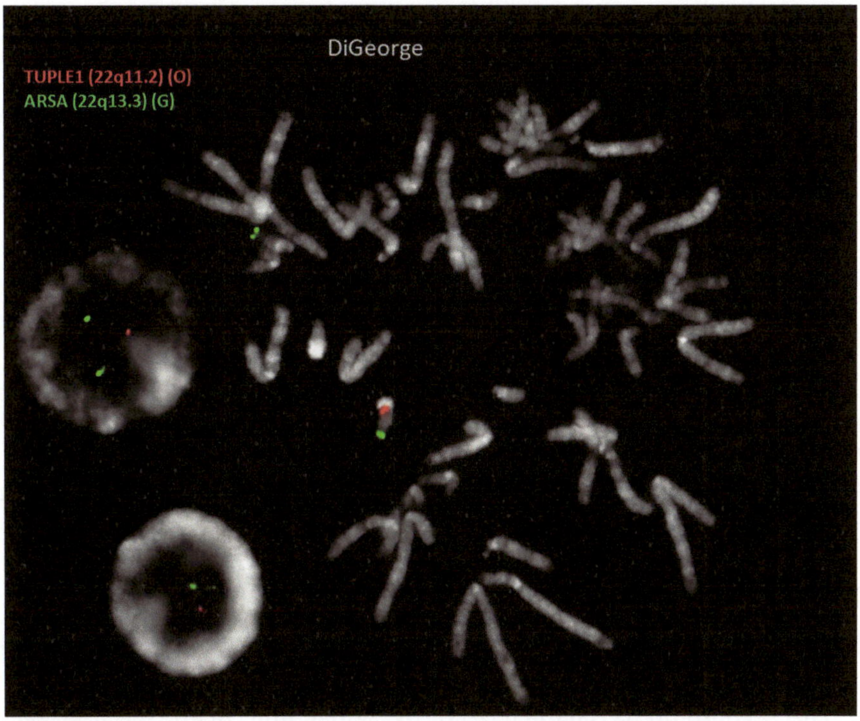

FIG. 9.5.1 FISH showed a deletion of *TUPLE1* in all metaphase cells examined. ISCN: ish del(22)(q11.2)(TUPLE1-)[10]

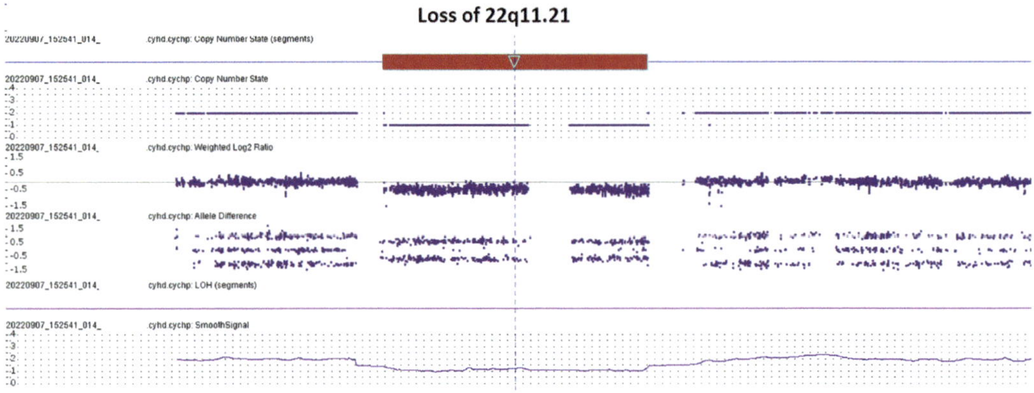

FIG. 9.5.2 CMA identified a deletion of 2.55 Mb DNA from the 22q11.2 region. ISCN: arr[hg19] 22q11.21(18,916,843-21,465,659)x1 arr[hg19] 22q13.1(38,132,597-38,187,935)x3

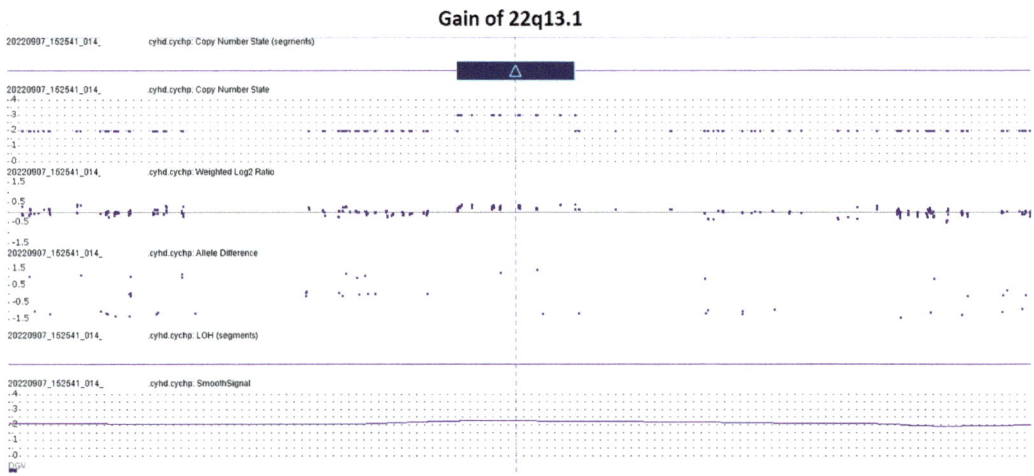

FIG. 9.5.3 CMA identified a gain of 55 kb DNA from the 22q13.1 region.

CMA also revealed a copy number gain of 55 kb of DNA from chromosome 22 at band 22q13.1, overlapping 1 OMIM gene, a portion of the *TRIOBP* (609761), of uncertain clinical significance (Fig. 9.5.3).

Results with interpretations

FISH identified a deletion of *TUPLE1*, which was confirmed by CMA analysis. The results are consistent with the diagnosis of DiGeorge/VCF syndrome (22q11.2 deletion syndrome). This syndrome is characterized by congenital heart disease, particularly conotruncal malformations (tetralogy of Fallot, interrupted aortic arch, ventricular septal defect, and truncus arteriosus), immune deficiency, and learning difficulties (https://www.ncbi.nlm.nih.gov/books/NBK1523/).

The second CMA finding was a gain of 55 kb of DNA from 22q13.1 including the *TRIOBP* gene. Homozygous mutations in the *TRIOBP* gene are associated with autosomal recessive deafness-28 (DFNB28), which is a form of nonsyndromic sensorineural hearing loss. However, this patient has a partial duplication, and no exact similar gain has been reported previously in individuals with the disease. Also, this gene is not directly relevant to the reason for the patient's referral. This analysis cannot determine the exact location or the orientation of the duplicated region. This duplication may be present, in the same or opposite orientation, or it may be inserted into a different region of the genome. Depending on the location, this duplication may or may not disrupt the function of any of the involved genes. Similar gains have been observed among healthy individuals in the Database of Genomic Variants (DGVs). As such, the clinical significance of this variation is unclear (https://omim.org/entry/609823).

Clinical correlation, genetic counseling, and continued surveillance of the literature regarding the clinical relevance of these CNVs are recommended.

Future testing recommendation

A thorough clinical assessment of this patient and genetic counseling are recommended. A proportion of DiGeorge/VCF-associated deletions of chromosome 22 are inherited from a parent. Clinical expression in a carrier parent may vary with features, often subtle, that are associated with the deletion. Parental VCF microdeletion FISH analysis is indicated.

Case 9.6 Contiguous gene syndrome with a deletion of 1q43q44

Clinical indication

A 15-day-old newborn baby girl presented with rocker bottom foot, microcephaly, corpus callosum, and agenesis.

Test ordered

- Chromosome analysis: Routine blood
- Chromosome microarray (CMA)

Laboratory test performed

Chromosome analysis and microarray methods were described in Chapter 1.

Test results

Chromosome analysis identified a terminal deletion of chromosome 1 at band 1q43 (Fig. 9.6.1). Chromosomal Microarray Analysis (CMA) revealed a copy number loss of 5.9 Mb of DNA from the long arm of chromosome 1 at band 1q43q44, including 24 OMIM genes of clinical significance (Fig. 9.6.2).

Results with interpretations

The terminal deletion of 1q43 identified by karyotyping was confirmed by CMA. These results are consistent with a diagnosis of Chromosome 1q43-q44 deletion syndrome (OMIM: #612377). This syndrome is characterized by moderate to severe mental retardation, limited or no speech, and variable but characteristic facial features, including a round face, prominent forehead, flat nasal bridge, hypertelorism, epicanthal folds, and low-set ears. Other features may include hypotonia, poor growth, microcephaly, agenesis of the corpus callosum, seizures, cardiac, gastroesophageal, and urogenital anomalies. Evidence from the literature suggests that almost 70% of children with similar deletions have a foot anomaly. The phenotype is variable, and not all features are observed in all patients, which may be

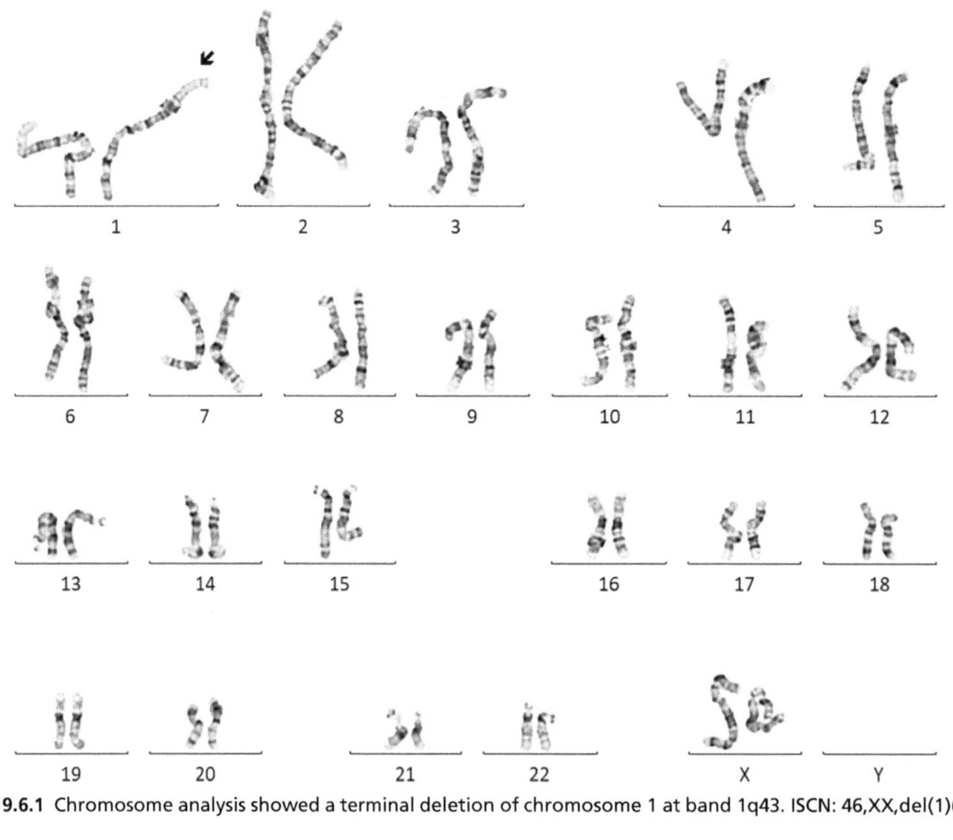

FIG. 9.6.1 Chromosome analysis showed a terminal deletion of chromosome 1 at band 1q43. ISCN: 46,XX,del(1)(q43)

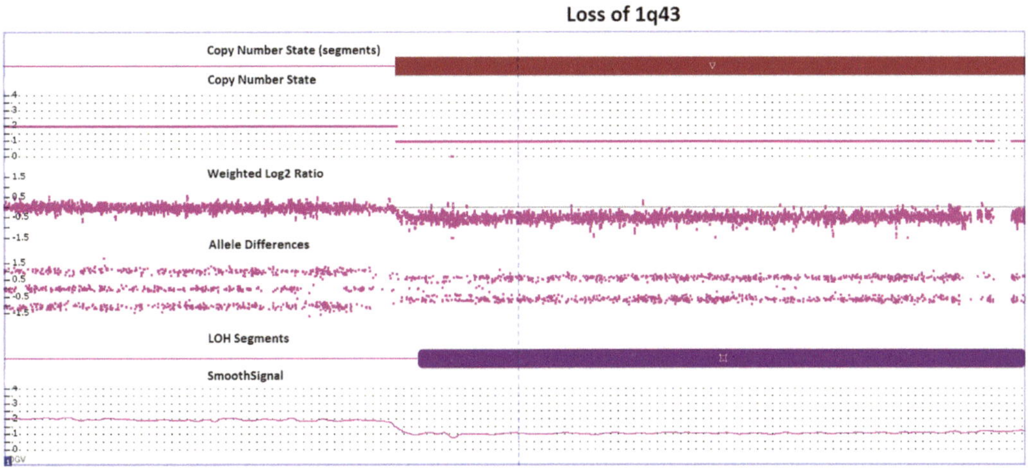

FIG. 9.6.2 CMA identified a loss of 5.9 Mb DNA from the 1q43q44 region. ISCN: arr[hg19] 1q43q44(243,338,560-249,224,684)x1

explained in some cases by incomplete penetrance or variable expressivity (https://omim.org/entry/612337, https://www.rarechromo.org/media/information/Chromosome%20%201/1q4%20deletions%20from%201q42%20and%20beyond%20FTNW.pdf) [24].

Future testing recommendation

A thorough clinical assessment of this patient and genetic counseling are recommended. Parental chromosome analysis is also indicated to rule out a balanced translocation involving 1q in either of the parents.

Summary of key learning points

1. CNVs may cause diseases with blended phenotypes.
2. Disease manifestation may be different between recurrent and nonrecurrent CNVs affecting the same critical region.
3. Interpretation of CNV requires careful correlation between the genomic content and the phenotypes of the patient.

References

[1] J.R. Lupski, Genomic disorders: structural features of the genome can lead to DNA rearrangements and human disease traits, Trends Genet. 14 (10) (1998) 417–422.

[2] J.R. Lupski, Genomic disorders ten years on, Genome Med. 1 (4) (2009) 42.

[3] P. Stankiewicz, J.R. Lupski, Structural variation in the human genome and its role in disease, Annu. Rev. Med. 61 (2010) 437–455.

[4] R.D. Schmickel, Contiguous gene syndromes: a component of recognizable syndromes, J. Pediatr. 109 (2) (1986) 231–241.

[5] W. Bi, et al., Reciprocal crossovers and a positional preference for strand exchange in recombination events resulting in deletion or duplication of chromosome 17p11.2, Am. J. Hum. Genet. 73 (6) (2003) 1302–1315.

[6] K.S. Chen, et al., Homologous recombination of a flanking repeat gene cluster is a mechanism for a common contiguous gene deletion syndrome, Nat. Genet. 17 (2) (1997) 154–163.

[7] L. Potocki, et al., Characterization of Potocki-Lupski syndrome (dup(17)(p11.2p11.2)) and delineation of a dosage-sensitive critical interval that can convey an autism phenotype, Am. J. Hum. Genet. 80 (4) (2007) 633–649.

[8] L. Potocki, et al., Molecular mechanism for duplication 17p11.2—the homologous recombination reciprocal of the Smith-Magenis microdeletion, Nat. Genet. 24 (1) (2000) 84–87.

[9] P.J. Hastings, et al., Mechanisms of change in gene copy number, Nat. Rev. Genet. 10 (8) (2009) 551–564.

[10] F. Zhang, et al., The DNA replication FoSTeS/MMBIR mechanism can generate genomic, genic and exonic complex rearrangements in humans, Nat. Genet. 41 (7) (2009) 849–853.

[11] B. Yuan, et al., Nonrecurrent 17p11.2p12 rearrangement events that result in two concomitant genomic disorders: the PMP22-RAI1 contiguous gene duplication syndrome, Am. J. Hum. Genet. 97 (5) (2015) 691–707.

[12] B. Yuan, et al., Nonrecurrent PMP22-RAI1 contiguous gene deletions arise from replication-based mechanisms and result in Smith-Magenis syndrome with evident peripheral neuropathy, Hum. Genet. 135 (10) (2016) 1161–1174.

[13] C.W. Wu, et al., Complicated diagnosis and treatment of HA20 due to contiguous gene deletions involving 6q23.3, J. Clin. Immunol. 41 (6) (2021) 1420–1423.

[14] M. Meins, et al., Partial trisomy of chromosome 22 resulting from an interstitial duplication of 22q11.2 in a child with typical cat eye syndrome, J. Med. Genet. 40 (5) (2003) e62.

[15] Z. Ou, et al., Microduplications of 22q11.2 are frequently inherited and are associated with variable phenotypes, Genet. Med. 10 (4) (2008) 267–277.

[16] M. Descartes, et al., Distal 22q11.2 microduplication encompassing the BCR gene, Am. J. Med. Genet. A 146A (23) (2008) 3075–3081.

[17] J. Coppinger, et al., Identification of familial and de novo microduplications of 22q11.21-q11.23 distal to the 22q11.21 microdeletion syndrome region, Hum. Mol. Genet. 18 (8) (2009) 1377–1383.

[18] K. Shimojima, K. Imai, T. Yamamoto, A de novo 22q11.22q11.23 interchromosomal tandem duplication in a boy with developmental delay, hyperactivity, and epilepsy, Am. J. Med. Genet. A 152A (11) (2010) 2820–2826.

[19] J. Wincent, et al., Sixteen new cases contributing to the characterization of patients with distal 22q11.2 microduplications, Mol. Syndromol. 1 (5) (2010) 246–254.

[20] A. Ujfalusi, et al., 22q13 microduplication syndrome in siblings with mild clinical phenotype: broadening the clinical and behavioral spectrum, Mol. Syndromol. 11 (3) (2020) 146–152.

[21] M. Johannessen, et al., A 22q13.33 duplication harbouring the SHANK3 gene: does it cause neuropsychiatric disorders? BMJ Case Rep. 12 (11) (2019).

[22] M.I. Srebniak, et al., Interstitial 6q21q23 duplication—variant of variable phenotype and incomplete penetrance or benign duplication? Mol. Cytogenet. 9 (2016) 43.

[23] L. Desch, et al., 6q16.3q23.3 duplication associated with Prader-Willi-like syndrome, Mol. Cytogenet. 8 (2015) 42.

[24] A.M. Mohamed, et al., Cytogenomic characterization of 1q43q44 deletion associated with 4q32.1q35.2 duplication and phenotype correlation, Mol. Cytogenet. 11 (2018) 57.

10

Thrombosis

Xia Li
SONORA QUEST LABORATORIES, PHOENIX, AZ, UNITED STATES

Background

Deep vein thrombosis (DVT) refers to a medical condition that occurs when a blood clot forms in a deep vein. These clots usually develop in the lower leg, thigh, or pelvis, but can also occur in the arm. The clots can break off and travel through the bloodstream to the lungs, causing a blockage called pulmonary embolism (PE). DVT can happen to anybody and can cause serious illness, disability, and in some cases, death. With the advancement of genetic testing, these conditions are preventable and treatable if discovered early (CDC.gov).

DVT is the most common venous thromboembolism (VTE) and is a multifactorial disease influenced by genetic, environmental, and circumstantial risk factors. The c.*97G>A variant (previously referred to as G20210A) in the *Factor II* gene is a genetic risk factor for VTE. Carriers of a mutation in the *prothrombin* (*factor II*) or *factor V* gene have a two- to fourfold greater risk for VTE than subjects without the mutations. Homozygotes for the c. *97G>A variant are rare with the annual risk of VTE in homozygotes at 1.1% per year. Individuals who carry both a *97G>A variant in the *Factor II* gene and a c. 1601G>A (p.Arg534Gln) variant in the *Factor V* gene (commonly referred to as *Factor V Leiden*) have an approximately 20-fold increased risk for VTE. Risks are likely to be even higher in more complex genotype combinations involving *Factor II* c.*97G>A variant and *Factor V Leiden* [1,2]. Additional risk factors may be associated with deficiency of protein C, protein S, or antithrombin III-, age, male sex, personal or family history of DVT, smoking, surgery, prolonged immobilization, malignant neoplasm, tamoxifen treatment, raloxifene treatment, oral contraceptive use, hormone replacement therapy, and pregnancy. Management of thrombotic risk and thrombotic events should follow established guidelines and fit the clinical circumstance [3–5].

Methylenetetrahydrofolate Reductase (*MTHFR*) Deficiency is the most common genetic cause of elevated levels of homocysteine in the plasma (hyperhomocysteinemia). Genetic variations in the *MTHFR* gene including 677C>T and 1298A>C are associated with high homocysteinemia and possibly venous thrombosis [6,7]. Therefore, these two variants were included in the coagulation panel. In this chapter, a case will be illustrated on how a patient can benefit from genetic testing of the coagulation panel.

Cases in Laboratory Genetics and Genomics (LGG) Practice. https://doi.org/10.1016/B978-0-323-99622-8.00013-7

Case 10.1 Deep vein thrombosis

Clinical indication

A 20-year-old female presented with obesity complicating pregnancy in the second trimester and a history of DVT in a prior pregnancy.

Test ordered

- Coagulation PCR panel

Laboratory test performed

The coagulation test is performed by the TaqMan assay, which uses real-time PCR to detect single nucleotide polymorphisms. There are three genes involved in this assay: *prothrombin* (*factor II*) c.*97G>A (G20210A), *factor V Leiden* c.1601G>A (p.Arg534Gln), and *MTHFR* c.1298A>C(p.Gln429Ala) and c.667C>T (p.Arg223Trp). The methods followed the manufacturer's protocol (refer to TaqMan SNP Genotyping assay from Thermo Fisher).

Test results

Real-time PCR showed negative results (wild type) for *factor II* 20210G>A, heterozygous for *factor V (Leiden)* mutation, homozygous for C667T mutation in *MTHFR*, and wild type for C1298T in *MTHFR* (Fig. 10.1.1).

Results with interpretations

The patient has a history of venous thrombosis. Her genotype for *factor II* was a wild type. Her genotype for *factor V* was a heterozygous Leiden mutation, and for *MTHFR* was homozygous for C677T. Heterozygous mutation for *factor V* is associated with activated protein C resistance and increased risk for venous thrombosis. The genotype of C677T in *MTHFR* occurs in 1.5%–15% of the population and is associated with increased plasma homocysteine levels, a risk factor for arteriosclerotic coronary disease, and venous thrombosis. Recently, the benefit of *MTHFR* testing has come into question because of conflicting results on an association between elevated homocysteine levels and the risk of venous thrombosis or the risk of coronary heart disease. Since the patient has a history of thrombosis, clinical correlation with her genotype results from the coagulation panel will help the physician to make decisions on clinical care.

Future testing and recommendations

Both *Factor II* and *Factor V* Leiden thrombophilia are inherited in an autosomal dominant manner. Heterozygosity for the 20210G>A or 1601G>A variant results in an increased risk for thrombosis; homozygosity for either of these two variants confer a higher risk for thrombosis than heterozygosity. Because of the high prevalence of *factor II* and *factor V*

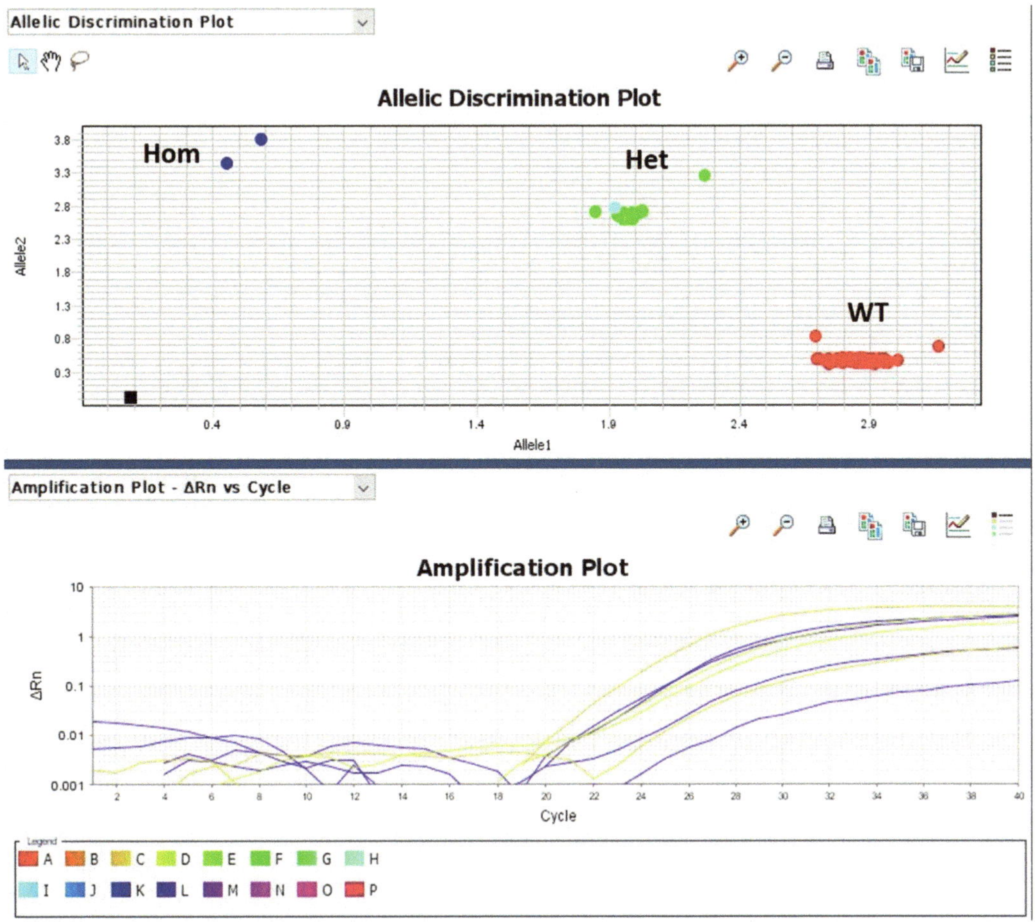

FIG. 10.1.1 Allelic discrimination plot and amplification plot from coagulation panel by real-time PCR for this run including this patient (*blue*: homozygous mutation; *green*: heterozygous; *red*: wild type).

Leiden allele in the general population, the genetic status of both parents and/or the reproductive partner of an affected individual needs to be evaluated before information regarding potential risks to sibs or offspring can be provided. Once the variants have been identified in a family member, prenatal testing for pregnancies at increased risk and preimplantation genetic testing is possible. Genetic counseling is recommended.

Summary of key learning points

- DVT is a multifactorial disease influenced by genetic, environmental, and circumstantial risk factors.
- DVT is associated with genes such as *factor II* G20210G>A and *factor V Leiden* c.1601G>A, *MTHFR* 677C>T, and 1298G>A.

- TaqMan real-time PCR genotyping assay is used to identify the variants in the coagulation panel.
- Risk of *Factor II*, *Factor V*, and *MTHFR* variants for sibs and offspring should be considered when proband has known mutations.

References

[1] J.L. Kujovich, Factor V Leiden thrombophilia, in: M.P. Adam, G.M. Mirzaa, R.A. Pagon, S.E. Wallace, L.J.H. Bean, K.W. Gripp, A. Amemiya (Eds.), GeneReviews® [Internet], University of Washington, Seattle, Seattle, WA, 1993–2023. 20301542.

[2] J.L. Kujovich, Prothrombin thrombophilia, in: M.P. Adam, G.M. Mirzaa, R.A. Pagon, S.E. Wallace, L.J.H. Bean, K.W. Gripp, A. Amemiya (Eds.), GeneReviews® [Internet], University of Washington, Seattle, Seattle, WA, 1993–2023. 20301327.

[3] S. Safavi-Abbasi, et al., Thrombophilia due to factor V and factor II mutations and formation of a dural arteriovenous fistula: case report and review of a rare entity, Skull Base 18 (2) (2008) 135–143.

[4] J. Emmerich, et al., Combined effect of factor V Leiden and prothrombin 20210A on the risk of venous thromboembolism—pooled analysis of 8 case-control studies including 2310 cases and 3204 controls. Study group for pooled-analysis in venous thromboembolism, Thromb. Haemost. 86 (3) (2001) 809–816.

[5] P. Simioni, et al., Risk for subsequent venous thromboembolic complications in carriers of the prothrombin or the factor V gene mutation with a first episode of deep-vein thrombosis, Blood 96 (10) (2000) 3329–3333.

[6] L. Dean, M. Kane, Medical genetics summaries expert reviewers, in: V.M. Pratt, S.A. Scott, M. Pirmohamed, et al. (Eds.), Medical Genetics Summaries [Internet], National Center for Biotechnology Information (US), Bethesda, MD, 2012. Available from: https://www.ncbi.nlm.nih.gov/books/NBK436916/.

[7] R. Clarke, et al., Homocysteine and coronary heart disease: meta-analysis of MTHFR case-control studies, avoiding publication bias, PLoS Med. 9 (2) (2012) e1001177.

11

Pharmacogenomics

Xia Li

SONORA QUEST LABORATORIES, PHOENIX, AZ, UNITED STATES

Background

Pharmacogenetics refers to the role of genetics in drug responses. It studies individual gene-drug interactions, usually one or two genes that have a dominant effect on drug response. Pharmacogenomics (PGx) refers to the science that allows us to predict responses to drugs based on an individual's genetic makeup. It studies genomic influences on drug response, often using high-throughput data such as sequencing, SNP chip, expression, and proteomics. The goals of PGx include maximizing drug efficacy, minimizing drug toxicity, predicting patients who will respond to intervention, and aiding in new drug development [1]. In the field of neuropsychiatry, mental illness is the leading cause of bad health in many developing countries because medications can fail to bring about remission in more than half of patients with either depression or mental disorders [2,3]. It is known that acetaminophen(APAP) is widely used as an over-the-counter fever reducer and pain reliever. However, the variability among patients with different age groups in both efficacy and toxicity has limited the use of this drug for pediatric patients. The toxic metabolite, *N*-acetyl-p-benzoquinone imine (NAPQI) is the cause of adverse effects resulting from an overdose of acetaminophen. There are a few genes involved in the metabolism of APAP including *CYP2D6*, *CYP1A2*, *CYP2A6*, *CYP2E1*, and *CYP3A4*. Knowing the genotypes of these genes by PGx testing will benefit patients in managing drug usage [4].

The advancement of DNA sequencing has evolved rapidly within the last 50 years. The clinical utility of PGx has been shown to maximize the drug efficacy and "tailor" for individualized drug therapy [5]. There are many panels on the market focusing on psychiatry, cardiology, pain management, or oncology, and specific genes can be tested to guide physicians for prescription drugs. The Clinical Pharmacogenetics Implementation Consortium, a collaboration between the PGRN and the PharmGKB database, or a similar European consortium, such as the Dutch Pharmacogenetics Working Group, provides evidence-based guidelines for decision-making regarding alerts for any specific medication (https://cpicpgx.org/). PGx has become an important field of precision medicine, which signals the moving of health care from a one-size-fits-all approach to a more targeted approach. The technologies used in the clinical diagnostic lab for PGx testing include Sanger sequencing, genotyping, multiplex PCR, and NGS, etc. The case shown

Cases in Laboratory Genetics and Genomics (LGG) Practice. https://doi.org/10.1016/B978-0-323-99622-8.00009-5

in this chapter will illustrate how PGx is performed and how the results will benefit the patient in the long term with various medications.

Case 11.1 Overdose acetaminophen (APAP)

Clinical indication

A 30-year-old female was admitted with acute liver injury presumed due to the accidental overuse of APAP. The patient reported that she had severe migraine and took APAP 1500 (3× 500 mg tabs) every 4 h. She had nausea and vomited before being presented to the ER.

Test ordered

- PGx panel

Laboratory test performed

PGx is the study of genetic variation as it relates to drug response. PGx studies involve testing samples for multiple variants in drug metabolism enzyme (DME) and transporter genes. The assay uses procedures from Thermo Fisher for a sample-to-result PGx workflow solution using the QuantStudio 12K Flex Real-Time PCR System. Variant testing uses TaqMan SNP Genotyping Assay or TaqMan Copy Number Assays targeting DME genes. The TaqMan Drug Metabolism Genotyping Assay collection has ~2700 assays that detect potentially causative polymorphisms in 221 drug metabolism enzymes and associated transporter genes. TaqMan Copy Number Assays examine copy number variation (CNV) in DME genes. Array-based assays detect alleles, including all common and rare variants with known clinical significance at analytical sensitivity and specificity >99%.

Test results

The PGx assay detected the genotypes from 22 genes with 120 SNPs and two copy number variations from *CYP2D6* using the TaqMan Genotyper software. The haplotypes and predicted phenotypes were annotated by Translational software (Translational Software, Inc.). The gene names, genotypes, or haplotypes if available, and predicted phenotypes are shown in Fig. 11.1.1. The alerts for drugs and dosing guidance were also listed in the PGx report (data not shown).

Results with interpretations

The PGx results showed all genotypes for the genes and SNPs that were tested. Based on the genotyping data using TaqMan Genotyper software, translational software converted the genotyping data into haplotypes if available and predicted the phenotype for each gene. The report generated contains alternates for specific medications and dosing

Pharmacogenetic Test Summary

ABCB1	3435C>T C/C	Variant Allele Not Present
ABCB1	1236T>C C/C	Homozygous Mutant - Variant Allele Present
ABCB1	2677G>T G/G	Variant Allele Not Present
CYP1A2	*1F/*1V	Normal Metabolizer- Possible Inducibility
CYP2B6	*1/*1	Normal Metabolizer
CYP2C	g.96405502G>A G/A	High Sensitivity
CYP2C19	*1/*2	Intermediate Metabolizer
CYP2C9	*1/*2	Intermediate Metabolizer
CYP2D6	*1/*4	Normal Metabolizer
CYP3A4	*1/*1	Normal Metabolizer
CYP3A5	*1/*3	Intermediate Metabolizer
CYP4F2	1347G>A A/G	Heterozygous for the A allele (rs2108622)
DPYD	Activity Score: 2	Normal Metabolizer
IFNL3	rs12979860 C/T	Heterozygous for rs12979860 T allele
NAT2	c.857G>A G/A	Heterozygous for rs1799931 A allele
NAT2	c.364G>A G/G	Homozygous for rs4986996 G allele
NAT2	c.341T>C T/T	Homozygous for rs1801280 T allele
NAT2	c.191G>A G/G	Homozygous for rs1801279 G allele
NAT2	c.590G>A G/A	Heterozygous for rs1799930 A allele
NUDT15	*1/*1	Normal Metabolizer
RARG	rs2229774 C/C	Normal Function
SLC28A3	rs7853758 C/C	Normal Function
SLCO1B1	521T>C T/T	Normal Function
TPMT	*1/*1	Normal Metabolizer
UGT1A1	*1/*1	Normal Metabolizer
UGT1A6	rs17863783 G/G	Normal Metabolizer
UGT2B15	*1/*2	Intermediate Metabolizer
VKORC1	-1639G>A G/A	Intermediate Warfarin Sensitivity
VKORC1	c.3730G>A G/G	Homozygous for rs7294 C allele
WBP2NL	c.63-2604G>A A/A	Homozygous for rs5758550 A allele

FIG. 11.1.1 PGx results showed genotypes and phenotypes from a customized panel with 22 genes, 120 SNPs, and 2 copy number variations of *CYP2D6* for the patient.

guidance. For this patient, the normal dosing guidance APAP for adults is two pills every 6h (500mg per pill) (https://www.getreliefresponsibly.com/use-pain-medicine-safely/adult-acetaminophen-dosing). This patient had overdosed on APAP, which was resolved after she was treated in hospital. The PGx results will guide physicians in the future, including those for psychiatric clinical treatment.

Future testing and recommendation

It is recommended that this patient should bring her PGx report and show her doctor or pharmacist when she receives a prescription or over-the-counter medications. When the patient takes medication, always follow the physician's guidance regarding dosage instruction.

Summary of key learning points

- There are many clinical applications of pharmacogenomics.
- PGx testing can reduce drug toxicity and maximize drug efficacy.
- Individuals' genetic variation may affect drug response.
- Many technologies are used to identify genes associated with drug metabolism.
- Always follow physicians' guidance when taking medications.

References

[1] T.P. Aneesh, et al., Pharmacogenomics: the right drug to the right person, J. Clin. Med. Res. 1 (4) (2009) 191–194.

[2] F.J. McMahon, T.R. Insel, Pharmacogenomics and personalized medicine in neuropsychiatry, Neuron 74 (5) (2012) 773–776.

[3] J. Garcia-Gonzalez, et al., Pharmacogenetics of antidepressant response: a polygenic approach, Prog. Neuro-Psychopharmacol. Biol. Psychiatry 75 (2017) 128–134.

[4] A.E. Krasniak, et al., Pharmacogenomics of acetaminophen in pediatric populations: a moving target, Front. Genet. 5 (2014) 314.

[5] R.M. Weinshilboum, L. Wang, Pharmacogenomics: precision medicine and drug response, Mayo Clin. Proc. 92 (11) (2017) 1711–1722.

PART II

Hematologic malignancies

12

Chronic myeloid leukemia

Xia Li and Guang Liu
SONORA QUEST LABORATORIES, PHOENIX, AZ, UNITED STATES

Background

The discovery of the Philadelphia chromosome (Ph) as a hallmark of chronic myeloid leukemia (CML) in 1960 by Peter Nowell provided evidence for a genetic marker that links to cancer. Philadelphia chromosome is the chromosome abnormality that causes CML, acute myeloid leukemia (AML), mixed phenotype acute leukemia, and B-lymphoblastic leukemia (B-ALL). It is most commonly seen in CML [1].

CML, also known as chronic myelogenous leukemia, is a type of cancer that starts in the blood-forming cells of the bone marrow and invades the blood. About 15% of leukemias in adults are CML (https://www.cancer.org/cancer/chronic-myeloid-leukemia.html).

CML is a myeloproliferative neoplasm (MPN) characterized by granulocytes as a major clinical presentation. It arises from a hematopoietic stem cell and is characterized by a balanced translocation called t(9;22)(q34;q11.2) [2]. The derivative chromosome 22 containing a fusion of *BCR::ABL1* is called Philadelphia chromosome [3–10]. Molecular diagnostics of *BCR::ABL1* fusion are routinely done in clinical genetics laboratories. There are various assays for detecting *BCR::ABL1* fusion including karyotyping, real-time PCR, FISH, and NGS. Each of these techniques has a different analytical sensitivity. For typical *BCR::ABL1* fusion transcripts, there are major (e13a2 and e14a2, called p210) and minor (e1a2, called p190) transcripts depending on the variable breakpoints, which form the hybrid proteins. The atypical transcripts include e1a3, e13a3, e6a2, and e19a2 (p230). p210, p190, and p230 are named based on the molecular weight of the hybrid proteins. From the molecular investigation, p210 is commonly seen in CML, occasionally in AML or Philadelphia chromosome-positive acute lymphoblastic leukemia (Ph+ ALL); p190 is more common in B-cell ALL and occasionally in AML, but rarely observed in CML. p230 is generated by the fusion of the near-complete *BCR* gene with the *ABL1* gene and is considered a molecular diagnostic marker for neutrophilic-chronic myeloid leukemia (CML-N) [11,12].

The National Comprehensive Cancer Network (NCCN) guidelines for CML recommend that quantitative RT-PCR (qPCR) using an international scale (IS) for *BCR::ABL1* p210 be performed in the initial workup to establish the baseline. This applies to disease monitoring as well. FISH, cytogenetics, and NGS can also be used as alternative tests for initial diagnostic purposes. However, for monitoring response to tyrosine kinase inhibitors (TKIs) therapy, conventional cytogenetics and qPCR are the recommended assays [13–16].

Cases in Laboratory Genetics and Genomics (LGG) Practice. https://doi.org/10.1016/B978-0-323-99622-8.00020-4

In CML TKI therapeutic monitoring, *BCR::ABL1* transcript $\leq$10% by qPCR (IS) at 3- and 6-month initiation of TKI therapy indicates early molecular response (EMR), while *BCR::ABL1* transcript $\leq$0.1% corresponds to a major molecular response (MMR). In patients who experience a 10-fold increase in *BCR::ABL1* transcript levels after the initial achievement of an MMR, evaluation of *ABL1* kinase domain mutations, which results in a secondary loss of response to TKI therapy, is recommended [13,17,18].

BCR::ABL1 quantitative real-time PCR for p190 is also an RNA-based test and quantifies *BCR::ABL1* transcript p190, the minor breakpoint cluster region (e1a2) using a quantitative, reverse transcription PCR (qPCR). The pattern has been mainly associated with Philadelphia chromosome-positive acute lymphoblastic leukemia (Ph+ ALL), although it can be seen in rare cases of CML. Results are reported as a percent ratio of *BCR::ABL1* p190 to control gene *ABL1*. In minimal residual disease (MRD) assessment of Ph+ ALL during TKI therapy, an increasing level of p190 transcript may indicate a poor initial response or a secondary loss of response to TKI. Evaluation of *ABL1* kinase domain mutations in recurrent Ph+ ALL is recommended [13,19].

ABL1 kinase domain mutation analysis is a test designed for the detection of mutations in imatinib-treated CML patients with acquired resistance. The early detection of mutations should provide clinical benefits by allowing for earlier intervention. Candidates for the *ABL1* kinase domain mutation analysis include those who fail to respond to Gleevec (imatinib) and other TKI therapy or are in an accelerated phase/blast crisis. Most of the labs use RT-PCR and the Sanger sequencing method for mutation detection. The analysis includes the detection of all mutations recommended by guidelines, including common mutations such as T315I, Y253H, E255K/V, F359V/C/I, F317L/V/I/C, T315A, and V299L. Alternatively, NGS is used to detect these mutations with a higher sensitivity than those of Sanger sequencing [19].

In most clinical diagnostics laboratories, only typical transcripts are available for testing. In our lab, the NGS assay can detect 35 different transcripts, including both typical and atypical forms. The following cases demonstrate how these transcripts are detected by different assays, and what limitation each technology may have when evaluating the concordance of the test results.

Case 12.1 Chronic myeloid Leukemia (CML) with t(1;9;22;15)(p32;q34;q11.2;q25)

Clinical indication

A 66-year-old male presented to the clinic for an initial evaluation of leukocytosis, hyperuricemia, and splenomegaly. He complained of fatigue and easier bruising tendency over the past months. He also has suffered from painful "lumps" and myalgias (muscle aches and pain) throughout his body. Bone marrow aspirate smears showed a predominance of granulocytes including neutrophils, monocytes, basophils, and rare promyelocytes and blasts. These findings were suggestive of myeloproliferative neoplasm.

Test ordered

- Chromosome analysis of the bone marrow
- FISH: MPN panel including *BCR::ABL1* probes
- *BCR::ABL1* quantitative real-time PCR
- NGS Hematology Molecular Profile

Laboratory test performed

Chromosome analysis and FISH methods were described in Chapter 1.

Real-time PCR tests were performed according to the following methods. *BCR::ABL1* quantitative real-time PCR for p210 is an RNA-based test, designed to detect the *BCR::ABL1* transcript p210 associated with the major breakpoint cluster region (e2a2 and e3a2) in CML, or Ph+ ALL, using a quantitative reverse transcription PCR (qPCR). The percent ratio of *BCR::ABL1* to control gene ABL adjusted with the international scale (IS) is reported as *BCR::ABL1*/*ABL1* % (IS).

BCR::ABL1 quantitative real-time PCR for p190 is also an RNA-based test and quantifies *BCR::ABL1* transcript p190, the minor breakpoint cluster region (e1a2) using a quantitative, reverse transcription PCR (qPCR). The sensitivity of both p210 and p190 assays is a 4.5-log reduction in comparison with the universal baseline, which meets the requirement of NCCN guidelines [12,13]. The limit of detection for both assays is 0.003%.

The NGS Hematology Molecular Profile assay was performed. This assay is designed to detect mutations present in any blood or bone marrow samples. The test detects single-nucleotide variants (SNVs) and small insertions/deletions (In/Dels) as well as full gene alterations and translocations in a select group of genes. The 74 genes are selected based on the actionability of mutations identified in those genes using currently available evidence from national and international guidelines and literature. Actionability is defined as information a clinician might find useful to aid in the diagnosis, prognosis, and/or treatment strategy for a patient. The results of the test should be correlated with clinical findings. This test was designed for the detection and annotation of somatic variants and is not intended to be a germline test. For myeloid neoplasms, some germline mutations in genes such as *CEBPA*, *RUNX1*, *ETV6*, *GATA2*, and *TP53* may be detected by the assay. When the allele frequency is approximately 50%, the possibility of these being germline mutations cannot be excluded.

Test results

Chromosome analysis was performed on bone marrow. Of the 20 cells examined, all exhibited a four-way translocation involving 1p, 9q, 22q, and 15q (Fig. 12.1.1).

FISH for MPN panel was performed on interphase nuclei using probes localized to the *ABL1* (9q34.12), *BCR* (22q11), *D8Z1* (8cen), and *D20S108* (20q12), *D13S319* (13q14.3), and *LAMP1* (13q34) gene regions. Two hundred nuclei were examined for each probe. The results demonstrated *BCR::ABL1* rearrangements in 200/200 (100.0%) of the cells scored, consistent with a diagnosis of chronic myeloid leukemia with *BCR::ABL1* rearrangement (Fig. 12.1.2).

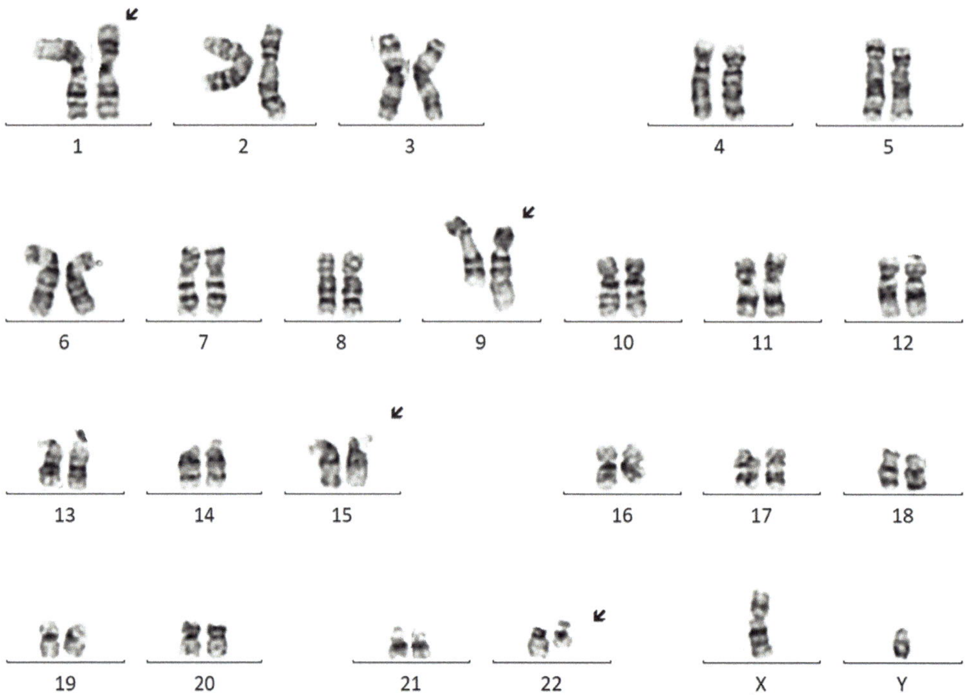

FIG. 12.1.1 The karyotype of the bone marrow showed a four-way translocation involving 1p, 9q, 22q, and 15q. ISCN: 46,XY,t(1;9;22;15)(p32;q34;q11.2;q25)[20]

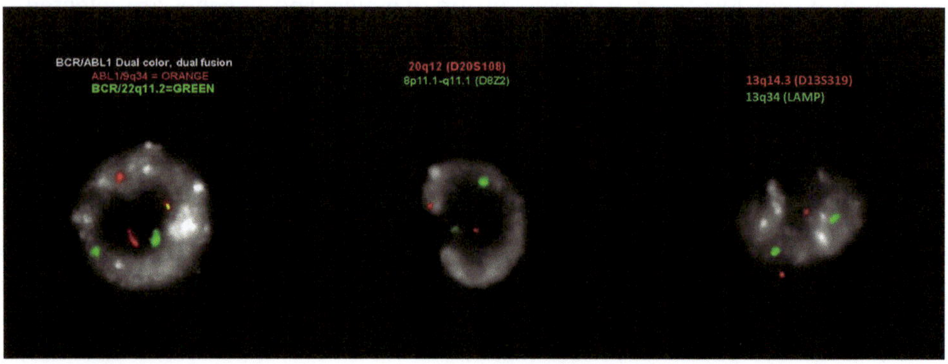

FIG. 12.1.2 FISH for the MPN panel was positive for *BCR::ABL1* rearrangement in 100% of cells. ISCN: nuc ish (D8Z2D,20S108)x2[200/200],(ABL1,BCR)x3(ABL1 con BCRx1)[184/200]/(ABL1,BCR)x3(ABL1 con BCRx2)[16/200],(D13S319,LAMP1)x2[200/200]

Then *BCR::ABL1* quantitative real-time PCR designed for detecting p210 and p190 was performed. The results were positive for the p210 transcript (Fig. 12.1.3).

Finally, the NGS Hematology profile was performed and detected p210 transcript (e14a2) (Fig. 12.1.4).

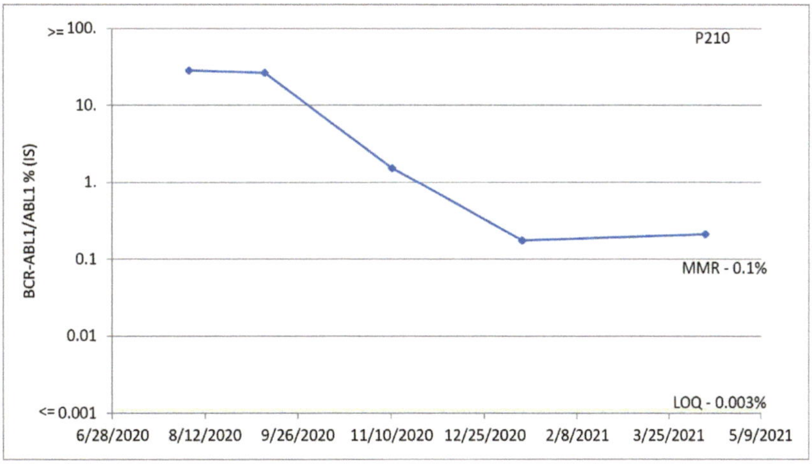

FIG. 12.1.3 Trend and chart for p210 by qPCR showed positive results for *BCR::ABL1* rearrangement.

Relevant Biomarkers

Tier	Genomic Alteration	Relevant Therapies (In this cancer type)	Clinical Trials
IA	*BCR-ABL1 fusion*	**asciminib** [1] **bosutinib** [1, 2] **dasatinib** [1, 2] **imatinib*** [1, 2] **interferon alpha-2b** [2] **nilotinib** [1, 2] ponatinib	10
	Diagnostic significance: Chronic Myeloid Leukemia		

Public data sources included in relevant therapies: FDA1, NCCN, **EMA**2, ESMO
Public data sources included in prognostic and diagnostic significance: NCCN, ESMO
Tier Reference: Li et al. *Standards and Guidelines for the Interpretation and Reporting of Sequence Variants in Cancer: A Joint Consensus Recommendation of the Association for Molecular Pathology, American Society of Clinical Oncology, and College of American Pathologists.* J Mol Diagn. 2017 Jan;19(1):4-23.
*** Includes biosimilars**

Variant Details

Gene Fusions (RNA)

Genes	Variant ID	Locus
BCR-ABL1	BCR-ABL1.B14A2.1	chr22:23632600 - chr9:133729451

FIG. 12.1.4 NGS demonstrated fusion of *BCR::ABL1* with transcript for p210 (e14a2).

Results with interpretations

Karyotype revealed a four-way translocation involving chromosomes 1p, 9q, 22q, and 15q. FISH revealed one fusion, two orange, and two green signals, which confirmed this four-way translocation by karyotype. qPCR results demonstrated that p210 was positive, also concordant with the NGS finding showing the p210 transcript (e14a2). Overall, all tests performed had concordant results, which are consistent with a diagnosis of CML. The patient was treated with dasatinib, which targets Ph+ CML.

Future testing and recommendations

BCR::ABL1 qPCR and chromosome analysis on bone marrow or leukemic blood for monitoring the minimal residual disease are recommended.

Case 12.2 Chronic myeloid leukemia (CML) with t(9;22;17)(q34;q11.2;q24)

Clinical indication

A 72-year-old male presented with a history of hypertension. He was diagnosed with high white blood cells, and was briefly placed on hydroxyurea. The patient reported feeling pain all over his body, particularly in the chest, abdomen, flank, and shoulders, which worsened with movement. He had generalized weakness and shortness of breath. Laboratory data were remarkable for markedly elevated WBC and renal insufficiency. CT chest/abdomen/pelvis showed splenomegaly.

Test ordered

- Chromosome analysis of the bone marrow
- FISH: AML and MPN panel including *BCR::ABL1* probes
- *BCR::ABL1* quantitative real-time PCR
- NGS Hematology Molecular Profile

Laboratory test performed

Chromosome analysis, FISH, qPCR, and NGS on bone marrow were described previously in this chapter, case 1.

Test results

Chromosome analysis was performed on the bone marrow. Of the 22 cells examined, all exhibited a three-way translocation involving the long arms of chromosomes 9, 22, and 17, the possible secondary translocation involving chromosomes 9 and 17, concordant with the concurrent FISH signal patterns (Fig. 12.2.1).

FISH for AML and MPN panels was performed on interphase nuclei using probes localized to the *D5S721* (5p15.2), *EGR1* (5q31), *D7Z1* (7cen), *D7S486* (7q31), *D8Z2* (8cen), *RUNX1T1* (8q22), *ABL1* (9q34.12), *KMT2A* (11q23), *D13S319* (13q14.3), *LAMP1* (13q34), PML (15q24.1), *CBFB* (16q22.1), *RARA* (17q21.1), *D20S108* (20q12), *RUNX1* (21q22.3), and *BCR* (22q11) gene regions. Two hundred nuclei were examined, and the results demonstrated an atypical *BCR::ABL1* rearrangement in 173/200 (86.5%) of the cells scored

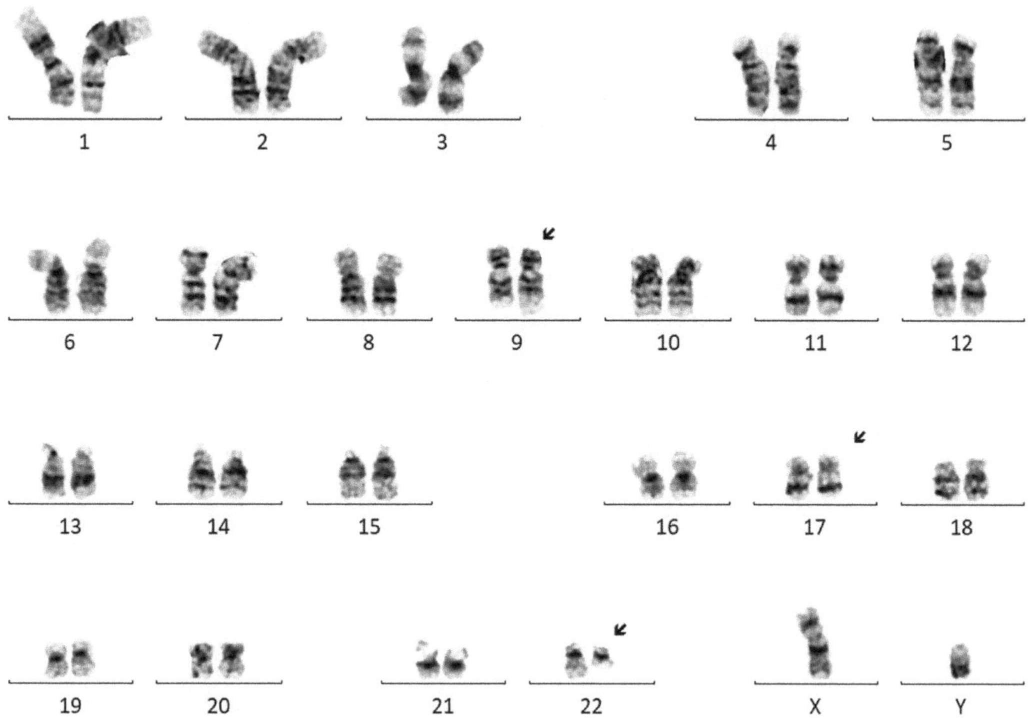

FIG. 12.2.1 The karyotype of the bone marrow showed a three-way translocation involving 9q, 22q, and 17q. ISCN: 46, XY,der(9)t(9;22;17)(q34;q11.2;q24)t(9;22)(q34;q12),der(17)t(9;22;17)(q34;q11.2;q24)?t(9;22)(q34;q12),der(22)?t(9;22)(q34;q12)[22]

(Fig. 12.2.2). Metaphase FISH using *BCR::ABL1* probes confirmed the three-way translocation (Fig. 12.2.3).

Then *BCR::ABL1* quantitative real-time PCR designed for detecting p210 and p190 was performed. The results were positive for p210 transcript (Fig. 12.2.4).

Finally, the NGS Hematology profile was performed and detected p210 transcript (e13a2) (Fig. 12.2.5).

Results with interpretations

Chromosome analysis identified a three-way translocation involving 9q, 22q, and 17q.

FISH analysis of the interphase cells identified an atypical *BCR::ABL1* rearrangement. FISH analysis on metaphase cells confirmed the three-way translocation. These results are concordant with the cytogenetics finding and are consistent with a diagnosis of CML.

BCR::ABL1 q-PCR testing revealed a fusion product for *BCR::ABL1* p210, and so were the results from NGS testing. The patient was treated with dasatinib, which targets Ph + CML.

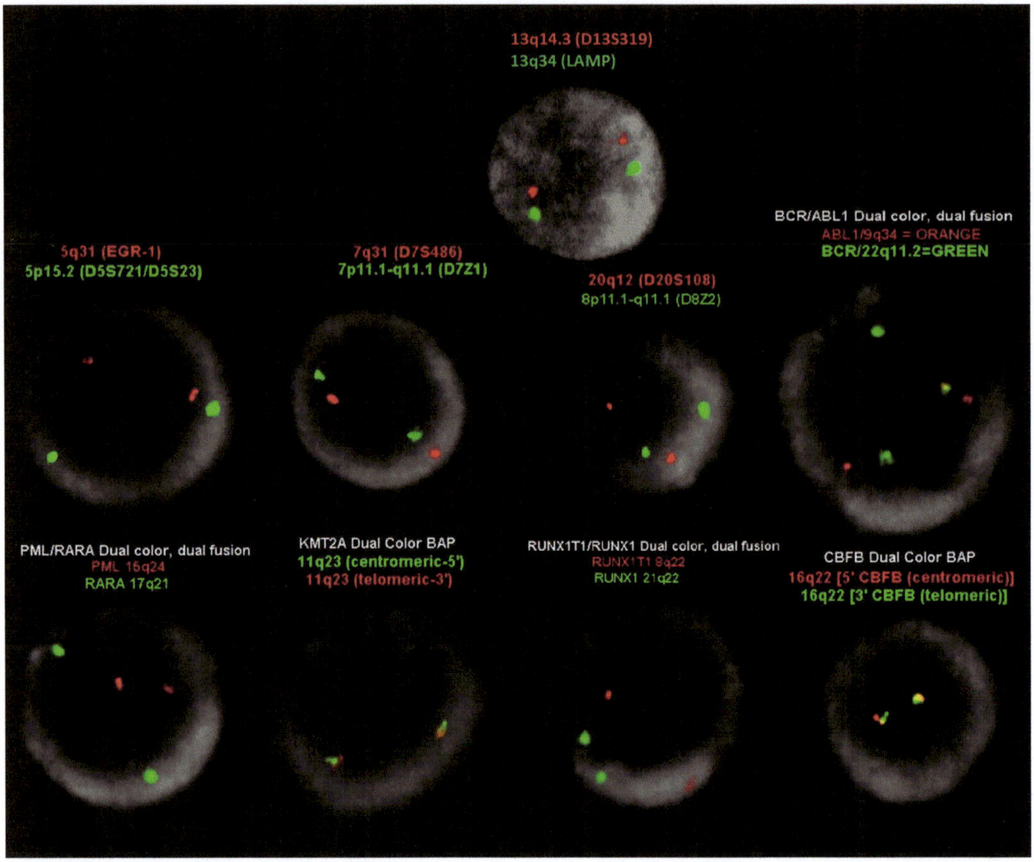

FIG. 12.2.2 FISH for the AML panel and probes for 13q showed positive results for *BCR::ABL1* rearrangement in 86.5% of the cells examined. ISCN: nuc ish(D5S721,EGR1)x2[198/200],(CEP7,D7S486)x2[197/200],(D8Z2,D20S108)x2[197/200],(RUNX1T1,RUNX1)x2[200/200],(ABL1,BCR)x3(ABL1 con BCRx1)[173/200],(KMT2Ax2)[200/200], (D13S319,LAMP1)x2 [198/200],(PML,RARA)x2[200/200],(CBFBx2)[197/200]

Future testing and recommendations

BCR::ABL1 qPCR and chromosome analysis on the bone marrow or leukemic blood for monitoring the minimal residual disease are recommended.

Case 12.3 Chronic myeloid leukemia (CML) with t(9;22)(q34;q11.2)inv(22)

Clinical indication

A 33-year-old male presented with leukocytosis with neutrophilia, myelocytes, and metamyelocytes. However, he did not show peripheral blood circulating blasts by flow, which was suggestive of possibly CML, chronic phase, although other subtypes cannot be excluded.

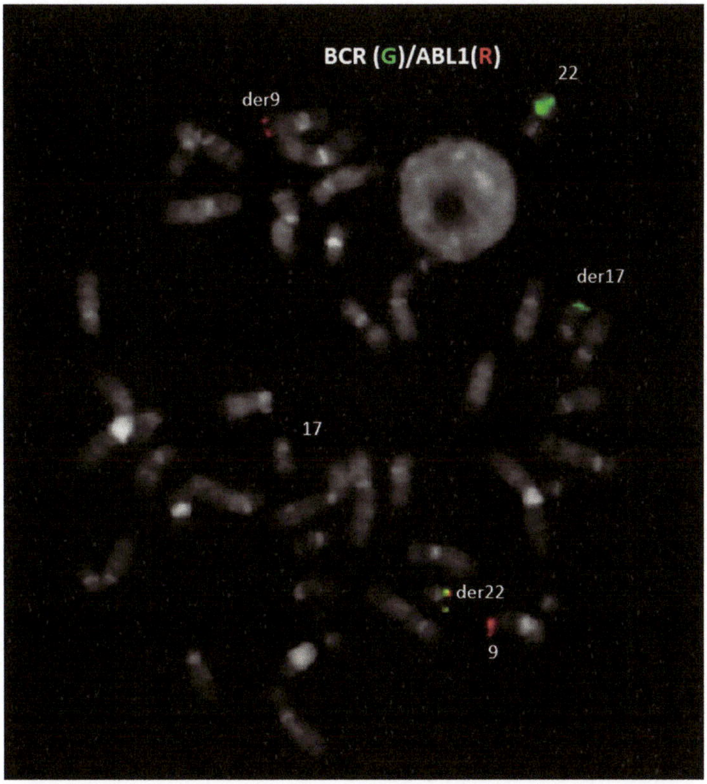

FIG. 12.2.3 FISH for *BCR::ABL1* on metaphase cells showed a three-way translocation involving 9q, 22q, and 17q.

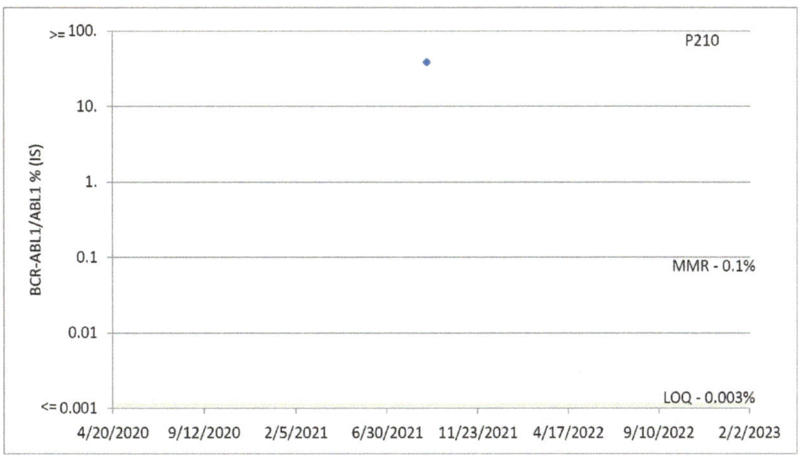

FIG. 12.2.4 Trend and chart for p210 by qPCR showed positive results for *BCR::ABL1* rearrangement.

Relevant Biomarkers

Tier	Genomic Alteration	Relevant Therapies (In this cancer type)	Clinical Trials
IA	*BCR-ABL1 fusion*	**bosutinib** [1,2] **dasatinib** [1,2] **imatinib** [1,2] **interferon alpha-2b** [2] **nilotinib** [1,2]	33

Prognostic significance: None
Diagnostic significance: Chronic Myeloid Leukemia

Public data sources included in relevant therapies: FDA1, NCCN, **EMA**2, ESMO
Public data sources included in prognostic and diagnostic significance: NCCN, ESMO
Tier Reference: Li et al. *Standards and Guidelines for the Interpretation and Reporting of Sequence Variants in Cancer: A Joint Consensus Recommendation of the Association for Molecular Pathology, American Society of Clinical Oncology, and College of American Pathologists.* J Mol Diagn. 2017 Jan;19(1):4-23.

Variant Details

Gene Fusions (RNA)

Genes	Variant ID	Locus
BCR-ABL1	BCR-ABL1.B13A2	chr22:23631808 - chr9:133729451

FIG. 12.2.5 NGS demonstrated the fusion of *BCR::ABL1* with a transcript for p210 (e13a2).

Test ordered

- Chromosome analysis of the bone marrow
- FISH: AML and MPN panels including *BCR::ABL1* probes
- *BCR::ABL1* quantitative real-time PCR
- NGS Hematology Molecular Profile

Laboratory test performed

Chromosome analysis, FISH, qPCR, and NGS on the bone marrow were described previously in this chapter, case 1.

Test results

For chromosome analysis, of the 21 cells examined, all exhibited a balanced translocation involving the long arms of chromosomes 9 and 22 with a paracentric inversion of chromosome 22 (Fig. 12.3.1).

FISH for AML and MPN panels was performed on interphase nuclei using probes localized to the *D5S721* (5p15.2), *EGR1* (5q31), *D7Z1* (7cen), *D7S486* (7q31), *D8Z2* (8cen), *RUNX1T1* (8q22), *ABL1* (9q34.12), *KMT2A* (11q23), *D13S319* (13q14), *LAMP1* (13q34), *PML* (15q24.1), *CBFB* (16q22.1), *RARA* (17q21.1), *D20S108* (20q12), *RUNX1* (21q22.3), and *BCR* (22q11) gene regions. Two hundred nuclei were examined, and the results demonstrated *BCR::ABL1* rearrangements in 177/200 (88.5%) of the cells scored (Fig. 12.3.2). Metaphase FISH confirmed the translocation involving *BCR::ABL1* fusion (the

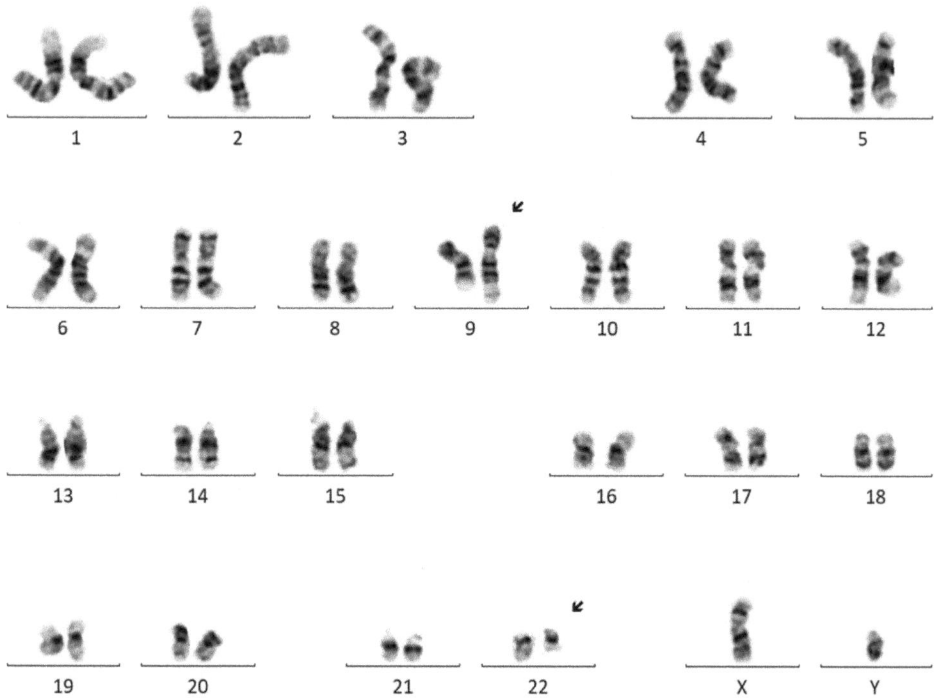

FIG. 12.3.1 The karyotype of the bone marrow showed a *BCR::ABL1* translocation with a paracentric inversion of 22q. ISCN: 46,XY,der(9)t(9;22)(q34;q11.2)inv(22)(q11.2q13),der(22)t(9;22)[21]

Philadelphia chromosome) on derivative chromosome 22, and the second fusion on derivative 9 with inv(22) (Fig. 12.3.3).

Then *BCR::ABL1* quantitative real-time PCR designed for detecting p210 and p190 was performed. The results were positive for p210 transcript (Fig. 12.3.4).

Finally, the NGS Hematology profile was performed and detected p210 transcripts (e14a2) (Fig. 12.3.5).

Results with interpretations

Chromosome analysis identified a *BCR::ABL1* translocation with a paracentric inversion of chromosome 22q. FISH analysis on interphase cells identified an atypical *BCR::ABL1* rearrangement. Metaphase FISH showed a signal pattern, consistent with those of karyotype and interphase FISH signal patterns. *BCR::ABL1* q-PCR testing revealed a fusion product for *BCR::ABL1* p210, and so were the results from NGS testing. These results are concordant with the cytogenetics finding and are consistent with a diagnosis of CML. The patient was treated with dasatinib, which targets Ph+ CML.

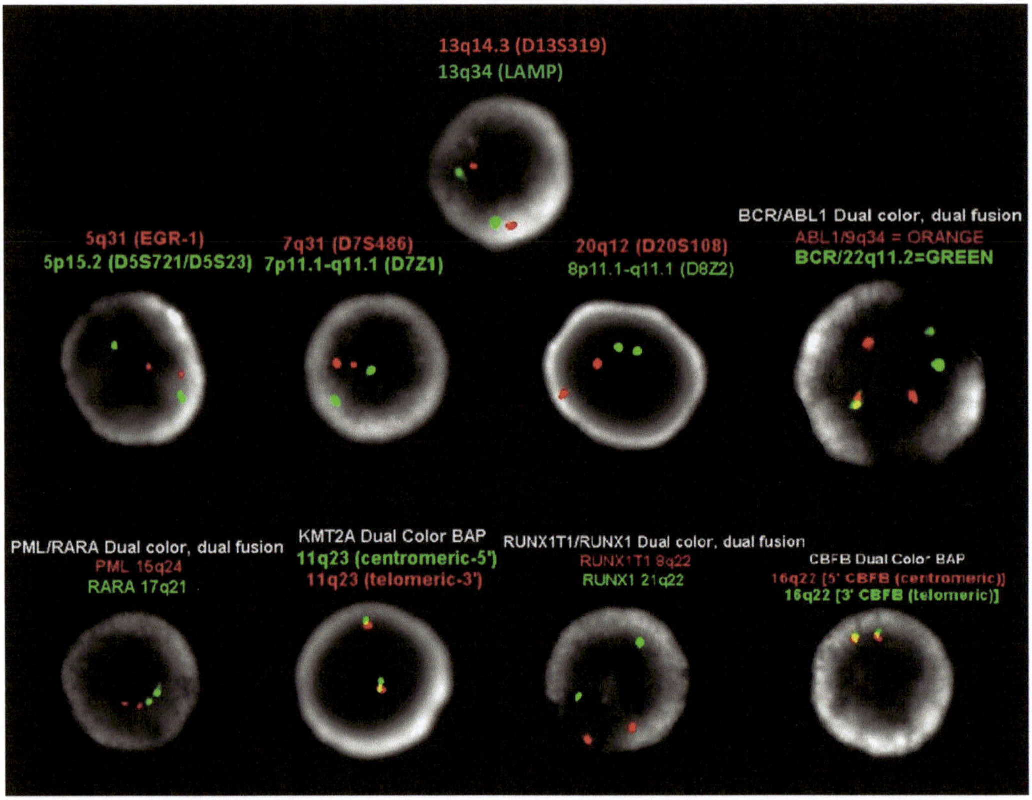

FIG. 12.3.2 FISH for the AML panel and del(13) probes was positive for *BCR::ABL1* rearrangement in 88.5% of the cells examined. ISCN: nuc ish(D5S721,EGR1)x2[198/200],(CEP7,D7S486)x2[197/200],(D8Z2,D20S108)x2[199/200],(RUNX1T1,RUNX1)x2[200/200],(ABL1,BCR)x3(ABL1 con BCRx2)[78/200]/(ABL1,BCR)x3(ABL1 con BCRx1)[99/200],(KMT2Ax2)[198/200], (D13S319,LAMP1)x2[198/200], (PML,RARA)x2[200/200],(CBFBx2)[200/200]

Future testing and recommendations

BCR::ABL1 qPCR and chromosome analysis on the bone marrow or leukemic blood for monitoring the minimal residual disease are recommended.

Case 12.4 Chronic myeloid leukemia and acute lymphoblastic leukemia with t(9;22)(q34;q11.2)

Clinical indication

A 42-year-old male presented with a history of CML for nearly 10 years. He did not respond well to the initial treatment. After multiple relapses, he had the first and then the second allogeneic stem cell transplant. He was diagnosed with recurrent Philadelphia chromosome-positive acute lymphoblastic leukemia, status posts second allogeneic stem cell transplant. Since that time, the patient has failed with the treatment of ponatinib, gemcitabine, venetoclax, and more recently single agent asciminib.

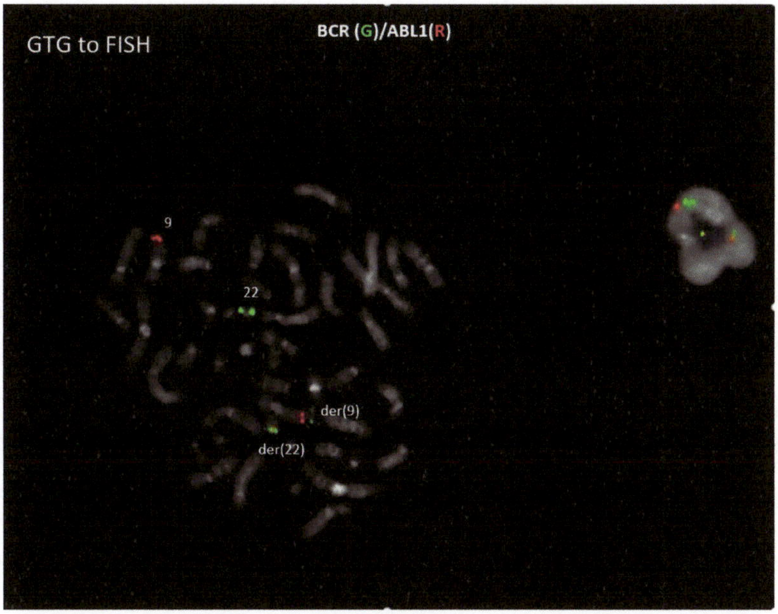

FIG. 12.3.3 Metaphase FISH confirmed the translocation involving *BCR::ABL1* fusion (the Philadelphia chromosome) on derivative chromosome 22, and the second fusion on derivative 9 with inv(22).

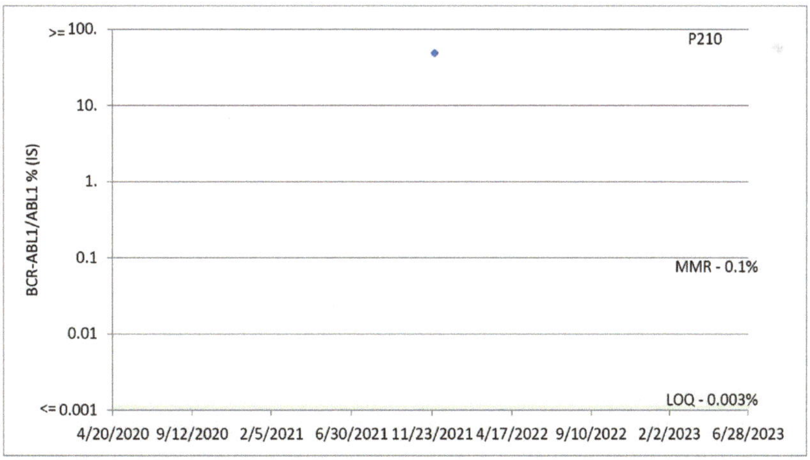

FIG. 12.3.4 RT-PCR demonstrated a fusion of *BCR::ABL1* with a transcript of p210 (e14a2).

Test ordered

- Chromosome analysis of the bone marrow
- FISH: AML and MPN panels including *BCR::ABL1* probes, plus *FGFR1, PDGFRA*, and *PDGFRB* probes
- *BCR::ABL1* quantitative real-time PCR
- NGS Hematology Molecular Profile

Relevant Biomarkers

Tier	Genomic Alteration	Relevant Therapies (In this cancer type)	Clinical Trials
IA	*BCR-ABL1 fusion*	**bosutinib** [1,2] **dasatinib** [1,2] **imatinib*** [1,2] **interferon alpha-2b** [2] **nilotinib** [1,2]	29

Prognostic significance: None
Diagnostic significance: Chronic Myeloid Leukemia

Public data sources included in relevant therapies: **FDA**1, NCCN, **EMA**2, ESMO
Public data sources included in prognostic and diagnostic significance: NCCN, ESMO
Tier Reference: Li et al. *Standards and Guidelines for the Interpretation and Reporting of Sequence Variants in Cancer: A Joint Consensus Recommendation of the Association for Molecular Pathology, American Society of Clinical Oncology, and College of American Pathologists.* J Mol Diagn. 2017 Jan;19(1):4-23.
*** Includes biosimilars**

Variant Details

DNA Sequence Variants

Gene	Amino Acid Change	Coding	Locus	Allele Frequency	Variant Effect
BCOR	p.(R1661*)	c.4981C>T	chrX:39911649	47.80%	nonsense

Gene Fusions (RNA)

Genes	Variant ID	Locus
BCR-ABL1	BCR-ABL1.B14A2.1	chr22:23632600 - chr9:133729451

FIG. 12.3.5 NGS detected a fusion of *BCR::ABL1* with a transcript of p210 (e14a2) and a *BCOR* c.4981C>T, p.(R1661*) mutation.

Laboratory test performed

Chromosome analysis, FISH, qPCR, and NGS on bone marrow were described previously in this chapter, case 1.

Test results

The initial chromosome analysis identified the Philadelphia chromosome in all 20 cells examined (Fig. 12.4.1). After relapse, the karyotype showed trisomy 8 in addition to the Philadelphia chromosome (Fig. 12.4.2).

FISH for AML and MPN panels as well as *FGFR1, PDGFRA,* and *PDGFRB* probes was performed on interphase nuclei using probes localized to the *SCFD2, FIP1L1, LNX1, PDGFRA* (4q12), *D5S721* (5p15.2), *EGR1* (5q31), probes localized to the 5' and 3' ends of the *PDGFRB* (5q33.1), *D7Z1* (7cen), *D7S486* (7q31), probes localized to the 5' and 3' ends of the *FGFR1* (8p12.23-p11.22), *D8Z2* (8cen), *RUNX1T1* (8q22), *ABL1* (9q34.12), *KMT2A* (11q23), *D13S319* (13q14.3), and *LAMP1* (13q34), *PML* (15q24.1), *CBFB* (16q22.1), *RARA* (17q21.1), *D20S108* (20q12), *RUNX1* (21q22.3), and *BCR* (22q11) gene regions. Two hundred nuclei were examined, and the results demonstrated a *BCR::ABL1* rearrangement in 197/200 (98.5%), trisomy 8 in 180/200 (90.0%), and gain of *FGFR1* in 183/200 (91.5%) of the cells scored (Fig. 12.4.3).

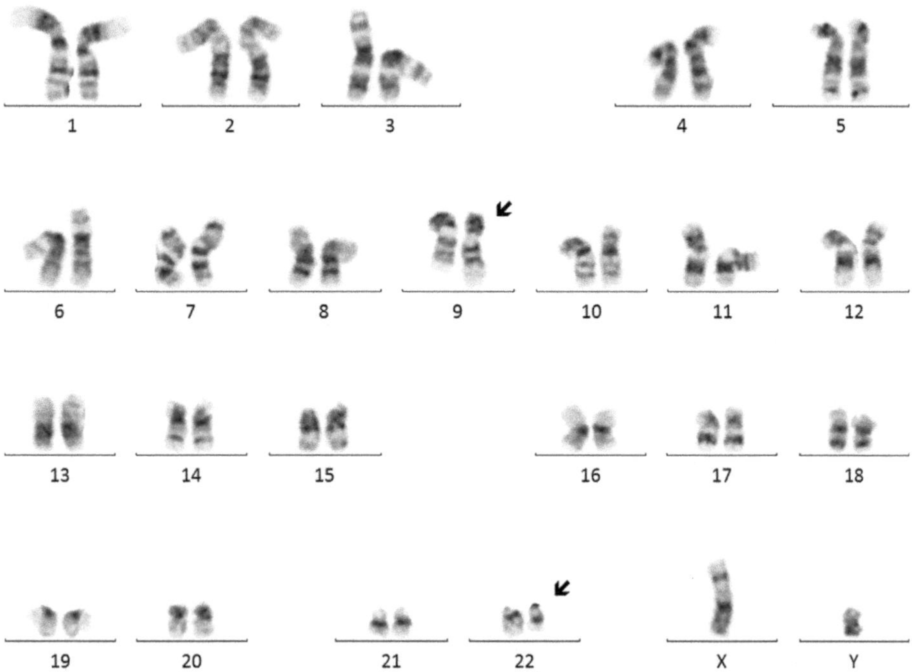

FIG. 12.4.1 The karyotype of the bone marrow showed a balanced translocation involving 9q and 22q, resulting in a Philadelphia chromosome. ISCN: 46,XY,t(9;22)(q34;q11.2)[4]/47,idem,+8[16]

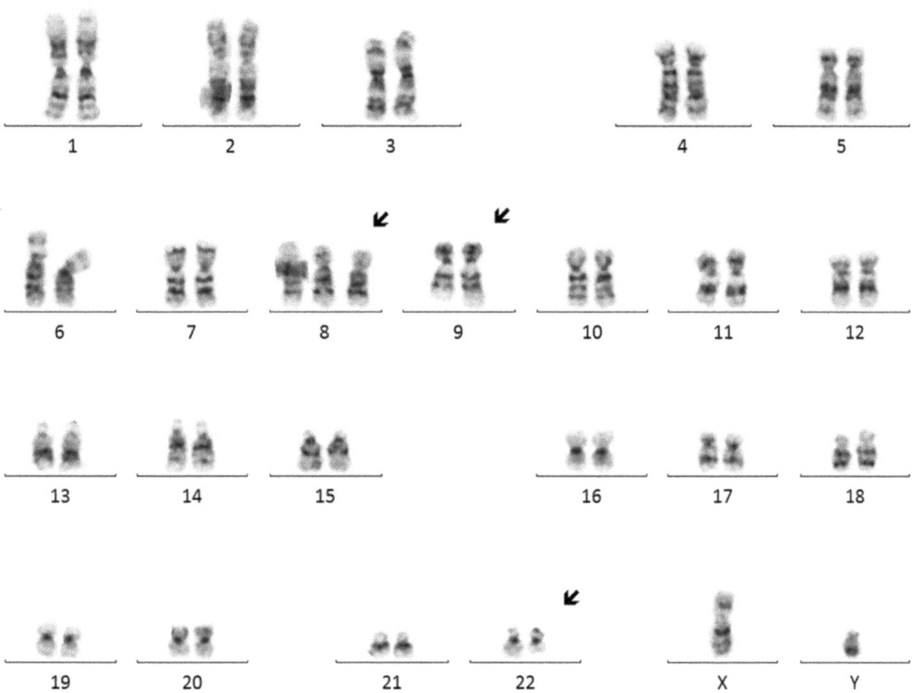

FIG. 12.4.2 The karyotype of the bone marrow showed a balanced translocation involving 9q and 22q, resulting in a Philadelphia chromosome. In addition, there was a trisomy 8, indicating an accelerated phase of the disease. ISCN: 46, XY,t(9;22)(q34;q11.2)[4]/47,idem,+8[16]

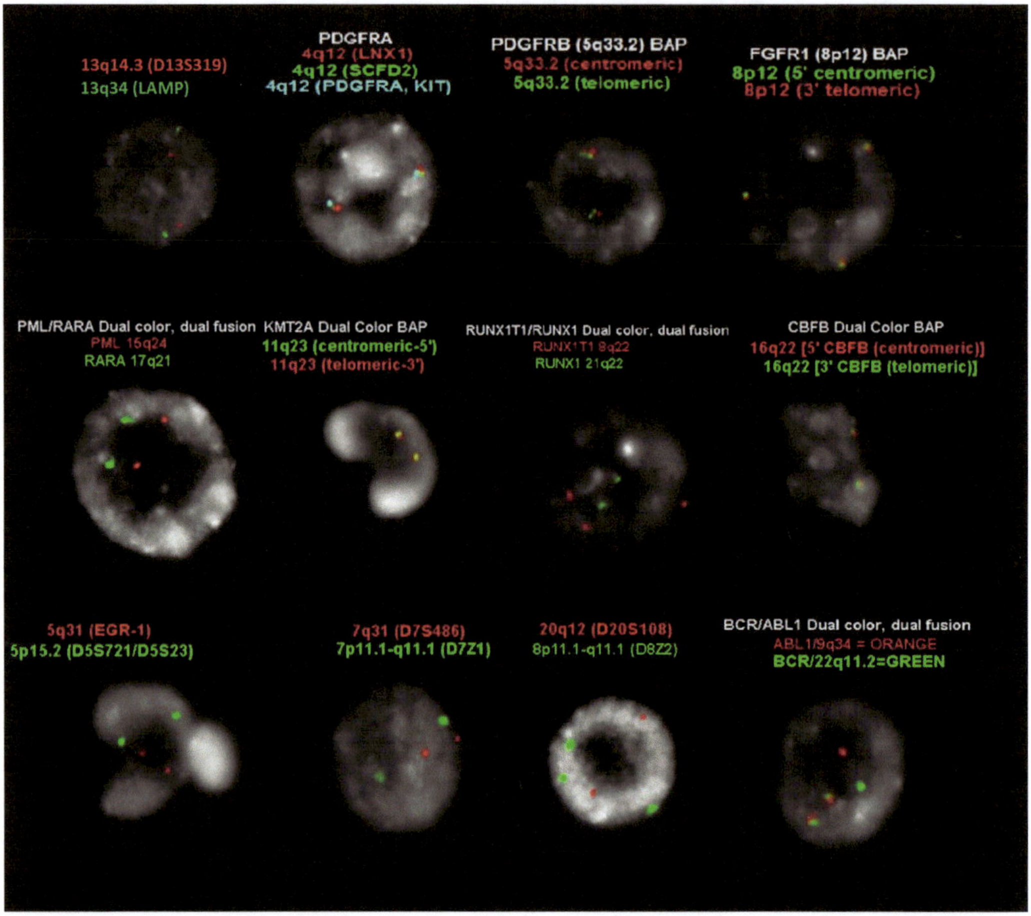

FIG. 12.4.3 FISH for the AML and MPN panels, *FGFR1*, *PDGFRA*, and *PDGFRB* probes showed positive results for *BCR::ABL1* rearrangement in 98.5% of cells, trisomy 8 in 90% of the cells, and gain of *FGFR1* in 91.5% of the cells examined. ISCN: nuc ish(SCFD2,FIP1L1,LNX1,PDGFRAx2)[199/200],(D5S721,EGR1)x2[200/200], (PDGFRBx2)[200/200],(CEP7, D7S486)x2[199/200],(FGFR1x3)[183/200],(D8Z2x3,D20S108x2)[180/200],(RUNX1T1x3,RUNX1x2)[173/200],(ABL1, BCR)x3(ABL1 con BCRx2)[197/200],(KMT2Ax2)[199/200],(D13S319,LAMP1)x2[200/200],(PML,RARA)x2[200/200], (CBFBx2)[198/200]

Then *BCR::ABL1* quantitative real-time PCR designed for detecting p210 and p190 was performed. The results were positive for p210 transcript (Fig. 12.4.4).

Finally, the NGS Hematology profile was performed and detected p210 transcripts (e14a2) and *ABL1* p.(T315I) resistance mutation (Fig. 12.4.5).

Results with interpretations

Chromosome analysis identified a *BCR::ABL1* translocation and trisomy 8 in CML accelerated phase. FISH analysis on interphase cells identified a typical *BCR::ABL1* rearrangement and trisomy 8, concordant with cytogenetic findings. *BCR::ABL1* q-PCR testing

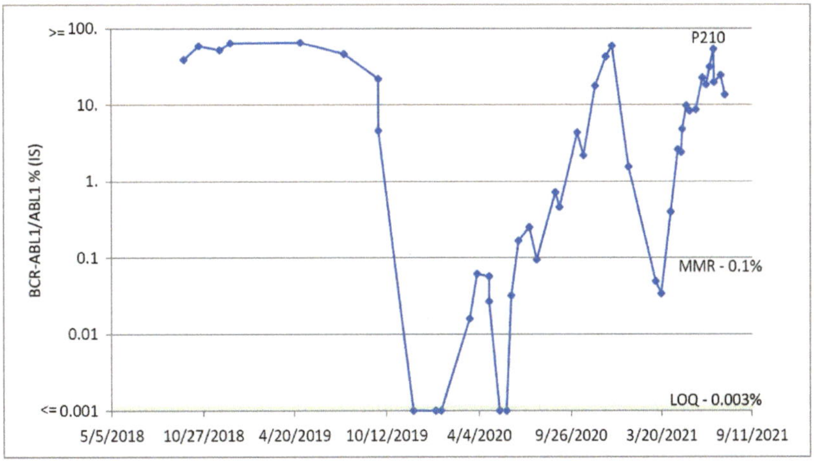

FIG. 12.4.4 Trend and chart for p210 by qPCR showed positive results for *BCR::ABL1* rearrangement.

Relevant Biomarkers

Tier	Genomic Alteration	Relevant Therapies (In this cancer type)	Clinical Trials
IA	*BCR-ABL1 fusion*	**asciminib** **imatinib** **interferon alpha-2b** ponatinib	6
	Diagnostic significance: Chronic Myeloid Leukemia		
IA	*ABL1 p.(T315I) c.944C>T*	**asciminib** **ponatinib** allogeneic stem cells omacetaxine	4
	Prognostic significance: NCCN: Poor		

Alerts informed by public data sources: Contraindicated, Resistance

ABL1 p.(T315I) c.944C>T	Contraindicated: bosutinib, dasatinib, nilotinib Resistance: bosutinib, dasatinib, imatinib, nilotinib

Variant Details

DNA Sequence Variants

Gene	Amino Acid Change	Coding	Locus	Allele Frequency	Variant Effect
ABL1	p.(T315I)	c.944C>T	chr9:133748283	40.46%	missense
ASXL1	p.(E635Rfs*15)	c.1900_1922delAGAGAGGC GGCCACCACTGCCAT	chr20:31022402	15.58%	frameshift Deletion
RUNX1	p.(A409Rfs*188)	c.1225_1232delGCCTCGGC	chr21:36164642	40.33%	frameshift Deletion

Gene Fusions (RNA)

Genes	Variant ID	Locus
BCR-ABL1	BCR-ABL1.B14A2.1	chr22:23632600 - chr9:133729451

FIG. 12.4.5 NGS demonstrated the fusion of *BCR::ABL1* with a transcript for p210 (e14a2) and *ABL1* kinase domain mutation (p.T315I).

revealed a fusion product for *BCR::ABL1* p210, and so did the results from NGS testing. NGS also detected *ABL1* T315I, which is a resistance mutation to the first generation TKIs. These results are consistent with a diagnosis of CML with no remission. According to the pathology finding, the results also confirmed that the patient was transformed to B-ALL after the second allogeneic transplant.

Future testing and recommendations

BCR::ABL1 qPCR and chromosome analysis on the bone marrow or leukemic blood for monitoring the minimal residual disease are recommended.

Summary of key learning points

- The Philadelphia chromosome is the hallmark of CML, and it can be involved in complex chromosome translocations.
- Quantitative real-time PCR (q-PCR) is the gold standard for testing of *BCR::ABL1* fusion recommended by NCCN.
- The importance of the q-PCR sensitivity in determining the minimal residual disease (MRD) level is described.
- For detection of different transcripts, NGS can identify p210, p190, p230, and many other transcripts while q-PCR in most labs only detects p210 and p190.
- *ABL1* kinase domain mutations are associated with resistance to imatinib treatment. For patients who do not respond to the first line of TKI therapy, this test can provide patients with an alternative drug.

References

[1] G.A. Koretzky, The legacy of the Philadelphia chromosome, J. Clin. Invest. 117 (8) (2007) 2030–2032.

[2] E. Sabattini, et al., WHO classification of tumours of haematopoietic and lymphoid tissues in 2008: an overview, Pathologica 102 (3) (2010) 83–87.

[3] P.C. Nowell, D.A. Hungerford, Chromosome studies in human leukemia. II. Chronic granulocytic leukemia, J. Natl. Cancer Inst. 27 (1961) 1013–1035.

[4] P.C. Nowell, D.A. Hungerford, Chromosome studies in human leukemia. IV. Myeloproliferative syndrome and other atypical myeloid disorders, J. Natl. Cancer Inst. 29 (1962) 911–931.

[5] P.C. Nowell, Discovery of the Philadelphia chromosome: a personal perspective, J. Clin. Invest. 117 (8) (2007) 2033–2035.

[6] D.A. Hungerford, P.C. Nowell, Chromosome studies in human leukemia. III. Acute granulocytic leukemia, J. Natl. Cancer Inst. 29 (1962) 545–565.

[7] K.H. Rothfels, L. Siminovitch, An air-drying technique for flattening chromosomes in mammalian oells grown in vitro, Stain. Technol. 33 (2) (1958) 73–77.

[8] I.M. Tough, et al., Cytogenetic studies in chronic myeloid leukaemia and acute leukaemia associated with monogolism, Lancet 1 (7174) (1961) 411–417.

[9] I.M. Tough, et al., Cytogenetic studies on bone-marrow in chronic myeloid leukaemia, Lancet 1 (7286) (1963) 844–846.

[10] E.A. Bruford, et al., HUGO Gene Nomenclature Committee (HGNC) recommendations for the designation of gene fusions, Leukemia 35 (11) (2021) 3040–3043.

[11] C.L. Sawyers, Chronic myeloid leukemia, N. Engl. J. Med. 340 (17) (1999) 1330–1340.

[12] Z.J. Kang, et al., The Philadelphia chromosome in leukemogenesis, Chin. J. Cancer 35 (2016) 48.

[13] M.W. Deininger, et al., Chronic myeloid leukemia, version 2.2021, NCCN clinical practice guidelines in oncology, J. Natl. Compr. Cancer Netw. 18 (10) (2020) 1385–1415.

[14] N.P. Shah, Ponatinib: targeting the T315I mutation in chronic myelogenous leukemia, Clin. Adv. Hematol. Oncol. 9 (12) (2011) 925–926.

[15] T. Hughes, et al., Monitoring CML patients responding to treatment with tyrosine kinase inhibitors: review and recommendations for harmonizing current methodology for detecting BCR-ABL transcripts and kinase domain mutations and for expressing results, Blood 108 (1) (2006) 28–37.

[16] J.D. Khoury, et al., The 5th edition of the World Health Organization classification of haematolymphoid tumours: myeloid and histiocytic/dendritic neoplasms, Leukemia 36 (7) (2022) 1703–1719.

[17] D. Jones, et al., Kinase domain point mutations in Philadelphia chromosome-positive acute lymphoblastic leukemia emerge after therapy with BCR-ABL kinase inhibitors, Cancer 113 (5) (2008) 985–994.

[18] C.L. Sawyers, B. Druker, Tyrosine kinase inhibitors in chronic myeloid leukemia, Cancer J. Sci. Am. 5 (2) (1999) 63–69.

[19] S. Soverini, et al., Next-generation sequencing for BCR-ABL1 kinase domain mutation testing in patients with chronic myeloid leukemia: a position paper, J. Hematol. Oncol. 12 (1) (2019) 131.

13

Myeloid/lymphoid neoplasms with eosinophilia and tyrosine kinase gene fusions

Dongbin Xu[a], Guang Liu[b], Xia Li[b], and Hanyin Cheng[a]
[a]HEMATOLOGICS, INC., SEATTLE, WA, UNITED STATES [b]SONORA QUEST LABORATORIES, PHOENIX, AZ, UNITED STATES

Background

Myeloid/lymphoid neoplasms with eosinophilia and tyrosine kinase gene fusions (MLN-TK) are a family of myeloid or lymphoid neoplasms resulting from tyrosine kinase gene rearrangement [1]. These neoplasms are commonly associated with eosinophilia and sensitive to tyrosine kinase inhibitors. They can present as myeloproliferative neoplasms (MPN), myelodysplastic/myeloproliferative neoplasms (MDS/MPN), acute myeloid leukemia (AML), mixed-phenotype acute leukemia (MPAL), B- or T-acute lymphoblastic leukemia/lymphoma (ALL). The defining genetic lesions include rearrangements involving tyrosine kinase genes *PDGFRA*, *PDGFRB*, *FGFR1*, *JAK2*, *FLT3*, and *ETV6*::*ABL1* fusion as well as some less frequently observed tyrosine kinase fusions, such as *ETV6*::*FGFR2*, *ETV6*::*LYN*, *ETV6*::*NTRK3*, *RANBP2*::*ALK*, *BCR*::*RET*, and *FGFR1OP*::*RET* [1].

Myeloid/lymphoid neoplasms with *PDGFRA* rearrangement commonly result from an interstitial microdeletion in 4q12 that leads to *FIP1L1*::*PDGFRA* gene fusion. Such deletions are typically a little larger than 800 Kb and therefore are cytogenetically cryptic. In rare cases, the *PDGFRA* rearrangement may present as a fusion with other partner genes such as *ETV6*, *BCR*, *CDK5RAP2*, *KIF5B*, *STRN*, and *TNKS2* [2–7]. Eosinophilia is a common feature in most, but not all cases, with many presenting hypereosinophilia (absolute eosinophil count $\geq 1.5 \times 10^9$/L). Elevated serum tryptase ($\geq$12 ng/mL) can be a strong indicator for *FIP1L1*::*PDGFRA* fusion [8]. Patients often have increased mast cells presenting as loose clusters in the bone marrow [9,10]. Blasts level in bone marrow or peripheral blood can be applied to distinguish between chronic phase (<20% blasts) and blast phase ($\geq$20% blasts) diseases. The kinase inhibitor, imatinib, is recommended in first-line therapy of this disease either in the chronic or blast phase due to its remarkable efficacy [3,11]. These neoplasms have a strong male predominance and mostly occur in patients within the age range of 25–55 [1].

Myeloid/lymphoid neoplasms with *PDGFRB* rearrangement can present with fusions involving more than 30 partner genes, with *ETV6*::*PDGFRB* as the most frequently

Cases in Laboratory Genetics and Genomics (LGG) Practice. https://doi.org/10.1016/B978-0-323-99622-8.00004-6

reported gene fusion that results from a translocation between chromosomes 5 and 12, t(5;12)(q32;p13.2). Besides prominent eosinophilia, other features include neutrophilia, monocytosis, mastocytosis, and myelofibrosis. These cases may resemble chronic myelomonocytic leukemia (CMML) with eosinophilia and can be treated effectively by tyrosine kinase inhibitor therapy using imatinib [12–14].

Please note that *ETV6::PDGFRB* and other *PDGFRB* rearrangements involving *EBF1, SSBP2, TNIP1, ZEB2,* and *ATF7IP* have been reported in *BCR::ABL1*-like B-lymphoblastic leukemia (Ph-like B-ALL) [15–17]. However, these Ph-like B-ALL cases generally do not present eosinophilia, monocytosis, or myelofibrosis [18].

Myeloid/lymphoid neoplasms with *FGFR1* rearrangement may manifest as chronic myeloid neoplasms or a blast crisis of B-cell, T-cell, myeloid or mixed-phenotype neoplasms [1]. At least 14 partner genes have been reported to form fusion with *FGFR1*. The most common *FGFR1* rearrangement is *ZMYM2::FGFR1* gene fusion resulting from chromosome translocation t(8;13)(p11.2;q12) that is often seen in T-cell lymphoblastic leukemia/lymphoma (T-ALL). *CNTRL::FGFR1* resulting from t(8;9)(p11.2;q33) and *FGFR1OP::FGFR1* from t(6;8)(q27;p11.2) have been reported in chronic myelomonocytic leukemia (CMML). *BCR::FGFR1* resulting from t(8;22)(p11.2;q11.2) was seen in MPN with basophilia, resembling chronic myeloid leukemia (CML) [19–26].

Myeloid/lymphoid neoplasms with *JAK2* rearrangement may present as fusions with different partner genes, among which *PCM1::JAK2* is the most common fusion resulting from chromosome translocation between chromosomes 8 and 9, t(8;9)(p22;p24) [12]. These cases showed apparent male predominance [12,27]. *PCM1::JAK2* cases may present unique "typical triad" bone marrow features including hypercellularity with eosinophilia, large aggregates of immature erythroid precursors, and myelofibrosis [18]. Please note that *JAK2* rearrangements can be associated with de novo B-ALL. In contrast, those B-ALL cases generally lack myeloid/myeloproliferative features and eosinophilia [17,27].

Myeloid/lymphoid neoplasms with *FLT3* rearrangement are rare and can manifest as myeloid sarcoma with MPN features in the bone marrow or T-ALL with associated eosinophilia. The most common fusion in this category is *ETV6::FLT3* resulting from chromosome translocation t(12;13)(p13;q12). Other fusion partner genes include *BCR, ZMYM2, TRIP11, SPTBN, GOLGB1, CCDC88C, ZBTB44,* and *MYO18A* [28,29]. Please note that myeloid neoplasms with the acquisition of *FLT3* fusion secondary to other primary genetic defining factors are excluded from this category [30].

Myeloid/lymphoid neoplasm with *ETV6::ABL1* fusion is rare and may present hypercellular bone marrow with increased myeloid to erythroid ratio and prominent eosinophilia with or without basophilia as seen in CML [1]. Due to the orientation of *ETV6* (12p13) and *ABL1* (9q34) in the original chromosomes, the *ETV6::ABL1* fusion cannot be formed by a balanced translocation. Translocation with inversion or insertion is required to form the fusion [31]. Please note that *ETV6::ABL1* fusion has been reported in B-ALL and other neoplasms [32]. It is necessary to separate myeloid/lymphoid neoplasm with *ETV6::ABL1* fusion from B-ALL or other neoplasms.

In addition to the aforementioned major categories, the last category from this family is myeloid/lymphoid neoplasms with other tyrosine kinase fusion genes, which currently include *ETV6::FGFR2*, *ETV6::LYN*, *ETV6::NTRK3*, *RANBP2::ALK*, *BCR::RET*, and *FGFR1OP::RET* [1,33–37].

Case 13.1 Myeloid/lymphoid neoplasm (MLN) with *FGFR1* rearrangement

Clinical indication

A 55-year-old male presented with enlarged axillary lymph nodes, night sweats, weight loss, and early satiety for about 2 months. Flow cytometry from the left axillary lymph node biopsy showed an abnormal T-cell population. The cells expressed cytoplasmic CD3, CD2, CD4, CD5, partial TdT, and partial CD56. The cells were negative for surface CD3, CD8, CD10, and CD34. A partial loss of CD7 and a subpopulation co-expressing CD4/CD8 were observed. The flow data were suggestive of an abnormal T-cell population with features consistent with T-acute lymphoblastic leukemia/lymphoma. Flow cytometry from bone marrow showed an aberrant, immature T-cell population comprising 5% of total events.

Test ordered

- Chromosome analysis of the bone marrow
- FISH: ALL panel, MPN panel plus *PDGFRA*, *PDGFRB*, and *FGFR1* probes
- NGS Hematology Molecular Profile

Laboratory test performed

Chromosome analysis, FISH, and NGS methods were described in Chapter 12.

Test results

Firstly, the cytogenetic analysis revealed an abnormal chromosome complement with two related clones; clonal evolution was evident. Clone 1 with 13 cells exhibited a translocation between the short arm of chromosome 8 and the long arm of chromosome 13 (Fig. 13.1.1). Clone 2 with two cells showed a trisomy 21 in addition to the t(8;13) translocation (Fig. 13.1.2). The remaining five cells appeared to be chromosomally normal.

Secondly, FISH for ALL and MPN panels was performed on interphase nuclei using probes localized to the chromosome 4, 10, and 17 centromeric regions, *ABL1* (9q34.12), *BCR* (22q11), *KMT2A* (11q23), *ETV6* (12p13), *RUNX1* (21q22.3), *D8Z2* (8cen), *D20S108* (20q12) *D13S319* (13q14.3), and *LAMP1* (13q34) gene regions. Two hundred nuclei were examined, and the results were within normal limits for the laboratory's established background rates (Fig. 13.1.3).

Next, FISH also was performed on interphase nuclei using probes localized to *SCFD2/FIP1L1/LNX1/PDGFRA* (4q12), the 3′ and 5′ ends of the *PDGFRB* (5q33.1), the 5′ and 3′ ends of the *FGFR1* (8p12.23-p11.22) gene region (supplied by Abbott Molecular, Inc.). Two hundred nuclei were examined, and the results were positive for *FGFR1* rearrangement in 162/200 (81.0%) of the cells scored (Fig. 13.1.4).

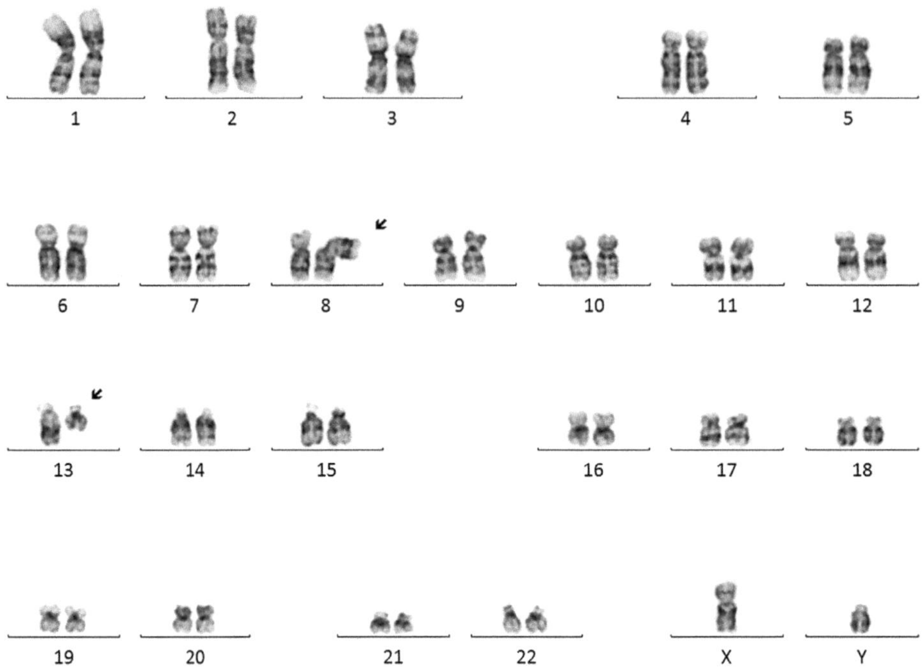

FIG. 13.1.1 Chromosome analysis showed the abnormal clone 1 with t(8;13). ISCN: 46,XY,t(8;13)(p11.2;q12)[cp13]/47,idem,+21[cp2]/46,XY[5]

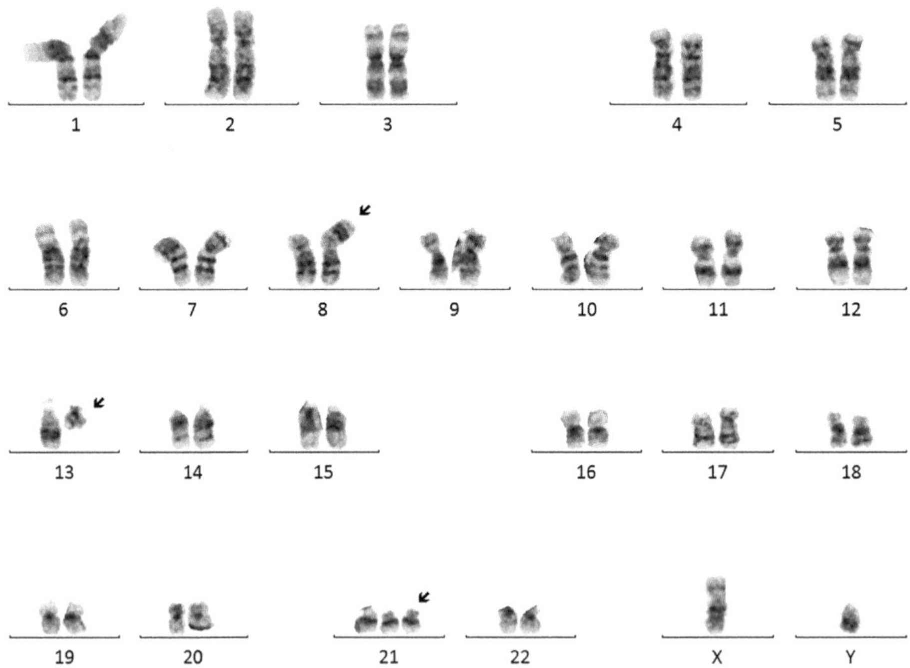

FIG. 13.1.2 Chromosome analysis revealed the abnormal clone 2 with t(8;13) and trisomy 21.

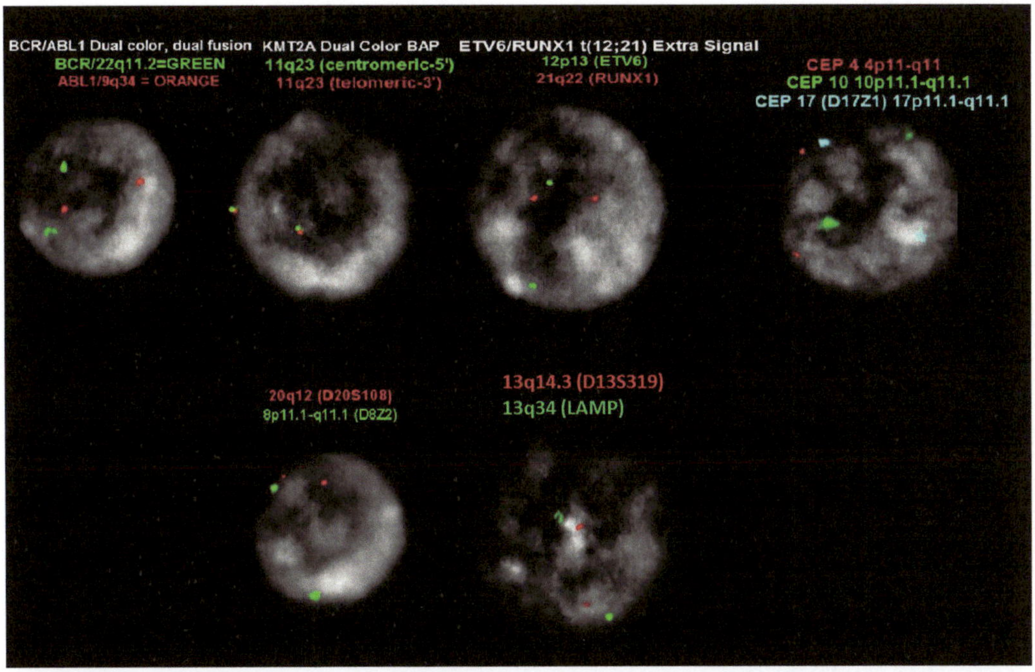

FIG. 13.1.3 FISH was negative for ALL and MPN panels. ISCN: nuc ish(CEP4,CEP10,CEP17)x2[200/200],(D8Z2,D20S108)x2[199/200],(ABL1,BCR)x2[200/200],(KMT2Ax2)[200/200],(ETV6,RUNX1)x2[199/200],(D13S319,LAMP1)x2[200/200]

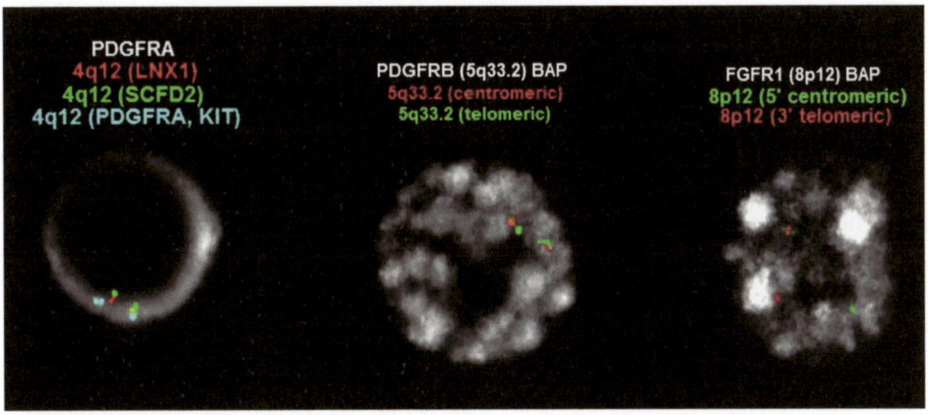

FIG. 13.1.4 FISH was positive for an *FGFR1* rearrangement and was negative for *PDGFRA* and *PDGFRB* rearrangements. ISCN: nuc ish(FGFR1x2)(5′FGFR1 sep 3′FGFR1x1)[162/200], (SCFD2,FIP1L1,LNX1,PDGFRAx2)[200/200], (PDGFRBx2)[199/200]

Finally, NGS Hematology Molecular Profile was conducted, and the results showed a fusion of *ZMYM2::FGFR1* and mutations of *RUNX1* and *DNMT3A* (Fig. 13.1.5).

Results with interpretations

Among the tests performed for this patient, chromosome analysis, FISH, and NGS were concordant, all of which detected t(8;13) or *ZMYM2::FGFR1* rearrangement. The t(8;13)

Relevant Biomarkers

Tier	Genomic Alteration	Relevant Therapies (In this cancer type)	Clinical Trials
IA	*ZMYM2-FGFR1 fusion*	midostaurin midostaurin + chemotherapy pemigatinib pemigatinib + chemotherapy ponatinib ponatinib + chemotherapy	2
	Diagnostic significance: Myeloid/Lymphoid Neoplasms with Eosinophilia		

Public data sources included in relevant therapies: FDA1, NCCN, **EMA2**, ESMO
Public data sources included in prognostic and diagnostic significance: NCCN, ESMO
Tier Reference: Li et al. *Standards and Guidelines for the Interpretation and Reporting of Sequence Variants in Cancer: A Joint Consensus Recommendation of the Association for Molecular Pathology, American Society of Clinical Oncology, and College of American Pathologists.* J Mol Diagn. 2017 Jan;19(1):4-23.

Prevalent cancer biomarkers without relevant evidence based on included data sources
*RUNX1 p.(P325Tfs*275) c.972_973insA, DNMT3A p.(R736H) c.2207G>A*

Variant Details

DNA Sequence Variants

Gene	Amino Acid Change	Coding	Locus	Allele Frequency	Variant Effect
DNMT3A	p.(R736H)	c.2207G>A	chr2:25463286	42.40%	missense
RUNX1	p.(P325Tfs*275)	c.972_973insA	chr21:36164902	40.98%	frameshift Insertion

Gene Fusions (RNA)

Genes	Variant ID	Locus
ZMYM2-FGFR1	ZMYM2-FGFR1.Z17F10	chr13:20633702 - chr8:38275891

FIG. 13.1.5 NGS showed the fusion of *ZMYM2::FGFR1* as well as *RUNX1* and *DNMT3A* mutations.

or *ZMYM2::FGFR1* has been reported in myeloid/lymphoid neoplasms with *FGFR1* rearrangement, in which trisomy 21 is the most common secondary cytogenetic abnormality. It is often associated with an adverse clinical prognosis. The t(8;13) or *ZMYM2::FGFR1* fusion is the most common translocation in this neoplasm and is associated with a high incidence of T-ALL, which is consistent with the patient's clinicopathologic findings.

Future testing and recommendations

Chromosome analysis, FISH, or NGS can be ordered to monitor disease progression.

Case 13.2 Myeloid/lymphoid neoplasm (MLN) with *PDGFRA* rearrangement (*LNX1* Deletion by FISH)

Clinical indication

A 30-year-old male with T-lymphoblastic lymphoma was admitted for acute relapse of the myeloproliferative neoplasm. He had a past medical history of peripheral neuropathy, chronic anemia, and thrombocytopenia. A bone marrow biopsy showed 12% blasts with the presence of eosinophilia.

Test ordered

- Chromosome analysis of the bone marrow
- FISH: *FIP1L1::PDGFRA* rearrangement
- NGS Hematology Molecular Profile

Laboratory test performed

Chromosome analysis, FISH, and NGS methods were described in Chapter 12.

Test results

Chromosome analysis was performed. A normal male karyotype was identified (Fig. 13.2.1).

FISH was performed with probes specific for the *SCFD2, FIP1L1, LNX1,* and *PDGFRA* gene regions. Two hundred nuclei were examined, and the results were positive for a rearrangement involving *FIP1L1::PDGFRA* in 192/200 (96.0%) of the cells scored (Fig. 13.2.2).

Finally, NGS detected *FIP1L1::PDGFRA* fusion as well as *TET2* and *U2AF1* mutations (Fig. 13.2.3).

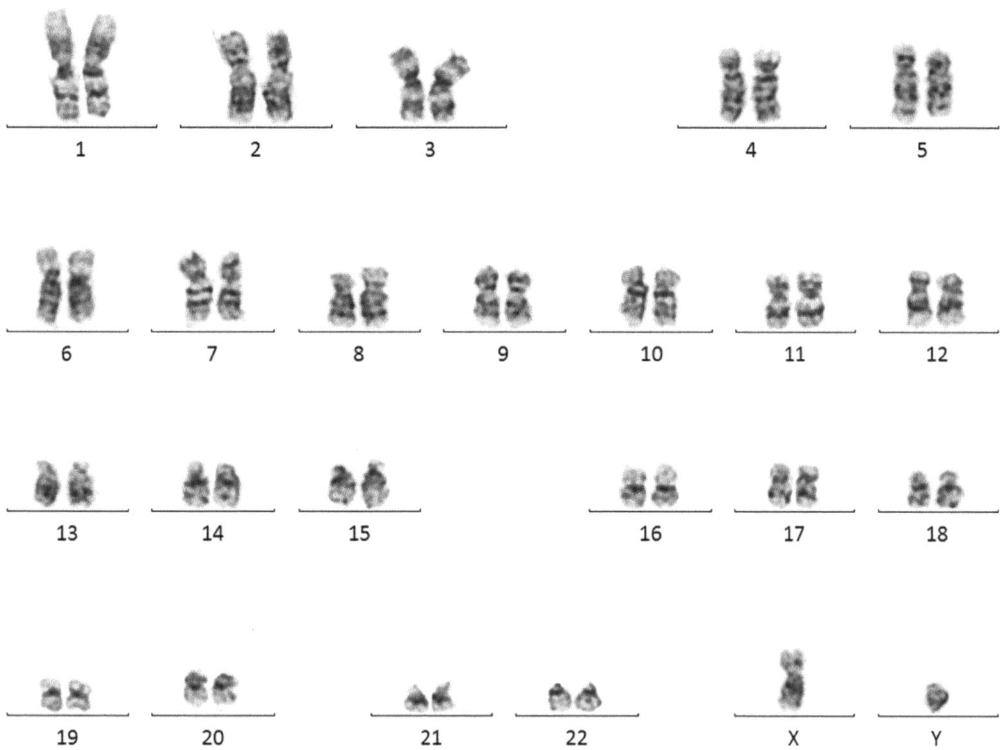

FIG. 13.2.1 Chromosome analysis showed a normal male karyotype. ISCN: 46,XY[21]

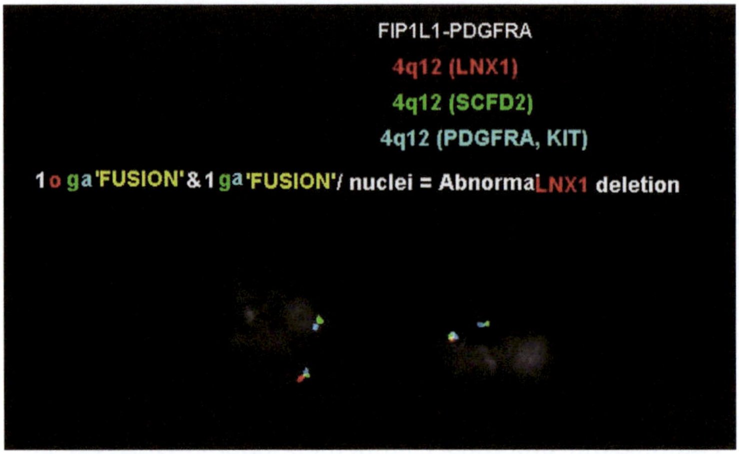

FIG. 13.2.2 FISH for *FIP1L1::PDGFRA* rearrangement was positive for a deletion of *LNX1* resulting in *FIP1L1::PDGFRA* fusion. ISCN: nuc ish(SCFD2x2,FIP1L1x2,LNX1x1,PDGFRAx2)(FIP1L1 con PDGFRAx1)[192/200]

Relevant Biomarkers

Tier	Genomic Alteration	Relevant Therapies (In this cancer type)	Clinical Trials
IA	*FIP1L1-PDGFRA fusion*	imatinib	1
	Diagnostic significance: Myeloid/Lymphoid Neoplasms with Eosinophilia		

Public data sources included in relevant therapies: FDA1, NCCN, **EMA**2, ESMO
Public data sources included in prognostic and diagnostic significance: NCCN, ESMO
Tier Reference: Li et al. *Standards and Guidelines for the Interpretation and Reporting of Sequence Variants in Cancer: A Joint Consensus Recommendation of the Association for Molecular Pathology, American Society of Clinical Oncology, and College of American Pathologists.* J Mol Diagn. 2017 Jan;19(1):4-23.

Prevalent cancer biomarkers without relevant evidence based on included data sources
U2AF1 p.(S34F) c.101C>T, TET2 p.(Q729) c.2185C>T*

Variant Details

DNA Sequence Variants

Gene	Amino Acid Change	Coding	Locus	Allele Frequency	Variant Effect
TET2	p.(Q729*)	c.2185C>T	chr4:106157284	32.25%	nonsense
U2AF1	p.(S34F)	c.101C>T	chr21:44524456	32.32%	missense

Gene Fusions (RNA)

Genes	Variant ID	Locus
FIP1L1-PDGFRA	FIP1L1-PDGFRA.F10P12del47.1	chr4:54266006 - chr4:55141055

FIG. 13.2.3 NGS detected *FIP1L1::PDGFRA* fusion as well as *TET2* and *U2AF1* mutations.

Results with interpretations

FISH and NGS results were concordant with the identification of a *FIP1L1::PDGFRA* fusion. Most MLN-TK cases associated with *PDGFRA* rearrangements have a cytogenetically cryptic deletion of 4q12 resulting in *FIP1L1::PDGFRA*. This patient has T-cell lymphoma, which is one of the many disorders possessing this rearrangement and can be treated by TKI therapy, particularly imatinib.

Future testing and recommendations

Since *PDGFRA* rearrangements are cytogenetically cryptic, FISH or NGS can be ordered to monitor disease progression.

Case 13.3 Myeloid/lymphoid neoplasm (MLN) with *PDGFRA* rearrangement (*CHIC2* Deletion by CMA)

Clinical indication

A 21-year-old male presented with hypereosinophilia. The blood counts from the lab showed anemia (hemoglobin 8.8 gm/dL), thrombocytopenia (platelets 93×10^3/microliter), and leukocytosis (white blood cells 52.7×10^3/microliter) with eosinophilia (42%).

Test ordered

- Hematopathology consultation
- Immunophenotypic analysis of bone marrow by flow cytometry
- Chromosome analysis of the bone marrow
- FISH: MPN panel and AML panels
- Chromosome microarray as a reflex test
- *FIP1L1::PDGFRA* quantitative real-time PCR as a reflex test

Laboratory test performed

Morphology studies on the peripheral blood smear, bone marrow aspirate smear, blood clot section, and trephine biopsy were performed as part of the hematopathology consultation.

Flow cytometry analysis on bone marrow aspirate was performed at first, as described by Wells et al. [38]. Briefly, cytometric analysis was performed on bone marrow aspirate after lysis with NH_4Cl erythrocyte. A standardized panel of monoclonal antibodies was used to characterize populations of myeloid and monocytic lineages using the combinations and sources of reagents as described [38]. Specimens were analyzed on a FACS Calibur flow cytometer (Becton Dickinson Immunocytometry Systems, San Jose, CA, USA). Data acquired were analyzed with WinList software (Verity Software, Topsham, ME, USA).

FISH for AML and MPN panels was performed on bone marrow aspirate using a deletion/fusion probe set (supplied by Oxford Gene Technology, Inc.) covering *FIP1L1*, *CHIC2*, and *PDGFRA* loci. Two hundred nuclei were analyzed for each probe.

Chromosome analysis on bone marrow aspirate was performed using a culture condition optimized for precursor and myeloid cell growth.

A microarray study was performed on DNA extracted from the bone marrow aspirate as a reflex test after reporting cytogenetics results and FISH results. Genomic DNA was isolated from the whole bone marrow aspirate. Oligonucleotide array-based comparative genomic hybridization (array CGH) and SNP (single-nucleotide polymorphism) analyses were performed using the Agilent SurePrint G3 Cancer CGH+SNP microarray with 25 KB overall median probe spacing (Design based on UCSC hg19 [NCBI Build 37, February 2009]). The array features 180,000 genomic loci, including ~20,000 CGH and ~60,000 SNP probes, which are cancer-associated. The assay was performed on the extracted DNA sample and referenced to a normal male DNA sample.

FIP1L1::PDGFRA quantitative real-time PCR was performed on RNA extracted from the bone marrow aspirate as a reflex test after reporting FISH results. Briefly, RNA is isolated from the sample provided and converted into cDNA using reverse transcriptase and amplified by real-time polymerase chain reaction (RQ-PCR). *FIP1L1::PDGFRA* transcript levels were reported as a ratio of fusion gene transcript/*ABL1* reference gene transcript. This test is designed to detect *FIP1L1::PDGFRA* fusion transcript with a sensitivity level of >1 in 10e4 transcripts (0.01%).

Test results

The peripheral blood smear showed increased eosinophils and basophils. Subsets of the neutrophils, eosinophils, and basophils have hypo granular cytoplasm. The marrow aspirate smears have 25% eosinophils, 2% basophils, and <1% blasts. The trephine biopsy has 90–100% cellularity. Flow cytometric analysis of the bone marrow aspirate revealed abnormal maturing neutrophilic cells, increased eosinophils (to 44% of nonerythroid cells), and increased basophils (to 3.1%) but no increase in myeloblasts (0.1%). The neutrophilic cells were predominantly mature and abnormally lacked CD36 and CD64.

The cytogenetic analysis revealed a normal male karyotype. FISH analysis using the probe set that covers the *FIP1L1-CHIC2-PDGFRA* genomic region detected a deletion of *CHIC2* in 84.5% of nuclei (Fig. 13.3.1A and Fig. 13.3.1B). FISH revealed no evidence of the following gene rearrangements: *PDGFRB, FGFR1, BCR::ABL1, JAK2, KMT2A, RUNX1::RUNX1T1, PML::RARA*, or *CBFB::MYH11* (data not shown). Microarray test on bone marrow revealed a microdeletion (about 849 kb) in chromosome band 4q12 (Fig. 13.3.1C). Quantitative RT-PCR on this diagnostic bone marrow specimen detected the *FIP1L1::PDGFRA* fusion transcript with a normalized copy number (NCN) at 0.702 (Fig. 13.3.1D).

Results with interpretations

Flowcytometric study on the bone marrow and morphologic analysis on both bone marrow and peripheral blood concordantly revealed prominent eosinophilia in this case. It was no surprise that chromosome analysis and other FISH tests for AML or MPN returned a normal result, while the FISH using the probes covering the *FIP1L1-CHIC2-PDGFRA* loci

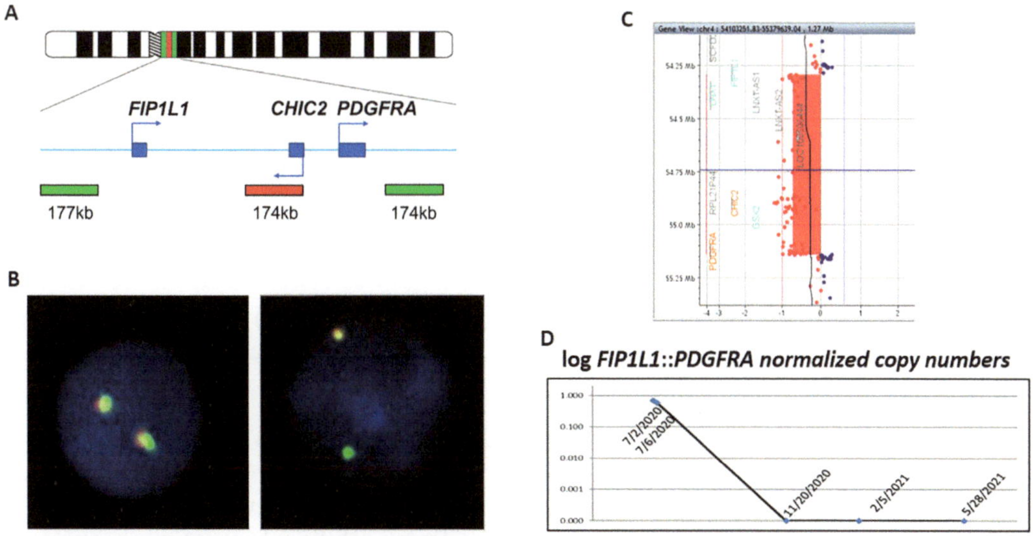

FIG. 13.3.1 FISH, microarray, and RT-PCR showed abnormal results with *CHIC2* deletion resulting in the fusion of *FIP1L1::PDGFRA*. (A) Diagram of the FISH probe set that covers the *FIP1L1-CHIC2-PDGFRA* genomic region. (B) A normal interphase nucleus with two fusion signals *(left)*. An interphase nucleus with an abnormal signal pattern: one fusion and one *green (right)*. (C) Microarray data showed an 849 kb deletion covering 3′-*FIP1L1*, whole *CHIC2* locus, and 5′-*PDGFRA*. (D) Quantitative RT-PCR data showed this patient's diagnostic specimen and follow-up specimens. ISCN: arr[GRCh37] 14q12(54292027_55141033)x1

revealed a loss of *CHIC2*. This FISH probe set was designed to specifically detect a small deletion of the *CHIC2* locus. The probe covering the *CHIC2* locus in the middle of this genomic region is produced with red fluorochrome, and the probes covering *FIP1L1* and *PDGFRA* at both sides of this genomic region are produced with green fluorochrome. A chromosome with intact *FIP1L1-CHIC2-PDGFRA* loci will be hybridized by both red and green probes adjacent to each other and will appear as a typical fusion signal. A chromosome carrying a *CHIC2* deletion can only be hybridized by two green probes adjacent to each other, but lacks a red probe signal, thus appearing as one green signal. In this case, 84.5% of nuclei examined present such deletion signal patterns, one fusion signal from a normal chromosome 4 and one green signal from a chromosome 4 carrying the *CHIC2* deletion. A microarray test can be an effective approach to detecting such small deletions. The microarray data verified an 849 kb deletion covering 3′-*FIP1L1*, whole *CHIC2* locus, and 5′-*PDGFRA*, resulting in the fusion of 5′-*FIP1L1* with 3′-*PDGFRA*. The formation of an in-frame fusion of the *FIP1L1* gene with the *PDGFRA* gene leads to constitutive activation of the *PDGFRA* kinase domain in humans [39,40]. RT-PCR test is a defining approach to verify the production of *FIP1L1::PDGFRA* fusion transcript. The RT-PCR assay also helped to establish a patient-specific baseline for future monitoring. The value of normalized copy numbers (NCNs) was 0.702 in the diagnostic specimen, which has been used as the baseline for monitoring along the patient treatment courses (Fig. 13.3.1D).

Future testing and recommendations

It has been reported that almost all patients with *FIP1L1::PDGFRA* fusion are responsive to the kinase inhibitor imatinib. Therefore, treatment with imatinib has been recommended as the definitive first-line therapy for patients with this gene fusion by the WHO [3,11]. For this patient, the NCN values for all follow-up specimens have been detected as 0.0000, reduced by 100.0% (log reduction >3.5) in comparison with the patient's baseline specimen.

Case 13.4 Myeloid/lymphoid neoplasm (MLN) with *PDGFRB* rearrangement

Clinical indication

A 46-year-old man presented with a headache and blurred vision. He had a medical history of hyperleukocytosis and leukocytosis. Bone marrow biopsy was inconclusive. The patient has been largely asymptomatic with no fevers, chills, chest pain, shortness of breath, abdominal discomfort, or changes in bowel or bladder habits. Labs obtained were concerning for WBC of 107.8, 87.1% neutrophils with 2.1% lymphocytes and 5.5% eosinophils. A concern of MPN or AML was indicated.

Test ordered

- Chromosome analysis of the bone marrow
- FISH: *PDGFRB* rearrangement

Laboratory test performed

Chromosome analysis and FISH methods were described in Chapter 12.

Test results

Chromosome analysis of the bone marrow was performed initially. Of the 20 cells analyzed, three exhibited a balanced translocation involving 5q and 12p likely resulting in *ETV6::PDGFRB* fusion. The remaining 17 cells appear to be chromosomally normal (Fig. 13.4.1).

FISH was performed on interphase nuclei using probes localized to the 5′ and 3′ ends of the *PDGFRB* (5q33.1) gene region. One hundred nuclei were examined, and the results demonstrated a rearrangement involving *PDGFRB* in 47/100 (23.5%) of the cells scored (Fig. 13.4.2).

Results with interpretations

This t(5;12) was noted in two prior studies which was concordant with the *PDGFRB* rearrangement seen in the concurrent FISH analysis. It is consistent with the diagnosis of a persistent myeloid/lymphoid neoplasm with *PDFGRB* rearrangement. Correlation with other clinical, laboratory, or pathology data is essential.

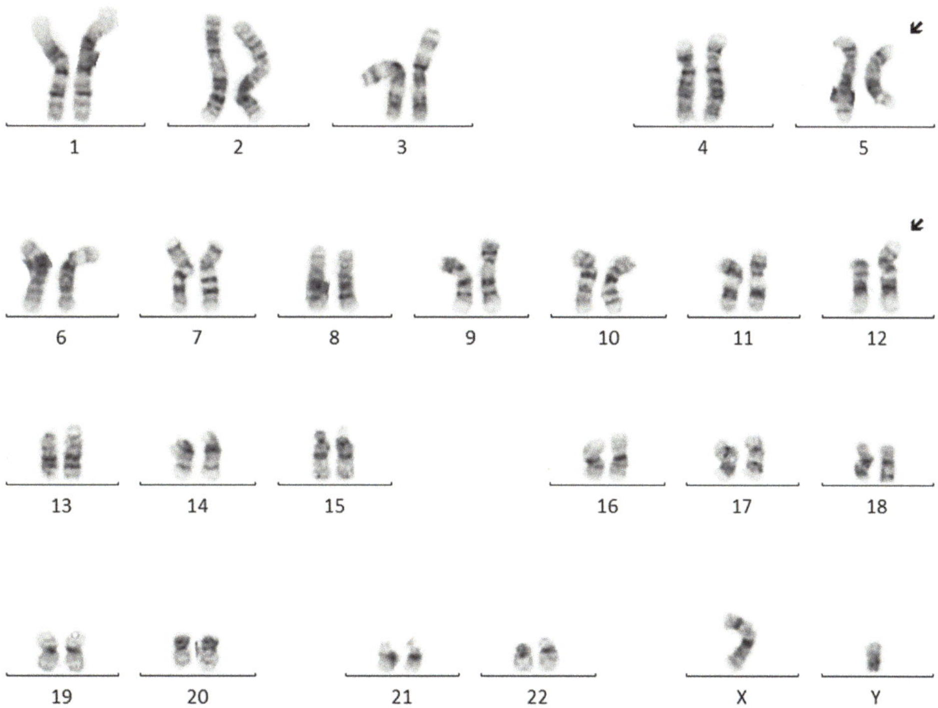

FIG. 13.4.1 Chromosome analysis showed an abnormal male karyotype with t(5;12). ISCN: 46,XY,t(5;12)(q33;p13)[3]/46,XY[17]

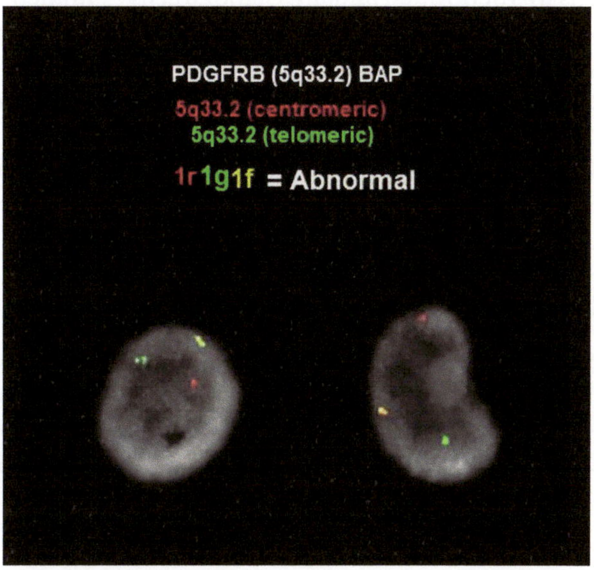

FIG. 13.4.2 FISH for *PDGFRB* break-apart probe showed positive results for *PDGFRB* rearrangement. ISCN: nuc ish(PDGFRBx2)(3′PDGFRB sep 5′PDGFRBx1)[47/200]

Future testing and recommendations

Chromosome analysis and FISH can be ordered to monitor disease progression.

Case 13.5 Myeloid/lymphoid neoplasm (MLN) with a variant *PDGFRB* rearrangement

Clinical indication

A 13-year-old female with a history of Ph-like pre-B ALL was treated on AALL1131 with Gleevec, off study, off therapy, for 2 months. She had headaches and blurry vision but denied any fever, nausea, vomiting, diarrhea, or bleeding symptoms. There was concern about a relapse of the disease.

Test ordered

- Chromosome analysis of the bone marrow
- FISH: *PDGFRB* rearrangement

Laboratory test performed

Chromosome analysis and FISH methods were described in Chapter 12.

Test results

Chromosome analysis was performed initially. Of the 20 cells examined, 8 exhibited multiple structural chromosome abnormalities including a derivative chromosome with del(2q) and translocation involving 2p&18q, unbalanced translocations involving 5q&22q, and 11q&22q. The remaining 12 cells appear to be chromosomally normal (Fig. 13.5.1).

FISH was performed on interphase nuclei using probes localized to the 5′ and 3′ ends of the *PDGFRB* (5q33.1) gene region. Two hundred nuclei were examined, and the results were positive for rearrangements involving *PDGFRB* in 192/200 (96.0%) of the cells scored (Fig. 13.5.2).

Results with interpretations

Chromosome analysis revealed a complex karyotype with rearrangements involving chromosomes 5, 11, and 22. Rearrangement of *PDGFRB* on 5q33 was also seen in the concurrent FISH testing, concordant with t(5;22) from karyotyping (a variant of *PDGFRB* rearrangement). These results are consistent with the diagnosis of myeloid/lymphoid neoplasm with a variant *PDGFRB* rearrangement. Clinicopathologic correlation is advised. These patients will benefit from imatinib therapy.

Future testing and recommendations

Chromosome analysis and FISH can be ordered to monitor disease progression.

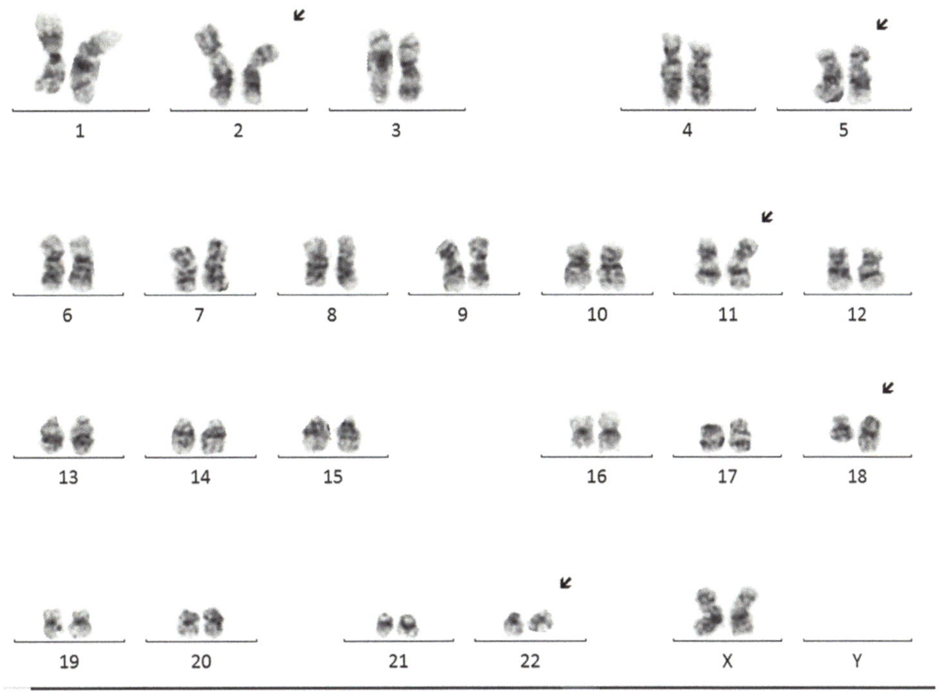

FIG. 13.5.1 Chromosome analysis revealed an abnormal complex karyotype including t(5;22). ISCN: 46,XX,der(2)del(2)(q11.2q21)t(2;18)(p13;q12),der(5)t(5;22)(q33;q13),der(11)t(11;22)(q24;q12)t(5;22),der(22)t(11;22),der(18)t(2;18)[8]/46,XX[12]

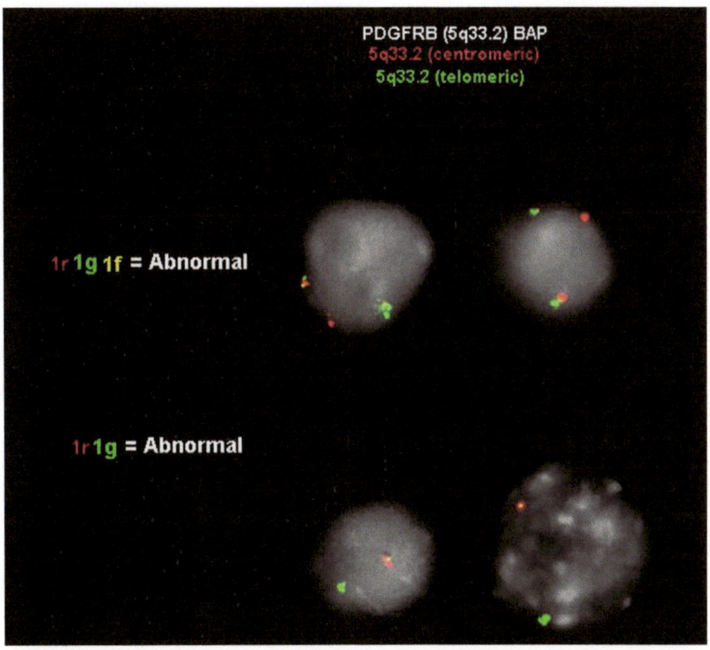

FIG. 13.5.2 FISH for *PDGFRB* break-apart probe showed positive results for *PDGFRB* rearrangement. ISCN: nuc ish(PDGFRBx2)(3′PDGFRB sep 5′PDGFRBx1)[167/200]/nuc ish(PDGFRBx1)(3′PDGFRB sep 5′PDGFRBx1)[25/200]

Case 13.6. Myeloid/lymphoid neoplasm with *JAK2* rearrangement

Clinical indication

A 78-year-old male has anemia (hemoglobin 8.1 gm/dL), thrombocytopenia (platelets 66×10^3/microliter), and eosinophilia (33% eosinophils).

Test ordered

- Chromosome analysis of the bone marrow
- FISH: MDS and MPN panels

Laboratory test performed

FISH was performed on bone marrow aspirate with an MDS panel and MPN panel including a break-apart probe set (supplied by Oxford Gene Technology, Inc.) covering the *JAK2* locus at chromosome band 9p24.1. Two hundred nuclei were analyzed for each probe set.

Chromosome analysis of bone marrow aspirate was performed using the culture condition optimized for precursor and myeloid cell growth.

Metaphase FISH using the *JAK2* break-apart probe set was performed as a reflex assay to evaluate if the *JAK2* arrangement involves the translocation between chromosomes 8 and 9.

Test results

An initial interphase FISH study was performed with probes from an MDS panel. The results revealed no evidence for 5q deletion, 7q deletion, 20q deletion, trisomy 8, or loss of the *TP53* gene in 17p. An additional FISH study was performed with probes from an MPN panel that includes *FIP1L1/CHIC2/PDGFRA*, *PDGFRB*, *FGFR1*, *BCR/ABL1*, and *JAK2* probe sets. The findings from the MPN panel revealed a rearrangement of the *JAK2* gene, but no evidence of aberrations involving other loci tested. Analysis of the *JAK2* break-apart probe set detected a typical rearrangement signal pattern (one red signal for *5′JAK2*, one green signal for *3′JAK2*, and one fusion signal for an intact *JAK2*) in 162/200 (81%) of the nuclei analyzed (Fig. 13.6.1A and B). Chromosome analysis revealed a translocation between chromosomes 8 and 9, t(8;9)(p22;p24), with a breakpoint at 8p22 and a breakpoint at 9p24 (Fig. 13.6.1C). Metaphase FISH analysis using the *JAK2* break-apart probe was performed as a reflex assay and verified transposition of *5′JAK2* from 9p into 8p (Fig. 13.6.1D).

Results with interpretations

The results from chromosome analysis, interphase, and metaphase FISH analyses provided evidence supporting a translocation t(8;9)(p22;p24) involving *JAK2* gene rearrangement. These results supported a diagnosis of myeloid/lymphoid neoplasm with *JAK2*

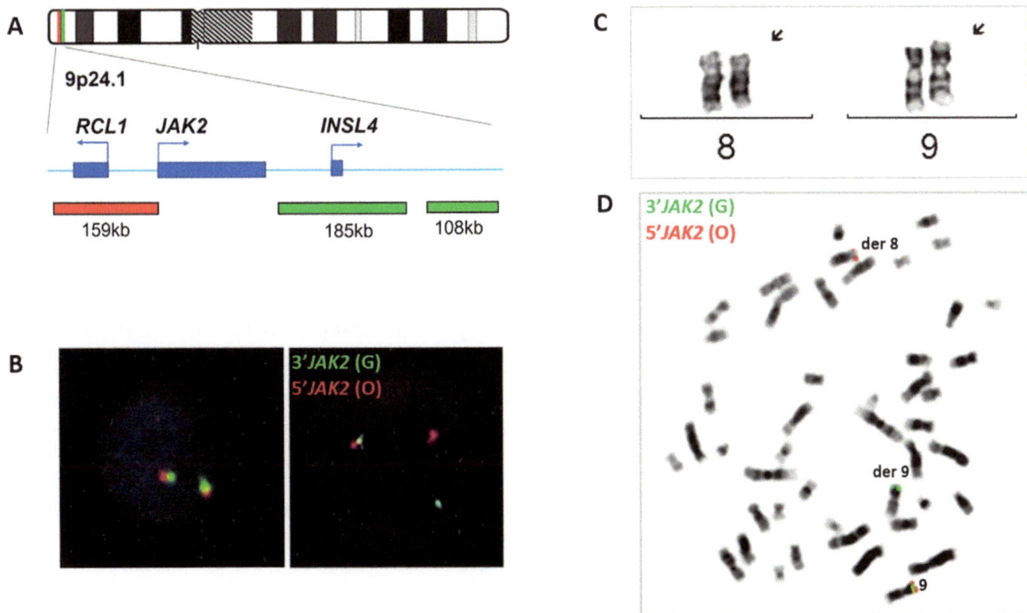

FIG. 13.6.1 Diagram of the *JAK2* break-apart FISH probe set that covers the *5'JAK2* and *3'JAK2* genomic regions (A). A normal interphase nucleus showed two fusion signals *(left)*, and an interphase nucleus showed an abnormal signal pattern: one fusion, one *green* signal, and one *red* signal separated (B). G-banded chromosomes demonstrated t(8;9)(p22;p24) (C). Metaphase FISH verified the presence of *5'JAK2* (*red*) in the derivative chromosome 8 at 8p22 and retained only *3'JAK2* (*green*) with the missing *5'JAK2* (*red*) in the derivative chromosome 9. An intact fusion signal (*yellow*) was observed in the normal chromosome 9 (D).

rearrangement. The finding of a reciprocal balanced translocation by karyotyping analysis combined with metaphase FISH finding of transposition of part of *JAK2* from chromosome 9 to chromosome 8 strongly supports the formation of *PCM1::JAK2* gene fusion that results from this translocation. Based on the orientation of these two genes in relevant chromosomes, the effective fusion with *5'PCM1* and *3'JAK2* most likely located on the derivative chromosome 9.

PCM1 gene at chromosome band 8p22 is the most common partner gene among the known partners involved with *JAK2* rearrangements, such as *BCR* and *ETV6* [41,42]. *PCM1::JAK2*-positive patients may present characteristic histopathologic features in bone marrow, such as hypercellularity with eosinophilia, large aggregates of immature erythroid precursors, and myelofibrosis [18]. Patients with *JAK2* fusions, but no prominent myeloid/myeloproliferative component or eosinophilia, may be defined as *BCR::ABL1*-like B-ALL [17,27].

Future testing and recommendations

FISH test with the *JAK2* probe can be applied for future disease monitoring. Quantitative RT-PCR for *PCM1::JAK2* fusion, or NGS testing if available, can be ordered to further verify the fusion transcript and can be an ideal monitoring approach with high sensitivity. It has been reported that myeloid/lymphoid neoplasm with *JAK2* rearrangement may be resistant to first-generation and second-generation tyrosine kinase inhibitors such as imatinib, bosutinib, dasatinib, and nilotinib. These neoplasms may respond to the *JAK2* inhibitor, ruxolitinib, for 1–2 years. Therefore, early allogeneic hematopoietic stem cell transplantation is recommended for eligible patients [13,25,43,44].

Assessment of tyrosine gene fusions including *JAK2* gene rearrangement should be considered for patients presenting eosinophilia. *PCM1::JAK2* fusion resulting from chromosome translocation t(8;9)(p22;p24) is the most common *JAK2* gene rearrangement associated with myeloid and/or lymphoid neoplasms. Please note that chromosome morphology changes can be subtle and cryptic for this translocation t(8;9) if banding resolution is not optimal. FISH testing with appropriate probes and molecular assays such as RT-PCR, or RNA fusion NGS test can help to verify the translocation and the gene fusions involving the *JAK2* gene.

Summary of key learning points

- According to the fifth edition of WHO classifications, evaluation for *PDGFRA* rearrangement has been recommended for patients presenting with unexplained eosinophilia, particularly when a conventional cytogenetic study reveals a normal karyotype.
- Patients with myeloid/lymphoid neoplasms with *FGFR1, PDGFRA, PDGFRB, or JAK2* rearrangements can present as a myeloproliferative neoplasm, acute myeloid leukemia, T- or B-lymphoblastic leukemia/lymphoma, or mixed-phenotype acute leukemia.
- The most common mechanism contributing to the *FIP1L1::PDGFRA* gene fusion is an interstitial microdeletion around 800 kb over the region containing the *CHIC2* locus at chromosome band 4q12.
- FISH and microarray can be effective approaches to detect fusions. RNA sequencing can be a powerful diagnostic approach if the sequencing cost and turnaround time become practicable in clinical applications.
- Myeloid/lymphoid neoplasms with *PDGFRA or PDGFRB* rearrangement can be treated effectively by tyrosine kinase inhibitor therapy using imatinib.
- Not all abnormalities of chromosome 8p11-12 involve the *FGFR1* gene. Therefore, FISH and/or molecular studies may be required for confirmation.

References

[1] J.D. Khoury, et al., The 5th edition of the World Health Organization Classification of haematolymphoid tumours: myeloid and histiocytic/dendritic neoplasms, Leukemia 36 (7) (2022) 1703–1719.

[2] E.J. Baxter, et al., The t(4;22)(q12;q11) in atypical chronic myeloid leukaemia fuses BCR to PDGFRA, Hum. Mol. Genet. 11 (12) (2002) 1391–1397.

[3] A.M. Safley, et al., Molecular and cytogenetic characterization of a novel translocation t(4;22) involving the breakpoint cluster region and platelet-derived growth factor receptor-alpha genes in a patient with atypical chronic myeloid leukemia, Genes Chromosom. Cancer 40 (1) (2004) 44–50.

[4] J. Score, et al., Identification of a novel imatinib responsive KIF5B-PDGFRA fusion gene following screening for PDGFRA overexpression in patients with hypereosinophilia, Leukemia 20 (5) (2006) 827–832.

[5] C. Walz, et al., Transient response to imatinib in a chronic eosinophilic leukemia associated with ins (9;4)(q33;q12q25) and a CDK5RAP2-PDGFRA fusion gene, Genes Chromosom. Cancer 45 (10) (2006) 950–956.

[6] C.E. Curtis, et al., Two novel imatinib-responsive PDGFRA fusion genes in chronic eosinophilic leukaemia, Br. J. Haematol. 138 (1) (2007) 77–81.

[7] Z.R. Chalmers, et al., Comprehensive genomic profiling identifies a novel TNKS2-PDGFRA fusion that defines a myeloid neoplasm with eosinophilia that responded dramatically to imatinib therapy, Blood Cancer J. 5 (2015), e278.

[8] A.D. Klion, et al., Elevated serum tryptase levels identify a subset of patients with a myeloproliferative variant of idiopathic hypereosinophilic syndrome associated with tissue fibrosis, poor prognosis, and imatinib responsiveness, Blood 101 (12) (2003) 4660–4666.

[9] A.D. Klion, et al., Molecular remission and reversal of myelofibrosis in response to imatinib mesylate treatment in patients with the myeloproliferative variant of hypereosinophilic syndrome, Blood 103 (2) (2004) 473–478.

[10] A. Pardanani, et al., FIP1L1-PDGFRA fusion: prevalence and clinicopathologic correlates in 89 consecutive patients with moderate to severe eosinophilia, Blood 104 (10) (2004) 3038–3045.

[11] J. Cools, et al., A tyrosine kinase created by fusion of the PDGFRA and FIP1L1 genes as a therapeutic target of imatinib in idiopathic hypereosinophilic syndrome, N. Engl. J. Med. 348 (13) (2003) 1201–1214.

[12] G. Tang, et al., Hematopoietic neoplasms with 9p24/JAK2 rearrangement: a multicenter study, Mod. Pathol. 32 (4) (2019) 490–498.

[13] J. Schwaab, et al., Response to tyrosine kinase inhibitors in myeloid neoplasms associated with PCM1-JAK2, BCR-JAK2 and ETV6-ABL1 fusion genes, Am. J. Hematol. 95 (7) (2020) 824–833.

[14] C.E. Curtis, et al., A novel ETV6-PDGFRB fusion transcript missed by standard screening in a patient with an imatinib responsive chronic myeloproliferative disease, Leukemia 21 (8) (2007) 1839–1841.

[15] K. Kobayashi, et al., ATF7IP as a novel PDGFRB fusion partner in acute lymphoblastic leukaemia in children, Br. J. Haematol. 165 (6) (2014) 836–841.

[16] S.L. Ondrejka, et al., PDGFRB-rearranged T-lymphoblastic leukemia/lymphoma occurring with myeloid neoplasms: the missing link supporting a stem cell origin, Haematologica 99 (9) (2014) e148–e151.

[17] K.G. Roberts, et al., Targetable kinase-activating lesions in Ph-like acute lymphoblastic leukemia, N. Engl. J. Med. 371 (11) (2014) 1005–1015.

[18] O. Pozdnyakova, et al., Myeloid/lymphoid neoplasms associated with eosinophilia and rearrangements of PDGFRA, PDGFRB, or FGFR1 or with PCM1-JAK2, Am. J. Clin. Pathol. 155 (2) (2021) 160–178.

[19] A. Reiter, et al., Consistent fusion of ZNF198 to the fibroblast growth factor receptor-1 in the t(8;13) (p11;q12) myeloproliferative syndrome, Blood 92 (5) (1998) 1735–1742.

[20] D. Smedley, et al., The t(8;13)(p11;q11-12) rearrangement associated with an atypical myeloproliferative disorder fuses the fibroblast growth factor receptor 1 gene to a novel gene RAMP, Hum. Mol. Genet. 7 (4) (1998) 637–642.

[21] S. Xiao, et al., FGFR1 is fused with a novel zinc-finger gene, ZNF198, in the t(8;13) leukaemia/lymphoma syndrome, Nat. Genet. 18 (1) (1998) 84–87.

[22] B.J. Bain, S.H. Fletcher, Chronic eosinophilic leukemias and the myeloproliferative variant of the hypereosinophilic syndrome, Immunol. Allergy Clin. N. Am. 27 (3) (2007) 377–388.

[23] C.C. Jackson, L.J. Medeiros, R.N. Miranda, 8p11 myeloproliferative syndrome: a review, Hum. Pathol. 41 (4) (2010) 461–476.

[24] K.R. Kumar, et al., Myeloid and lymphoid neoplasm with abnormalities of FGFR1 presenting with trilineage blasts and RUNX1 rearrangement: a case report and review of literature, Am. J. Clin. Pathol. 143 (5) (2015) 738–748.

[25] A. Reiter, J. Gotlib, Myeloid neoplasms with eosinophilia, Blood 129 (6) (2017) 704–714.

[26] T. Wang, et al., Identification of a novel TFG-FGFR1 fusion gene in an acute myeloid leukaemia patient with t(3;8)(q12;p11), Br. J. Haematol. 188 (1) (2020) 177–181.

[27] B.J. Bain, S. Ahmad, Should myeloid and lymphoid neoplasms with PCM1-JAK2 and other rearrangements of JAK2 be recognized as specific entities? Br. J. Haematol. 166 (6) (2014) 809–817.

[28] H. Zhang, et al., Two myeloid leukemia cases with rare FLT3 fusions, Cold Spring Harb. Mol. Case Stud. 4 (6) (2018).

[29] A.K. Chao, et al., Fusion driven JMML: a novel CCDC88C-FLT3 fusion responsive to sorafenib identified by RNA sequencing, Leukemia 34 (2) (2020) 662–666.

[30] G. Tang, et al., Myeloid/lymphoid neoplasms with FLT3 rearrangement, Mod. Pathol. 34 (9) (2021) 1673–1685.

[31] M. Zaliova, et al., Characterization of leukemias with ETV6-ABL1 fusion, Haematologica 101 (9) (2016) 1082–1093.

[32] J. Yao, et al., Myeloid/lymphoid neoplasms with eosinophilia/basophilia and ETV6-ABL1 fusion: cell-of-origin and response to tyrosine kinase inhibition, Haematologica 106 (2) (2021) 614–618.

[33] S. Rottgers, et al., ALK fusion genes in children with atypical myeloproliferative leukemia, Leukemia 24 (6) (2010) 1197–1200.

[34] P. Ballerini, et al., RET fusion genes are associated with chronic myelomonocytic leukemia and enhance monocytic differentiation, Leukemia 26 (11) (2012) 2384–2389.

[35] Y. Maesako, et al., inv(2)(p23q13)/RAN-binding protein 2 (RANBP2)-ALK fusion gene in myeloid leukemia that developed in an elderly woman, Int. J. Hematol. 99 (2) (2014) 202–207.

[36] N. Telford, et al., Myeloproliferative neoplasm with eosinophilia and T-lymphoblastic lymphoma with ETV6-LYN gene fusion, Blood Cancer J. 6 (2016), e412.

[37] T. Carll, et al., Diagnosis and treatment of mixed phenotype (T-myeloid/lymphoid) acute leukemia with novel ETV6-FGFR2 rearrangement, Blood Adv. 4 (19) (2020) 4924–4928.

[38] D.A. Wells, et al., Myeloid and monocytic dyspoiesis as determined by flow cytometric scoring in myelodysplastic syndrome correlates with the IPSS and with outcome after hematopoietic stem cell transplantation, Blood 102 (1) (2003) 394–403.

[39] E.H. Stover, et al., Activation of FIP1L1-PDGFRalpha requires disruption of the juxtamembrane domain of PDGFRalpha and is FIP1L1-independent, Proc. Natl. Acad. Sci. U. S. A. 103 (21) (2006) 8078–8083.

[40] F. Vega, et al., Hematolymphoid neoplasms associated with rearrangements of PDGFRA, PDGFRB, and FGFR1, Am. J. Clin. Pathol. 144 (3) (2015) 377–392.

[41] K. Stewart, et al., Neutrophilic myelofibrosis presenting as Philadelphia chromosome negative BCR non-rearranged chronic myeloid leukemia, Am. J. Hematol. 34 (1) (1990) 59–63.

[42] A. Reiter, et al., The t(8;9)(p22;p24) is a recurrent abnormality in chronic and acute leukemia that fuses PCM1 to JAK2, Cancer Res. 65 (7) (2005) 2662–2667.

[43] J. Schwaab, et al., Limited duration of complete remission on ruxolitinib in myeloid neoplasms with PCM1-JAK2 and BCR-JAK2 fusion genes, Ann. Hematol. 94 (2) (2015) 233–238.

[44] D.M. Mattis, S.A. Wang, C.M. Lu, Contemporary classification and diagnostic evaluation of hypereosinophilia, Am. J. Clin. Pathol. 154 (3) (2020) 305–318.

14

Myelodysplastic/myeloproliferative neoplasms

Xia Li and Guang Liu

SONORA QUEST LABORATORIES, PHOENIX, AZ, UNITED STATES

Background

Myelodysplastic/myeloproliferative neoplasms (MDS/MPNs) are a group of diseases in which the bone marrow makes an excess of white blood cells. MDS/MPNs have features of both myelodysplastic and myeloproliferative neoplasms. They are characterized by the following features: (1) persistent monocytosis $>1x10^9/L$ in the peripheral blood (PB); (2) negative result for presence of a Philadelphia (Ph) chromosome and *BCR::ABL1* fusion gene; (3) absence of rearrangement of *PDGFRA* or *PDGFB*; (4) fewer than 20% blasts in the PB and bone marrow (BM), and (5) dysplasia involving one or more myeloid lineages [1,2]. There are different types of MDS/MPNs including Chronic Myelomonocytic Leukemia (CMML), atypical Chronic Myelogenous Leukemia (aCML), *BCR::ABL1* negative, Juvenile Myelomonocytic Leukemia (JMML), and Myelodysplastic/myeloproliferative Neoplasms, unclassifiable (MDS/MPN-UC). There are many challenges in diagnosing this group of diseases. For example, excluding other causes of monocytosis, cytogenetic abnormalities, and somatic mutations may help identify clonal hematopoiesis ruling out reactive processes in CMML [1,2].

Genetic markers are very important in making an accurate diagnosis of this group of diseases. Clonal cytogenetics abnormalities are found in 20%–40% of patients with CMML, but none are specific [3]. Common cytogenetic abnormalities include trisomy 8, -Y, -7/del(7q), structural abnormalities of 12p, trisomy 21, and del(20q). These are used effectively for risk stratification in CMML patients. Gene mutations identified from CMML include *TET2* (~60%), *ASXL1* (~40%), *SRSF2* (~50%), *RUNX1* (~15%), *RAS* (~30%), and *CBL* (~15%) [3]. In aCML, karyotypic abnormalities are reported in up to 80% of patients. The most common abnormalities are +8, del(20q), as well as abnormalities of chromosomes 13, 4, 17, 19, and 12 [1]. *JAK2* V617F mutations were reported in some aCML cases [4]. Monosomy 7 is seen in about 25% of JMML patients, other abnormalities in 10%, and a normal karyotype in 65% [5]. RAS mutations such as *NRAS*, *KRAS*, *NF1*, and *PTPN11* are also reported [6,7]. There is no cytogenetic or molecular genetic finding specific to MDS/MPN-UC. *BCR::ABL1* fusion, *PDGFRA*, *PDGFRB*, or *FGFR1* rearrangement or with isolated del(5q), t(3;3)(q21;q26.3), or inv(3)(q21q26) is excluded from this category [1].

Cases in Laboratory Genetics and Genomics (LGG) Practice. https://doi.org/10.1016/B978-0-323-99622-8.00016-2

In clinical diagnostic labs, FISH, karyotyping, RT-PCR, or NGS is commonly used to identify the chromosomal aberrations in MDS/MPN. In this chapter, two cases will be illustrated to demonstrate how genetic testing can benefit the diagnosis and risk stratification of MDS/MPN patients.

Case 14.1 Chronic myelomonocytic leukemia (CMML)

Clinical indication

A 68-year-old male presented with no energy, night sweats, and lower jaw pain. He was found to have elevated white blood cell count, anemia, and thrombocytopenia. A bone marrow biopsy was then performed. The findings in the bone marrow were a morphological diagnosis match for myelodysplastic/myeloproliferative neoplasm, best classified as chronic myelomonocytic leukemia-1 (CMML-1).

Test ordered

- Chromosome analysis of the bone marrow
- FISH: MDS and MPN panels
- *JAK2* V617F Mutation Detection, Qualitative
- NGS Hematology Molecular Profile

Laboratory test performed

Chromosome analysis, FISH, RT-PCR, and NGS on bone marrow were performed (methods were detailed in Chapter 12).

The *JAK2* (mRNA reference sequence: NM_004972.3) qualitative real-time PCR allelic discrimination analysis is specifically designed to detect the point mutation c.1849G>T (p.Val617Phe). This mutation assay cannot accurately detect mutations in samples comprising less than or equal to 2% of the mutant allele. Mutations outside the codon covered by this test will not be detected.

Test results

Chromosome analysis was performed on the bone marrow. All 20 cells examined exhibited a translocation between the long arms of chromosome 3 and the short arm of chromosome 12 (Fig. 14.1.1).

FISH for MDS and MPN panels was performed on interphase nuclei using probes (Abbott Molecular, Inc.) localized to the *D5S721* (5p15.2), *EGR1* (5q23-31), *D7Z1* (7cen), *D7S468* (7q31), *ABL1* (9q34.12), *BCR* (22q11), *D8Z1* (8cen), *D20S108* (20q12), *D13S319* (13q14.3), and *LAMP1* (13q34) gene regions. Two hundred nuclei were examined, and the results were within normal limits for the laboratory's established background rate (Fig. 14.1.2).

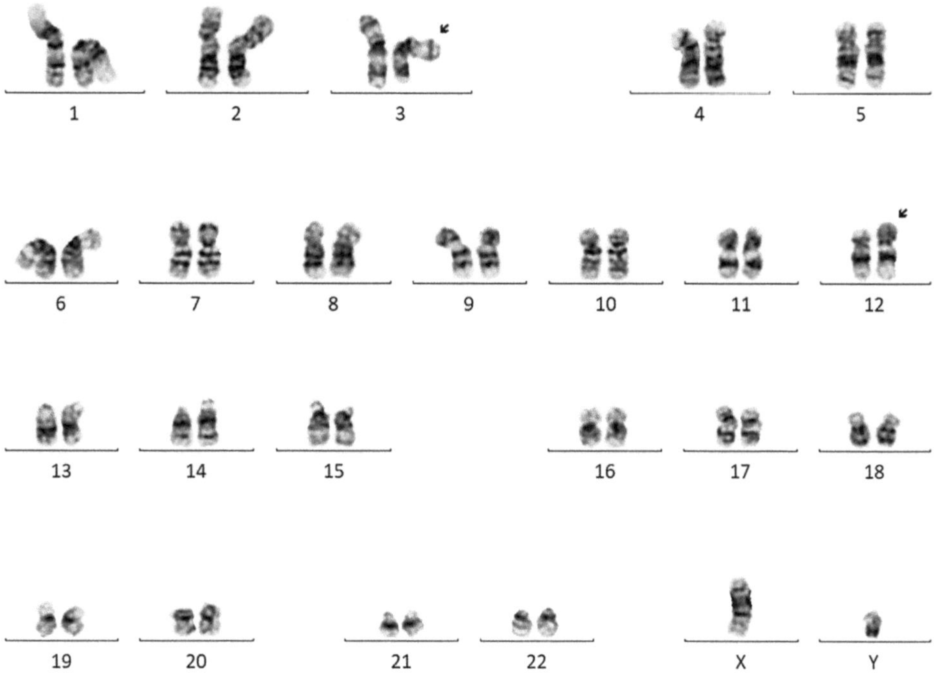

FIG. 14.1.1 The karyotype of the bone marrow showed a t(3;12) translocation. ISCN: 46,XY,t(3;12)(q26.2;p13)[20]

Qualitative *JAK2* V617F Mutation Detection was conducted. No mutation was detected (data not shown).

NGS Hematology Molecular Profile was performed, and the results showed a fusion of *ETV6::MECOM* and mutations of *ASXL1* and *SRSF2* (Fig. 14.1.3).

Results with interpretations

The translocation t(3;12)(q26.2;p13), resulting in the fusion of the *MECOM* gene at 3q26.2 with the *ETV6* gene at 12p13, was consistent with the concurrent NGS finding (*ETV6::MECOM* fusion). NGS results also showed *ASXL1* and *SRSF2* mutations. FISH was negative for MPN and MDS panels, including *BCR::ABL1* fusion probes. The t(3;12) is a recurrent abnormality seen in myeloid neoplasms including MDS, MPN, MDS/MPN and has been associated with a poor prognosis [https://atlasgeneticsoncology.org/haematological/1280/t(3;12)(q26;q21)] [8,9]. Many patients with CMML harbor somatic mutations, most frequently in *TET2* (found in 58% of cases), *SRSF2* (in 46%), and *ASXL1* (in 40%) [1]. Therefore, the results of karyotyping, FISH, and NGS from this patient are consistent with a myelodysplastic/myeloproliferative neoplasm (CMML) in the context of the pathology report. Clinicopathologic correlation is advised.

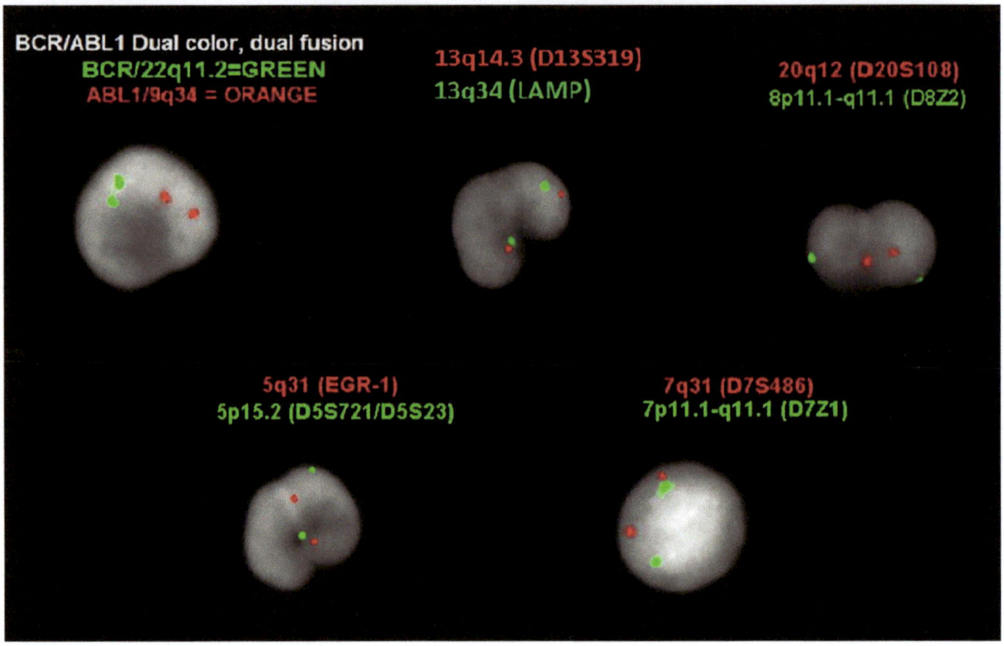

FIG. 14.1.2 FISH for MDS and MPN panel was negative for all probes examined. ISCN: nuc ish(D5S721,EGR1)x2 [200/200],(D7Z1,D7S468)x2[200/200],(D8Z2,D20S108)x2[200/200],(ABL1,BCR)x2[200/200],(D13S319,LAMP1)x2 [200/200]

Relevant Biomarkers

Tier	Genomic Alteration	Relevant Therapies (In this cancer type)	Clinical Trials
IA	*SRSF2 p.(P95H) c.284C>A* **Prognostic significance:** NCCN: Poor **Diagnostic significance:** None	None	3
IA	*ASXL1 p.(Q882*) c.2644C>T* **Prognostic significance:** NCCN: Poor **Diagnostic significance:** None	None	1

Public data sources included in relevant therapies: FDA1, NCCN, **EMA**2, ESMO
Public data sources included in prognostic and diagnostic significance: NCCN, ESMO
Tier Reference: Li et al. *Standards and Guidelines for the Interpretation and Reporting of Sequence Variants in Cancer: A Joint Consensus Recommendation of the Association for Molecular Pathology, American Society of Clinical Oncology, and College of American Pathologists.* J Mol Diagn. 2017 Jan;19(1):4-23.

Prevalent cancer biomarkers without relevant evidence based on included data sources
ETV6-MECOM fusion

Variant Details

DNA Sequence Variants

Gene	Amino Acid Change	Coding	Locus	Allele Frequency	Variant Effect
SRSF2	p.(P95H)	c.284C>A	chr17:74732959	53.10%	missense
ASXL1	p.(Q882*)	c.2644C>T	chr20:31023159	50.40%	nonsense

Gene Fusions (RNA)

Genes	Variant ID	Locus
ETV6-MECOM	ETV6-MECOM.E2M2	chr12:11905513 - chr3:168861620

FIG. 14.1.3 NGS demonstrated the fusion of *ETV::MECOM* and mutations of *ASXL1* and *SRSF2* (*MIR636*).

Future testing and recommendations

Chromosome analysis and NGS can be ordered to monitor disease progression and treatment efficacy in the future.

Case 14.2 Myelodysplastic/myeloproliferative neoplasm with ring sideroblasts and thrombocytosis (MDS/MPN-RS-T)

Clinical indication

A 63-year-old gentleman presented with macrocytic anemia and thrombocytosis. He complained about having decreased exercise tolerance, fatigue, hypertension, hypothyroidism, neutrophilic leukocytosis, and monocytosis. Bone marrow biopsy and aspirate showed trilineage hematopoiesis with erythroid hyperplasia and megakaryocytic hyperplasia consistent with myelodysplastic/myeloproliferative neoplasm with ring sideroblasts present. The patient also had thrombocytosis. Findings were consistent with those of myelodysplastic/myeloproliferative neoplasm with ring sideroblasts and thrombocytosis (MDS/MPN-RS-T). He was treated with Jakafi (Ruxolitinib).

Test ordered

- Chromosome analysis of the bone marrow
- FISH: MPN panel
- FISH: del(5q) and del(7q)
- NGS Hematology Molecular Profile

Laboratory test performed

Chromosome analysis, FISH, and NGS on bone marrow were performed (detailed methods were described previously in Chapter 12).

Test results

Chromosome analysis identified 4 related clones in 18 of the 20 metaphase cells analyzed. Specifically, clone 1 with 8 cells showed a solitary deletion 5q (Fig. 14.2.1). Clone 2 with three cells showed a rearrangement of chromosome 13, monosomy for chromosomes 16 and 18, trisomy 19, a dicentric chromosome involving 19q, a deletion 20q, a tricentric chromosome involving 21p&21q&22p, and presence of 1–2 marker chromosomes in addition to del(5q) (Fig. 14.2.2). Clone 3 with four cells showed, in addition to the aberrations in clone 2, loss of the Y chromosome, a rearrangement of 12p, trisomy 20, an unbalanced rearrangement involving 21q&21p, a rearrangement of 22q, and the presence of a marker chromosome. There was no evidence in this clone of the dicentric (19;19) and the tricentric (21;21;22) (Fig. 14.2.3). Clone 4 with three cells showed, in addition to the deletion 5q in clone 1, trisomy 21 (Fig. 14.2.4). The deletion 5q was detected in a previous chromosome study

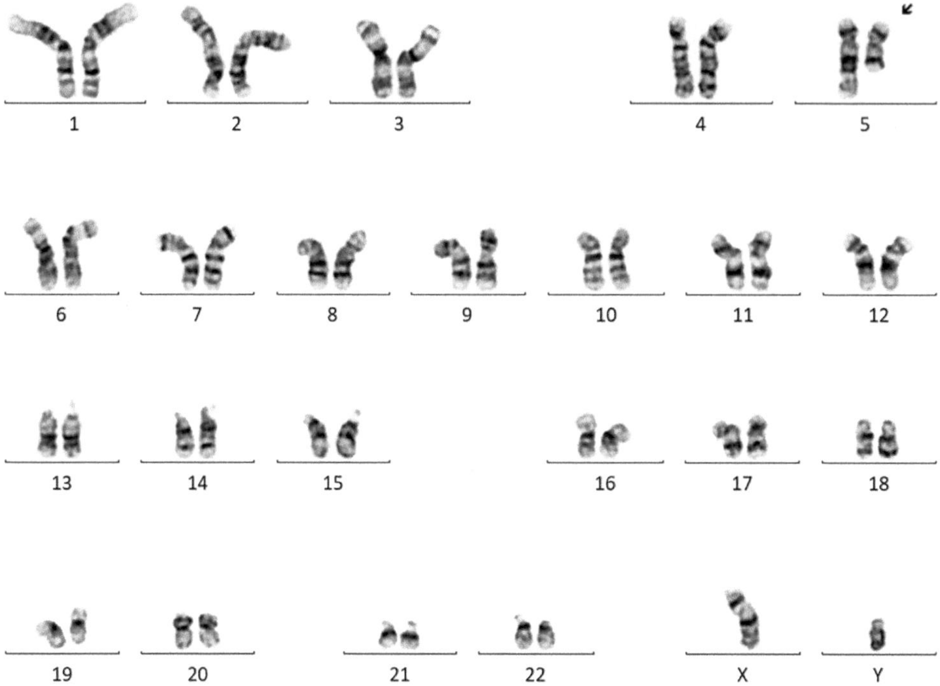

FIG. 14.2.1 The karyotype of the bone marrow showed clone 1 from a complex karyotype including del(5q). ISCN: 46, XY,del(5)(q22q35)[8]/46~47,sl,dic(13;?)(p11.2;?),-16,-18,+19,dic(19;19)(p13.2;p13.2)x2,del(20)(q11.2q13.3),trc(21;21;22)(p11.2;q22;p11.2),+1~2mar[cp3]/44~45,sdl,-Y,add(12)(p11.2),-dic(19;19),+20,-trc(21;21;22),+der(21;21)(q22;p11.2),add(22)(p11.2),+mar[cp4]/47,sl,+21[3]/46,XY[2]

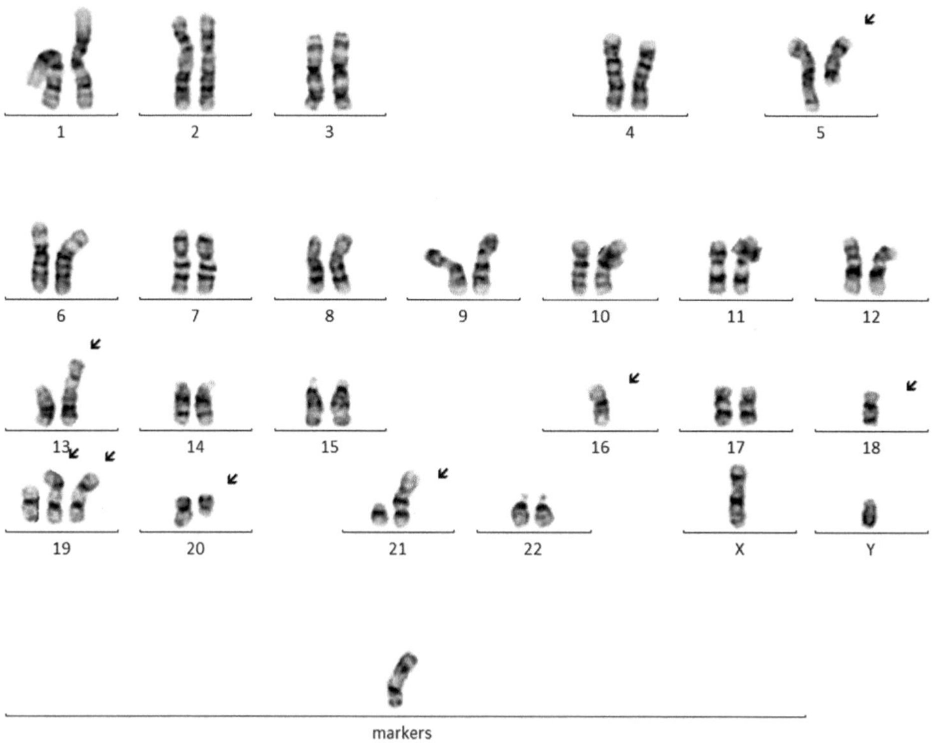

FIG. 14.2.2 The karyotype of the bone marrow showed clone 2 from a complex karyotype including many chromosome abnormalities.

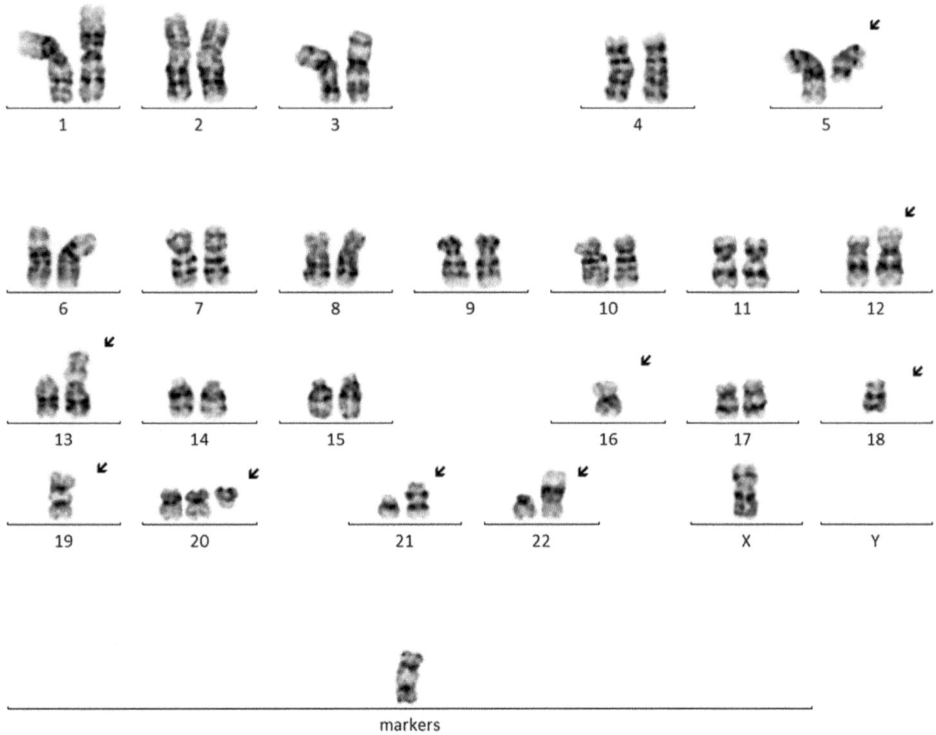

FIG. 14.2.3 The karyotype of the bone marrow showed clone 3 from a complex karyotype including many chromosome abnormalities.

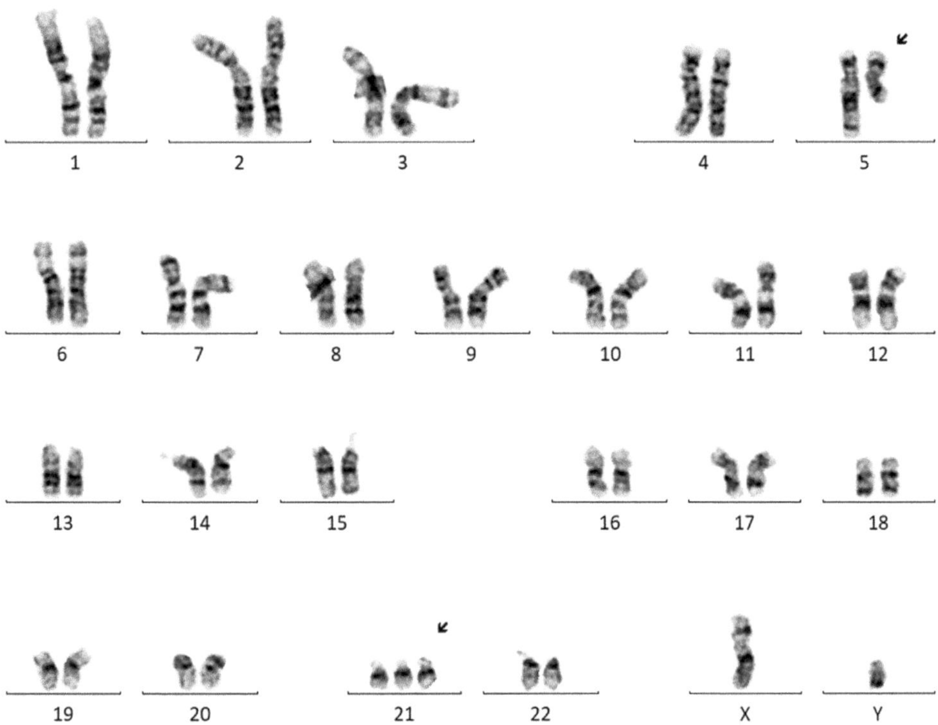

FIG. 14.2.4 The karyotype of the bone marrow showed clone 4 with del(5q) and trisomy 21.

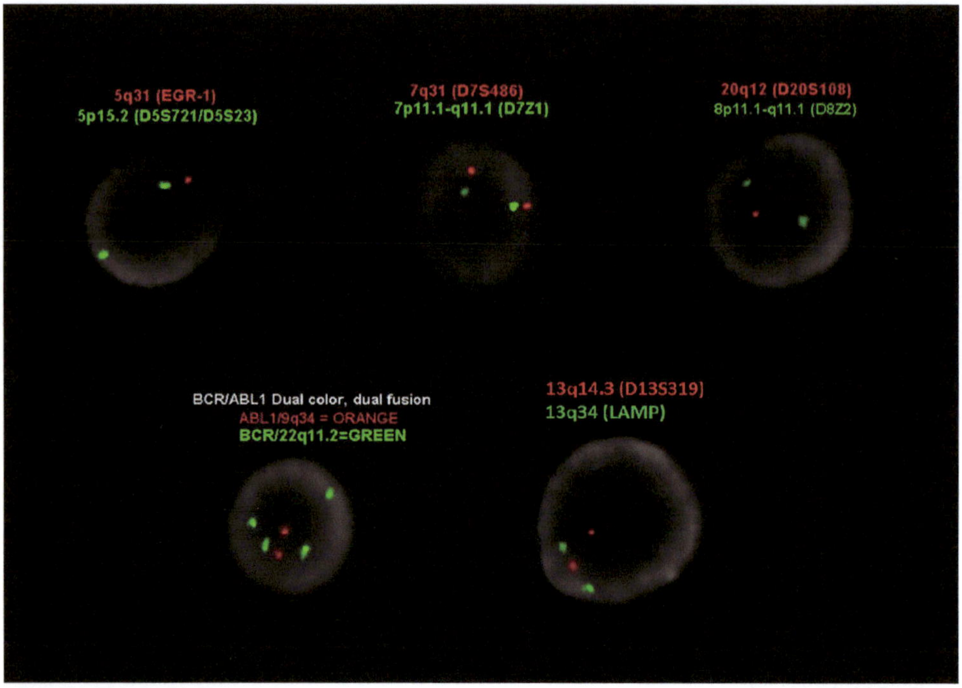

FIG. 14.2.5 FISH for MPN panel, del(5q)-5, and del(7q)/-7 showed gain of 22q (BCR), del(5q), and del(20q). ISCN: nuc ish(D5S721x2,EGR1x1)[150/200],(D7Z1,D7S468)x2[197/200], (D8Z2x2,D20S108x1)[70/200],(ABL1x2,BCRx3-4)[113/200],(D13S319,LAMP1)x2[195/200]

indicating persistent/recurrent disease. There were additional abnormalities in the present study compatible with clonal evolution. In addition to MDS/MPN, karyotypes such as this can be seen in acute myeloid leukemia (AML) and therapy-related MDS/AML. In the Revised International Prognostic Scoring System (IPSS-R) scoring system for MDS risk assessment, this result was associated with a very poor prognosis.

FISH for the MPN panel was performed on interphase nuclei using probes (Abbott Molecular, Inc.) localized to the *D5S721*, *EGR1*, *D7Z1* (7cen), *D7S468* (7q31), *ABL1* (9q34.12), *BCR* (22q11), *D8Z1* (8cen), *D20S108* (20q12), *D13S319* (13q14.3), and *LAMP1* (13q34) gene regions. Two hundred nuclei were examined, and the results demonstrated gains of 22q in 113/200 (56.5%) of the cells scored, deletion of 20q in 70/200 (35%), and deletion of 5q in 150/200 (75%) of the cells scored (Fig. 14.2.5).

Finally, NGS revealed two mutations including *JAK2* p.(V617F) and *TP53* p.(P152R) (Fig. 14.2.6).

Results with interpretations

Chromosome analysis showed a complex karyotype with clonal evolution. del(5q), del(20q), and gain of 22q were also seen from the current FISH testing. These findings are concordant with a persistent/recurrent myeloid neoplasm such as myelodysplasia (MDS) and/or myeloproliferative neoplasm (MPN) in the context of the pathology

Relevant Biomarkers

Tier	Genomic Alteration	Relevant Therapies (In this cancer type)	Clinical Trials
IA	*JAK2 p.(V617F) c.1849G>T*	None	3
	Prognostic significance: NCCN: Intermediate **Diagnostic significance:** Polycythemia Vera, Essential Thrombocythemia, Primary Myelofibrosis		
IA	*TP53 p.(P152R) c.455C>G*	None	0
	Prognostic significance: MIPSS-ET: Intermediate to High		

Public data sources included in relevant therapies: FDA1, NCCN, EMA2, ESMO
Public data sources included in prognostic and diagnostic significance: NCCN, ESMO
Tier Reference: Li et al. *Standards and Guidelines for the Interpretation and Reporting of Sequence Variants in Cancer: A Joint Consensus Recommendation of the Association for Molecular Pathology, American Society of Clinical Oncology, and College of American Pathologists.* J Mol Diagn. 2017 Jan;19(1):4-23.

Variant Details

DNA Sequence Variants

Gene	Amino Acid Change	Coding	Locus	Allele Frequency	Variant Effect
JAK2	p.(V617F)	c.1849G>T	chr9:5073770	19.80%	missense
TP53	p.(P152R)	c.455C>G	chr17:7578475	37.22%	missense

FIG. 14.2.6 NGS demonstrated mutations of *JAK2* and *TP53*.

report and clinical history. NGS revealed two mutations including *JAK2* p.(V617F) and *TP53* p.(P152R); both are associated with an intermediate prognosis for MDS/MPN.

Future testing and recommendations

Chromosome analysis and NGS can be ordered to monitor disease progression and treatment efficacy in the future.

Summary of key learning points

- MDS/MPNs have features of both myelodysplastic syndromes and myeloproliferative neoplasms.
- All MDS/MPN cases should be absent of *BCR::ABL1* fusion, with no *PDGFRA*, *PDGFRB*, or *FGFR1* rearrangement.
- FISH, karyotyping, RT-PCR, and NGS are the common technologies used to identify genetic abnormalities in MDS/MPN.

References

[1] D.A. Arber, et al., The 2016 revision to the World Health Organization classification of myeloid neoplasms and acute leukemia, Blood 127 (20) (2016) 2391–2405.

[2] J.D. Khoury, et al., The 5th edition of the World Health Organization classification of haematolymphoid tumours: myeloid and histiocytic/dendritic neoplasms, Leukemia 36 (7) (2022) 1703–1719.

[3] M.M. Patnaik, A. Tefferi, Cytogenetic and molecular abnormalities in chronic myelomonocytic leukemia, Blood Cancer J. 6 (2016), e393.

[4] T. Gao, et al., A rare atypical chronic myeloid leukemia BCR-ABL1 negative with concomitant JAK2 V617F and SETBP1 mutations: a case report and literature review, Ther. Adv. Hematol. 11 (2020). p. 2040620720927105.

[5] C.M. Niemeyer, et al., Chronic myelomonocytic leukemia in childhood: a retrospective analysis of 110 cases. European Working Group on Myelodysplastic Syndromes in Childhood (EWOG-MDS), Blood 89 (10) (1997) 3534–3543.

[6] L.E. Side, et al., Mutations of the NF1 gene in children with juvenile myelomonocytic leukemia without clinical evidence of neurofibromatosis, type 1, Blood 92 (1) (1998) 267–272.

[7] A.C. de Vries, C.M. Zwaan, M.M. van den Heuvel-Eibrink, Molecular basis of juvenile myelomonocytic leukemia, Haematologica 95 (2) (2010) 179–182.

[8] A. Ronaghy, et al., Myeloid neoplasms associated with t(3;12)(q26.2;p13) are clinically aggressive, show myelodysplasia, and frequently harbor chromosome 7 abnormalities, Mod. Pathol. 34 (2) (2021) 300–313.

[9] E. Padron, et al., ETV6 and signaling gene mutations are associated with secondary transformation of myelodysplastic syndromes to chronic myelomonocytic leukemia, Blood 123 (23) (2014) 3675–3677.

15 Myelodysplastic neoplasms

Xia Li and Guang Liu
SONORA QUEST LABORATORIES, PHOENIX, AZ, UNITED STATES

Background

Myelodysplastic neoplasms (MDS) consist of a group of blood disorders in which immature blood cells in the bone marrow do not mature or become healthy blood cells. In a patient with MDS, the blood stem cells (immature cells) do not become mature red blood cells, white blood cells, or platelets in the bone marrow. These immature blood cells, called blasts, either die in the bone marrow or die soon after they enter the blood. This reduces the room available for healthy blood cells to form in the bone marrow. When there are fewer healthy blood cells, infection, anemia, or easy bleeding is more likely to occur [1].

There are different types of myelodysplastic neoplasms, and they are diagnosed based on certain changes in the blood cells and bone marrow. These types include MDS with single lineage dysplasia (MDS-SLD), MDS with ring sideroblasts (MDS-RS), MDS with multilineage dysplasia (MDS-MLD), MDS with excess blasts (MDS-EB), MDS with excess blasts and erythroid predominance (MDS-EB/EP), MDS with excess blasts and fibrosis (MDS-EB/F), MDS with isolated del(5q) [MDS del(5q)], MDS unclassifiable (MDS-U), Childhood MDS, and Refractory cytopenia of childhood (MDS-RCC) [2]. An accurate diagnosis of MDS should include persistent and clinically unexplained cytopenia, a morphologic dysplasia of hematopoietic elements, and cytogenetic and/or molecular genetic evidence of clonal hematopoiesis. In addition, morphologic assessment of bone marrow and blood, flow cytometry immunophenotyping, and cytogenetics and molecular genetics testing can help with the classification, prognosis, and risk stratification of the diseases [1]. Finally, next-generation sequencing (NGS) has played an important role in the identification of mutations associated with MDS, making diagnosis and treatment options more accessible. The new WHO classification has included mutations of *SF3B1*, in the definition of a single MDS subtype, MDS with ring sideroblasts [2]. *TP53* mutation is associated with complex karyotype, therapy-related disease, and adverse prognosis in all MDS disease subtypes [3]. The number of genes and mutations in MDS identified by NGS is so large, that these can now be organized into a limited number of subtypes and implicated cellular modes of action such as RNA-splicing factors, epigenetics regulators, cohesin components, transcription factors, DNA damage response, and signal transduction molecules [4,5].

From the clinical and therapeutic points of view, the options for targeted treatment of MDS are rapidly increasing. For example, the drug luspatercept has been used in patients with *SF3B1* mutation, and other specific therapies including drugs targeting aberrant DNA

Cases in Laboratory Genetics and Genomics (LGG) Practice. https://doi.org/10.1016/B978-0-323-99622-8.00025-3

methylation and chromatin remodeling, modulating/activating the immune system to enhance tumor-specific cellular immune responses, etc., are under development. The integration of precision medicine into practice has become a reality for patients with MDS and/or other cancers [6]. It is known that 4%–10% of children and young adults with MDS or AML and 4% of adults with AML carry inherited damaging mutations in cancer susceptibility genes. These genes include *CEBPA*, *GATA2*, *RUNX1*, *ANKRD26*, *BRCA1*, and *MSH6*. The most recently identified genes such as *DDX41*, *SAMD9*, and *SAMD9L* are involved in leukemogenesis and deepen our understanding of the basic biology of MDS/AML [7]. Testing for germline mutations is indicated for families with prior known medical history.

In clinical diagnostic labs, the most common technologies for MDS patients include karyotype, which will define clonal evolution for abnormal cells, FISH, which looks for specific gain or loss as well as chromosome translocations, and NGS, which will elucidate all mutations and fusions associated with MDS. These tests will help to classify disease subtypes, prognosis, risk stratification, and therapy options. In this chapter, a few cases will be illustrated using the diagnostic technologies mentioned above. The results of the testing will aid in the decision-making process for treating MDS patients.

Case 15.1 Myelodysplastic neoplasms with excess blasts-2 (MDS EB-2)

Clinical indication

A 72-year-old female presented with pancytopenia. She reported fatigue, difficulty swallowing, and some chest pain.

Test ordered

- Chromosome analysis of the bone marrow
- NGS Hematology Molecular Profile

Laboratory test performed

Chromosome analysis and NGS methods were described in Chapters 1 and 12.

Test results

Chromosome analysis showed a complex karyotype with three abnormal clones identified. Of the eight cells examined, seven were abnormal, and clonal evolution was evident. All seven abnormal cells exhibited del(5q) and del(7q) (Fig. 15.1.1). Three cells also had a dicentric chromosome involving 1q and 17q with unknown material inserted to 1p and add(2q) (Fig. 15.1.2). Another three cells had additional abnormalities, including gains of chromosomes 1, 4, 6, 9, 14, and 21 (Fig. 15.1.3). The remaining cell appears to be chromosomally normal.

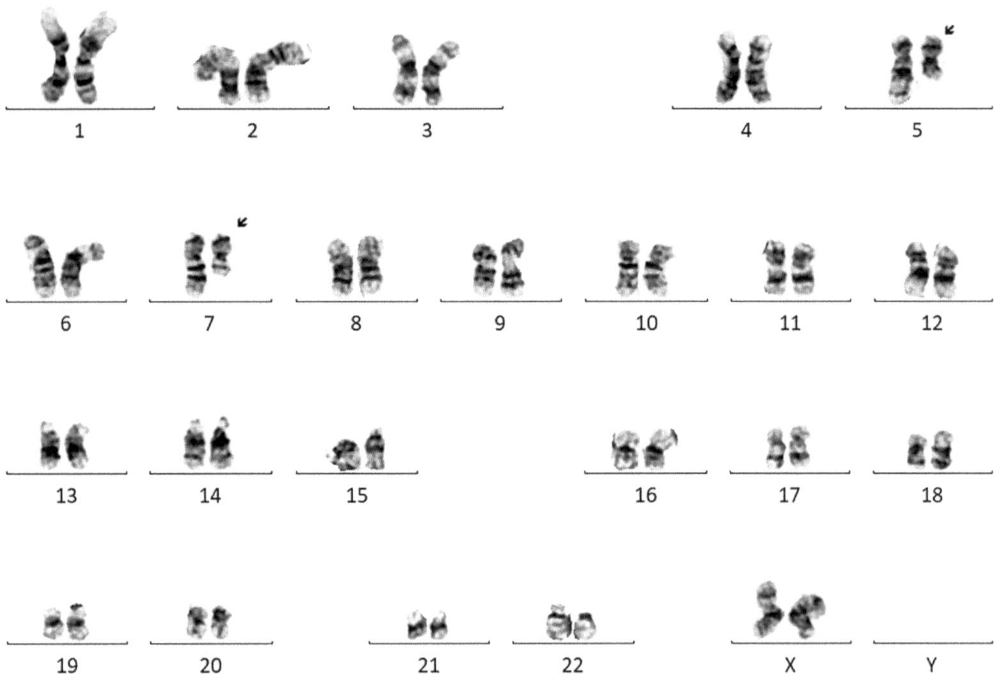

FIG. 15.1.1 Chromosome analysis of the bone marrow showed clone 1 from a complex karyotype including del(5q) and del(7q). ISCN: 46,XX,del(5)(q13q33),del(7)(q21q36)[1]/45,idem,dic(1;17)(p12;p11.2)ins(1;?)(p12;?),add(2)(q33)[3]/52,idem,+1,+4,+6,+9,+14,+21[3]/46,XX[1]

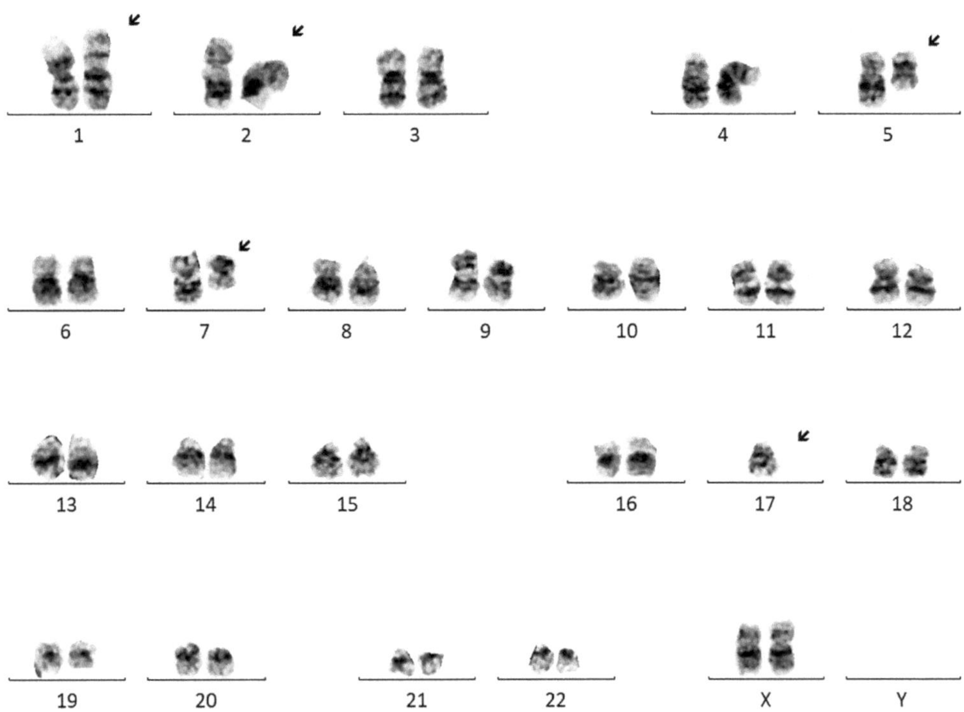

FIG. 15.1.2 Chromosome analysis of the bone marrow showed clone 2 from a complex karyotype including a dicentric chromosome involving 1q&17q with a possible insertion and add(2q) in addition to del(5q) and del(7q).

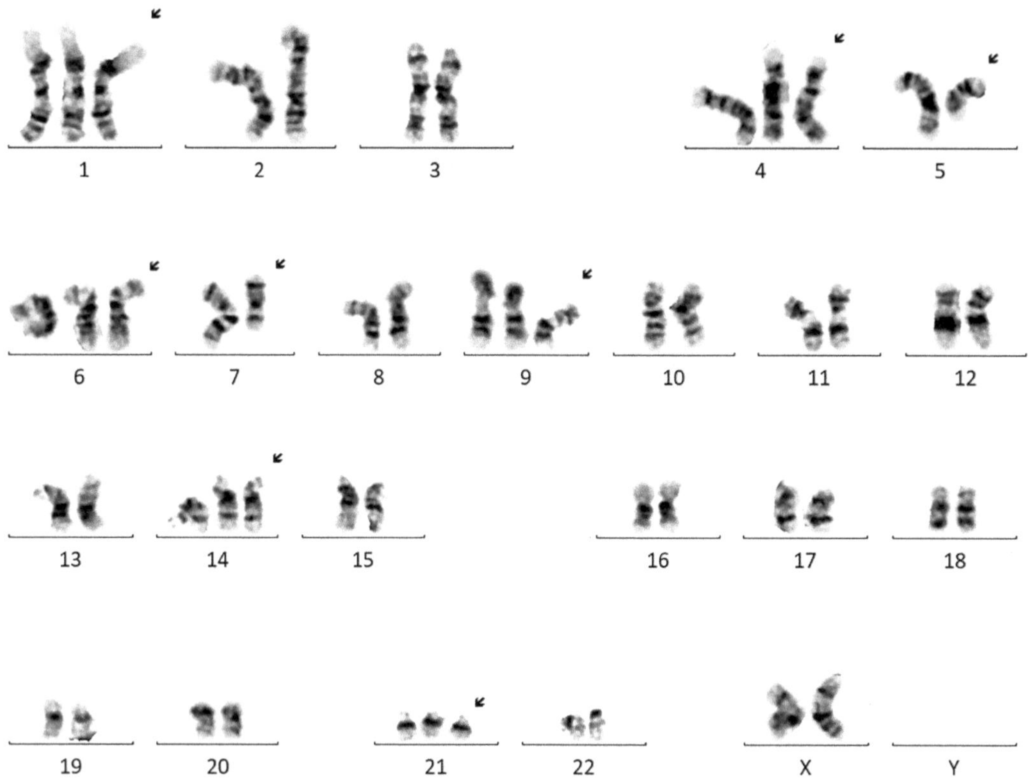

FIG. 15.1.3 Chromosome analysis of the bone marrow showed clone 3 from a complex karyotype including gains of chromosomes 1, 4, 6, 9, 14, and 21 in addition to del(5q) and del(7q).

NGS Hematology Molecular Profile was performed. The results were negative for DNA mutations and RNA fusions (data now shown).

Results with interpretations

del(5q) and del(7q) are commonly seen in myeloid disorders. These results are consistent with a diagnosis of MDS EB-2. Previous karyotype results from bone marrow performed 3 years ago showed del(5q) and del(7q). The accumulation of novel abnormalities in this specimen could represent sampling or could signify an increase in disease aggression. According to the revised International Prognostic Scoring System (IPSS-R) for MDS, this is a complex karyotype with >3 abnormalities. Therefore, it was classified as a very poor prognosis.

Future testing and recommendations

Regular follow-up and clinicopathologic correlation are recommended. Chromosome analysis can be ordered for monitoring disease progression and treatment efficacy in the future.

Case 15.2 Myelodysplastic neoplasms: Refractory anemia with ring sideroblasts (RARS)

Clinical indication

An 84-year-old female presented with normochromic, normocytic anemia, and erythroid hyperplasia. The morphology report showed that she had ring sideroblasts in her bone marrow.

Test ordered

- Chromosome analysis of the bone marrow
- FISH: MDS panel
- NGS Hematology Molecular Profile

Laboratory test performed

Chromosome analysis, FISH, and NGS methods were described previously in Chapters 1 and 12.

Test results

Chromosome analysis exhibited an abnormal karyotype with two distinct clones. Clone 1 with two cells had a balanced translocation involving Xp and 18q (Fig. 15.2.1). Clone 2 with three cells exhibited trisomy 1, a whole-arm translocation involving 1q and 16p, and del(20q) (Fig. 15.2.2).

FISH for MDS panel was performed on interphase nuclei using probes localized to the *D5S721* (5p15.2), *EGR1* (5q23-31), *D7Z1* (7cen), *D7S486* (7q31), *D8Z2* (8cen), and *D20S108* (20q12) gene regions. Two hundred nuclei were examined, and the results were within normal limits for the laboratory's established background rates (Fig. 15.2.3).

Finally, the NGS Hematology Molecular Profile showed results with mutations from *SF3B1*, *ASXL1*, and *BCOR*. The therapeutic drug for *SF3B1* mutation p.(R625L) c.1874G>T was shown in the report (Fig. 15.2.4).

Results interpretations

Of the 25 cells analyzed, 5 were abnormal, and 2 distinct clones were present. The remaining 15 cells appeared to be chromosomally normal.

Clone 1 (two cells) showed a balanced translocation between the short arm of the X chromosome and the long arm of chromosome 18. This clone was observed previously in 2020, indicating a persistent disease.

Clone 2 (three cells) showed a 20q deletion; a derivative chromosome consisting of 1q and 16p resulting in a gain of 1q and loss of 16q. der(1;16)(q10;p10) is a rare but non-random abnormality found in MDS and AML. Collectively, these findings are consistent with persistent MDS.

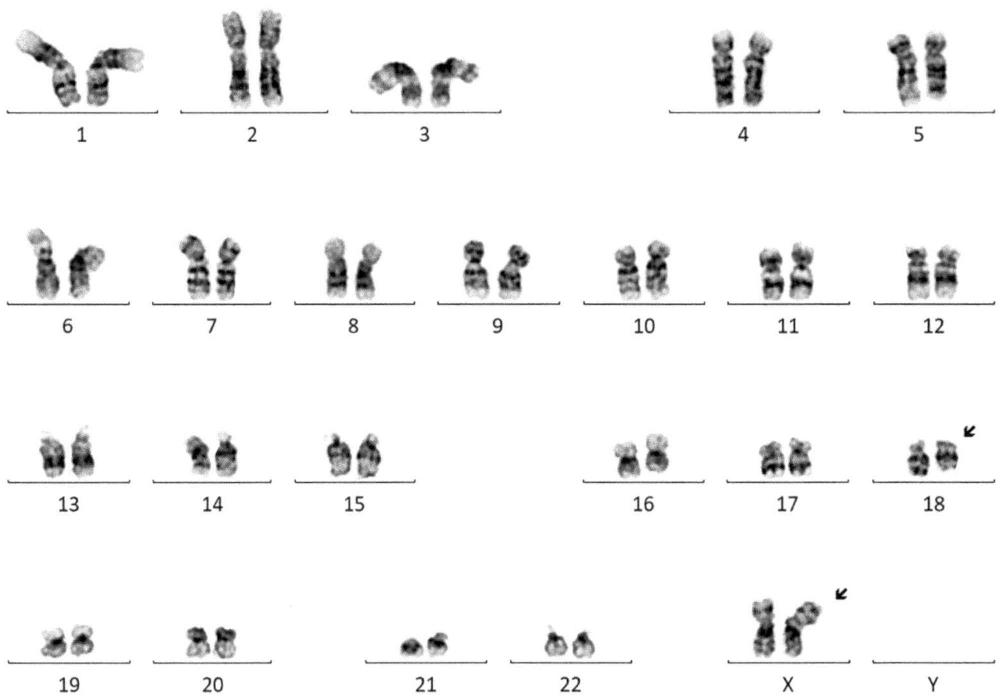

FIG. 15.2.1 The karyotype of this patient showed clone 1 with a balanced translocation involving Xp and 18q. ISCN: 46,XX,t(X;18)(p22.1;q21)[2]/46,XX,+1,der(1;16)(q10;p10),del(20)(q11.2)[3]/46,XX[15]

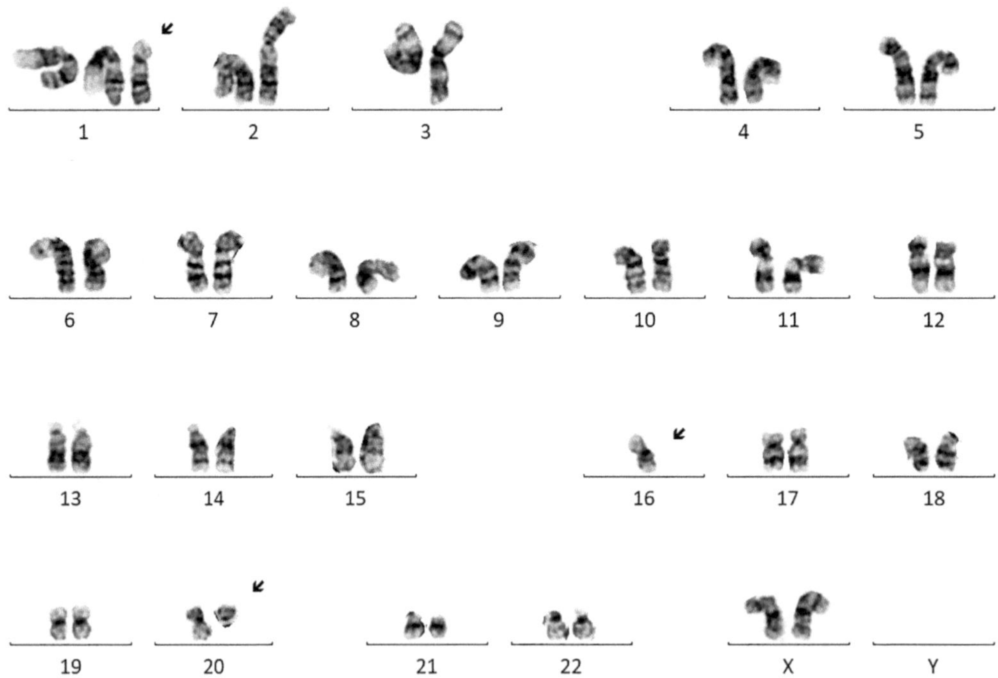

FIG. 15.2.2 The karyotype of this patient showed clone 2 with trisomy 1, a whole-arm translocation involving 1 and 16p, and del(20q).

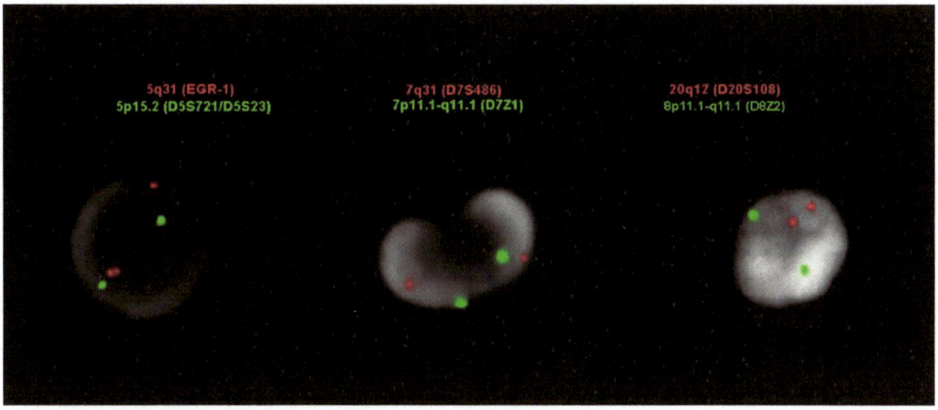

FIG. 15.2.3 FISH for the MDS panel showed negative results for the probes examined. ISCN: nuc ish(D5S721,EGR1)x2[200/200],(CEP7,D7S486)x2[200/200],(D8Z2,D20S108)x2[200/200]

Relevant Biomarkers

Tier	Genomic Alteration	Relevant Therapies (In this cancer type)	Clinical Trials
IA	*SF3B1 p.(R625L) c.1874G>T* **Prognostic significance:** NCCN: Favorable **Diagnostic significance:** None	luspatercept	2
IA	*ASXL1 p.(E635Rfs*15) c.1900_1922delAGAGAGGCGGC CACCACTGCCAT* **Prognostic significance:** NCCN: Poor **Diagnostic significance:** None	None	1
IA	*BCOR p.(Q1653*) c.4957C>T* **Prognostic significance:** NCCN: Poor **Diagnostic significance:** None	None	0

Public data sources included in relevant therapies: **EMA**1, **FDA**2, ESMO, NCCN
Public data sources included in prognostic and diagnostic significance: ESMO, NCCN
Tier Reference: Li et al. *Standards and Guidelines for the Interpretation and Reporting of Sequence Variants in Cancer: A Joint Consensus Recommendation of the Association for Molecular Pathology, American Society of Clinical Oncology, and College of American Pathologists.* J Mol Diagn. 2017 Jan;19(1):4-23.

Variant Details

DNA Sequence Variants

Gene	Amino Acid Change	Coding	Locus	Allele Frequency	Variant Effect
SF3B1	p.(R625L)	c.1874G>T	chr2:198267483	42.00%	missense
ASXL1	p.(E635Rfs*15)	c.1900_1922delAGAGAGGC GGCCACCACTGCCAT	chr20:31022403	4.08%	frameshift Deletion
BCOR	p.(Q1653*)	c.4957C>T	chrX:39913158	5.55%	nonsense

FIG. 15.2.4 NGS results revealed *SF3B1*, *ASXL1*, and *BCOR* mutations. The NCCN recommended therapy was shown.

NGS Hematology Molecular Profile has identified mutations from three genes: *SF3B1*, *ASXL1*, and *BCOR*. According to NCCN guidelines, *SF3B1* p.(R625L) c.1874G>T is associated with a favorable prognosis in MDS, and *ASXL1* p.(E635Rfs*15)c.1900_1922del AGAGAGGCGGCCACCACTGCCAT, and *BCOR* p.(Q1653*) c.4957C>T are associated with poor prognosis. Furthermore, luspatercept is a recommended drug for MDS-RS patients based on ESMO recommendations [8,9].

Future testing and recommendations

Regular follow-up and clinicopathologic correlation are recommended. Chromosome analysis and NGS can be ordered to monitor disease progression and treatment efficacy in the future.

Case 15.3 High-grade myelodysplastic neoplasms (MDS)

Clinical indication

A 70-year-old male had a history of coronary artery disease, congestive heart failure, and diabetes. His labs showed severe anemia.

Test ordered

- Chromosome analysis of the bone marrow
- NGS Hematology Molecular Profile

Laboratory test performed

Chromosome analysis and NGS methods were described previously in Chapters 1 and 12.

Test results

Of the 20 cells examined, 18 were abnormal, and clonal evolution was evident. All abnormal cells exhibited multiple numerical and structural chromosome abnormalities, including an unbalanced translocation involving 5q and 14q resulting in del(5q), del(7q), a pericentric inversion of chromosome 7, del(12p), monosomy 14, del(20q), and a marker chromosome (Fig. 15.3.1); 16 of these also had trisomy 8 (Fig. 15.3.2). The remaining two cells appear to be chromosomally normal.

NGS Hematology Molecular profile revealed *TP53* p.(H214R) c.641A>G, *TP53* p.(Y220C) c.659A>G and *BCOR* p.(W1218*) c.3654_3655delGG mutations (Fig. 15.3.3).

Results with interpretations

Chromosome analysis revealed a complex karyotype with two related clones, and clonal evolution was evident. del(5q), del(7q), del(12p), and del(20q) were all seen in myeloid disorders. These results are consistent with a diagnosis of high-grade myeloid neoplasms. The degree of karyotypic complexity indicates a more aggressive disease process.

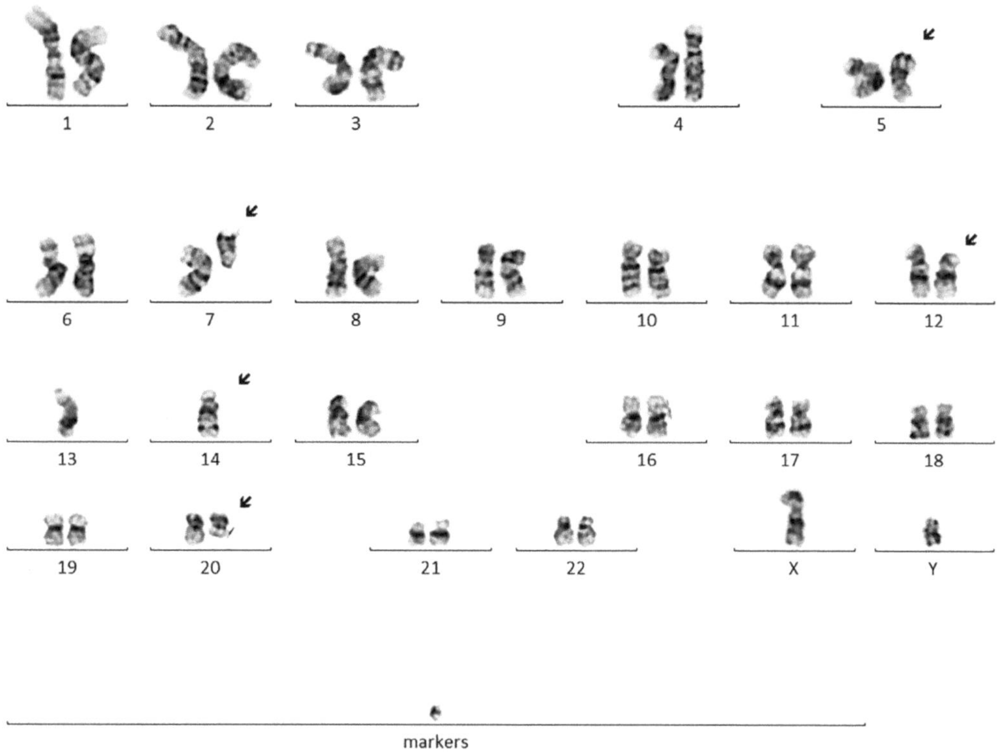

FIG. 15.3.1 The karyotype of this patient showed clone 1 in two cells with an unbalanced translocation involving 5q and 14q, del(7q), a pericentric inversion of chromosome 7, del(12p), monosomy 14, del(20q), and a marker chromosome. ISCN: 46,XY,der(5)t(5;14)(q13;q11.2),der(7)del(7)(q11.2)inv(7)(p11.2q11.2),del(12)(p11.2),-14,del(20)(q11.2q13.3),+mar[2]/46,idem,+8,-mar[16]/46,XY[2]

NGS results with *TP53* and *BCOR* mutations are associated with a poor prognosis in MDS based on the NCCN guideline. These two *TP53* mutations are also resistant to lenalidomide therapy.

Future testing and recommendations

Regular follow-up and clinicopathologic correlation are recommended. Chromosome analysis and NGS can be ordered to monitor disease progression and treatment efficacy in the future.

Case 15.4 Myelodysplastic neoplasms with excess blasts (MDS EB-1) transforming to AML

Clinical indication

A 75-year-old male had a recent diagnosis of myelodysplastic neoplasm with excess blasts (EB-1). The present bone marrow biopsy was performed for restaging in preparation for chemotherapy and possible stem cell transplant. The peripheral blood smear showed

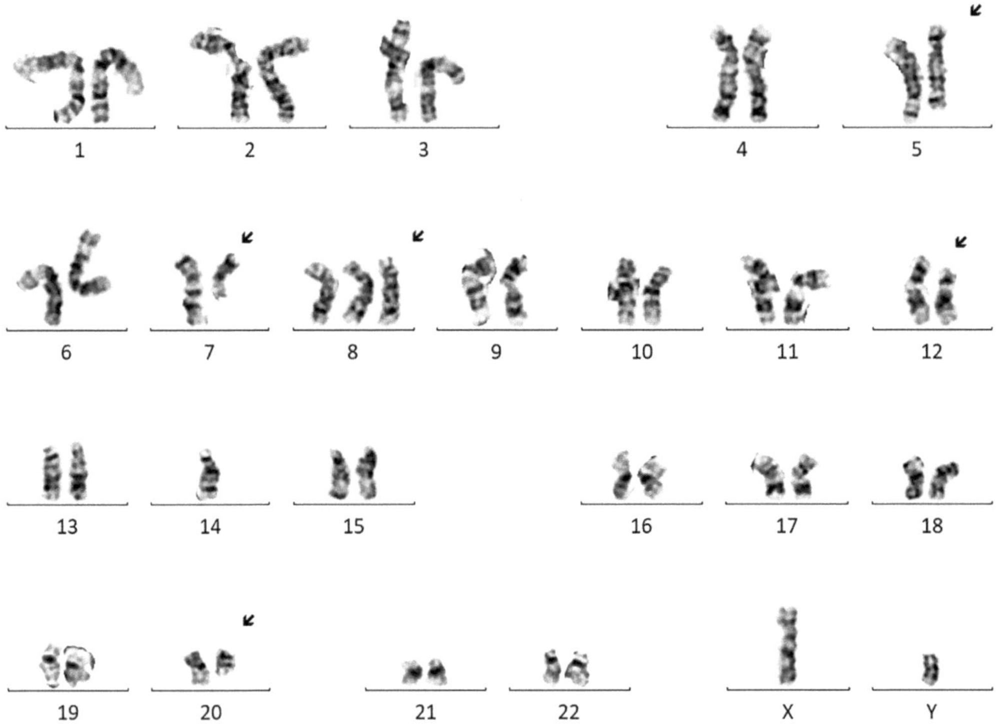

FIG. 15.3.2 The karyotype of this patient showed clone 2 in 16 cells with trisomy 8 and loss of the marker chromosome in addition to the abnormalities seen in clone 1.

pancytopenia with circulating blasts (10%). A flow cytometry analysis on the marrow aspirate showed 35% blasts. The marrow aspirate revealed 43% blasts. The patient was diagnosed with AML rising from MDS.

Test ordered

- Chromosome analysis of the bone marrow
- FISH: AML panel
- NGS Hematology Molecular Profile

Laboratory test performed

Chromosome analysis, FISH, and NGS methods were described previously in Chapters 1 and 12.

Test results

Of the 20 cells examined, 17 exhibited an abnormal chromosome complement with two unrelated clones. Clone 1 with 12 cells showed a monosomy 18 and one to two marker

Relevant Biomarkers

Tier	Genomic Alteration	Relevant Therapies (In this cancer type)	Clinical Trials
IA	*TP53 p.(H214R) c.641A>G, TP53 p.(Y220C) c.659A>G* **Prognostic significance:** NCCN: Poor **Diagnostic significance:** None	None	8
IA	*BCOR p.(W1218*) c.3654_3655delGG* **Prognostic significance:** NCCN: Poor **Diagnostic significance:** None	None	0

Public data sources included in relevant therapies: FDA[1], NCCN, EMA[2], ESMO
Public data sources included in prognostic and diagnostic significance: NCCN, ESMO
Tier Reference: Li et al. *Standards and Guidelines for the Interpretation and Reporting of Sequence Variants in Cancer: A Joint Consensus Recommendation of the Association for Molecular Pathology, American Society of Clinical Oncology, and College of American Pathologists.* J Mol Diagn. 2017 Jan;19(1):4-23.

Alerts informed by public data sources: Contraindicated, Resistance

TP53 p.(H214R) c.641A>G, TP53 p.(Y220C) c.659A>G	lenalidomide

Public data sources included in alerts: FDA[1], NCCN, EMA[2], ESMO

Variant Details

DNA Sequence Variants

Gene	Amino Acid Change	Coding	Locus	Allele Frequency	Variant Effect
TP53	p.(Y220C)	c.659A>G	chr17:7578190	42.80%	missense
TP53	p.(H214R)	c.641A>G	chr17:7578208	44.07%	missense
BCOR	p.(W1218*)	c.3654_3655delGG	chrX:39923052	58.06%	nonsense

FIG. 15.3.3 NGS results revealed *TP53* and *BCOR* mutations. The NCCN indicated a poor prognosis for MDS.

chromosomes (Fig. 15.4.1). Clones 2 with five cells revealed a deletion of 1q (Fig. 15.4.2). The remaining three cells appeared to be chromosomally normal.

FISH for AML panel was performed on interphase nuclei using probes localized to the *D5S721* (5p15.2), *EGR1* (5q31), *D7Z1* (7cen), *D7S486* (7q31), *D8Z2* (8cen), *RUNX1T1* (8q22), *ABL1* (9q34.12), *KMT2A* (11q23), *PML* (15q24.1), *CBFB* (16q22.1), *RARA* (17q21.1), *D20S108* (20q12), *RUNX1* (21q22.3), and *BCR* (22q11) gene regions. Two hundred nuclei were examined, and the results were within normal limits for the laboratory's established background rates (Fig. 15.4.3).

NGS Hematology Molecular Profile has identified *FLT3*-ITD with 35% allele frequency, *NPM1* p.(W288Cfs*12) with 3.5% allele frequency, and *STAG2* p.(M1200Nfs*4) with 75.30% allele frequency (Fig. 15.4.4).

Results with interpretations

Chromosome analysis identified two unrelated clones, not specific to MDS or AML. FISH for AML was normal. NGS findings with FMS-like tyrosine kinase III (*FLT3*)-ITD and *NPM1* were commonly seen in AML.

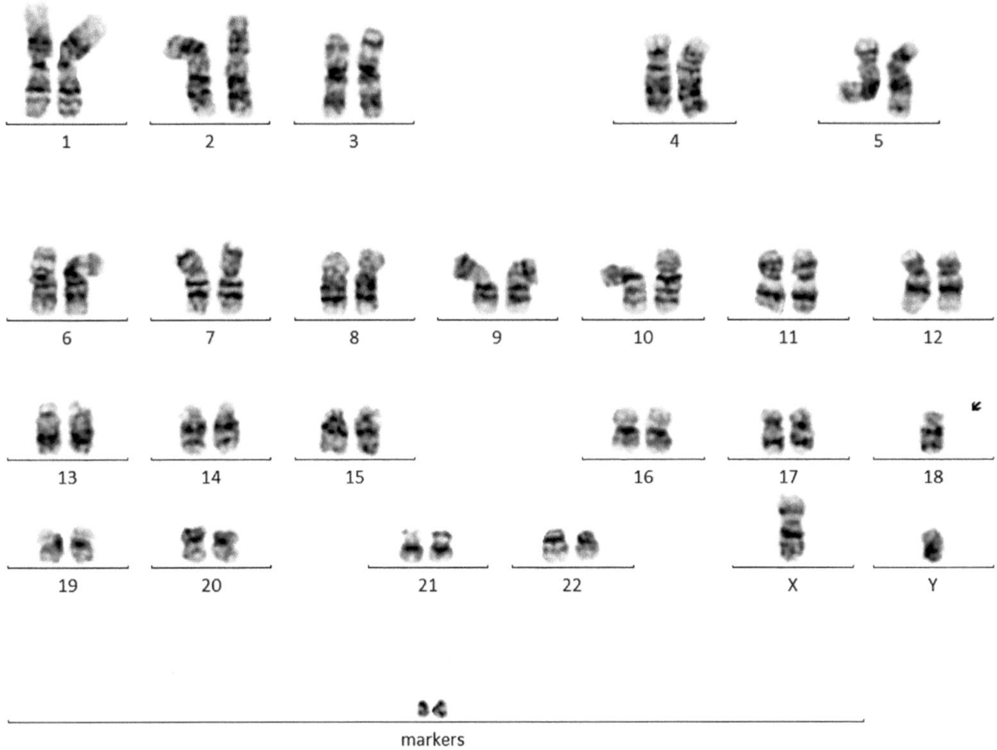

FIG. 15.4.1 Chromosome analysis of the bone marrow showed clone 1 with loss of chromosome 18 and one to two marker chromosomes. ISCN: 46-47,XY,-18,+1-2mar[cp12]/46,XY,del(1)(q21q25)[5]/46,XY[3]

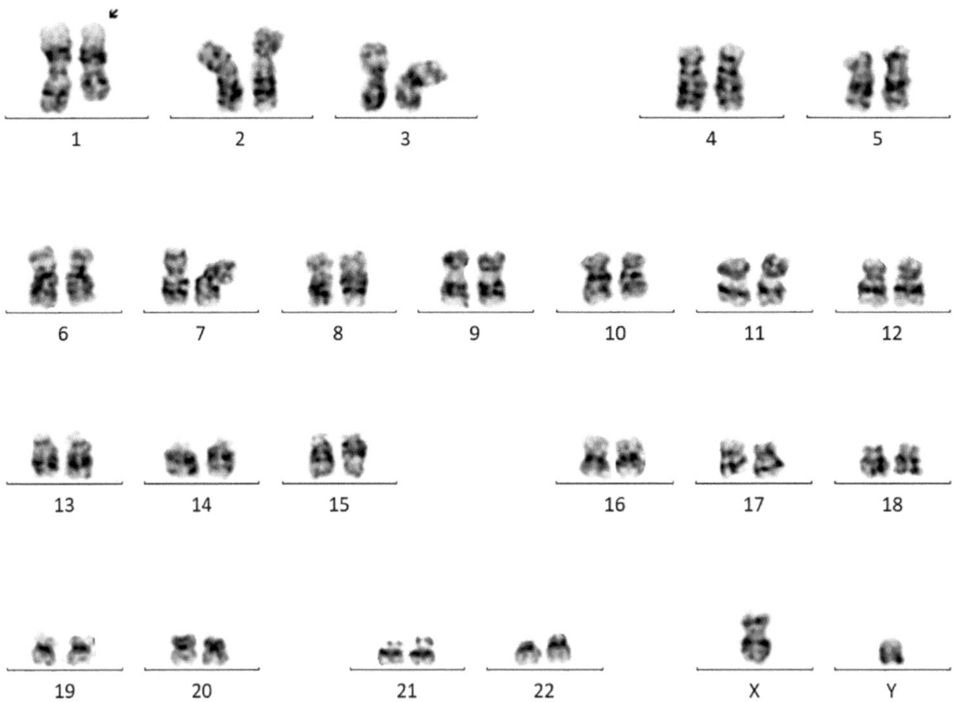

FIG. 15.4.2 Chromosome analysis of the bone marrow showed clone 2 with del(1q).

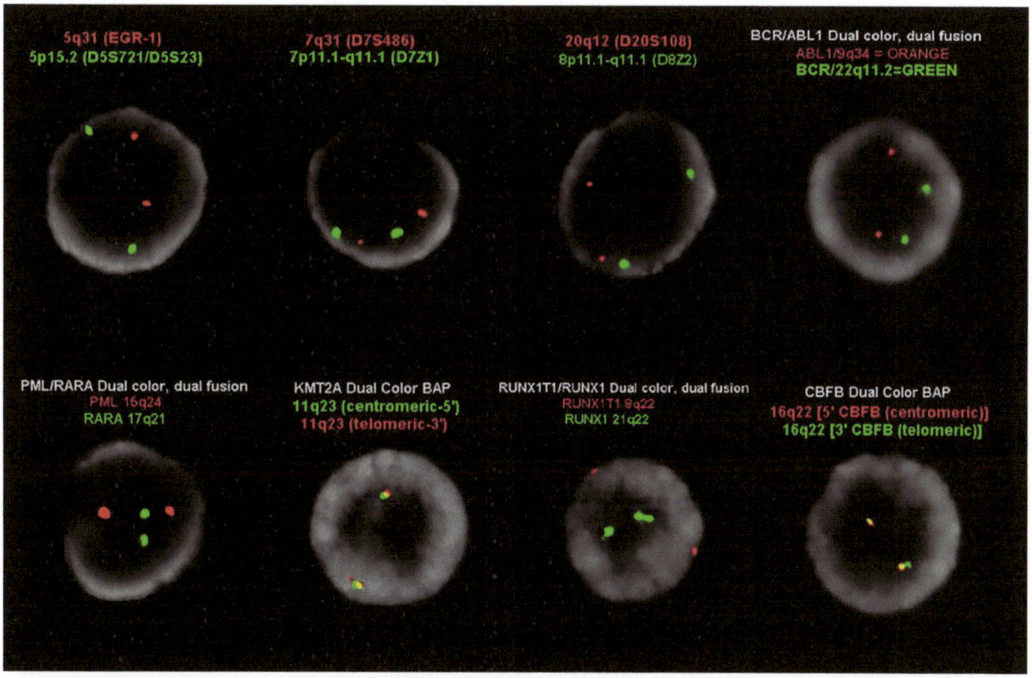

FIG. 15.4.3 FISH for the AML panel showed normal signal patterns for all probes tested. ISCN: nuc ish(D5S721,EGR1)x2 [200/200],(CEP7,D7S486)x2[199/200],(D8Z2,D20S108)x2[199/200],(RUNX1T1,RUNX1)x2[200/200],(ABL1,BCR)x2 [200/200],(KMT2Ax2)[200/200],(PML,RARA)x2[200/200],(CBFBx2)[196/200]

The incidence and impact of *FLT3* in myelodysplastic neoplasm (MDS) and chronic myelomonocytic leukemia (CMML) are unknown. A study from Daver et al. showed that 12/1232 (0.95%) MDS patients had *FLT3* mutations, and the overall survival did not differ significantly for *FLT3*-mutated versus *FLT3*-nonmutated patients [10]. However, *FLT3* mutations have been identified in one-third of AML and have multiple small-molecule inhibitors (midostaurin and gilteritinib) [11]. The patient is currently on consolidation therapy with decitabine, gilteritinib, and venetoclax, and is doing well.

Future testing and recommendations

Regular follow-up and clinicopathologic correlation are recommended. Chromosome analysis and NGS can be ordered to monitor disease progression and treatment efficacy in the future.

Case 15.5 Myelodysplastic neoplasms: Refractory cytopenia with multilineage dysplasia (MDS-RCMD)

Clinical indication

The patient was an 81-year-old male and was noted to be mildly neutropenic and thrombocytopenic. His hemoglobin was 136 g/L, white cell count was 3.1×10^9/L, the neutrophil count was 1.8×10^9/L, and platelet count was 113×10^9/L. Bone marrow biopsy showed

Relevant Biomarkers

Tier	Genomic Alteration	Relevant Therapies (In this cancer type)	Clinical Trials
IA	*FLT3 ITD mutation* Mut/WT Ratio: 0.036 **Prognostic significance:** ELN 2017: Favorable to Intermediate **Diagnostic significance:** None	**gilteritinib** [1,2] **midostaurin + chemotherapy** [1,2] sorafenib + chemotherapy venetoclax + chemotherapy	44
IA	*NPM1 p.(W288Cfs*12) c.863_864insCATG* **Prognostic significance:** ELN 2017: Favorable to Intermediate **Diagnostic significance:** Acute Myeloid Leukemia	None	8
IIC	*STAG2 p.(M1200Nfs*4) c.3596_3597insT* **Prognostic significance:** None **Diagnostic significance:** None	None	3

Public data sources included in relevant therapies: FDA1, NCCN, **EMA**2, ESMO
Public data sources included in prognostic and diagnostic significance: NCCN, ESMO
Tier Reference: Li et al. *Standards and Guidelines for the Interpretation and Reporting of Sequence Variants in Cancer: A Joint Consensus Recommendation of the Association for Molecular Pathology, American Society of Clinical Oncology, and College of American Pathologists.* J Mol Diagn. 2017 Jan;19(1):4-23.

Variant Details

DNA Sequence Variants

Gene	Amino Acid Change	Coding	Locus	Allele Frequency	Variant Effect
NPM1	p.(W288Cfs*12)	c.863_864insCATG	chr5:170837545	35.02%	frameshift Insertion
FLT3	p.(W603_E604insDYVDFREYEYDLKW)	c.1810_1811insACTACGTTGATTTCAGAGAATATGAATATGATCTCAAATGGG	chr13:28608245	3.50%	nonframeshift Insertion
STAG2	p.(M1200Nfs*4)	c.3596_3597insT	chrX:123227884	75.30%	frameshift Insertion

FIG. 15.4.4 NGS results revealed *FLT3*-ITD, *NPM1*, and *STAG2* mutations. The NCCN indicated therapy drug and prognosis information for MDS transforming to AML.

hypercellular marrow with moderate dyserythropoiesis and dysgranulopoiesis, which was consistent with the diagnosis of refractory cytopenia with multilineage dysplasia.

Test ordered

- Chromosome analysis of the bone marrow
- FISH: MDS panel
- NGS Hematology Molecular Profile

Laboratory test performed

Chromosome analysis, FISH, and NGS methods were described in Chapters 1 and 12.

Test results

Firstly, FISH for MDS panel was performed on interphase nuclei using probes localized to the *D5S721* (5p15.2), *EGR1* (5q23-31), *D7Z1* (7cen), *D7S486* (7q31), *D8Z2* (8cen), and *D20S108* (20q12) gene regions. Two hundred nuclei were examined, and the results were positive for deletion 20q in 134/200 (67.0%) of the cells scored (Fig. 15.5.1).

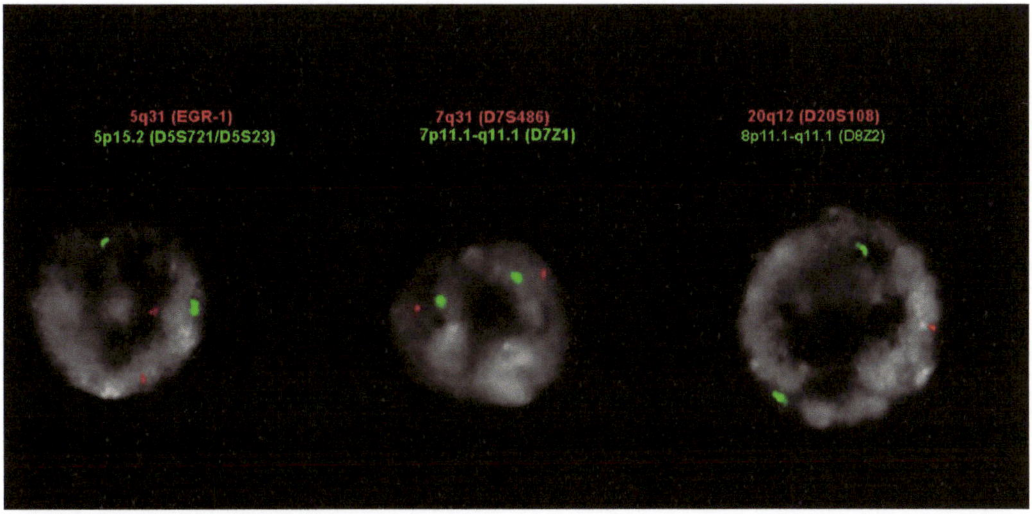

FIG. 15.5.1 FISH for MDS panel was positive for deletion 20q. ISCN: nuc ish(D5S721,EGR1)x2[200/200],(CEP7,D7S486)x2 [199/200],(D8Z2x2,D20S108x1)[134/200]

Secondly, NGS Hematology Molecular Profile was conducted, and the results were negative (data not shown).

Finally, chromosome analysis revealed an isochromosome on the long arm of chromosome 20 with loss of an interstitial segment of 20q11-q13 in 15 of the 20 metaphase cells examined (Fig. 15.5.2). The remaining five cells were chromosomally normal.

Results with interpretations

The finding of an isochromosome of the long arm of chromosome 20 with loss of interstitial material [ider(20q)] was consistent with the concurrent FISH finding. The ider(20q) is a rare but recurrent abnormality in myelodysplastic neoplasm (MDS) and acute myeloid leukemia (AML) [12,13]. The chromosome result is consistent with myelodysplastic neoplasm (MDS) in the context of the pathology report for this patient. The prognosis of patients with ider(20q) seems to be poor compared to patients with del(20q) [14]. In conventional cytogenetics, ider(20q) is a small metacentric chromosomal marker associated with the monosomy of chromosome 20. Therefore, an inexperienced cytogenetic technologist could read a result of 46,XY,-20,+mar for this type of case. Accurate identification of this abnormality is important for the patient's diagnosis, prognosis, and selection of the treatment strategy.

Future testing and recommendations

Regular follow-up and clinicopathologic correlation are recommended. Chromosome analysis and FISH can be ordered to monitor disease progression and treatment efficacy in the future.

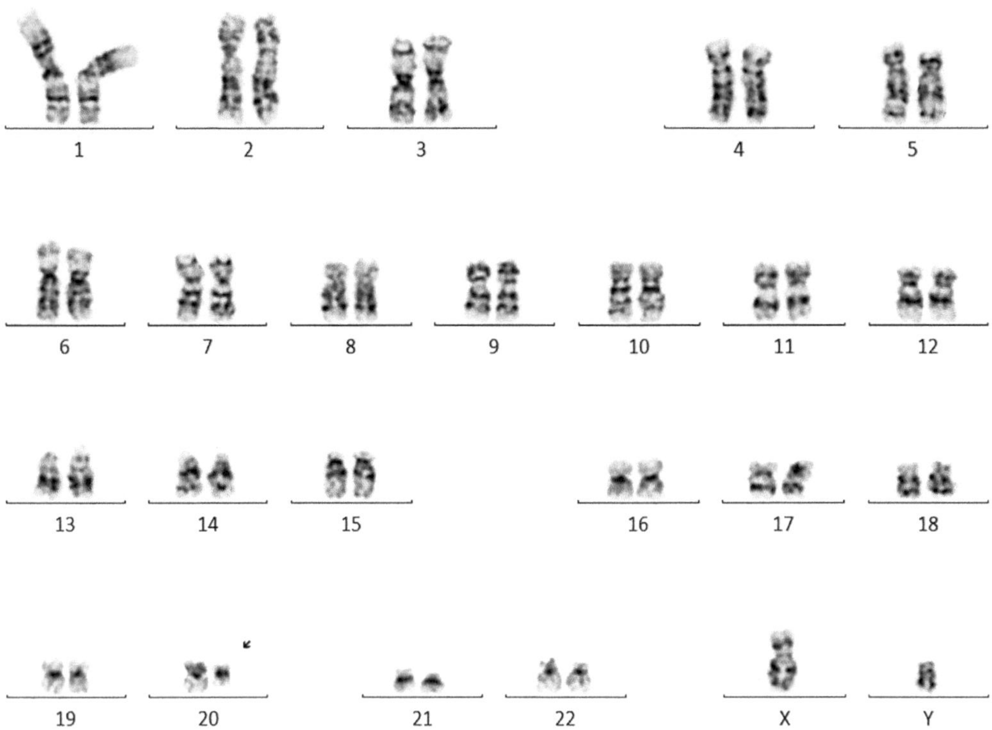

FIG. 15.5.2 The karyotype of bone marrow from the patient showed an ider(20q). ISCN: 46,XY,ider(20)(q10)del(20)(q11q13)[15]/46,XY[5]

Summary of key learning points

- MDS can be diagnosed through morphology, flow, and cytogenetics analysis.
- del(5q), del(7q), del(20q), and +8 are the most common findings in MDS.
- Karyotypes can be very complex in MDS. The results will help with prognosis, risk stratification, and treatment option.
- NGS can identify DNA mutations or fusions. Some of these, such as *SF3B1*, may provide prognosis or therapy information.

References

[1] R.P. Hasserjian, Myelodysplastic syndrome updated, Pathobiology 86 (1) (2019) 7–13.

[2] J.D. Khoury, et al., The 5th edition of the World Health Organization classification of haematolymphoid tumours: myeloid and histiocytic/dendritic neoplasms, Leukemia 36 (7) (2022) 1703–1719.

[3] T. Haferlach, The molecular pathology of myelodysplastic syndrome, Pathobiology 86 (1) (2019) 24–29.

[4] J.A. Kennedy, B.L. Ebert, Clinical Implications of Genetic Mutations in Myelodysplastic Syndrome, J. Clin. Oncol. 35 (9) (2017) 968–974.

[5] M.F. Zahid, et al., Insight into the molecular pathophysiology of myelodysplastic syndromes: targets for novel therapy, Eur. J. Haematol. 97 (4) (2016) 313–320.

[6] M. Tobiasson, A.O. Kittang, Treatment of myelodysplastic syndrome in the era of next-generation sequencing, J. Intern. Med. 286 (1) (2019) 41–62.

[7] J.E. Churpek, Familial myelodysplastic syndrome/acute myeloid leukemia, Best Pract. Res. Clin. Haematol. 30 (4) (2017) 287–289.

[8] P. Fenaux, et al., Myelodysplastic syndromes: ESMO Clinical Practice Guidelines for diagnosis, treatment and follow-up, Ann. Oncol. 25 Suppl 3 (2014) iii57–69.

[9] A.A. van de Loosdrecht, I. Mandac Smoljanovic, EHA endorsement of ESMO clinical practice guidelines for diagnosis, treatment, and follow-up for myelodysplastic syndromes, Hemasphere 6 (3) (2022), e695.

[10] N. Daver, et al., FLT3 mutations in myelodysplastic syndrome and chronic myelomonocytic leukemia, Am. J. Hematol. 88 (1) (2013) 56–59.

[11] V.E. Kennedy, C.C. Smith, FLT3 mutations in acute myeloid leukemia: key concepts and emerging controversies, Front. Oncol. 10 (2020), 612880.

[12] K. Saunders, et al., Isochromosome of a deleted 20q: a rare but recurrent chromosome abnormality in myelodysplastic syndromes, Cancer Genet. Cytogenet. 156 (2) (2005) 154–157.

[13] E. Wong, S. Juneja, Myelodysplastic syndrome with ider(20q) and prominent emperipolesis, Ann. Hematol. 93 (2) (2014) 341–342.

[14] N. Douet-Guilbert, et al., Fluorescence in situ hybridization characterization of ider(20q) in myelodysplastic syndrome, Br. J. Haematol. 143 (5) (2008) 716–720.

16

Acute myeloid leukemia (AML)

Dongbin Xu[a], Guang Liu[b], Xia Li[b], and Hanyin Cheng[a]
[a]HEMATOLOGICS, INC., SEATTLE, WA, UNITED STATES [b]SONORA QUEST LABORATORIES, PHOENIX, AZ, UNITED STATES

Background

Based on the fifth edition of the World Health Organization (WHO) Classification of Haematolymphoid Tumors published in 2022, acute myeloid leukemia (AML) falls into two families: AML with defining genetic abnormalities and AML defined by differentiation [1].

AML can be diagnosed regardless of myeloid blast percentage by detecting defining genetic abnormalities including gene fusions such as *PML*::*RARA*, *RUNX1*::*RUNX1T1*, *MYH11*::*CBFB*, *DEK*::*NUP214*, and *RBM15*::*MRTFA*; gene rearrangements involving *KMT2A*, *MECOM*, and *NUP98* as well as *NPM1* mutation. Please note that the 2022 International Consensus Classification (ICC) of myeloid neoplasms requires a 10% or higher blast percentage along with the defining genetic abnormalities to define AML [2].

AML can also be defined based on findings of *BCR*::*ABL1* fusion or *CEBPA* mutation along with at least 20% blasts. The defining *CEBPA* mutation can be either monoallelic in the basic leucine zipper (bZIP) region of the gene or a biallelic mutation based on the fifth edition of the World Health Organization Classification [1]. In the 2022 ICC classification, biallelic *CEBPA* mutations are not required to make this diagnosis [2].

Acute promyelocytic leukemia (APL) with *PML*::*RARA* fusion is characterized by the predominance of abnormal promyelocytes and fusion of the *RARA* gene on the chromosome at band 17q21.2 with the *PML* gene on the chromosome at band 15q24.1 or with other rare fusion partner genes. APL is frequently associated with coagulopathy and the risk of disseminated intravascular coagulation (DIC) [3].

APL cells may contain Auer rods and kidney-shaped or bilobed nuclei, with high-level myeloperoxidase expression. APL immunophenotypic features include negative expression of CD34 and HLA-DR, with characteristically high side scatter and forward scatter and positive expression of CD13, CD33, and CD117 [4–7].

APL diagnoses can be confirmed by defining *PML*::*RARA* fusion or other variant *RARA* gene fusions. Other chromosomal aberrations including 7q deletion and trisomy 8 can be seen in APL. *FLT3* mutations such as *FLT3*-ITD and *FLT3* p.D835, as well as other mutations, can be found in APL.

APL can be stratified into three risk categories based on WBC and platelet count: high-risk with WBC $\geq 10 \times 10^9$/L, intermediate-risk with WBC $<10 \times 10^9$/L and platelet count

Cases in Laboratory Genetics and Genomics (LGG) Practice. https://doi.org/10.1016/B978-0-323-99622-8.00006-X

$>40 \times 10^9$/L, and low-risk with WBC $<10 \times 10^9$/L and platelet count $\leq 40 \times 10^9$/L [8–11]. Accordingly, the therapy strategy is adjusted based on risk category [12].

Acute myeloid leukemia (AML) with *RUNX1::RUNX1T1* is characterized by a fusion of the *RUNX1* gene on the chromosome 21q22.1 and *RUNX1T1* gene on the chromosome 8q22, resulting from the translocation t(8;21)(q22;q22.1).

RUNX1 and *CBFB* encode components in the transcription factor complex called core binding factor (CBF). The *RUNX1::RUNX1T1* fusion leads to leukemogenesis by blocking myeloid differentiation. Besides t(8;21), other secondary chromosomal aberrations include deletion in chromosome 9q and loss of an X chromosome in females or loss of a Y chromosome in males [13].

Blasts in this AML category are large with abundant basophilic cytoplasm, often containing numerous azurophilic granules and perinuclear halo (clearing), with fewer monocytes and increased eosinophil [14,15]. Its immunophenotype is characterized by high CD34 expression and aberrant expression of CD19 and cytoplasmic CD79a, with abnormal neutrophil maturation. HLA-DR, CD13, myeloperoxidase, and PAX-5 are often expressed, with weak or no expression of CD33. Blasts are usually positive for myeloperoxidase [16–20].

Most patients in this category have higher rates of complete remission and long-term disease-free survival when treated with intensive consolidation therapy (e.g., high-dosage cytarabine) [21,22]. The presence of *KIT* p.D816 in adults correlates with a lower relapse-free survival rate, while hyperdiploidy or presence of del(9q) is associated with longer overall survival [23]. A high ($\geq$2) mutation burden is associated with inferior outcomes [24,25].

Patients who do not achieve major molecular remission after consolidation therapy are at a high risk for relapse and may benefit from allogeneic stem cell transplant therapy [26]. Detection of measurable residual disease either by flow cytometry or PCR-based techniques is associated with lower complete remission rates and shorter survival, even after transplantation [27].

Acute myeloid leukemia with *CBFB::MYH11* fusion is characterized by fusion of the *CBFB* gene on the chromosome 16q22 and the *MYH11* gene on the chromosome 16p13.1, resulting from either a pericentric inversion of a chromosome 16 or a translocation between two homologous chromosome 16.

Increased blasts with myelomonocytic differentiation can be seen in bone marrow smears. Marrow eosinophilia with abnormal eosinophils morphology is a feature of AML with *CBFB::MYH11*.

The flow cytometric study usually identifies two abnormal blast populations: one CD45-dim immature blast population with CD34 expression, and another more mature monocytic CD45-bright population that lacks CD34 expression but expresses monocytic antigens such as CD14, CD64 and lysozyme [1].

Besides inv(16) or t(16;16), other chromosomal abnormalities include trisomy 8, trisomy 21, trisomy 22, and/or deletion of 7q [13]. *CBFB::MYH11* is associated with a favorable prognosis, with long-term survival rate of ~50% in adults [22].

AML with *KMT2A* rearrangement is defined irrespective of the bone marrow blast percentage. More than 80 fusion partners have been identified involving *KMT2A* rearrangements [28]. The most common rearrangement is *KMT2A::MLLT3* fusion resulting from translocation between *KMT2A* on the chromosome 11q23.3 and *MLLT3* on the chromosome 9p21.3. Please note, some *KMT2A* rearrangements can be seen in acute lymphoblastic leukemia (ALL). *KMT2A* rearrangements can occur after treatment with topoisomerase II inhibitors, supporting diagnosis of myeloid neoplasm post cytotoxic therapy. *KMT2A::CREBBP fusion resulting from* a translocation t(11;16)(q23;p13) is highly indicative of such disease [29,30].

KMT2A rearrangements are more often seen in children than adults with AML [31,32]. The rearrangement results in a chimeric protein consisting of N-terminus of *KMT2A* and C-terminus of a partner protein [8].

Overall, *MLLT3, AFDN, ELL,* and *MLLT10* are most commonly detected in *KMT2A*-rearranged AML, resulting from the translocations t(9;11)(p22;q23), t(6;11)(q27;q23), t(11;19)(p13.1;q23), and t(10;11)(p12;q23), respectively [28,32–34].

The immunophenotype of leukemic cells with a *KMT2A* rearrangement varies based on cell lineage. The most often leukemic cells are of the monocytic lineage and express CD33, CD65, CD4, CD15, HLA-DR, and lysozyme. In children, *KMT2A::MLLT3* and *KMT2A::MLLT10* can present as an acute megakaryoblastic leukemia (AMKL) [35–37].

Cases with subpopulations of leukemic cells meeting the criteria for mixed phenotype acute leukemia (MPAL) should not be classified as AML. The finding of a partial tandem duplication (PTD) in *KMT2A,* common in adults AML, does not support the diagnosis of AML with *KMT2A* rearrangement. Some *KMT2A* rearrangements can be cytogenetically cryptic and can be detectable by FISH, RT-PCR, or RNA sequencing.

For AML with *KMT2A* rearrangement, the prognostic impact depends on the fusion partner. In adults, cases with *KMT2A::MLLT3* fusion have an intermediate risk of relapse, whereas other *KMT2A* rearrangements are associated with a poor prognosis [32–34,38]. In children, the prognostic impact of fusion partners varies based on treatment protocol. The *KMT2A* rearrangements involving *ABL1, AFDN, AFF1, MLLT1,* and *MLLT10* are commonly associated with a high risk of relapse, and *ELL, MLLT3, MLLT11,* and *SEPTIN6* with a standard risk [35].

AML with *MECOM* rearrangement is characterized by rearrangements involving the *MDS1* and *EVI1* complex locus (*MECOM*) on the chromosome 3q26.2, resulting from a paracentric inversion inv(3)(q21.3;q26.2) in chromosome 3q or a translocation t(3.3)(q21.3;q26.2) between two homologous chromosome 3. Such rearrangements lead to overexpression of *EVI1* but not *MDS1::EVI1* as well as functional *GATA2* haploinsufficiency due to the *GATA2* enhancer being hijacked to 3q26.2 from its original position 3q21.3 [39–41].

Multilineage dysplasia is common, often presenting small megakaryocytes with nonlubated or bilobed nucleus. Blasts are positive for CD34, CD33, CD13, CD117, and HLA-DR and often positive for CD38, but negative for CD7 [42]. High CD34 expression is more observed in inv(3) cases rather than in t(3;3) cases [43].

GATA2::MECOM is the most frequent rearrangement resulting from either inv(3) or t(3;3), but more than 30 partner genes including *MYC, ETV6, RUNX1* have been described, with some of those being cytogenetically cryptic [25,44–47]. In addition to chromosome karyotyping, fluorescence in situ hybridization (FISH) using break-apart probes for *MECOM* locus can be considered as assessing possible cryptic *MECOM* rearrangements [48,49].

Some myeloid neoplasm-related cytogenetic aberrations such as loss of 5q, loss of 7q, monosomy 7, and/or complex karyotype can be seen along with 3q26.2 rearrangements [50,51].

A hallmark of AML with *MECOM* rearrangement is *EVI1* overexpression in the absence of *MDS1::EVI1* transcript [45].

The rare cases with concurrent *BCR::ABL1* and *MECOM* rearrangement at presentation are best considered as having blast phase CML [1].

AML with *BCR::ABL1* fusion is a de novo AML with *BCR::ABL1* detected at initial diagnosis. More than 20% myeloid blasts in the bone marrow and/or peripheral blood and no evidence of CML are required to make this diagnosis [1].

It is challenging to distinguish AML with *BCR::ABL1* and CML in myeloid blast phase of CML. AML with *BCR::ABL1* is associated with a higher blast percentage (median: 13% vs 47%), a lower absolute basophil count or basophil percentage (median: 0% vs 2.5%), and a lower frequency of splenomegaly (25% vs 65%) [52,53]. AML with *BCR::ABL1* is rare and occurs mainly in adults with a male predominance [54].

About 70%–80% cases of AML with *BCR::ABL1* present p210 transcripts, and 50%–60% cases have additional chromosomal alterations [54]. Cryptic deletions within immunoglobulin and T-cell receptor genes are seen in almost all cases, with frequent microdeletions in *IKZF1* and/or *CDNK2A/B* genes, detectable by array comparative genomic hybridization [55]. These deletions are also seen in the lymphoid blast phase of CML and in MPAL with *BCR::ABL1*, but not in myeloid blast phase of CML [56].

Compared with the myeloid blast phase of CML, AML with *BCR::ABL1* shows a lower myeloid/erythroid ratio (median: 2.0 vs 4.8) and a lower frequency of basophils being higher than 2% (13% vs 53%) in bone marrow [52].

AML with *DEK::NUP214* fusion is characterized by fusion of *DEK* gene on the chromosome 6p22.3 and *NUP214* gene on the chromosome 9q34, resulting from a translocation t(6;9)(p23;q34). This rearrangement can be detected by chromosome analysis, FISH, or molecular approaches. AML with *DEK::NUP214* is rare. It comprises 0.6%–1.7% of AML in children and about 1% of AML in adults [57–60]. *FLT3*-ITD mutation presents in many cases and does not impact overall survival but has faster relapse [61]. AML with *DEK::NUP214* is associated with a poor prognosis and short overall survival [62].

AML with *NUP98* rearrangement is characterized by fusion of *NUP98* on chromosome 11p15.4 with various partner genes. *NUP98::KDM5A* and *NUP98::NSD1* are the two most frequently seen *NUP98* rearrangements, resulting from cytogenetically cryptic translocations t(11;12)(p15;p13) and t(5;11)(q35;p15.5), respectively. *NUP98* rearrangements are frequently found in childhood acute leukemias with erythroid differentiation

[63]. Among various *NUP98* rearrangements, *NUP98::KMD5A* and *NUP98::JARID1A* are most often associated with megakaryoblastic differentiation [64,65].

Assessment of *NUP98* rearrangement is recommended for AML cases with normal karyotype and *FLT3-ITD* mutation and/or *WT1* mutation and for pediatric acute megakaryoblastic leukemia. Loss of *RB1* is particularly associated with *NUP98::KDM5A* [66,67].

AML with *NUP98* rearrangement is associated with poor prognosis. *FLT3*-ITD makes prognosis even worse [67–69]. *NUP98* rearrangement has been found in almost half of pediatric refractory AML cases, suggesting allogeneic stem cell transplantation for *NUP98* rearranged patients who achieve remission [70].

AML with *RBM15::MRTFA* is defined by fusion of *RBM15* on the chromosome 1p13.3 and *MRTFA* on the chromosome 22q13.1 and is characterized by megakaryocytic differentiation. This type of AML accounts for less than 1% of all AML and 10%–12% of pediatric AMKL. It occurs more frequently in infants ($<$6 months) and young children ($\leq$3 years) without trisomy 21 (Down syndrome), with a female predominance [71].

AML with mutated *NPM1* is a myeloid neoplasm characterized by somatic mutations involving the nucleophosmin (*NPM1*) gene. Nearly 1/3 of adult AML patients have *NPM1* -mutation, most of them with normal karyotype at middle age [72–75]. Approximately 20% of AML in patients over 70 years old has *NPM1* mutation [76]. *NPM1* mutation was found in 6%–8% of pediatric AML [77–80].

NPM1 mutation is characterized by a four-base pair insertion in its exon 12. *NPM1* mutations result in a new nuclear export signal (NES) motif and loss of the tryptophan residue at codon 288 or 290, leading to aberrant localization of mutant *NPM1* to the cytoplasm [81–84]. Identification of cup-like nuclear morphology in more than 10% of blasts is highly specific for AML with *NPM1* mutation [85].

NPM1 mutation is an AML-defining genetic lesion. The finding of *NPM1* mutation overrides other myelodysplastic or myeloproliferative morphologic features and defines this AML diagnosis [86,87]. *NPM1* mutations are mostly ($>$99%) restricted in exon 12 [88]. More than 85% of these cases present a normal karyotype [73,89]. Patients with this diagnosis typically experienced a good response to induction therapy [38,73]. However, higher *NPM1* mutant expression is associated with increased risk of relapse and death post therapy [90–92].

The prognosis of *NPM1* mutated AML is influenced by other mutations. *NPM1* mutation positive AML without *FLT3*-ITD or with low *FLT3*-ITD (VAF $<$ 0.5) is classified as the favorable-risk category, while *NPM1* mutation with high *FLT3*-ITD (VAF $\geq$0.5) is intermediate risk [38], wild-type *NPM1* without *FLT3*-ITD or with low *FLT3*-ITD (VAF $<$ 0.5) is intermediate risk, and wild-type *NPM1* high *FLT3*-ITD (VAF $\geq$0.5) as adverse risk [93–98]. Triple positive for *NPM1*, *FLT3*-ITD, and *DNMT3A* mutations is associated with a particularly poor outcome [99–101]. Cytogenetic abnormalities with poor prognosis override favorable risk of *NPM1* mutation without *FLT3*-ITD or with low *FLT3*-ITD [102].

AML with *CEBPA* mutation is characterized by biallelic *CEBPA* mutation (bi*CEBPA*) or single *CEBPA* mutation in the basic leucine zipper (bZIP) region of the gene (smbZIP-*CEBPA*). More than 20% myeloid blasts in bone marrow or blood are required for this diagnosis. AML with *CEBPA* mutation accounts for ~5% of AML in children and 5%–11% in adults [103–105].

Among cases with bi*CEBPA*, up to 10% of patients have a germline N-terminal mutation [106]. Therefore, bi*CEBPA* may reflect a germline *CEBPA* variant, indicating referral to genetic counseling.

Please note, bi*CEBPA* was previously required for this AML subtype due to the association with a favorable prognosis [107–109]. Now, the bZIP single mutation is being included in this category with a favorable prognosis [103–105].

AML, myelodysplasia-related (AML-MR) can be defined with at least 20% blasts and carrying specific cytogenetic and molecular abnormalities associated with MDS, de novo or secondary. These cytogenetic abnormalities include complex karyotype with three or more abnormalities; 5q deletion or loss of 5q; monosomy 7, 7q deletion, or loss of 7q; 11q deletion; 12p deletion or loss of 12p; monosomy 13 or 14q deletion; 17p deletion or loss of 17p; isochromosome 17q; and isodicentric chromosome X idic(X)(q13) [1]. The recognized MDS-associated molecular abnormalities include mutations of eight genes: *SRSF2, SF3B1, U2AF1, ZRSR2, ASXL1, EZH2, BCOR*, and *STAG2* [110,111].

AML-MR is associated with a poor prognosis [112,113]. AML-MR patients often have *TP53* mutations and complex karyotype, indicative of a worse prognosis [114].

AML with other defined genetic alterations includes emerging or provisional AML subtypes with distinct genetic features. This diagnosis requires ≥20% myeloid blasts in bone marrow and/or blood. This category includes several subtypes: AML with *CBFA2T3::GLIS2,* AML with *KAT6A::CREBBP,* AML with *FUS::ERG,* AML with *MNX1::ETV6, and* AML with *NPM1::MLF1.*

CBFA2T3::GLIS2 is exclusively seen in childhood AML, particularly in infants, and often associated with non-Down-syndrome related AMKL [36,115,116]. Cases with this fusion often present RAM immunophenotype (strong CD56 expression and lack of HLA-DR and CD38) and poor prognosis [117–119]. AML with *KAT6A::CREBBP* is rare, but mostly seen in the neonatal period [120,121]. AML with *FUS::ERG* is mainly found in young adults [11,122]. AML with *MNX1::ETV6* accounts for less than 1% of overall pediatric AML, but up to one-third of infant (0–2 years old) AML [123,124].

Other rare fusions in this category include *RARG* (12q13) fusions such as *CPSF6::RARG,* t(12;12)(q13;q15) [125,126], *NUP98::RARG,* t(11;12)(p15;q13) [127]; *PML::RARG,* t(12;15)(q13;q22) [128], and *HNRNPC::RARG* [129]. These fusions are associated with APL-like features and resistance to all-trans retinoic acid-based therapy.

Besides AML with defining genetic abnormalities, AML can also be diagnosed without defined genetic abnormalities by presenting more than 20% of myeloblasts in peripheral blood or bone marrow based on morphology analysis. **AML defined by differentiation** includes several types distinguished by differentiation status. Details of these AML types can be found in the updated fifth edition of WHO classification [1]. In addition, **myeloid sarcoma** represents a unique tissue-based manifestation of AML or transformed MDS, MDS/MPN, or MPN.

In the genetic diagnostic lab, karyotyping, FISH, PCR, or next-generation sequencing (NGS) is used to identify chromosome abnormalities including gene rearrangements and mutations. The information from these testing will help patients with targeted therapy, predicting prognosis, and risk stratification. In this chapter, over 10 cases will be illustrated on how patients will benefit from these genetic testing.

Case 16.1 Acute myeloid leukemia (AML) with t(3;21)(q26.2; q22)/*RUNX1::MECOM* fusion

Clinical indication

This is a 70-year-old man with leukocytosis, anemia, and thrombocytopenia. The new diagnosis of acute myeloid leukemia with 31.89% blasts was made based on flow and morphology from the bone marrow.

Test ordered

- Chromosome analysis of the bone marrow
- FISH: AML panel
- NGS Hematology Molecular Profile

Laboratory test performed

Chromosome analysis, FISH, and NGS methods were described in Chapters 1 and 12.

Test results

All 20 metaphase cells analyzed exhibited a translocation between the long arms of chromosomes 3 and 21 (Fig. 16.1.1).

FISH for AML panel was performed on interphase nuclei using probes localized to the *D5S721* (5p15.2), *EGR1* (5q31), *D7Z1* (7cen), *D7S486* (7q31), *D8Z2* (8cen), *RUNX1T1* (8q22), *ABL1* (9q34.12), *KMT2A* (11q23), *PML* (15q24.1), *CBFB* (16q22.1), *RARA* (17q21.1), *D20S108* (20q12), *RUNX1* (21q22.3), and *BCR* (22q11) gene regions (supplied by Abbott Molecular, Inc.). Two hundred nuclei were analyzed for each probe, the results demonstrated three *RUNX1* signals in 200/200 (100%) of the cells scored (Fig. 16.1.2).

NGS Hematology Molecular Profile was conducted on bone marrow, and the results showed a *RUNX1::MECOM* (also called *EVI1*) fusion and several mutations including *ASXL1*, *NRAS*, and *WT1*. The relevant therapies for *ASXL1* mutations were also shown in the report (Fig. 16.1.3).

Results with interpretations

The results from chromosome analysis, FISH, and NGS were concordant for this patient. The t(3;21)(q26.2;q22) translocation, resulting in the fusion of the *MECOM* gene at 3q26

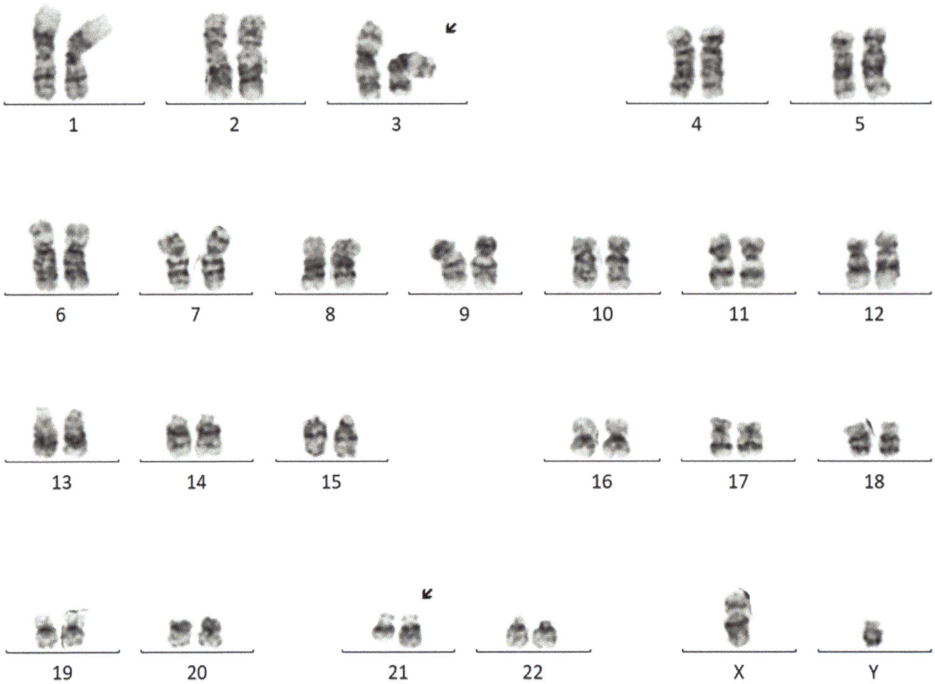

FIG. 16.1.1 The karyotype showed a translocation between chromosomes 3 and 21. ISCN: 46,XY,t(3;21)(q26.2;q22)[20]

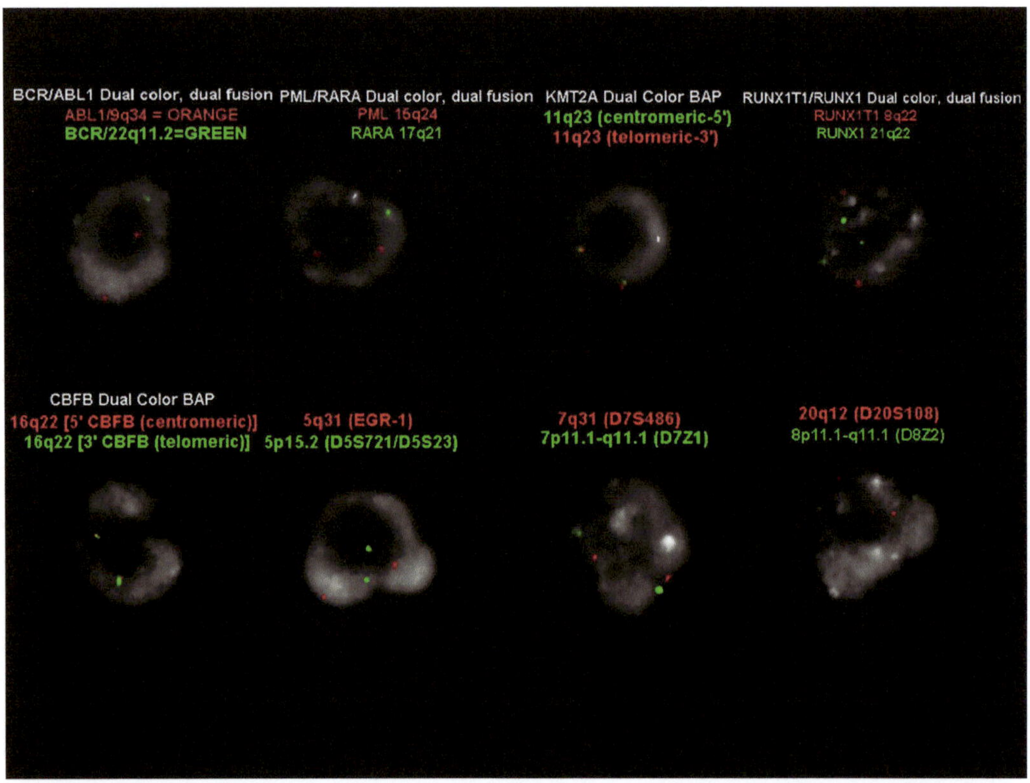

FIG. 16.1.2 FISH with AML panel was positive for three signals of *RUNX1*. ISCN: nuc ish(D5S721,EGR1)x2[200/200],(CEP7,D7S486)x2[200/200],(D8Z2,D20S108)x2[198/200],(RUNX1T1x2,RUNX1x3)[200/200],(ABL1,BCR)x2[200/200],(KMT2Ax2)[200/200],(PML,RARA)x2[200/200],(CBFBx2)[200/200]

Relevant Biomarkers

Tier	Genomic Alteration	Relevant Therapies (In this cancer type)	Clinical Trials
IA	*ASXL1 p.(E635Rfs*61) c.1900_1926delAGAGAGGCGGC CACCACTGCCATCGGAinsCGGAG , ASXL1 p.(E635Rfs*15) c.1900_1922delAGAGAGGCGGC CACCACTGCCAT* **Prognostic significance:** ELN 2017: Adverse	allogeneic stem cells azacitidine cytarabine cytarabine + daunorubicin cytarabine + daunorubicin + etoposide cytarabine + etoposide + idarubicin cytarabine + fludarabine + idarubicin + filgrastim cytarabine + idarubicin cytarabine + mitoxantrone decitabine gemtuzumab ozogamicin + chemotherapy venetoclax + chemotherapy	5
IIC	*NRAS p.(Q61H) c.183A>T*	None	1
IA	*RUNX1-MECOM fusion* **Diagnostic significance:** Acute Myeloid Leukemia	None	1
IIC	*WT1 p.(R374Gfs*18) c.1120_1120delCinsGGGGTATG*	None	1

Public data sources included in relevant therapies: FDA1, NCCN, **EMA**2, ESMO
Public data sources included in prognostic and diagnostic significance: NCCN, ESMO
Tier Reference: Li et al. *Standards and Guidelines for the Interpretation and Reporting of Sequence Variants in Cancer: A Joint Consensus Recommendation of the Association for Molecular Pathology, American Society of Clinical Oncology, and College of American Pathologists.* J Mol Diagn. 2017 Jan;19(1):4-23.

Prevalent cancer biomarkers without relevant evidence based on included data sources
RUNX1-RPL22 fusion

Variant Details

DNA Sequence Variants

Gene	Amino Acid Change	Coding	Locus	Allele Frequency	Variant Effect
NRAS	p.(Q61H)	c.183A>T	chr1:115256528	34.00%	missense
WT1	p.(R374Gfs*18)	c.1120_1120delCinsGGGGTATG	chr11:32417947	34.57%	frameshift Block Substitution
ASXL1	p.(E635Rfs*15)	c.1900_1922delAGAGAGGCGGCCACCACTGCCAT	chr20:31022402	12.59%	frameshift Deletion
ASXL1	p.(E635Rfs*61)	c.1900_1926delAGAGAGGCGGCCACCACTGCCATCGGAinsCGGAG	chr20:31022415	9.18%	frameshift Block Substitution

Gene Fusions (RNA)

Genes	Variant ID	Locus
RUNX1-MECOM	RUNX1-MECOM.R4M2	chr21:36206707 - chr3:169099312
RUNX1-RPL22	RUNX1-RPL22.R3R2	chr21:36231771 - chr1:6257733

FIG. 16.1.3 NGS showed the fusion of *RUNX1::MECOM* and three gene mutations including *NRAS, WT1*, and *ASXL1*.

and the *RUNX1* gene at 21q22, is a rare and recurrent chromosomal abnormality exhibited in <1% of AML and myelodysplastic syndrome (MDS), primarily in therapy-related MDS/AML or in the blastic crisis phase of chronic myelogenous leukemia [130,131]. This abnormality typically activates the *MECOM* gene, which has been implicated in the pathogenesis of AML and is associated with a poor prognosis [132,133].

Future testing and recommendations

Chromosome analysis, FISH, and NGS on bone marrow can be ordered to monitor the disease progression and treatment efficacy in the future.

Case 16.2 Acute myeloid leukemia (AML) with t(6;9)(p22;q34)/*DEK::NUP214* fusion

Clinical indication

The patient was a previously healthy 19-year-old male with 1 week of fevers, chills, night sweats, and vomiting. Peripheral blood smear showed 42% blasts by morphology. By flow cytometry, there was an increased blast population comprising 55% of all events, 36% of all events were CD34 positive and express CD45, C38, CD33 (dim), CD117, CD13, HLA-DR, CD64 (subset), CD11b (dim), and CD11c (dim). There was a partial aberrant expression of CD7 (dim). The blasts did not show expression CD5, CD19, CD10, CD20, CD2, CD3, CD4, CD8, CD14, CD56, MPO, TdT, or CD3. Collectively, these findings demonstrated an early myeloid population with monocytic features and were consistent with a diagnosis of AML.

Test ordered

- Chromosome analysis of the bone marrow
- FISH: AML panel
- NGS Hematology Molecular Profile

Laboratory test performed

Chromosome analysis, FISH, and NGS methods were described in Chapters 1 and 12.

Test results

Chromosome analysis revealed a translocation t(6;9)(p22;q34) resulting in the fusion of the *DEK* gene at 6p22 with the *NUP214* gene at 9q34 in all 20 metaphase cells examined (Fig. 16.2.1)

FISH for AML panel was performed on interphase nuclei using probes localized to the *D5S721* (5p15.2), *EGR1* (5q31), *D7Z1* (7cen), *D7S486* (7q31), *D8Z2* (8cen), *RUNX1T1* (8q22), *ABL1* (9q34.12), *KMT2A* (11q23), *PML* (15q24.1), *CBFB* (16q22.1), *RARA* (17q21.1), *D20S108* (20q12), *RUNX1* (21q22.3), and *BCR* (22q11) gene regions (supplied

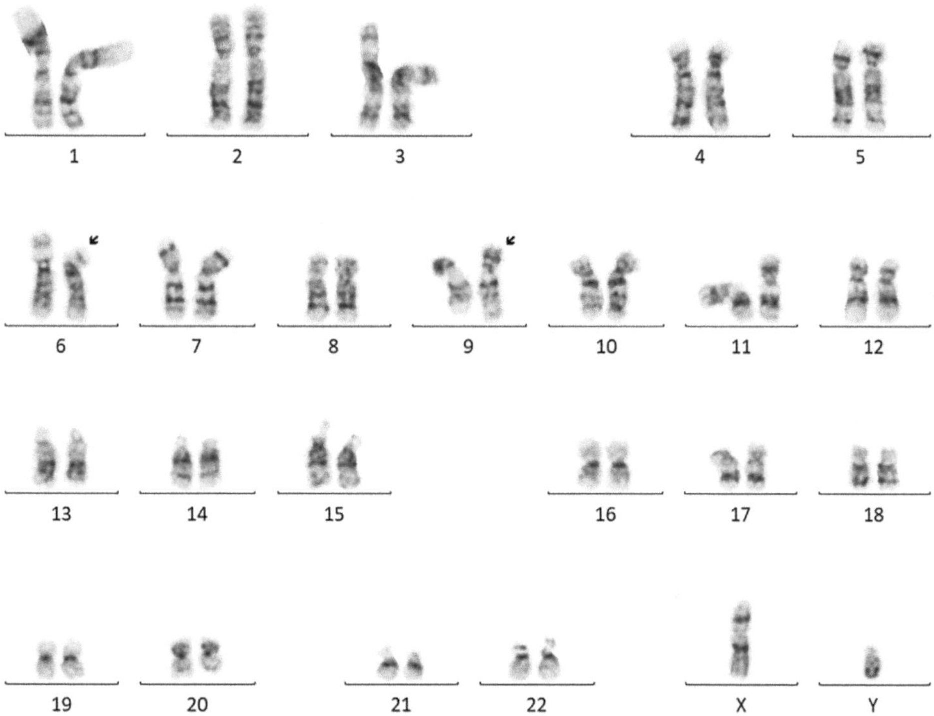

FIG. 16.2.1 The karyotype of bone marrow from the patient showed a t(6;9) translocation. ISCN: 46,XY,t(6;9)(p22;q34)[20]

by Abbott Molecular, Inc.). Two hundred nuclei were analyzed for these probes, and the results were negative for the AML panel (Fig. 16.2.2).

NGS Hematology Molecular Profile was conducted, and the results showed *DEK::NUP214* fusion and *FLT3*-ITD mutation (Fig. 16.2.3).

Results with interpretations

The t(6;9)(p22;q34) translocation identified by chromosome analysis and consistent with the concurrent NGS finding (*DEK::NUP214* fusion) is a recurrent abnormality observed in AML and has been associated with a poor clinical prognosis. The t(6;9) occurs in 0.7%–1.8% of AML cases and is predominantly detected in younger adults. This is a distinct entity in the World Health Organization (WHO) classification system and has been studied extensively [1,57,134]. AML with t(6;9) is an AML with $\geq$ 20% peripheral blood or bone marrow blasts with or without monocytic features. This is also concordant with this patient's pathologic findings. *FLT3*-ITD mutations are very common in AML with t(6;9), occurring in approximately 70% of cases. The presence of *FLT3*-ITD does not appear to negatively impact the outcome compared with the absence of *FLT3*-ITD [58]. The t(6;9) may be overlooked in chromosome analysis due to subtle chromosomal structural

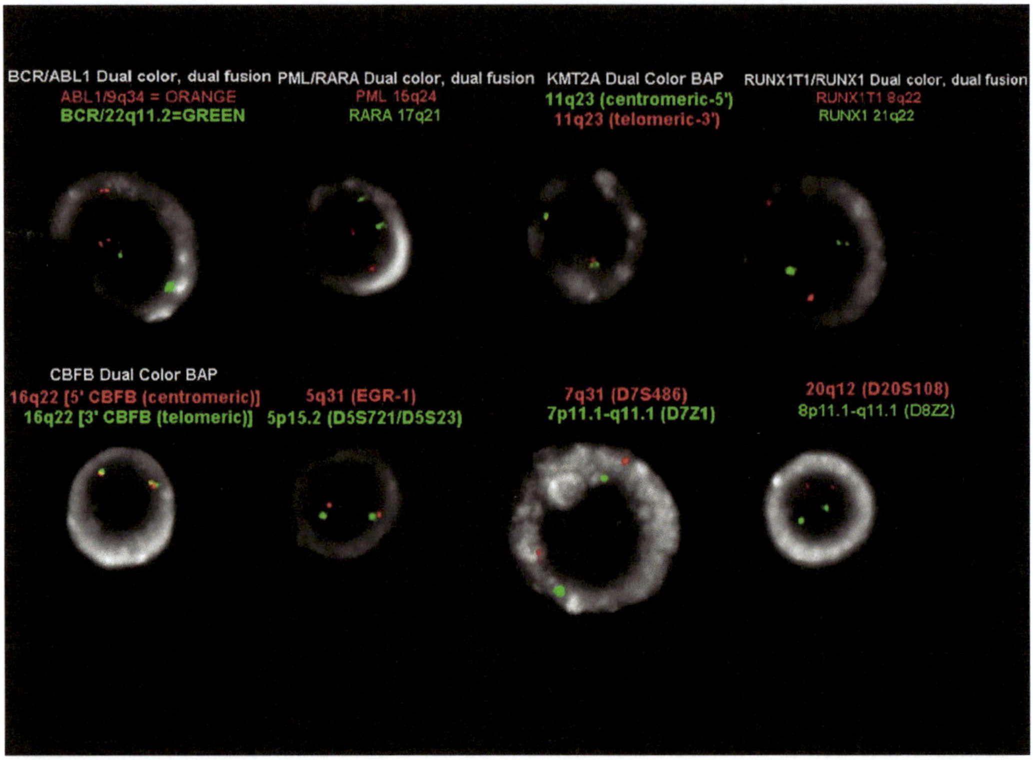

FIG. 16.2.2 FISH was negative for the AML panel. ISCN: nuc ish(D5S721,EGR1)x2[200/200],(CEP7,D7S486)x2[200/200],(D8Z2,D20S108)x2[200/200],(RUNX1T1,RUNX1)x2[200/200],(ABL1,BCR)x2[200/200],(KMT2Ax2)[200/200],(PML,RARA)x2[200/200],(CBFBx2)[200/200]

rearrangement (http://atlasgeneticsoncology.org/haematological/1014/t(6;9)(p22;q34)). Also, the FISH AML panel usually does not include this translocation. Therefore, NGS or RT-PCR testing is important to identify this rearrangement.

Future testing and recommendations

Chromosome analysis and NGS on bone marrow can be ordered to monitor the disease progression and treatment efficacy in the future.

Case 16.3 Acute myeloid leukemia (AML) with t(9;11)(p21;q23)/*KMT2A::MLLT3* fusion

Clinical indication

A 21-year-old female complained of nosebleeds and was found to have a population of immature monocytic cells in her peripheral blood. The bone marrow biopsy morphology report showed a predominant population of monoblasts. By flow cytometry, these cells

Relevant Biomarkers

Tier	Genomic Alteration	Relevant Therapies (In this cancer type)	Clinical Trials
IA	*FLT3 ITD mutation* **Prognostic significance:** ELN 2017: Intermediate to Adverse	**gilteritinib** [1, 2] **midostaurin + chemotherapy** [1, 2] azacitidine cytarabine + daunorubicin cytarabine + daunorubicin + etoposide cytarabine + etoposide + idarubicin cytarabine + fludarabine + idarubicin + filgrastim cytarabine + idarubicin cytarabine + mitoxantrone decitabine gemtuzumab ozogamicin + chemotherapy sorafenib sorafenib + chemotherapy venetoclax + chemotherapy	28
IA	*DEK-NUP214 fusion* **Prognostic significance:** ELN 2017: Adverse **Diagnostic significance:** Acute Myeloid Leukemia	allogeneic stem cells azacitidine cytarabine cytarabine + daunorubicin cytarabine + daunorubicin + etoposide cytarabine + etoposide + idarubicin cytarabine + fludarabine + idarubicin + filgrastim cytarabine + idarubicin cytarabine + mitoxantrone decitabine gemtuzumab ozogamicin + chemotherapy midostaurin + chemotherapy venetoclax + chemotherapy	10

Public data sources included in relevant therapies: FDA1, NCCN, **EMA**2, ESMO
Public data sources included in prognostic and diagnostic significance: NCCN, ESMO
Tier Reference: Li et al. *Standards and Guidelines for the Interpretation and Reporting of Sequence Variants in Cancer: A Joint Consensus Recommendation of the Association for Molecular Pathology, American Society of Clinical Oncology, and College of American Pathologists.* J Mol Diagn. 2017 Jan;19(1):4-23.

Variant Details

DNA Sequence Variants

Gene	Amino Acid Change	Coding	Locus	Allele Frequency	Variant Effect
FLT3	p.(V579_E598dup)	c.1792_1793insAGGTACAG GTGACCGGCTCCTCAGATA ATGAGTACTTCTACGTTGA TTTCAGAGAATATG	chr13:28608263	1.40%	nonframeshift Insertion

Gene Fusions (RNA)

Genes	Variant ID	Locus
DEK-NUP214	DEK-NUP214.D9N18	chr6:18236683 - chr9:134034770

FIG. 16.2.3 NGS showed the fusion of *DEK::NUP214* and *FLT3*-ITD mutation.

showed moderate side scatter and were positive for CD4, CD56, CD33, CD13, CD117, CD123, CD45 (dim), and HLA-DR while negative for B-cell markers, MPO, and cCD3. The collective findings indicate acute myeloid leukemia with monocytic differentiation.

Test ordered

- Chromosome analysis of the bone marrow
- FISH: AML panel
- NGS Hematology Molecular Profile

Laboratory test performed

Chromosome analysis, FISH, and NGS methods were described in Chapters 1 and 12.

Test results

Firstly, chromosome analysis revealed a translocation t(9;11)(p21;q23), resulting in the fusion of the *MLLT3* gene at 9p21 with the *KMT2A* gene at 11q23, in 19 of 20 metaphase cells examined (Fig. 16.3.1). The remaining cell appeared to be chromosomally normal.

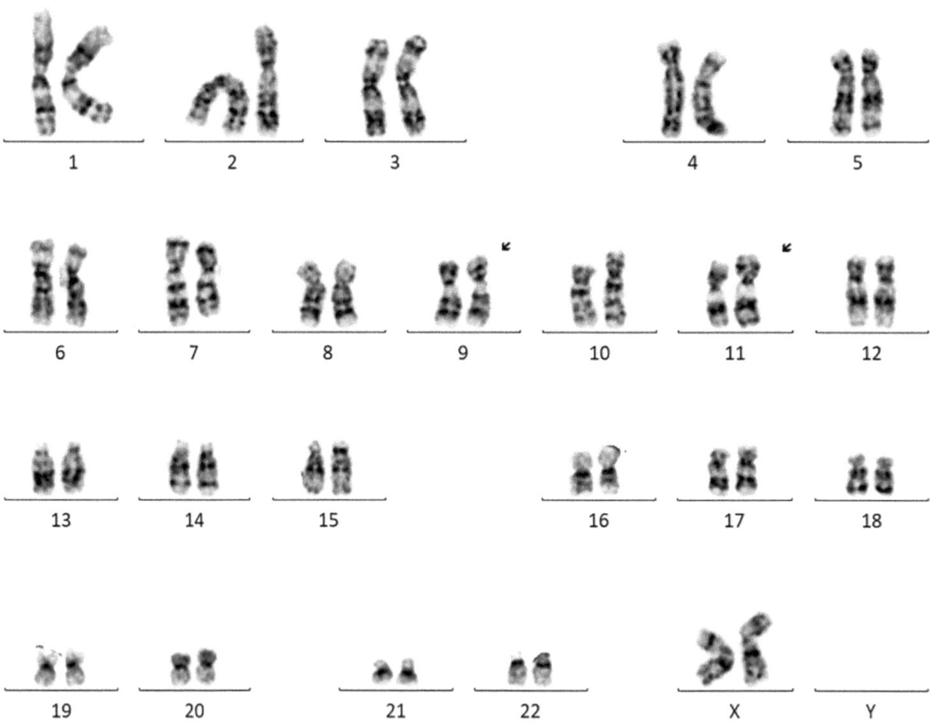

FIG. 16.3.1 The karyotype of bone marrow from the patient showed a t(9;11) translocation. ISCN: 46,XX,t(9;11)(p21;q23)[19]/46,XX[1]

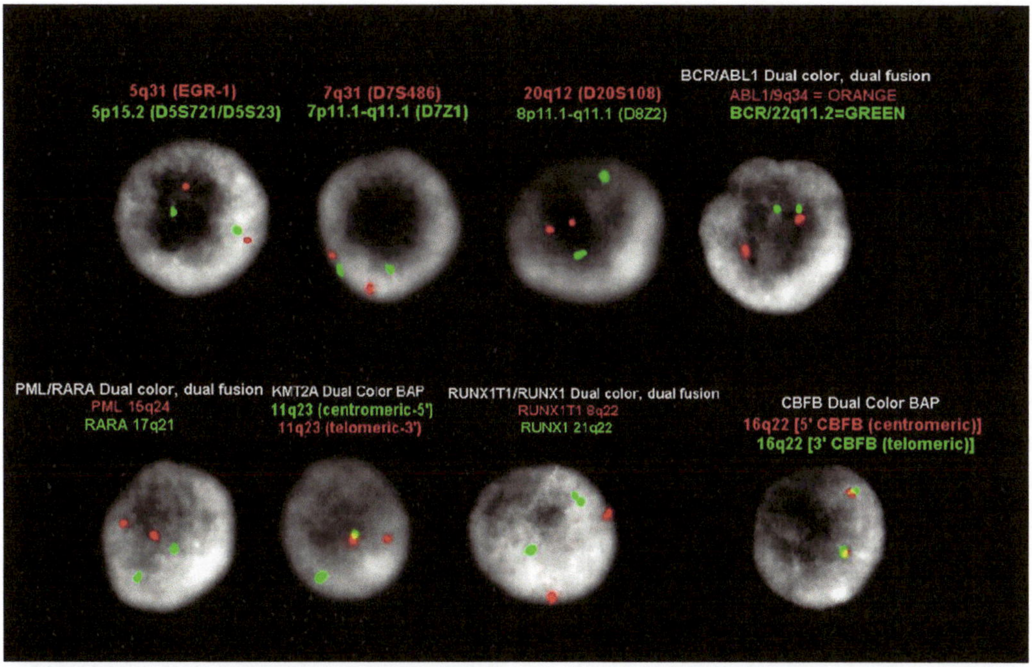

FIG. 16.3.2 FISH with AML panel was positive for *KMT2A* rearrangement. ISCN: nuc ish(D5S721,EGR1)x2[200/200], (CEP7,D7S486)x2[200/200], (D8Z2,D20S108)x2[198/200],(RUNX1T1,RUNX1)x2[200/200],(ABL1,BCR)x2[200/200], (KMT2Ax2)(5'KMT2A sep 3'KMT2Ax1)[132/200],(PML,RARA)x2[200/200],(CBFBx2)[200/200]

Next, FISH for the AML panel was performed on interphase nuclei using probes localized to the *D5S721* (5p15.2), *EGR1* (5q31), *D7Z1* (7cen), *D7S486* (7q31), *D8Z2* (8cen), *RUNX1T1* (8q22), *ABL1* (9q34.12), *KMT2A* (11q23), *PML* (15q24.1), *CBFB* (16q22.1), *RARA* (17q21.1), *D20S108* (20q12), *RUNX1* (21q22.3), and *BCR* (22q11) gene regions (supplied by Abbott Molecular, Inc.). Two hundred nuclei were examined, and the results demonstrated a rearrangement involving *KMT2A* in 132/200 (66.0%) of the cells scored (Fig. 16.3.2).

NGS Hematology Molecular Profile was performed, and the results showed a *KMT2A::MLLT3* fusion (Fig. 16.3.3).

Results with interpretations

The results of karyotyping, FISH, and NGS for this patient were concordant and consistent with t(9;11)(p21;q23) or *KMT2A::MLLT3* rearrangement. This is a distinct entity in the World Health Organization (WHO) classification system [1]. The t(9;11) can occur at any age but is more common in children and accounts for 9%–12% of pediatric cases of AML and 2% of adult AML cases [1]. Morphologically, it usually has monocytic features, which is also consistent with this patient's pathologic findings. This abnormality is generally associated with an intermediate prognosis in AML, which is superior to AML with other 11q23 (*KMT2A*)

Relevant Biomarkers

Tier	Genomic Alteration	Relevant Therapies (In this cancer type)	Clinical Trials
IA	*KMT2A-MLLT3 fusion* **Prognostic significance:** ELN 2017: Intermediate **Diagnostic significance:** Acute Myeloid Leukemia	allogeneic stem cells azacitidine cytarabine cytarabine + daunorubicin cytarabine + daunorubicin + etoposide cytarabine + etoposide + idarubicin cytarabine + fludarabine + idarubicin + filgrastim cytarabine + idarubicin cytarabine + mitoxantrone decitabine gemtuzumab ozogamicin + chemotherapy venetoclax + chemotherapy	14

Public data sources included in relevant therapies: FDA[1], NCCN, **EMA**[2], ESMO
Public data sources included in prognostic and diagnostic significance: NCCN, ESMO
Tier Reference: Li et al. *Standards and Guidelines for the Interpretation and Reporting of Sequence Variants in Cancer: A Joint Consensus Recommendation of the Association for Molecular Pathology, American Society of Clinical Oncology, and College of American Pathologists.* J Mol Diagn. 2017 Jan;19(1):4-23.

Variant Details

Gene Fusions (RNA)

Genes	Variant ID	Locus
KMT2A-MLLT3	KMT2A-MLLT3.K9M6	chr11:118355029 - chr9:20365742

FIG. 16.3.3 NGS showed the fusion of *KMT2A::MLLT3.*

translocations. The t(9;11) may easily be overlooked in chromosome analysis due to subtle chromosomal structural rearrangement (http://atlasgeneticsoncology.org/haematological/1001/t(9;11)(p21;q23)) [135,136]. Therefore, FISH, NGS, or RT-PCR testing is important to identify this rearrangement.

Future testing and recommendations

Chromosome analysis, FISH, or NGS can be ordered to monitor the disease progression and treatment efficacy in the future.

Case 16.4 Acute myeloid leukemia (AML) with t(8;21;21)(q22;p13;q22)/*RUNX1::RUNX1T1* fusion

Clinical indication

A 68-year-old male presented with a near syncopal episode and was found to have pancytopenia. He also had systemic mastocytosis observed on bone marrow biopsy. He denies any fevers, chills, chest pain, shortness of breath, palpitations, abdominal pain, nausea, vomiting, or diarrhea. He did note chronic, intermittent, moderate to severe, nonradiating, sharp, aching, right shoulder pain.

Test ordered

- Chromosome analysis of the bone marrow
- FISH: AML panel
- NGS Hematology Molecular Profile

Laboratory test performed

Chromosome analysis, FISH, and NGS methods are described in Chapters 1 and 12.

Test results

Firstly, chromosome analysis revealed a three-way translocation t(8;21;21)(q22;p13;q22), resulting in the fusion of the *RUNX1::RUNX1T1*, in 20 of 20 metaphase cells examined (Fig. 16.4.1). Fourteen cells also showed two copies of the derivative chromosome 21, which resulted from t(8;21;21) (Fig. 16.4.2). The karyotype also revealed del(9q) in one of the 20 cells analyzed (Fig. 16.4.3).

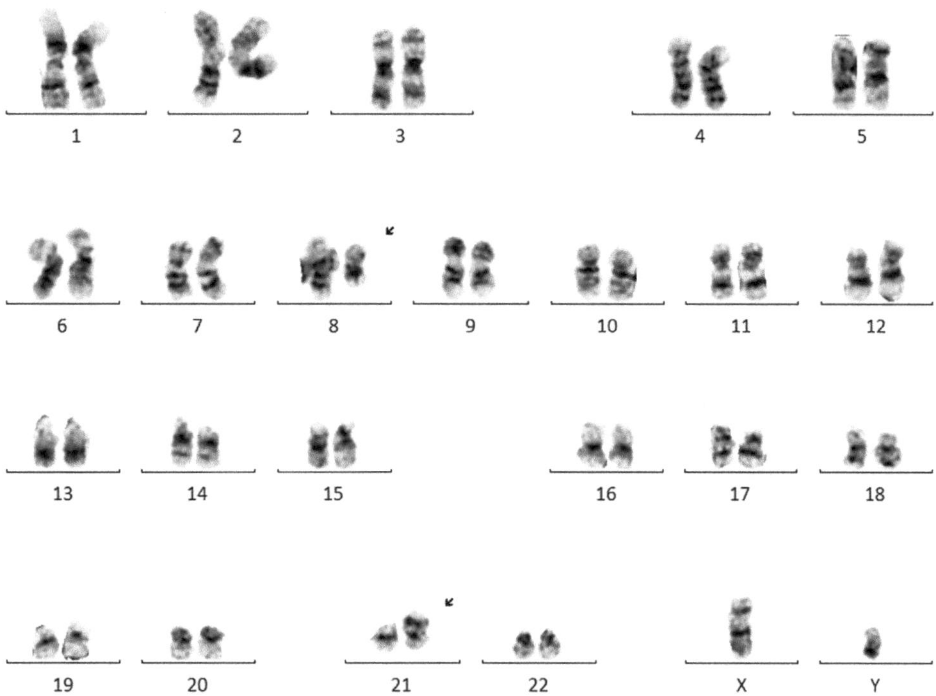

FIG. 16.4.1 Chromosome analysis revealed an abnormal clone 1 with a three-way translocation t(8;21;21). ISCN: 46, XY,t(8;21;21)(q22;p13;q22)[7]/46,idem,der(21)t(8;21;21)[14]/46,idem,del(9)(q13q22)[1]

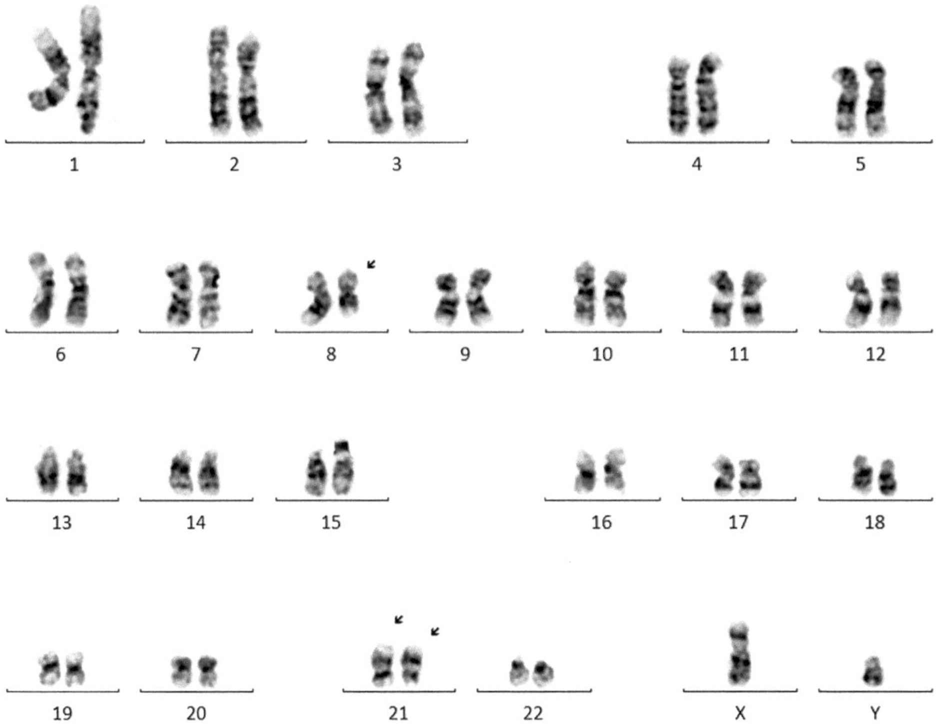

FIG. 16.4.2 Chromosome analysis showed an abnormal clone 2 with two copies of derivative chromosome 21, which resulted from a three-way translocation t(8,21;21).

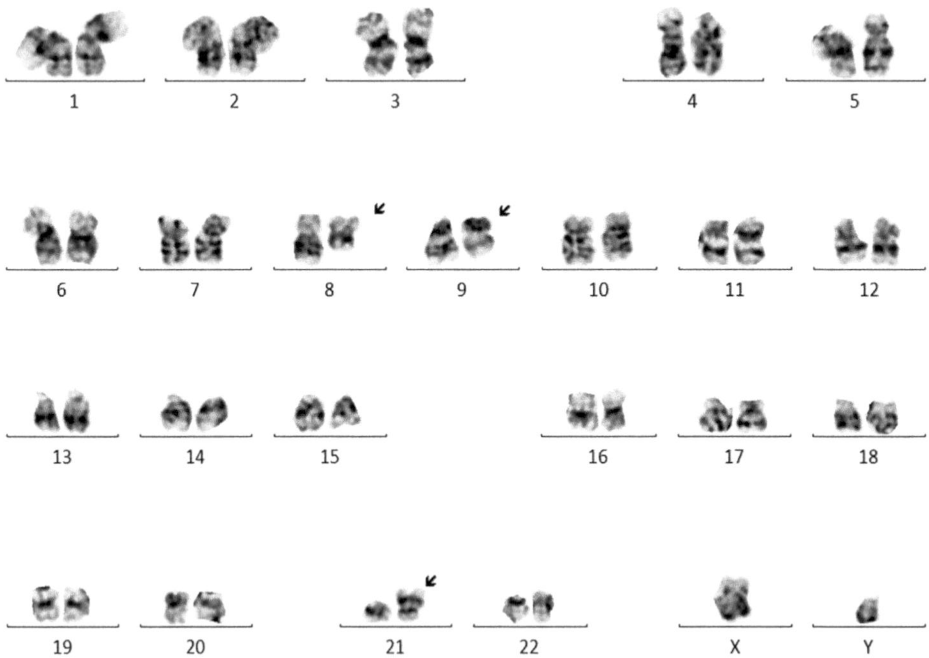

FIG. 16.4.3 Chromosome analysis showed an abnormal clone 3 with del(9q) in addition to a three-way translocation t(8;21;21).

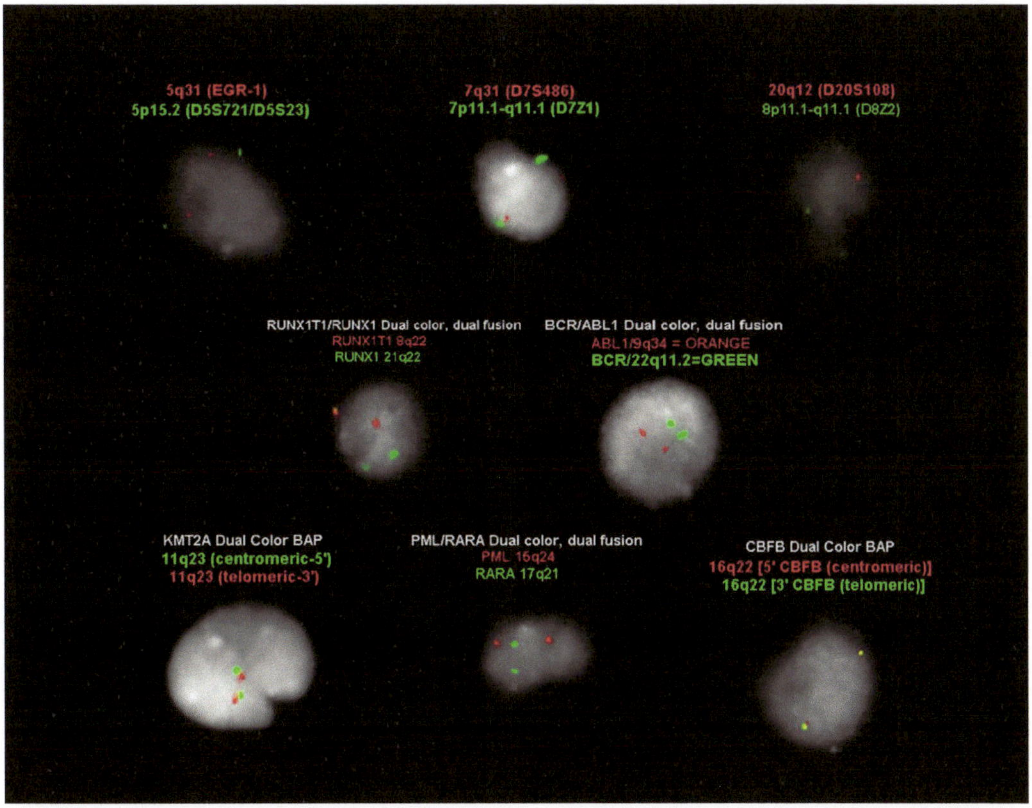

FIG. 16.4.4 FISH with AML panel was positive for an atypical *RUNX1::RUNX1T1* rearrangement. ISCN: nuc ish(D5S721, EGR1)x2[200/200],(CEP7,D7S486)x2[200/200],(D8Z2,D20S108)x2[200/200],(RUNX1T1x2,RUNX1x3)(RUNX1T1 con RUNX1x1)[164/200],(ABL1,BCR)x2[200/200],(KMT2Ax2)[200/200],(PML,RARA)x2[200/200],(CBFBx2)[200/200]

FISH for AML panel was also performed on interphase nuclei using probes localized to the *D5S721* (5p15.2), *EGR1* (5q31), *D7Z1* (7cen), *D7S486* (7q31), *D8Z2* (8cen), *RUNX1T1* (8q22), *ABL1* (9q34.12), *KMT2A* (11q23), *PML* (15q24.1), *CBFB* (16q22.1), *RARA* (17q21.1), *D20S108* (20q12), *RUNX1* (21q22.3), and *BCR* (22q11) gene regions. Two hundred nuclei were examined, and the results demonstrated an atypical *RUNX1T1::RUNX1* rearrangement in 164/200 (82.0%) of the cells scored. There was only one red signal indicating a possible partial deletion of *RUNX1T1* (Fig. 16.4.4).

NGS Hematology Molecular Profile was conducted, and the results showed a *RUNX1::RUNX1T1* fusion and *KIT* p.(D816V) c.2447A>T mutation (Fig. 16.4.5).

Results with interpretations

The results of karyotyping, FISH, and NGS for this patient were concordant with t(8;21;21) or *RUNX1::RUNX1T1* rearrangement. The morphology report concurred with a diagnosis of AML with the comments of hypercellular bone marrow (60%) with persistent/relapsed

Relevant Biomarkers

Tier	Genomic Alteration	Relevant Therapies (In this cancer type)	Clinical Trials
IA	*RUNX1-RUNX1T1 fusion* **Prognostic significance:** ELN 2017: Favorable **Diagnostic significance:** Acute Myeloid Leukemia	cytarabine + daunorubicin cytarabine + idarubicin cytarabine + mitoxantrone gemtuzumab ozogamicin + chemotherapy	6
IIC	*KIT p.(D816V) c.2447A>T*	None	2

Public data sources included in relevant therapies: FDA1, NCCN, EMA2, ESMO
Public data sources included in prognostic and diagnostic significance: NCCN, ESMO
Tier Reference: Li et al. *Standards and Guidelines for the Interpretation and Reporting of Sequence Variants in Cancer: A Joint Consensus Recommendation of the Association for Molecular Pathology, American Society of Clinical Oncology, and College of American Pathologists.* J Mol Diagn. 2017 Jan;19(1):4-23.

Variant Details

DNA Sequence Variants

Gene	Amino Acid Change	Coding	Locus	Allele Frequency	Variant Effect
KIT	p.(D816V)	c.2447A>T	chr4:55599321	34.35%	missense

Gene Fusions (RNA)

Genes	Variant ID	Locus
RUNX1-RUNX1T1	RUNX1-RUNX1T1.R3R3	chr21:36231771 - chr8:93029591

FIG. 16.4.5 NGS detected a fusion of *RUNX1::RUNX1T1* and *KIT* p.(D816V) c.2447A>T mutation.

systemic mastocytosis with associated hematological neoplasm represented by AML (87% BLASTS) (SM-AHN).

AML with t(8;21)(q22;q22) is the typical form presented as part of the group of AML with recurrent genetic abnormalities. It is commonly seen in AML-M2 and rarely in M1 or M4 (https://atlasgeneticsoncology.org/haematological/1019/t(8;21)(q22;q22)). This rearrangement has been seen in both children and adults but is more frequently observed in childhood AML. Complex t(8;21;Var) rearrangement involving a (variable) third chromosome has been described in 3% of AML patients. In terms of prognosis, complete remission (CR) in most cases (90%) can be expected with relatively long disease-free survival when treated with high-dose chemotherapy. NGS identified a *KIT* mutation in addition to the *RUNX1::RUNX1T1* fusion. The relapse-free survival (RFS) in *KIT*-mutated patients was inferior to those of unmutated patients. Based on subgroup analysis, *KIT* mutations had a prognostic impact in patients with *RUNX1::RUNX1T1* [137].

Future testing and recommendations

Chromosome analysis, FISH, or NGS can be ordered to monitor the disease progression and treatment efficacy in the future.

Case 16.5 Acute myeloid leukemia (AML) with *CEBPA* double mutations

Clinical indication

A 35-year-old male presented with muscle aches, progressive fatigue, self-limiting gum bleeding after brushing, intermittent hematemesis, nausea, and vomiting. Also noted intermittent chills and sweating along with headaches and watery eyes. Denied any chest pain, shortness of breath, dyspnea on exertion, neurological deficits, cough, or any abdominal pain. Evaluation of the bone marrow showed an increased blast cell population (68%). Flow cytometry also revealed an increased blast cell population that was positive for the CD34 dim subset, CD13, CD117, HLA-DR, CD38, CD45, and CD7 but negative for CD33, and CD10. AML was one of the differential diagnoses.

Test ordered

- Chromosome analysis of the bone marrow
- FISH: AML panel
- NGS Hematology Molecular Profile

Laboratory test performed

Chromosome analysis, FISH, and NGS methods were described previously in Chapters 1 and 12.

Test results

Of all 20 metaphases examined, no evidence for a chromosome abnormality was detected within the limits of the band level and technology utilized in this study (Fig. 16.5.1).

FISH for AML panel was also performed on interphase nuclei using probes localized to the *D5S721* (5p15.2), *EGR1* (5q31), *D7Z1* (7cen), *D7S486* (7q31), *D8Z2* (8cen), *RUNX1T1* (8q22), *ABL1* (9q34.12), *KMT2A* (11q23), *PML* (15q24.1), *CBFB* (16q22.1), *RARA* (17q21.1), *D20S108* (20q12), *RUNX1* (21q22.3), and *BCR* (22q11) gene regions. Two hundred nuclei were examined, and the results were within normal limits for the laboratory's established background rates. (Fig. 16.5.2).

Finally, NGS Hematology Molecular Profile was performed and showed a double mutation of *CEBPA* [p.(Q312dup) and p.(F77Wfs*82)] (Fig. 16.5.3).

Results with interpretations

This patient had normal results from chromosome analysis and FISH. However, NGS identified two mutations of *CEBPA*.

CEBPA mutations are seen in 6.86%–20.33% of AML patients [138]. Moreover, the frequencies of *CEBPA*^{sm} and *CEBPA*^{dm} are very similar in AML patients from Caucasian populations,

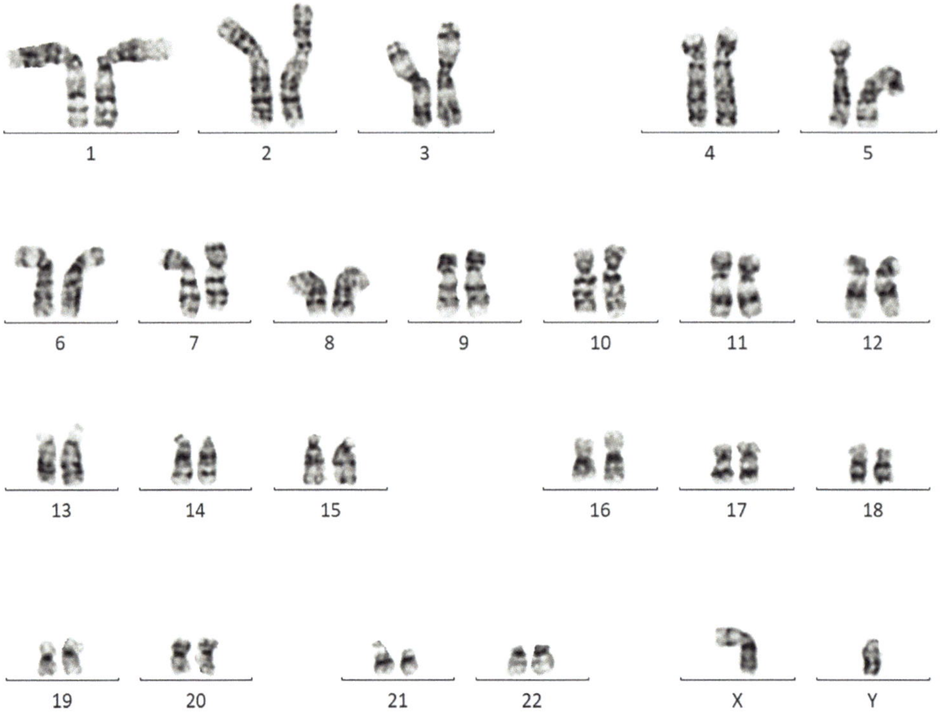

FIG. 16.5.1 Chromosome analysis showed a normal male karyotype. ISCN: 46,XY[20]

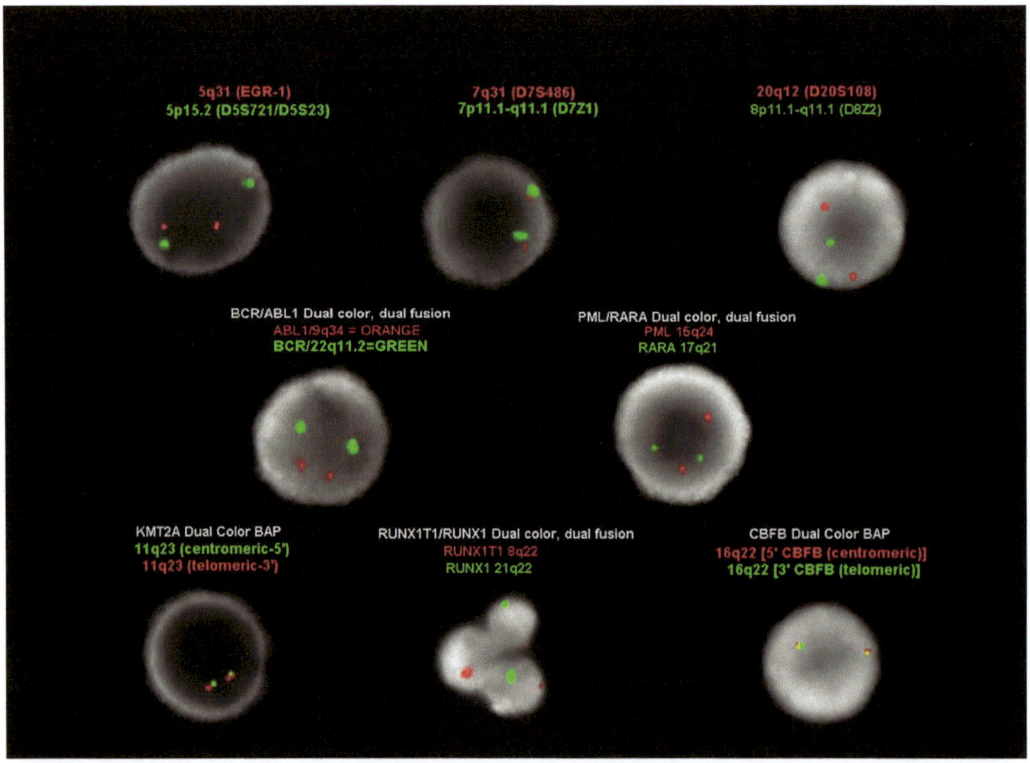

FIG. 16.5.2 FISH for the AML panel revealed normal results for all probes examined. ISCN: nuc ish(D5S721,EGR1)x2[200/200],(CEP7,D7S486)x2[200/200], (D8Z2,D20S108)x2[200/200],(RUNX1T1,RUNX1)x2[200/200], (ABL1,BCR)x2[200/200],(KMT2Ax2)[200/200], (PML,RARA)x2[200/200],(CBFBx2)[200/200]

Relevant Biomarkers

Tier	Genomic Alteration	Relevant Therapies (In this cancer type)	Clinical Trials
IA	*CEBPA p.(Q312dup) c.934_936dup, CEBPA p.(F77Wfs*82) c.229_232delTTCC*	cytarabine + daunorubicin cytarabine + idarubicin cytarabine + mitoxantrone gemtuzumab ozogamicin + chemotherapy	2

Prognostic significance: ELN 2017: Favorable
Diagnostic significance: Acute Myeloid Leukemia

Public data sources included in relevant therapies: FDA1, NCCN, **EMA**2, ESMO
Public data sources included in prognostic and diagnostic significance: NCCN, ESMO
Tier Reference: Li et al. *Standards and Guidelines for the Interpretation and Reporting of Sequence Variants in Cancer: A Joint Consensus Recommendation of the Association for Molecular Pathology, American Society of Clinical Oncology, and College of American Pathologists.* J Mol Diagn. 2017 Jan;19(1):4-23.

Variant Details

DNA Sequence Variants

Gene	Amino Acid Change	Coding	Locus	Allele Frequency	Variant Effect
CEBPA	p.(Q312dup)	c.934_936dup	chr19:33792384	25.96%	nonframeshift Insertion
CEBPA	p.(F77Wfs*82)	c.229_232delTTCC	chr19:33793088	29.76%	frameshift Deletion

FIG. 16.5.3 NGS results exhibited double mutations for *CEBPA*.

but the frequencies of *CEBPA*^{dm} are seen in more Asian populations [107,139]. Ho et al. studied 847 pediatric AML patients and evaluated the presence of *CEBPA* mutations. Clinical outcome was examined for the 847 patients with known *CEBPA* mutation status. The overall survival (OS), event-free survival (EFS), disease-free survival (DFS), and complete remission (CR) were more favorable in patients with *CEBPA* mutations than those without *CEBPA* mutations [140]. AML patients with *CEBPA*^{dm} are sensitive to chemotherapy and show favorable outcomes [107,109,138]. In this case, even though cytogenetics and FISH results were normal, the patient still benefited from NGS testing, which showed a favorable prognosis with *CEBPA*^{dm} mutations. The therapy drugs were also provided in the report (Fig. 16.5.3).

Future testing and recommendations

Since this patient had normal karyotype and negative FISH results, NGS can be ordered to monitor the disease progression and treatment efficacy in the future.

Case 16.6 Acute myeloid leukemia (AML) with a complex karyotype and *TP53* mutation

Clinical indication

The patient is a 66-year-old female with a past medical history of basal cell carcinoma, leukocytosis, monocytosis, anemia, and thrombocytopenia. A bone marrow biopsy was obtained for evaluation.

Test ordered

- Chromosome analysis of the bone marrow
- FISH: AML panel
- NGS Hematology Molecular Profile

Laboratory test performed

Chromosome analysis, FISH, and NGS methods were described previously in Chapters 1 and 12.

Test results

Chromosome analysis was performed initially. Of the 21 cells examined, 19 were abnormal and clonal evolution was evident. The stemline with two cells exhibited del(4q), monosomy 7, dup(11q), monosomy 20, and a marker chromosome (Fig. 16.6.1); in addition to the abnormalities from the stemline, sideline 1 with 7 cells showed add(5q), monosomy

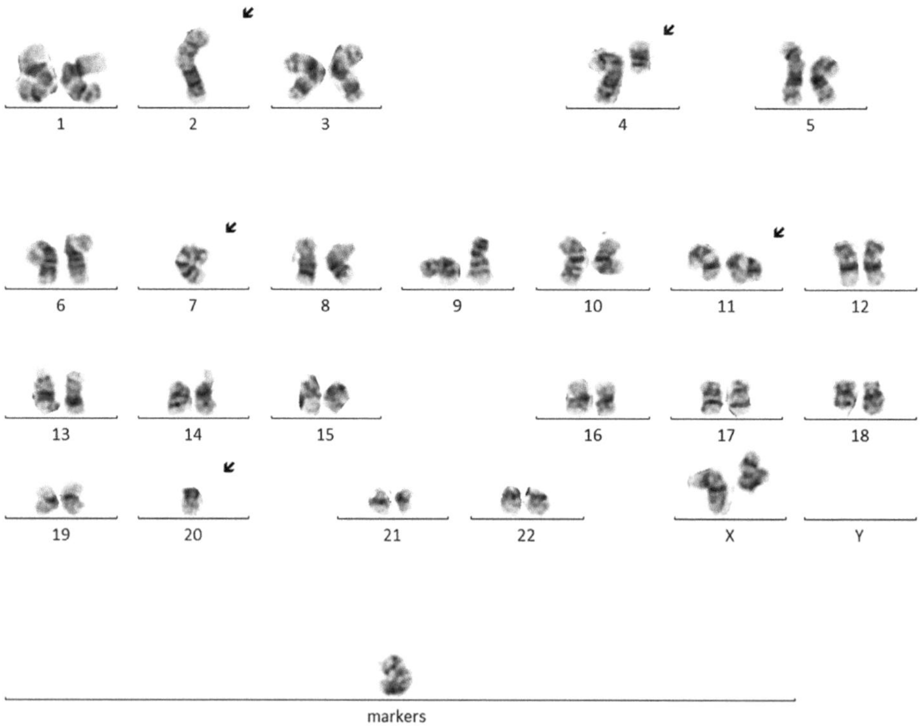

FIG. 16.6.1 Chromosome analysis showed an abnormal clone 1 with del(4q), monosomy 7 and 20, dup(11q), and a marker chromosome. ISCN: 45,XX,del(4)(q13q35),-7,dup(11)(q23q25),-20,+mar[2]/ 46~48,sl,add(5)(q13),-13,+3~5r [cp7]/ 46~50,sdl1,+8[cp8]/ 48,sdl2,add(2)(p10),-add(5),-19[2]/ 46,XX[2]

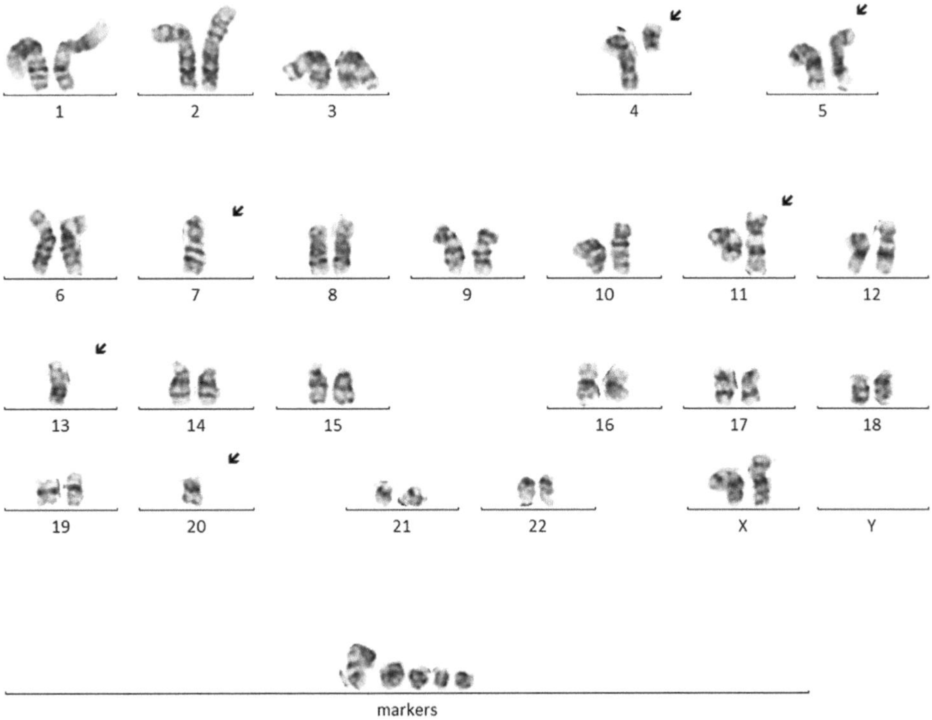

FIG. 16.6.2 Chromosome analysis showed an abnormal clone 2 with add(5q), monosomy 12, and 3–5 ring chromosomes in addition to the abnormalities seen in clone 1.

13, and 3–5 ring chromosomes (Fig. 16.6.2); sideline 2 with 8 cells revealed trisomy 8 (Fig. 16.6.3); sideline 3 with 2 cells had add(2p), monosomy 19 but without add(5q) (Fig. 16.6.4). The remaining two cells appear to be chromosomally normal.

FISH for AML panel was performed on interphase nuclei using probes localized to the *D5S721* (5p15.2), *EGR1* (5q31), *D7Z1* (7cen), *D7S486* (7q31), *D8Z2* (8cen), *RUNX1T1* (8q22), *ABL1* (9q34.12), *KMT2A* (11q23), *PML* (15q24.1), *CBFB* (16q22.1), *RARA* (17q21.1), *D20S108* (20q12), *RUNX1* (21q22.3), and *BCR* (22q11) gene regions. Two hundred nuclei were examined, and the results demonstrated multiple abnormalities (Fig. 16.6.5).

Deletion 5q was observed in 155/200 (77.5%) of the cells scored.

Deletion 7q was observed in 190/200 (95.0%) of the cells scored.

Trisomy 8 was observed in 102/200 (51.0%) of the cells scored.

A gain of 11q (*KMT2A*) was observed in 140/200 (70.0%) of the cells scored.

Deletion 20q was observed in 194/200 (97.0%) of the cells scored.

NGS Hematology Molecular Profile revealed a *TP53* p.(G187Vfs*60) c.560delG mutation. There was a fast-track treatment for AML patients with this mutation (Fig. 16.6.6).

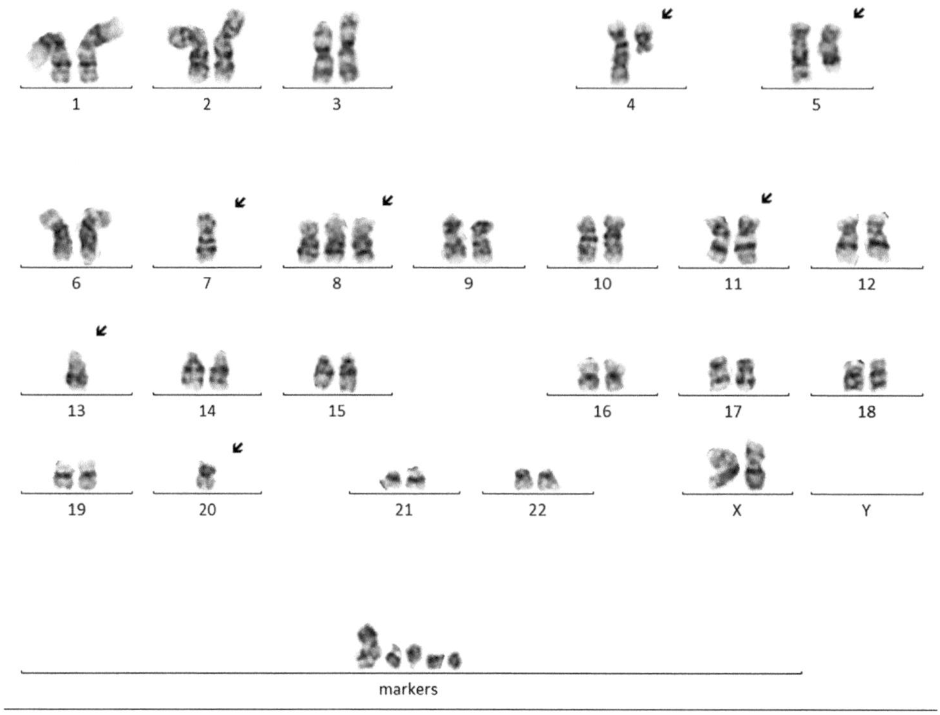

FIG. 16.6.3 Chromosome analysis showed an abnormal clone 3 with trisomy 8 in addition to the abnormalities seen in clone 2.

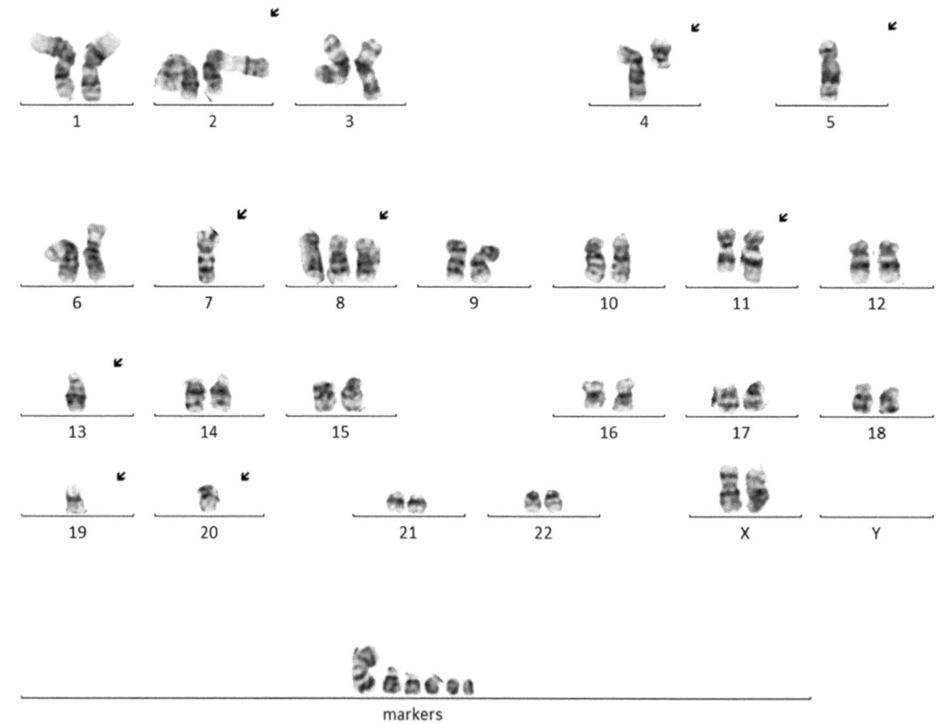

FIG. 16.6.4 Chromosome analysis showed an abnormal clone 4 with add(2p), monosomy 19, and without add(5q) in addition to the abnormalities seen in clone 3.

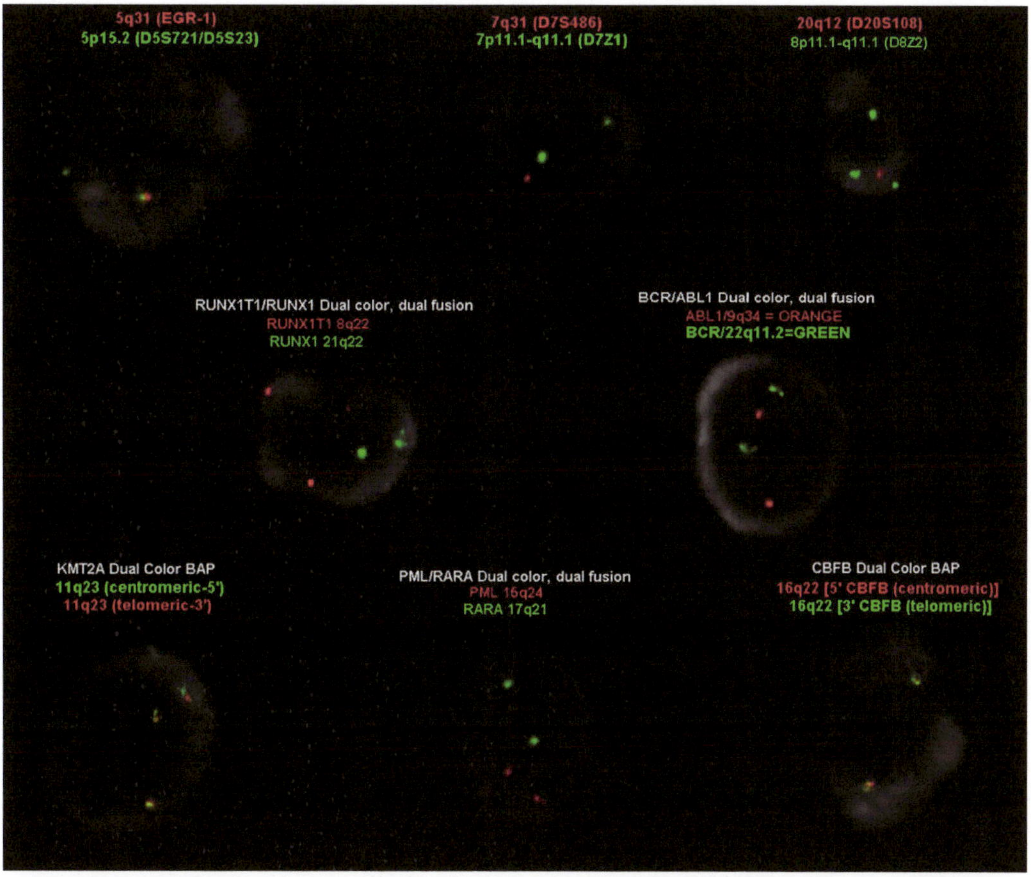

FIG. 16.6.5 FISH for AML panel identified deletion 5q, deletion 7q, trisomy 8, gain of 11q (*KMT2A*), and deletion 20q. ISCN: nuc ish(D5S721x2,EGR1x1)[155/200],(CEP7x2,D7S486x1)[190/200],(D8Z2x3,D20S108x1)[102/200]/(D8Z2x2,D20S108x1)[92/200],(RUNX1T1x3,RUNX1x2)[95/200],(ABL1,BCR)x2[200/200],(KMT2Ax3)[140/200],(PML,RARA)x2[200/200],(CBFBx2)[200/200]

Results with interpretations

Chromosome analysis identified four abnormal clones and clonal evolution was evident. The chromosome aberrations seen in this case were recurrent chromosome abnormalities associated with myeloid disorders. del(5q), del(7q), trisomy 8, a gain of 11q, and del(20q) were also seen from the concurrent FISH testing. Three or more FISH abnormalities and chromosome 7 anomalies indicate poor risk status in MDS and AML. These results are consistent with a diagnosis of AML. The degree of karyotypic complexity indicates a more aggressive disease process.

NGS Hematology Molecular Profile identified a *TP53* mutation, which provided the therapy option for a fast-track designation by the FDA (https://ir.aprea.com//news-releases/news-release-details/aprea-therapeutics-receives-fda-fast-track-designation).

Relevant Biomarkers

Tier	Genomic Alteration	Relevant Therapies (In this cancer type)	Clinical Trials
IA	*TP53 p.(G187Vfs*60) c.560delG* **Prognostic significance:** ELN 2017: Adverse **Diagnostic significance:** None	None	9

Public data sources included in relevant therapies: FDA1, NCCN, **EMA**2, ESMO
Public data sources included in prognostic and diagnostic significance: NCCN, ESMO
Tier Reference: Li et al. *Standards and Guidelines for the Interpretation and Reporting of Sequence Variants in Cancer: A Joint Consensus Recommendation of the Association for Molecular Pathology, American Society of Clinical Oncology, and College of American Pathologists.* J Mol Diagn. 2017 Jan;19(1):4-23.

Variant Details

DNA Sequence Variants

Gene	Amino Acid Change	Coding	Locus	Allele Frequency	Variant Effect
TP53	p.(G187Vfs*60)	c.560delG	chr17:7578288	75.64%	frameshift Deletion

Alerts Informed By Public Data Sources

Current FDA Information

Contraindicated   Not recommended   Resistance   Breakthrough   Fast Track

FDA information is current as of 2021-04-14. For the most up-to-date information, search www.fda.gov.

TP53 p.(G187Vfs*60) c.560delG

eprenetapopt + azacitidine, eprenetapopt + venetoclax + azacitidine

Cancer type: Acute Myeloid Leukemia   **Variant class:** TP53 mutation

Supporting Statement:

The FDA has granted Fast Track Designation to eprenetapopt in combination with azacitidine or azacitidine and venetoclax for:

- TP53 mutant acute myeloid leukemia (AML)
- TP53 mutant myelodysplastic syndrome (MDS)

Reference:

https://ir.aprea.com//news-releases/news-release-details/aprea-therapeutics-receives-fda-fast-track-designation

FIG. 16.6.6 NGS identified a *TP53* mutation.

Future testing and recommendations

Chromosome analysis, FISH, and NGS can be ordered to monitor the disease progression and treatment efficacy in the future.

Case 16.7 Acute promyelocytic leukemia (APL) with a typical *PML::RARA* fusion and *FLT3*-ITD mutation

Clinical indication

A 26-year-old male presented with a 2-week history of fever and bleeding gums. A bone marrow biopsy was then performed and revealed a hypercellular marrow with left-shifted maturation in the myeloid lineage with 77% blasts. Morphology showed an increased nuclear-cytoplasmic ratio and bilobed nuclei with sliding plate morphology. The morphologic and clinical findings were concerning for APL, and thus, the clinical team started ATRA (all-trans-retinoic acid) while awaiting final diagnosis. Flow cytometry revealed the blast population to be dim CD45, bright CD33, subset CD34, dim HLA-DR, dim CD13, CD117, subset CD56, and cytoplasmic MPO positive.

Test ordered

- FISH: AML panel including *PML::RARA* fusion probes
- Chromosome analysis of the bone marrow
- RT-PCR was conducted in a sendout lab
- NGS Hematology Molecular Profile

Laboratory test performed

Firstly, FISH for the AML panel was performed on interphase nuclei using probes localized to the *PML* (15q22-24) and *RARA* (17q21) gene regions (Abbott Molecular, Inc.) as a STAT case. Then, a full AML panel was also conducted for this patient.

Secondly, chromosome analysis of the bone marrow was performed. At the same time, RT-PCR was performed in another lab.

Finally, the NGS Hematology Molecular Profile assay was performed. In most clinical diagnostics laboratories, only typical transcripts are available for testing. In our lab, the NGS assay can detect 45 different transcripts for *RARA* gene rearrangements including common and variant fusion partners.

Test results

Firstly, FISH for AML panel was performed on interphase nuclei using probes localized to the *D5S721* (5p15.2), *EGR1* (5q31), *D7Z1* (7cen), *D7S486* (7q31), *D8Z2* (8cen), *RUNX1T1* (8q22), *ABL1* (9q34.12), *KMT2A* (11q23), *PML* (15q24.1), *CBFB* (16q22.1), *RARA* (17q21.1), *D20S108* (20q12), *RUNX1* (21q22.3), and *BCR* (22q11) gene regions. Two hundred nuclei were examined, and the results demonstrated a *PML::RARA* rearrangement in 187/200 (93.5%) of the cells scored and trisomy 8 in 172/200 (86.0%) of the cells scored. This *PML::RARA* gene rearrangement is commonly associated with acute promyelocytic leukemia (APL) (Fig. 16.7.1).

Secondly, chromosome analysis revealed a translocation between the long arms of chromosomes 15 and 17, and trisomy 8 in 15 of 20 the cells examined (Fig. 16.7.2). Five normal cells were observed

RT-PCR was conducted in another lab showing a fusion of *PML::RARA* (data not shown).

Lastly, NGS Hematology Molecular Profile was performed, and the results showed a fusion of *PML::RARA* and *FLT3*-ITD mutation (Fig. 16.7.3).

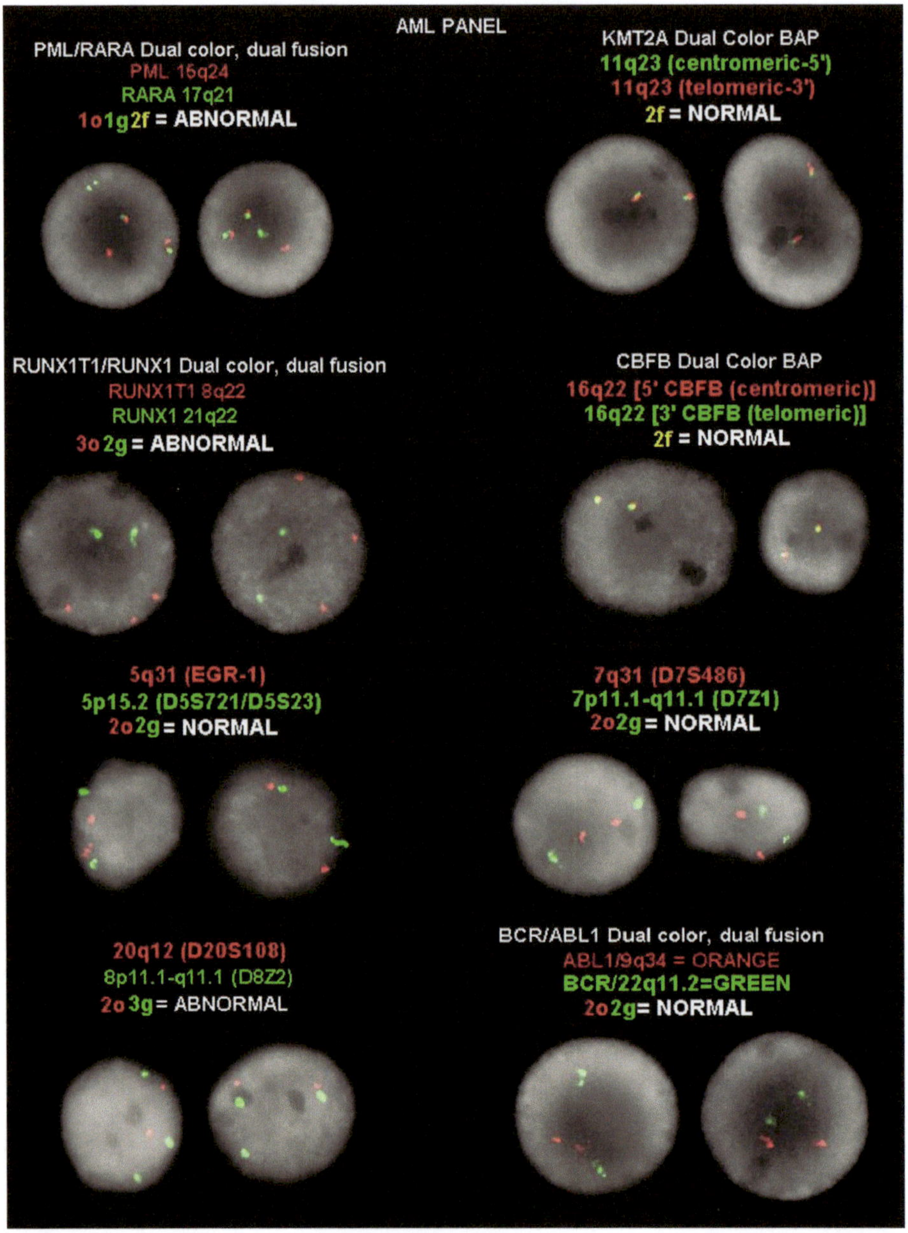

FIG. 16.7.1 FISH for the AML panel was positive for *PML::RARA* rearrangement and trisomy 8 (3 signals for *RUNX1T1*). ISCN: ish(D5S721,EGR1)x2[200/200],(CEP7,D7S486)x2[200/200],(D8Z2x3,D20S108x2)[166/200],(RUNX1T1x3,RUNX1x2)[172/200],(ABL1,BCR)x2[200/200],(KMT2Ax2)[200/200],(PML,RARA)x3(PML con RARAx2)[187/200],(CBFBx2)[200/200]

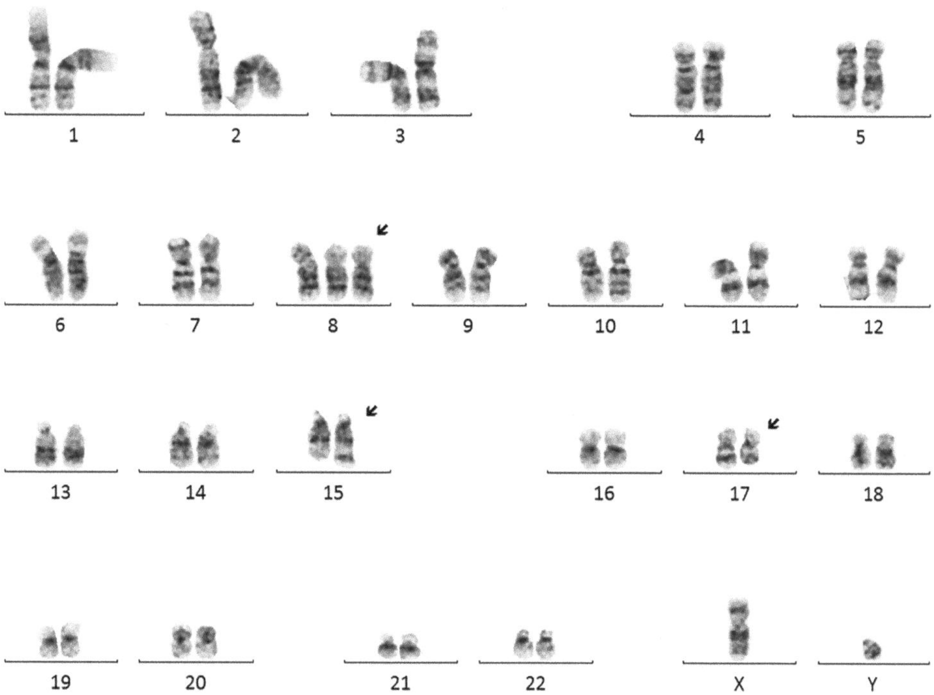

FIG. 16.7.2 The karyotype of the bone marrow showed a t(15;17) translocation and a trisomy 8. ISCN: 47,XY,+8,t(15;17)(q21;q21)[15]/46,XY[5]

Results with interpretations

The tests performed on this patient including chromosome analysis, FISH, and NGS were concordant, all of which detected t(15,17) or *PML::RARA* rearrangement. However, chromosome analysis and FISH also showed trisomy 8, which is the most common secondary change for APL patients. NGS showed an *FLT3*-ITD mutation in addition to the fusion of *PML::RARA*, which might affect risk stratification in patients with APL [141]. *FLT3*-ITD mutation with a high variant allele frequency (VAF), along with *PML::RARA* gene fusion and trisomy 8, could represent a poor prognostic for APL, especially in young patients [142]. The genetic investigation of patients with APL must be comprehensive to overcome the limits of each technique to have an appropriate characterization of the genome, a realistic prognosis, and personalized treatment.

APL is a subtype of AML in which abnormal promyelocytes predominate. APL accounts for 5%–8% of AML cases. The disease can occur at any age, but most APL patients are middle-aged adults. An important distinguishing feature of APL is the frequent association with disseminated intravascular coagulation (DIC) leading to high mortality and morbidity rate [2,143]. However, APL has a particular sensitivity to treatment with tretinoin and arsenic trioxide; therefore, a prompt diagnosis of APL is very important [144].

Relevant Biomarkers

Tier	Genomic Alteration	Relevant Therapies (In this cancer type)	Clinical Trials
IA	*FLT3 ITD mutation* Mut/WT Ratio: 0.563 **Prognostic significance:** ELN 2017: Intermediate to Adverse	**gilteritinib** [1, 2] **midostaurin + chemotherapy** [1, 2] azacitidine cytarabine + daunorubicin cytarabine + daunorubicin + etoposide cytarabine + etoposide + idarubicin cytarabine + fludarabine + idarubicin + filgrastim cytarabine + idarubicin cytarabine + mitoxantrone decitabine gemtuzumab ozogamicin + chemotherapy sorafenib sorafenib + chemotherapy venetoclax + chemotherapy	26
IA	*PML-RARA fusion* **Diagnostic significance:** Acute Promyelocytic Leukemia	**arsenic trioxide** [1, 2] **arsenic trioxide + tretinoin** [1, 2] anthracycline + arsenic trioxide anthracycline + arsenic trioxide + tretinoin arsenic trioxide + idarubicin + tretinoin cytarabine + daunorubicin + tretinoin gemtuzumab ozogamicin gemtuzumab ozogamicin + chemotherapy idarubicin + tretinoin	4

Public data sources included in relevant therapies: FDA1, NCCN, **EMA**2, ESMO
Public data sources included in prognostic and diagnostic significance: NCCN, ESMO
Tier Reference: Li et al. *Standards and Guidelines for the Interpretation and Reporting of Sequence Variants in Cancer: A Joint Consensus Recommendation of the Association for Molecular Pathology, American Society of Clinical Oncology, and College of American Pathologists.* J Mol Diagn. 2017 Jan;19(1):4-23.

Variant Details

DNA Sequence Variants

Gene	Amino Acid Change	Coding	Locus	Allele Frequency	Variant Effect
FLT3	p.(Y597_E598insDGSSDNEY FYVDFREY)	c.1792_1793insACGGCTCCT CAGATAATGAGTACTTCTAC GTTGATTTCAGAGAATATG	chr13:28608263	35.30%	nonframeshift Insertion
PML-RARA	PML-RARA.P3R3		chr15:74315749 - chr17:38504568		

FIG. 16.7.3 NGS demonstrated the fusion of *PML::RARA* and *FLT3*-ITD mutation.

This represents the most common STAT (urgent/rush) testing for clinical genetic/genomic labs.

An APL diagnosis requires APL cellular morphology and either translocation t(15:17)(q22;q21) by cytogenetics or promyelocytic leukemia (*PML*)/retinoic acid receptor alpha (*RARA*) oncoprotein by molecular testing. t(15;17) usually appears as the sole chromosomal abnormality, but complex translocations involving chromosomes 15 and 17 are also possible. The most common secondary change is trisomy 8 (seen in one-third of cases), followed by del(7q), del(9q), and ider(17)(q10)t(15,17) [145,146].

Mutations involving *FLT3*, including *FLT3*-ITD and *FLT3* tyrosine kinase domain (*FLT3*-TKD) mutations, occur in 30%–40% of APL. *FLT3*-ITD mutations are the most common, and patients with this mutation are more likely to have significantly higher white blood cell count at diagnosis, higher risk of induction deaths, and lower overall survival rates than those without *FLT3*-ITD mutations [141,142,147–152].

Future testing and recommendations

A repeat of FISH, NGS, qPCR, and chromosome analysis on bone marrow is recommended to confirm a complete response to the treatment and to monitor the minimal residual disease or a relapse of the disease.

Case 16.8 Acute promyelocytic leukemia (APL) with a variant *ZBTB16::RARA* fusion

Clinical indication

A 51-year-old male presented to the clinic due to fatigue and easy bruising. Morphologic examination of peripheral blood smear demonstrated some immature/blast-like cells. These cells were of predominantly intermediate to large size with oval to folded/convoluted nuclei, fine nuclear chromatin, small nucleoli, and scanty finely granular cytoplasm. Possible Auer rods were seen. Flow cytometry demonstrated that 35% blasts that were CD34(−), HLA-DR(−), CD33(+), CD13(+), CD117(+), and CD38(+). The findings were suspicious for APL.

Test ordered

- FISH: AML panel including *PML::RARA* duel color duel fusion probes
- FISH reflex to *RARA* break-apart probes
- Chromosome analysis of the bone marrow
- NGS Hematology Molecular Profile

Laboratory test performed

Chromosome analysis, FISH, qPCR, and NGS methods were described previously in Chapters 1 and 12.

Test results

Firstly, FISH for AML panel was performed on interphase nuclei using probes localized to the *D5S721* (5p15.2), *EGR1* (5q31), *D7Z1* (7cen), *D7S486* (7q31), *D8Z2* (8cen), *RUNX1T1* (8q22), *ABL1* (9q34.12), *KMT2A* (11q23), *PML* (15q24.1), *CBFB* (16q22.1), *RARA* (17q21.1),

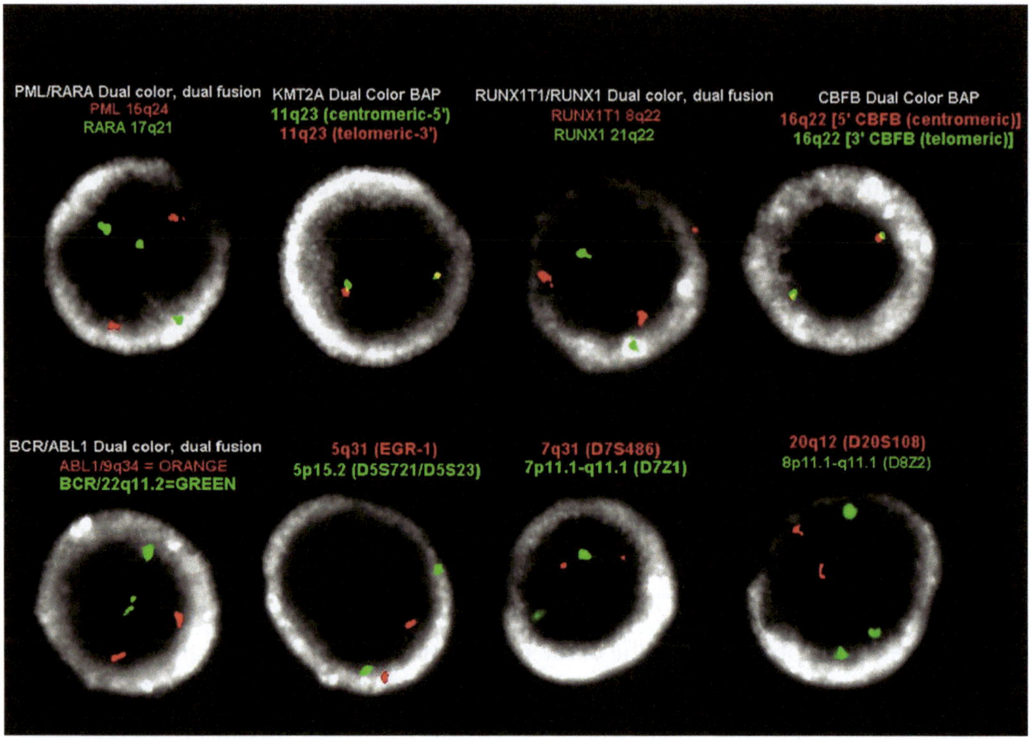

FIG. 16.8.1 FISH with AML FISH panel with *PML::RARA* dual-color dual-fusion probes showed negative for *PML::RARA* fusion, but three *RARA (green)* signals. nuc ish(D5S721,EGR1)x2[200/200],(CEP7,D7S486)x2[200/200],(D8Z2x3, D20S108x2)[68/200],(RUNX1T1x3,RUNX1x2)[73/200],(ABL1,BCR)x2[200/200],(KMT2Ax2)[200/200],(PMLx2,RARAx3)[173/200],(CBFBx2)[199/200]

D20S108 (20q12), *RUNX1* (21q22.3), and *BCR* (22q11) gene regions. Two hundred nuclei were examined, and the results demonstrated trisomy 8 in 68/200 (34.0%) of the cells scored and three copies of *RARA* in 173/200 (86.5%) of the cells scored (Fig. 16.8.1).

Secondly, FISH with *RARA* break-apart probes was performed on interphase nuclei, the results were positive for *RARA* rearrangement in 177/200 (88.5%) of the cells scored (Fig. 16.8.2).

Next, chromosome analysis was performed on bone marrow. All the 20 cells examined exhibited a translocation between the long arms of chromosomes 11 and 17 (Fig. 16.8.3). Two cells also had trisomy 8 in addition to t(11;17) (Fig. 16.8.4).

Finally, the NGS Hematology Molecular Profile was performed and detected fusion of *ZBTB16::RARA* in addition to *IDH2, TET2,* and *SRSF2* mutations (Fig. 16.8.5).

Results with interpretations

A subset of cases, often with morphological features resembling those of APL with t(15;17), show variant translocations involving *RARA*. The variant fusion partners include *ZBTB16* (previously called *PLZF*) at 11q23.2, *NUMA1* at 11q13.4, *NPM1* at 5q35.1, and *STAT5B* at

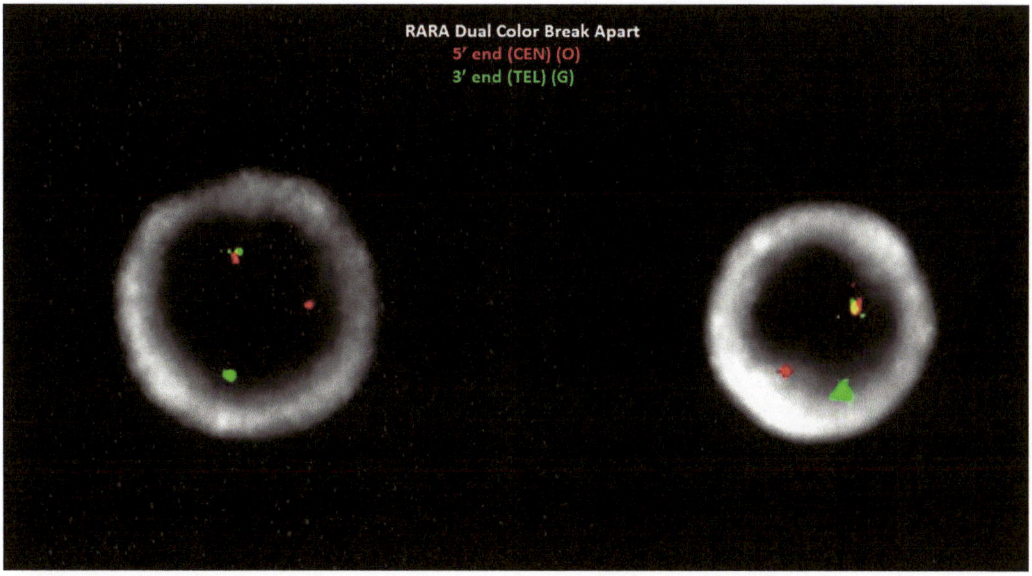

FIG. 16.8.2 FISH with *RARA* dual color break-apart probes was positive for *RARA* rearrangement. ISCN: nuc ish(RARA)x2(5'RARA sep 3'RARAx1)[177/200]

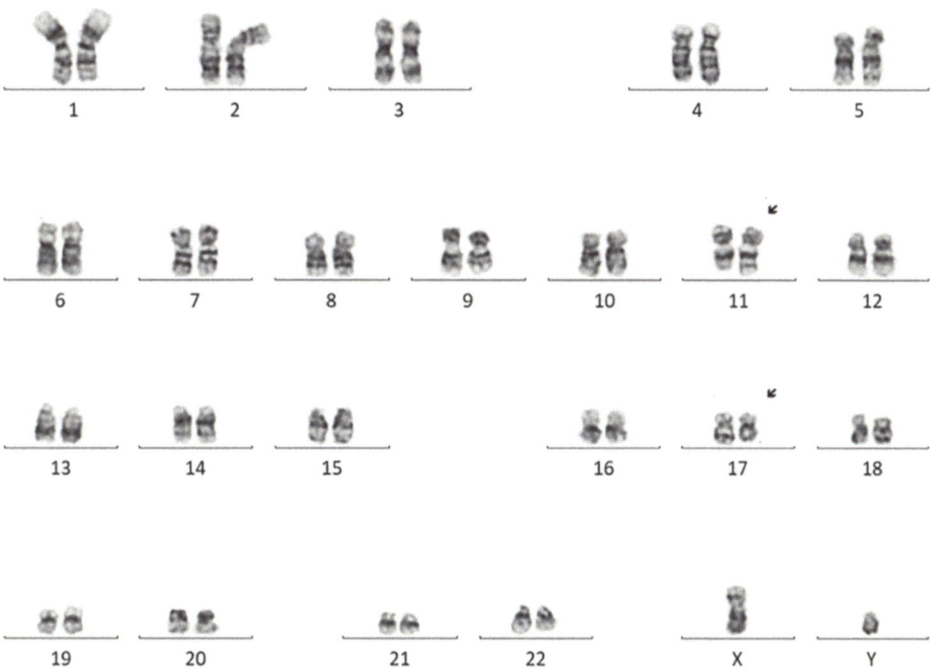

FIG. 16.8.3 The karyotype of the bone marrow showed an abnormal clone 1 with a t(11;17) translocation. ISCN: 46,XY, t(11;17)(q23;q21)[18]/47,idem,+8[2]

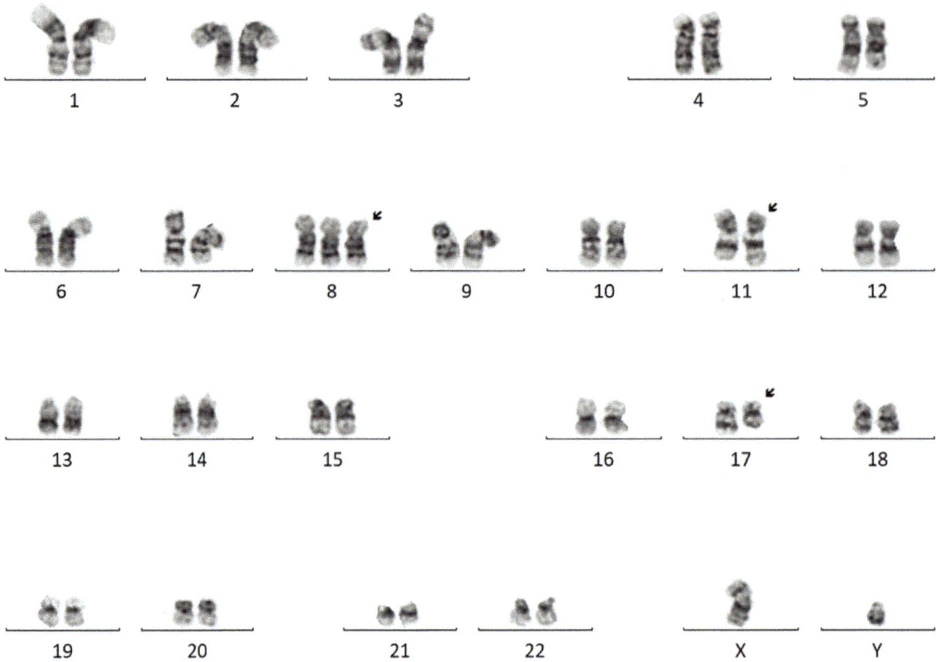

FIG. 16.8.4 The karyotype of the bone marrow showed an abnormal clone 2 with a t(11;17) translocation and trisomy 8.

Relevant Biomarkers

Tier	Genomic Alteration	Relevant Therapies (In this cancer type)	Clinical Trials
IA	*IDH2 p.(R140Q) c.419G>A*	**enasidenib** [1] azacitidine decitabine venetoclax + chemotherapy	6
IIC	*SRSF2 p.(P95T) c.283C>A*	None	3
IIC	*TET2 p.(V1718L) c.5152G>T*	None	2

Public data sources included in relevant therapies: FDA1, NCCN, **EMA**2, ESMO
Tier Reference: Li et al. *Standards and Guidelines for the Interpretation and Reporting of Sequence Variants in Cancer: A Joint Consensus Recommendation of the Association for Molecular Pathology, American Society of Clinical Oncology, and College of American Pathologists.* J Mol Diagn. 2017 Jan;19(1):4-23.

Prevalent cancer biomarkers without relevant evidence based on included data sources
ZBTB16-RARA fusion

Variant Details

DNA Sequence Variants

Gene	Amino Acid Change	Coding	Locus	Allele Frequency	Variant Effect
TET2	p.(V1718L)	c.5152G>T	chr4:106196819	50.73%	missense
IDH2	p.(R140Q)	c.419G>A	chr15:90631934	47.05%	missense
SRSF2	p.(P95T)	c.283C>A	chr17:74732960	55.15%	missense

Gene Fusions (RNA)

Genes	Variant ID	Locus
ZBTB16-RARA	ZBTB16-RARA.Z3R3	chr11:114027156 - chr17:38504568

FIG. 16.8.5 NGS demonstrated the fusion of *ZBTB16::RARA* and other gene mutations.

17q21.2 [148]. Some APL variants, including those with *ZBTB16::RARA* and *STAT5B::RARA* fusions, are resistant to tretinoin. Cases with these variant translocations should be diagnosed as APL with a variant *RARA* translocation [2,143–146].

The variant APL with t(11;17)(q23;q21) cases that are associated with the *ZBTB16::RARA* fusion gene has been reported as being resistant to all-trans-retinoic acid (ATRA). Therefore, differential diagnosis of variant APL with t(11;17)(q23;q12) from classical APL with t(15;17)(q22;q12)/*PML::RARA* is very important [149]. If FISH is negative for *PML::RARA* but three *RARA* signals are present or still have strong clinical suspicion, use *RARA* break-apart probes or karyotype to determine the partner. Our NGS hematology assay can detect *RARA* rearrangement with 16 different partners and a total of 45 transcripts, some of which cannot be identified by chromosome analysis or FISH due to cryptic or submicroscopic changes. Therefore, NGS analysis should be considered for APL variant cases. Also, NGS can identify other gene mutations. Molecular profiling of additional mutations may provide a platform to individualize therapeutic management in patients with this rare form of APL [150,151].

Future testing and recommendations

Since the patient had a variant *RARA* rearrangement (*ZBTB16::RARA* fusion), chromosome analysis, FISH using *RARA* break-apart probes, or NGS should be repeated to measure how this patient responds to treatment and to monitor the disease progression or relapse.

Case 16.9 Acute promyelocytic leukemia (APL) with a cryptic *PML::RARA* fusion

Clinical indication

A 32-year-old female presented with night sweats, pancytopenia, decreased platelets, and low fibrinogen. The blood smear consisted predominantly of normochromic, normocytic red blood cells. The bone marrow aspirate was cellular and pauciarticular. Most of the cells were abnormal promyelocytes and represent >90% of nucleated cells. Auer rods and APL cells were observed. The flow cytometry finding indicated APL.

Test ordered

- FISH: *PML::RARA* dual color dual fusion probes
- Chromosome analysis of the bone marrow
- NGS Hematology Molecular Profile
- RT-PCR: *PML::RARA* quantitative assay (send out)

Laboratory test performed

Chromosome analysis, FISH, and NGS methods were described previously in Chapters 1 and 12.

Test results

FISH for *PML::RARA* rearrangement was performed on interphase nuclei using probes localized to the *PML* (15q24.1) and *RARA* (17q21.1) gene regions. Two hundred nuclei were examined, and the results were negative for the rearrangement (Fig. 16.9.1A).

Then FISH with *RARA* break-apart probes was performed on interphase nuclei, and the results were negative for *RARA* rearrangement in all cells examined (Fig. 16.9.1B).

Next, chromosome analysis was performed on bone marrow. All 20 cells examined exhibited a normal female karyotype (Fig. 16.9.2).

Finally, the NGS hematology profile was performed and detected the fusion of *PML::RARA* in addition to *WT1* mutation (Fig. 16.9.3).

RT-PCR was concurrently performed in another lab and revealed the long fusion transcript of *PML::RARA* associated with t(15;17)(q24;q21) at a transcript level of 1478.933 with *ABL1* as the control gene (data not shown).

To further characterize these apparently discrepant results, a reanalysis of FISH using enhancement was performed on both metaphase and interphase cells. The enhanced FISH revealed a tiny cryptic insertion of *RARA* (green) into the *PML* (red) at chromosome 15q, which could not be appreciated in unenhanced images (Fig. 16.9.4A and B). This is a case of a cryptical *PML::RARA* fusion, likely due to a minor insertion, the lack of visibility on chromosome analysis, and the lack of detectability by conventional FISH.

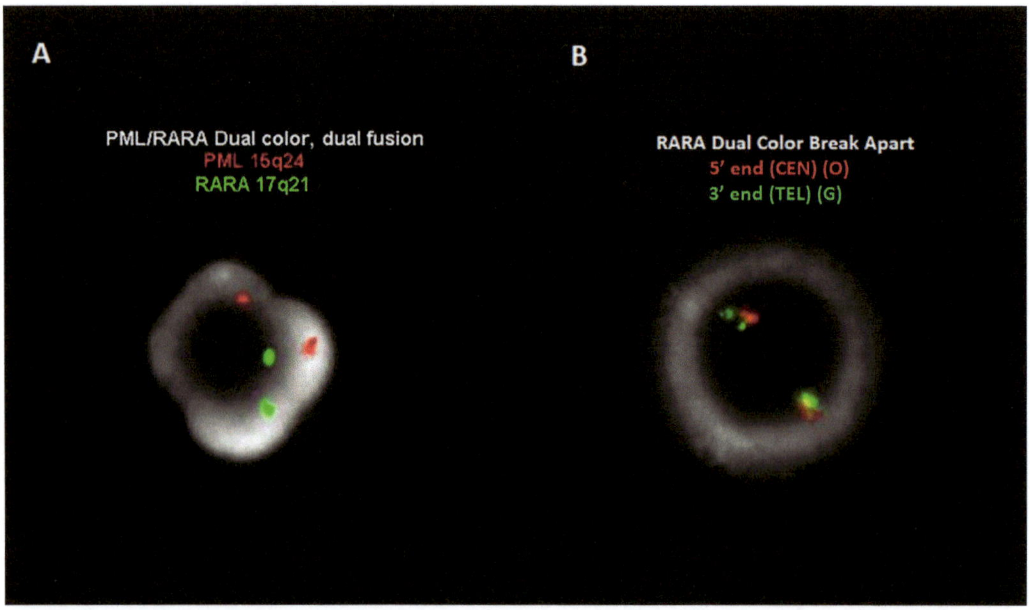

FIG. 16.9.1 FISH with *PML::RARA* dual-color dual-fusion probes and *RARA* dual-color break-apart probes both showed negative results for *PML::RARA* fusion (A) and *RARA* rearrangement (B). ISCN: nuc ish 15q24.1(PMLx2),17q21.1(RARAx2)[200].nuc ish 17q21.1(RARAx2)[200]

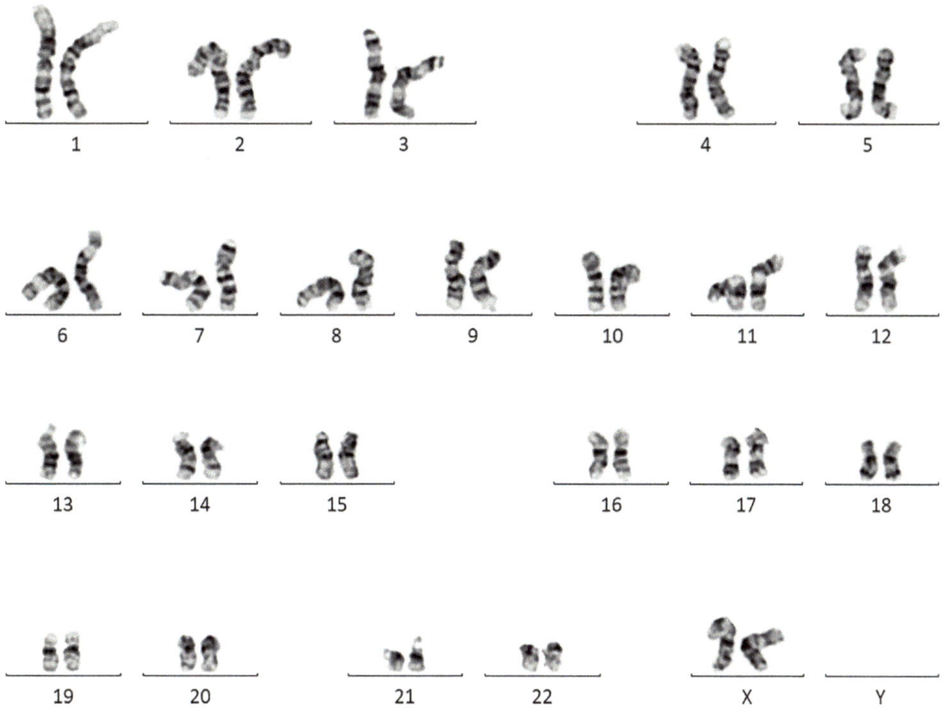

FIG. 16.9.2 Chromosome analysis revealed a normal female Karyotype, 46,XX[20]

Relevant Biomarkers

Tier	Genomic Alteration	Relevant Therapies (In this cancer type)	Clinical Trials
IA	*PML-RARA fusion* **Diagnostic significance:** Acute Promyelocytic Leukemia	**arsenic trioxide** [1, 2] **arsenic trioxide + tretinoin** [1, 2] anthracycline + arsenic trioxide anthracycline + arsenic trioxide + tretinoin arsenic trioxide + idarubicin + tretinoin cytarabine + daunorubicin + tretinoin gemtuzumab ozogamicin gemtuzumab ozogamicin + chemotherapy idarubicin + tretinoin	4
IIC	*WT1 p.(D469N) c.1405G>A*	None	1

Public data sources included in relevant therapies: FDA1, NCCN, **EMA**2, ESMO
Public data sources included in prognostic and diagnostic significance: NCCN, ESMO
Tier Reference: Li et al. *Standards and Guidelines for the Interpretation and Reporting of Sequence Variants in Cancer: A Joint Consensus Recommendation of the Association for Molecular Pathology, American Society of Clinical Oncology, and College of American Pathologists.* J Mol Diagn. 2017 Jan;19(1):4-23.

Variant Details

DNA Sequence Variants

Gene	Amino Acid Change	Coding	Locus	Allele Frequency	Variant Effect
WT1	p.(D469N)	c.1405G>A	chr11:32413560	12.81%	missense

Gene Fusions (RNA)

Genes	Variant ID	Locus
PML-RARA	PML-RARA.P6R3	chr15:74325755 - chr17:38504568

FIG. 16.9.3 NGS showed the fusion of *PML::RARA* and a *WT1* mutation.

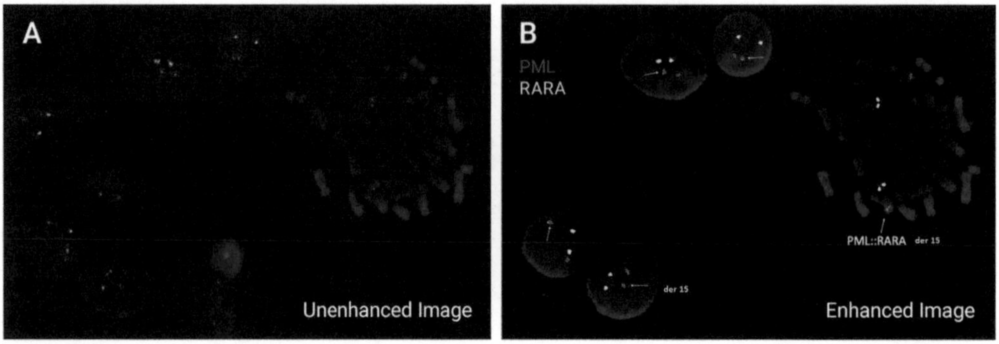

FIG. 16.9.4 FISH with *PML::RARA* dual-color dual-fusion probes showed negative results for *PML::RARA* fusion from the unenhanced image (A); FISH revealed a tiny cryptic insertion of *RARA (green)* into *PML (red)* from the enhanced image (B).

Results with interpretations

The conventional cytogenetics and FISH for *PML::RARA* rearrangement as well as *RARA* break-apart probes did not detect a fusion. In general, if morphology findings strongly suggest APL, then one should investigate further using NGS testing or metaphase FISH. In this case, NGS identified the *PML::RARA* rearrangement. Then reanalysis of FISH with enhancement revealed the cryptic fusion. Cryptic *PML::RARA* fusions have been reported in many studies; accurately identifying these cryptic fusions is crucial because patients with a typical *PML::RARA* fusion and with cryptic::*RARA* fusion both respond to targeted therapies including ATRA and ATO, with a favorable prognosis [153–156].

Our NGS hematology assay can detect *RARA* rearrangement with 16 different partners and a total of 45 transcripts, some of which cannot be identified by chromosome analysis or FISH due to cryptic or submicroscopic changes. Therefore, NGS analysis should be considered for APL variant cases.

Future testing and recommendations

NGS is superior to conventional cytogenetics and FISH in identifying fusions. It can be ordered for follow-up testing to monitor the disease progression or a relapse.

Case 16.10 Acute myeloid leukemia (AML) with jumping translocations

Clinical indication

A 45-year-old female presented with weakness and low blood counts. A bone marrow biopsy was collected. The blast level reached 34.35%. She was diagnosed with AML.

Test ordered

- Chromosome analysis of the bone marrow
- FISH: AML panel
- NGS Hematology Molecular Profile

Laboratory test performed

Chromosome analysis, FISH, and NGS methods were described previously in Chapters 1 and 12.

Test results

Chromosome analysis was performed initially. Of the 25 cells examined, 8 exhibited jumping translocations involving 7p with various partners that included 22q, 9q, and 1p (Figs. 16.10.1–16.10.3). The remaining 17 cells appear to be chromosomally normal.

FISH for AML panel was performed on interphase nuclei using probes localized to the *D5S721* (5p15.2), *EGR1* (5q31), *D7Z1* (7cen), *D7S486* (7q31), *D8Z2* (8cen), *RUNX1T1*

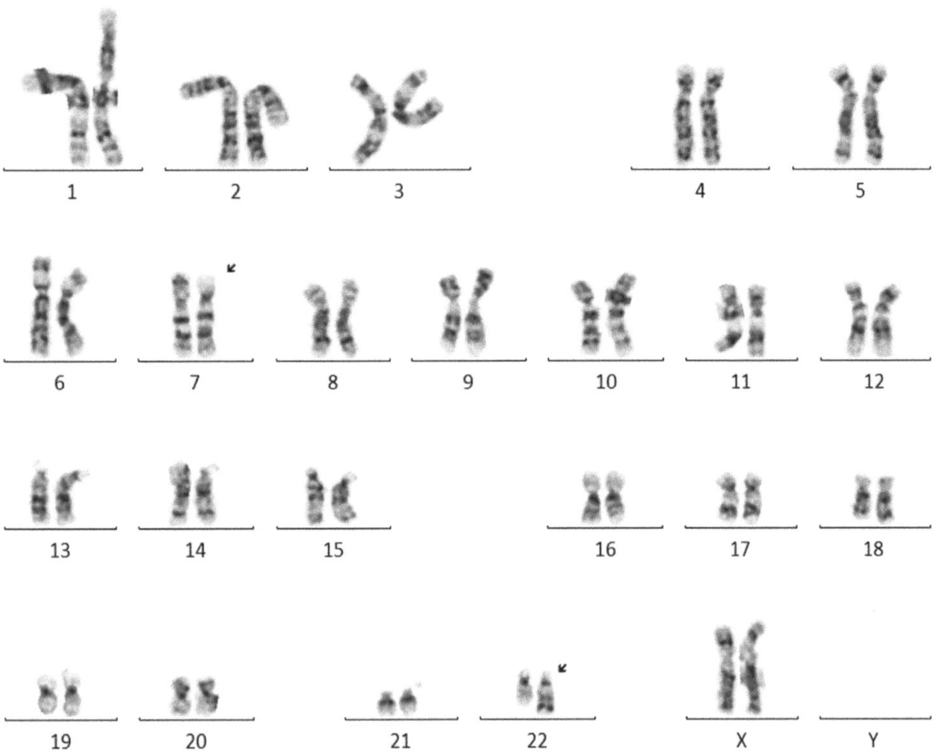

FIG. 16.10.1 Chromosome analysis showed the abnormal clone 1 with t(7;22). ISCN: 46,XX,t(7;22)(p12;q11.2)[5]/ 46,XX,t(7;9)(p12;q34)[2]/ 46,XX,t(1;7)(p12;p12)[1]/ 46,XX[17]

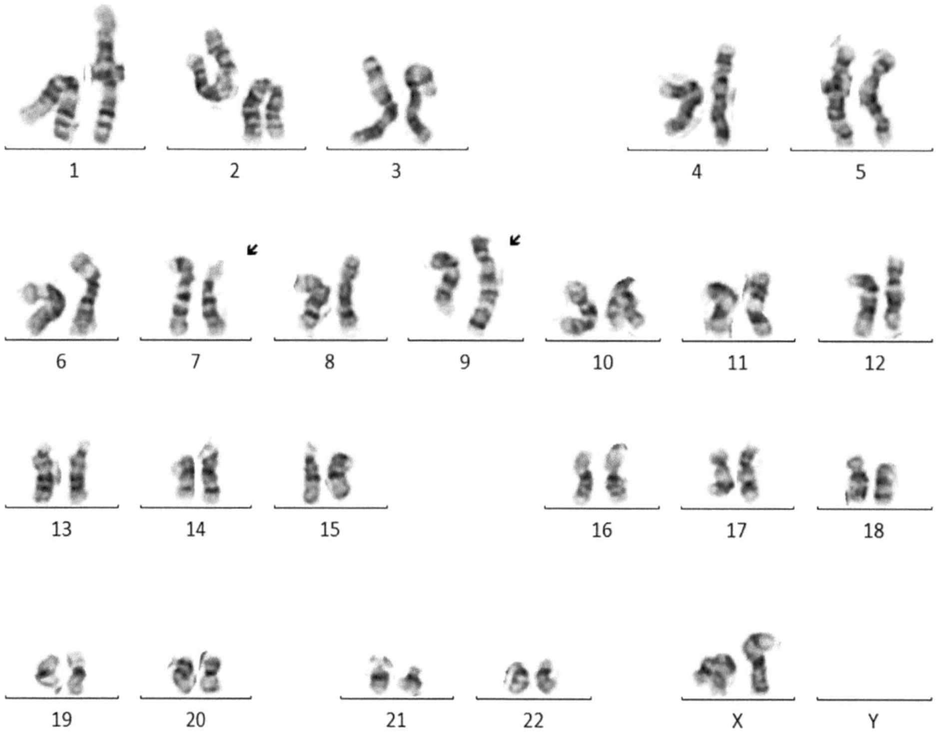

FIG. 16.10.2 Chromosome analysis showed the abnormal clone 2 with t(7;9).

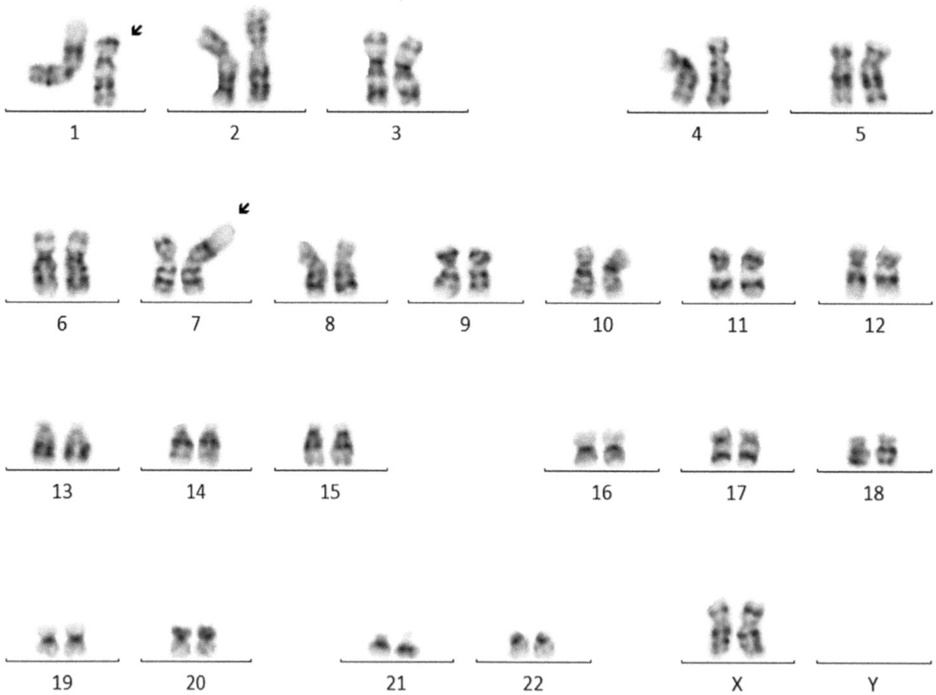

FIG. 16.10.3 Chromosome analysis showed the abnormal clone 3 with t(1;7).

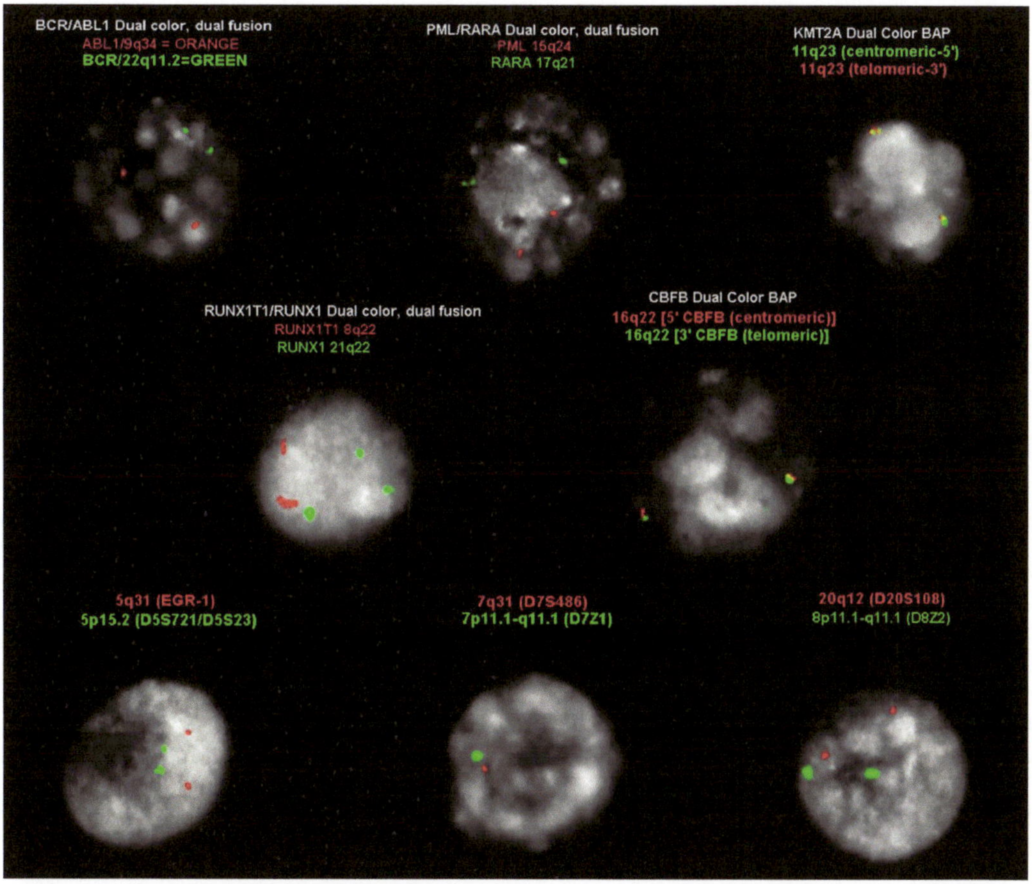

FIG. 16.10.4 FISH for AML panel identified monosomy 7 and gain of *RUNX1*. ISCN: nuc ish(D5S721,EGR1)x2[199/200],(CEP7,D7S486)x1[15/200],(D8Z2,D20S108)x2[200/200],(RUNX1T1x2,RUNX1x3)[9/200],(ABL1,BCR)x2[200/200],(KMT2Ax2)[200/200],(PML,RARA)x2[199/200],(CBFBx2)[198/200]

(8q22), *ABL1* (9q34.12), *KMT2A* (11q23), *PML* (15q24.1), *CBFB* (16q22.1), *RARA* (17q21.1), *D20S108* (20q12), *RUNX1* (21q22.3), and *BCR* (22q11) gene regions. Two hundred nuclei were examined, the results demonstrated monosomy 7 in 15/200 (7.5%) of the cells scored and three copies of *RUNX1* in 9/200 (4.5%) of the cells scored (Fig. 16.10.4).

NGS Hematology Molecular Profile identified *SF3B1*, *NRAS*, and *DNMT3A* mutations (Fig. 16.10.5).

Results with interpretations

The karyotyping and FISH results were consistent with a diagnosis of AML. Jumping translocations in myeloid malignancies are associated with treatment resistance and poor survival. The mutations identified from NGS were seen in myeloid disorders but were not

Relevant Biomarkers

Tier	Genomic Alteration	Relevant Therapies (In this cancer type)	Clinical Trials
IIC	*SF3B1 p.(K666N) c.1998G>T* **Prognostic significance:** None **Diagnostic significance:** None	None	3
IIC	*NRAS p.(G12C) c.34G>T* **Prognostic significance:** None **Diagnostic significance:** None	None	1

Public data sources included in relevant therapies: EMA[1], FDA[2], ESMO, NCCN
Public data sources included in prognostic and diagnostic significance: ESMO, NCCN
Tier Reference: Li et al. *Standards and Guidelines for the Interpretation and Reporting of Sequence Variants in Cancer: A Joint Consensus Recommendation of the Association for Molecular Pathology, American Society of Clinical Oncology, and College of American Pathologists.* J Mol Diagn. 2017 Jan;19(1):4-23.

Prevalent cancer biomarkers without relevant evidence based on included data sources
DNMT3A p.(R882C) c.2644C>T

Variant Details

DNA Sequence Variants

Gene	Amino Acid Change	Coding	Locus	Allele Frequency	Variant Effect
NRAS	p.(G12C)	c.34G>T	chr1:115258748	50.05%	missense
DNMT3A	p.(R882C)	c.2644C>T	chr2:25457243	48.57%	missense
SF3B1	p.(K666N)	c.1998G>T	chr2:198267359	47.95%	missense

FIG. 16.10.5 NGS detected *SF3B1*, *NRAS*, and *DNMT3A* mutations.

specific to AML. The degree of karyotypic complexity indicates a more aggressive disease process. Clinicopathologic correlation is advised.

Future testing and recommendations

Chromosome analysis, FISH, or NGS can be ordered to monitor the disease progression and treatment efficacy in the future.

Case 16.11 Acute myeloid leukemia (AML) with a complex karyotype and multiple mutations

Clinical indication

A 49-year-old female presented with the diagnosis of *FLT3*-positive acute myelogenous leukemia who had relapsed after a prior haploidentical stem cell transplant.

Test ordered

- Chromosome analysis of the bone marrow
- FISH: AML panel

- NGS Hematology Molecular Profile
- AML MRD assay (send out)
- *FLT3*-ITD MRD assay (send out)

Laboratory test performed

Chromosome analysis, FISH, and NGS methods were described previously in Chapters 1 and 12.

Test results

Chromosome analysis revealed 2 related abnormal clones from 10 out of 23 cells. These 10 cells were female chromosome complement, which was consistent with the patient's gender (Recipient Karyotype). Clone 1 with 8 cells exhibited a balanced translocation involving 3p and 12q, a dicentric chromosome involving 7q and 12q, add(14p), monosomy 17 possibly resulting in loss of *TP53*, and add(20q) (Fig. 16.11.1). Clone 2 with 2 cells had t(3;12), an isochromosome 13 for the long arm leading to a gain of 13q, add(14p), monosomy 17, and add(20q) (Fig. 16.11.2).

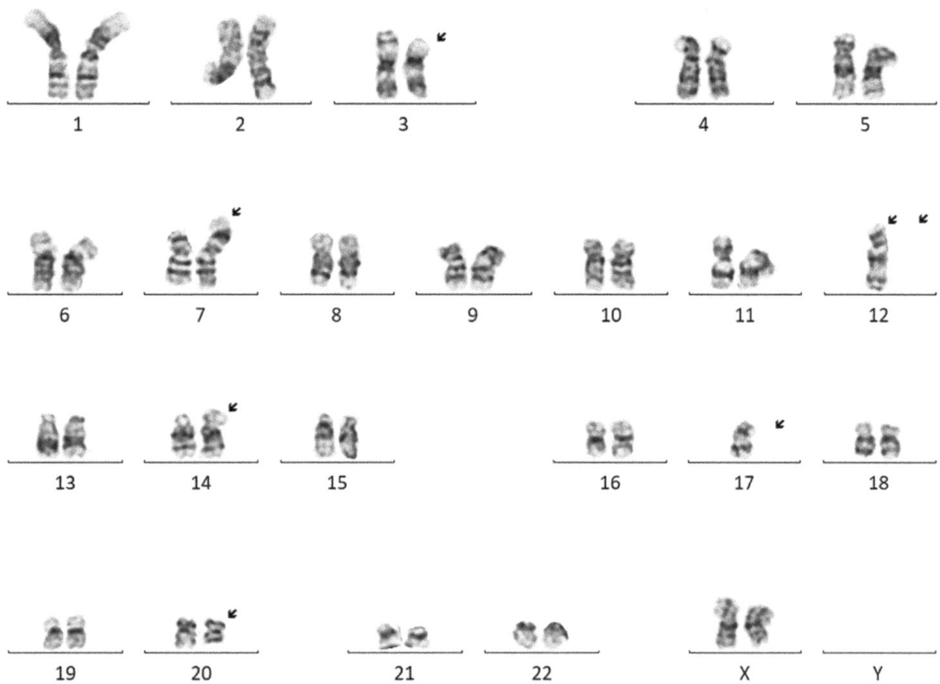

FIG. 16.11.1 Chromosome analysis showed the abnormal clone 1 with a complex karyotype including t(3;12) and other chromosomal abnormalities. ISCN: 44,XX,t(3;12)(p21;q24.1),dic(7;12)(p12;p11.2),add(14)(p11.2),-17, add(20)(q12)[8]/44,idem,-dic(7;12),+7,+12,i(13)(q10)[cp2]//46,XY[13]

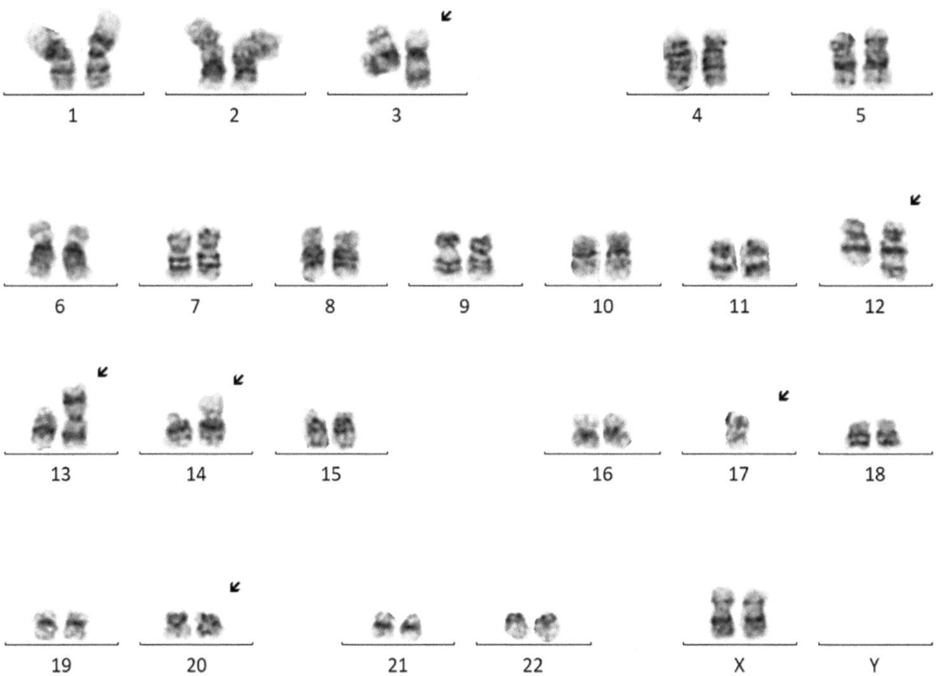

FIG. 16.11.2 Chromosome analysis showed the abnormal clone 2 with a complex karyotype with additional chromosomal abnormalities.

The sex chromosome complement, XY, of 13 metaphase cells examined is inconsistent with the patient's reported gender as a female. Based on the patient's history of bone marrow/stem cell transplant, these cells are therefore interpreted as being of donor origin.

FISH for the AML panel was performed on interphase nuclei using probes localized to the *D5S721* (5p15.2), *EGR1* (5q31), *D7Z1* (7cen), *D7S486* (7q31), *D8Z2* (8cen), *RUNX1T1* (8q22), *ABL1* (9q34.12), *KMT2A* (11q23), *PML* (15q24.1), *CBFB* (16q22.1), *RARA* (17q21.1), *D20S108* (20q12), *RUNX1* (21q22.3), and *BCR* (22q11) gene regions (supplied by Abbott Molecular, Inc.). Two hundred nuclei were examined, and the results demonstrated loss of one copy of *CBFB*, without evidence for rearrangement, in 166/200 (83.0%) of the cells scored (Fig. 16.11.3).

NGS identified several mutations including *FLT3*-ITD, *FLT3* p.(D835Y) (*FLT3*-TKD), *TP53* p.(R196*), and *RUNX1* p.(M78fs) (Fig. 16.11.4).

FLT3 MRD assay detected 159bp *FLT3*-ITD with 2.14E-0.3 variant read frequency (Fig. 16.11.5).

Results with interpretations

Two related abnormal clones with the sex chromosome complement, XX, were detected by chromosome analysis. The abnormal female chromosome complement was consistent with AML, status post bone marrow transplant, in the context of the pathology report. The concurrent FISH study also detected a deletion of *CBFB* (16q22), which was presumed to

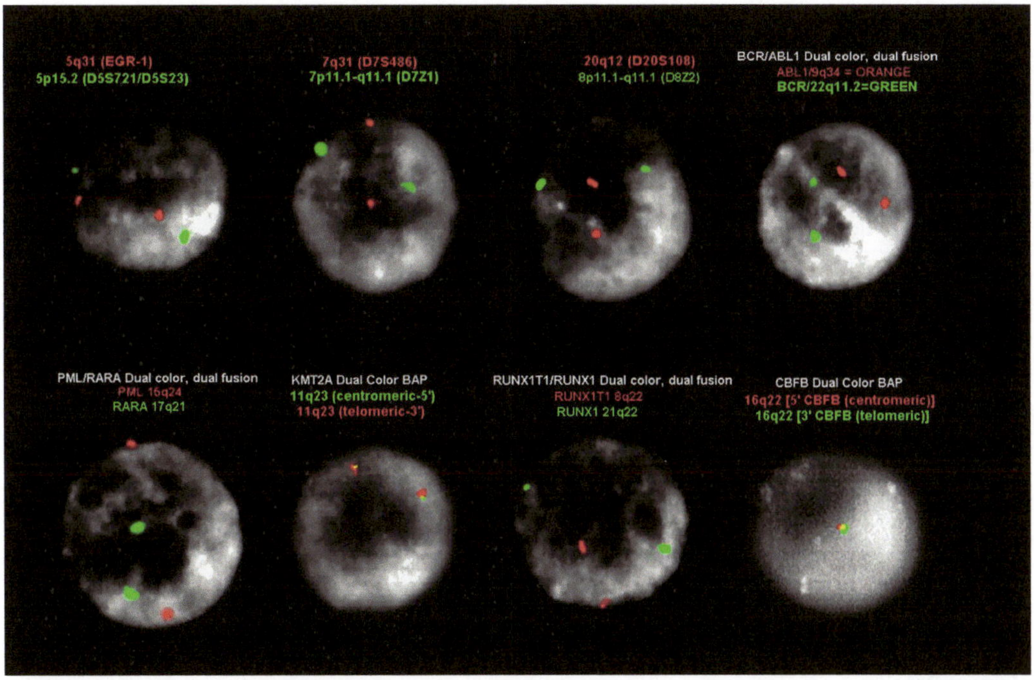

FIG. 16.11.3 FISH for the AML panel revealed a loss of one copy of CBFB in 83% of the cells scored. ISCN: nuc ish(D5S721,EGR1)x2[200/200],(CEP7,D7S486)x2[200/200],(D8Z2,D20S108)x2[200/200],(RUNX1T1,RUNX1)x2[200/200],(ABL1,BCR)x2[199/200],(KMT2Ax2)[200/200],(PML,RARA)x2[200/200],(CBFBx1)[166/200]

be cryptic at the conventional cytogenetic level. FISH also did not detect loss of 17p, suggesting that it might be located on one of the abnormal chromosomes. NGS detected *FLT3*-ITD, *FLT3*-TKD, *TP53*, and *RUNX1* mutations, concordant with the finding from *FLT3* MRD assay regarding *FLT3*-ITD. In the European Leukemia Net (ELN) scoring system for AML risk assessment, this combination of abnormalities is associated with an adverse prognosis. Clinicopathologic correlation is advised.

Future testing and recommendations

Chromosome analysis, FISH, or NGS can be ordered to monitor the disease progression and treatment efficacy in the future.

Case 16.12 Acute myeloid leukemia (AML) with t(10;11)(p12;q23)/*KMT2A::MLLT10* fusion

Clinical indication

A 48-year-old female was known to have a poor risk of refractory AML. She presented with weakness, cough, and lethargy. Her white blood count had been low, and the diagnosis was recurrent AML.

Relevant Biomarkers

Tier	Genomic Alteration	Relevant Therapies (In this cancer type)	Clinical Trials
IA	*FLT3 ITD mutation* **Prognostic significance:** ELN 2017: Intermediate to Adverse **Diagnostic significance:** None	**gilteritinib** [1, 2] **midostaurin + chemotherapy** [1, 2] sorafenib + chemotherapy venetoclax + chemotherapy	42
IA	*FLT3 p.(D835V) c.2504A>T* FLT3 TKD mutation **Prognostic significance:** None **Diagnostic significance:** None	**gilteritinib** [1, 2] **midostaurin + chemotherapy** [1, 2] venetoclax + chemotherapy	20
IA	*TP53 p.(R196*) c.586C>T* **Prognostic significance:** ELN 2017: Adverse **Diagnostic significance:** None	None	12
IA	*RUNX1 p.(M78fs) c.233_234insT* **Prognostic significance:** ELN 2017: Adverse **Diagnostic significance:** None	None	3

Public data sources included in relevant therapies: FDA1, NCCN, **EMA**2, ESMO
Public data sources included in prognostic and diagnostic significance: NCCN, ESMO
Tier Reference: Li et al. *Standards and Guidelines for the Interpretation and Reporting of Sequence Variants in Cancer: A Joint Consensus Recommendation of the Association for Molecular Pathology, American Society of Clinical Oncology, and College of American Pathologists.* J Mol Diagn. 2017 Jan;19(1):4-23.

Variant Details

DNA Sequence Variants

Gene	Amino Acid Change	Coding	Locus	Allele Frequency	Variant Effect
FLT3	p.(D835V)	c.2504A>T	chr13:28592641	7.88%	missense
FLT3	p.(R607_E608insRSNEYFYVDFREYEYDLKWEFPR)	c.1821_1822insAGATCCAATGAGTACTTCTACGTTGATTTCAGAGAATATGAATATGATCTCAAATGGGAGTTTCCAAGA	chr13:28608234	27.75%	nonframeshift Insertion
TP53	p.(R196*)	c.586C>T	chr17:7578263	5.65%	nonsense
RUNX1	p.(M78fs)	c.233_234insT	chr21:36259257	49.17%	frameshift Insertion

FIG. 16.11.4 NGS detected multiple mutations including *FLT3*-ITD.

Test Results:

Overall: POSITIVE for the *FLT3*-ITD Mutation.

***FLT3*-ITD Mutation(s) detected.**

ITD length	Variant Read Frequency
159bp	2.14E-03
69bp	2.14E-04
N/Abp	N/AE-N/A

FIG. 16.11.5 FLT3 MRD assay showed positive results for *FLT3*-ITD.

Test ordered

- Chromosome analysis of the bone marrow
- FISH: AML panel
- NGS Hematology Molecular Profile

Laboratory test performed

Chromosome analysis, FISH, and NGS methods were described previously in Chapters 1 and 12.

Test results

Chromosome analysis was performed initially. Of the 20 cells examined, 13 exhibited add(1p), a balanced translocation involving 2p and 6p, and a possible insertion of 10p to 11q, resulting in *KMT2A::MLLT10* Fusion (Fig. 16.12.1). The remaining seven cells appear to be chromosomally normal.

FISH for the AML panel was performed on interphase nuclei using probes localized to the *D5S721* (5p15.2), *EGR1* (5q31), *D7Z1* (7cen), *D7S486* (7q31), *D8Z2* (8cen), *RUNX1T1*

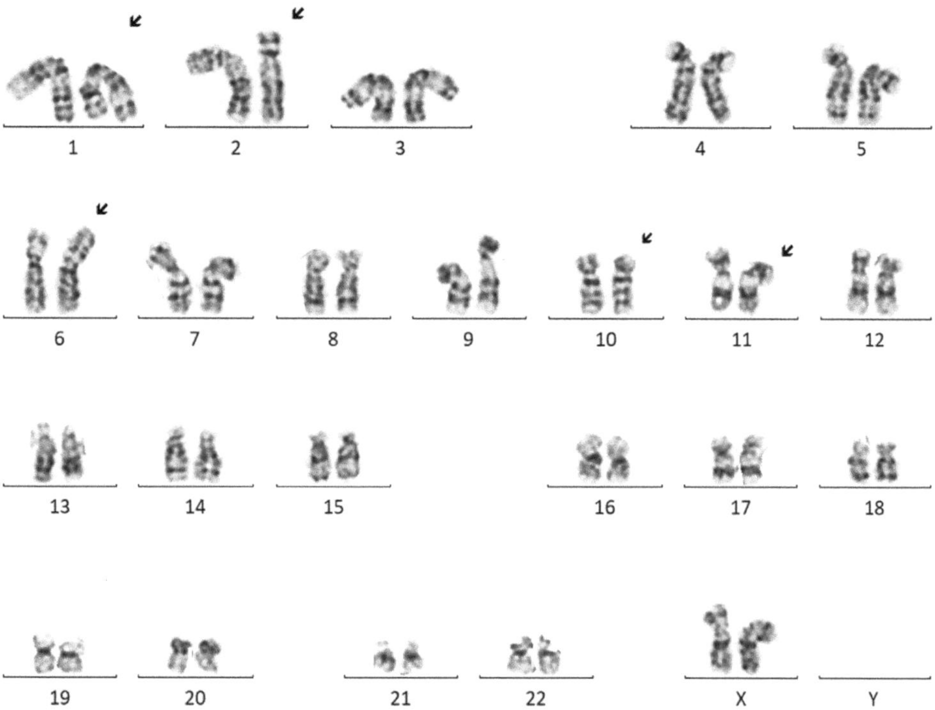

FIG. 16.12.1 Chromosome analysis showed an abnormal clone with a *KMT2A* rearrangement. ISCN: 46,XX,add(1)(p36.1),t(2;6)(p13;p21),?ins(11;10)(q23;p12p13)[13]/46,XX[7]

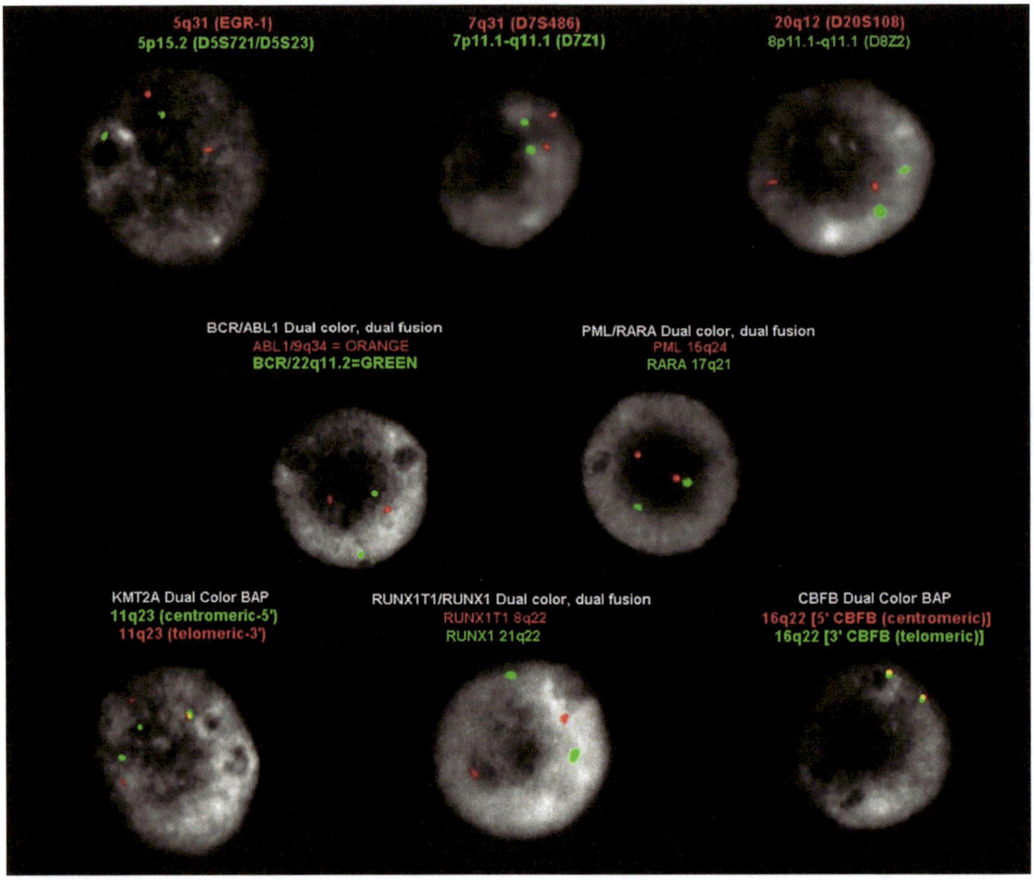

FIG. 16.12.2 FISH for the AML panel revealed a gain of *KMT2A* with a rearrangement involving two copies of *KMT2A* in 106/200 (53.0%) of the cells scored. ISCN: nuc ish(D5S721,EGR1)x2[200/200],(CEP7,D7S486)x2[196/200],(D8Z2, D20S108)x2[199/200],(RUNX1T1,RUNX1)x2[200/200],(ABL1,BCR)x2[200/200],(KMT2Ax3)(5'KMT2A sep 3'KMT2Ax2)[/200],(PML,RARA)x2[200/200],(CBFBx2)[199/200]

(8q22), *ABL1* (9q34.12), *KMT2A* (11q23), *PML* (15q24.1), *CBFB* (16q22.1), *RARA* (17q21.1), *D20S108* (20q12), *RUNX1* (21q22.3), and *BCR* (22q11) gene regions (supplied by Abbott Molecular, Inc.). Two hundred nuclei were examined, and the results demonstrated a gain of *KMT2A* with a rearrangement involving two *KMT2A* copies in 106/200 (53.0%) of the cells scored (Fig. 16.12.2).

NGS identified *KMT2A::MLLT10* fusion and a *DNMT3A* mutation (Fig. 16.12.3).

Results with interpretations

KMT2A rearrangement was seen from karyotyping, the concurrent FISH, and NGS testing. These results are consistent with a diagnosis of AML, and the abnormality is associated with an unfavorable prognosis. Clinicopathologic correlation is advised.

Relevant Biomarkers

Tier	Genomic Alteration	Relevant Therapies (In this cancer type)	Clinical Trials
IA	*KMT2A-MLLT10 fusion*	None	13
	Prognostic significance: ELN 2017: Adverse **Diagnostic significance:** None		

Public data sources included in relevant therapies: **FDA**1, NCCN, **EMA**2, ESMO
Public data sources included in prognostic and diagnostic significance: NCCN, ESMO
Tier Reference: Li et al. *Standards and Guidelines for the Interpretation and Reporting of Sequence Variants in Cancer: A Joint Consensus Recommendation of the Association for Molecular Pathology, American Society of Clinical Oncology, and College of American Pathologists.* J Mol Diagn. 2017 Jan;19(1):4-23.

Prevalent cancer biomarkers without relevant evidence based on included data sources
DNMT3A p.(R882H) c.2645G>A

Variant Details

DNA Sequence Variants

Gene	Amino Acid Change	Coding	Locus	Allele Frequency	Variant Effect
DNMT3A	p.(R882H)	c.2645G>A	chr2:25457242	13.40%	missense

Gene Fusions (RNA)

Genes	Variant ID	Locus
KMT2A-MLLT10	KMT2A-MLLT10.K8M4	chr11:118353210 - chr10:21875223

FIG. 16.12.3 NGS detected *KMT2A::MLLT10* fusion and a *DNMT3A* mutation.

Future testing and recommendations

Chromosome analysis, FISH, or NGS can be ordered to monitor the disease progression and treatment efficacy in the future.

Case 16.13 Acute myeloid leukemia (AML) with t(11;19)(q23;p13.3)/*KMT2A::MLLT1* fusion

Clinical indication

This is a 55-year-old female with a complex past medical history including breast cancer, complex regional pain syndrome, hypertension, type 2 diabetes, ulcerative colitis, gastroparesis, and sarcoidosis. She presented with shortness of breath. Her initial lab work showed abnormal results with a WBC of 5.3, Hgb of 5.7, and PLT of 16. Her peripheral blood flow showed a newly diagnosed therapy-related AML.

Test ordered

- Chromosome analysis of the bone marrow
- FISH: AML panel
- NGS Hematology Molecular Profile
- *FLT3*-ITD MRD assay (send out)

Laboratory test performed

Chromosome analysis, FISH, and NGS methods were described previously in Chapters 1 and 12.

Test results

Chromosome analysis revealed three related abnormal clones from all 20 cells examined, and clonal evolution was evident. Of the 20 cells examined, all exhibited t(11;19)(q23; p13.3)/*KMT2A::MLLT1* (Fig. 16.13.1); 17 cells also had two marker chromosomes in addition to t(11;19) (Fig. 16.13.2). The remaining two cells had an unbalanced translocation involving 1p and 11q with del(1p) (Fig. 16.13.3).

FISH for the AML panel was performed on interphase nuclei using probes localized to the *D5S721* (5p15.2), *EGR1* (5q31), *D7Z1* (7cen), *D7S486* (7q31), *D8Z2* (8cen), *RUNX1T1* (8q22), *ABL1* (9q34.12), *KMT2A* (11q23), *PML* (15q24.1), *CBFB* (16q22.1), *RARA* (17q21.1), *D20S108* (20q12), *RUNX1* (21q22.3), and *BCR* (22q11) gene regions (supplied by Abbott Molecular, Inc.). Two hundred nuclei were examined, and the results demonstrated *KMT2A* rearrangements in 192/200 (96.0%) of the the cells scored. This abnormality is associated with an unfavorable prognosis. The additional red signal represented the gain of *KMT2A* involving t(1;11) (Fig. 16.13.4).

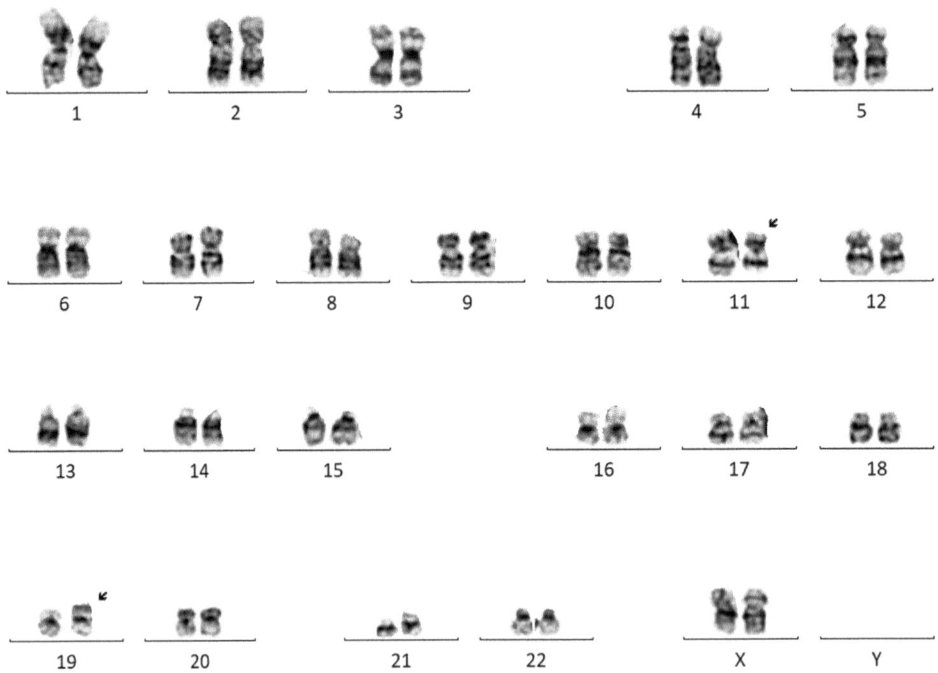

FIG. 16.13.1 Chromosome analysis showed the abnormal clone 1 with t(11;19). ISCN: 46,XX,t(11;19)(q23;p13.3)[1]/ 48,idem,+2mar[17]/ 46,idem,der(1)del(1)(p31p12)t(1;11)(p34.2;q23)[2]

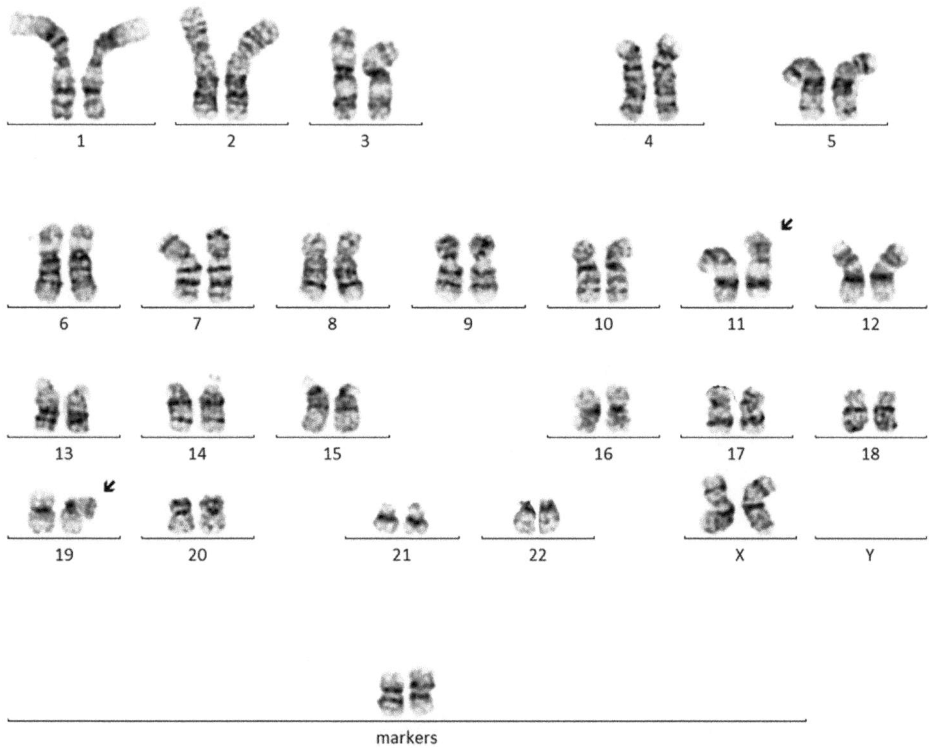

FIG. 16.13.2 Chromosome analysis showed the abnormal clone 2 with t(11;19) and two marker chromosomes.

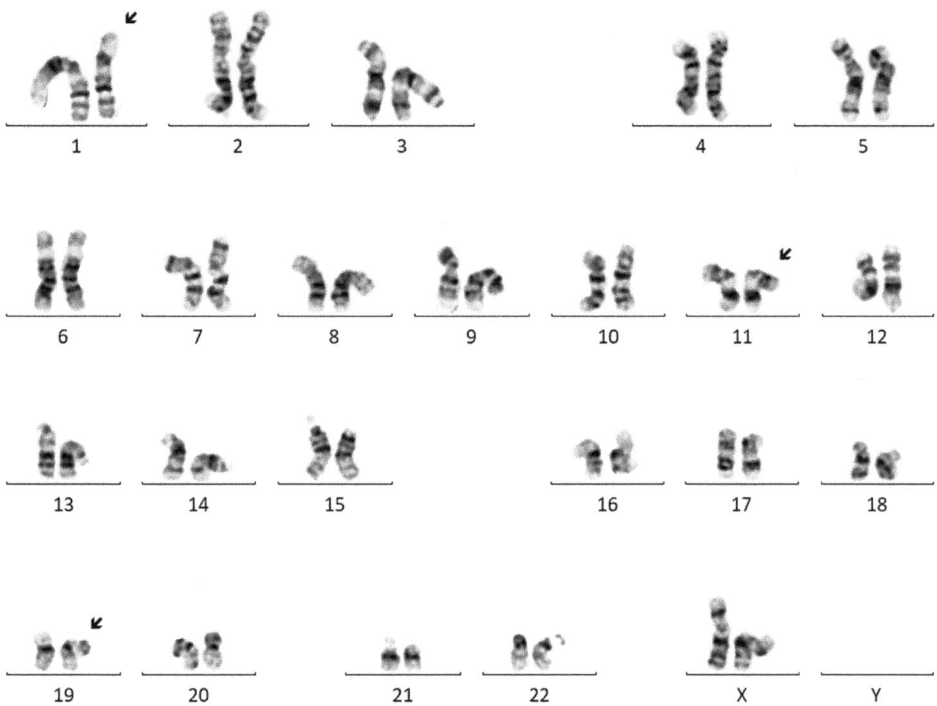

FIG. 16.13.3 Chromosome analysis showed the abnormal clone 2 with t(11;19), an unbalanced translocation involving 1p&11q and del(1p).

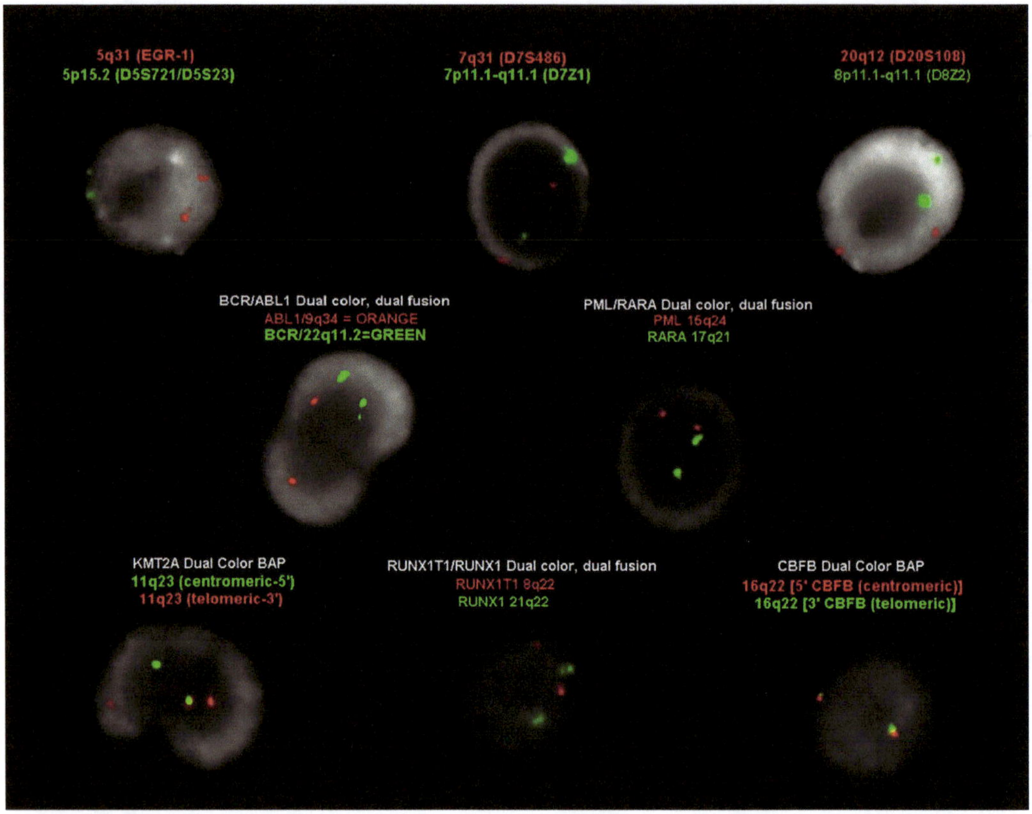

FIG. 16.13.4 FISH for the AML panel revealed a *KMT2A* rearrangement in 96% of the the cells scored. ISCN: nuc ish(D5S721,EGR1)x2[200/200],(CEP7,D7S486)x2[200/200],(D8Z2,D20S108)x2[200/200],(RUNX1T1,RUNX1)x2[200/200],(ABL1,BCR)x2[200/200],(KMT2Ax2)(5'KMT2A sep 3'KMT2Ax1)[136/200]/(5'KMT2Ax2,3'KMT2Ax3)(5'KMT2A con 3'KMT2Ax1)[56/200],(PML,RARA)x2[200/200],(CBFBx2)[200/200]

NGS identified a *KMT2A::MLLT1* fusion and *KRAS* p.(Q61H) c.183A>C mutation. The therapeutic drugs for this fusion were shown in the report (Fig. 16.13.5).

FLT3 MRD assay was negative for *FLT3*-ITD (data not shown).

Results with interpretations

Chromosome analysis identified a complex karyotype with t(11;19). *KMT2A* rearrangement was also seen from the concurrent FISH testing and NGS. These results are consistent with a diagnosis of therapy-related AML. The degree of karyotypic complexity indicates a more aggressive disease process. Clinicopathologic correlation is advised.

t(11;19)(q23;p13.3)/*KMT2A::MLLT1* has been reported in B-cell ALL, T-cell ALL, AML, and bi-phenotypic leukemia. Most cases are found in infants <1 year (congenital leukemia) with various phenotypes except in T-cell pediatric patients. Such a feature is particularly striking: most female cases exhibit a B-lineage or bi-phenotypic phenotype, and most male cases are M4/M5 cases. The prognosis is very poor (median <1 year) except

Relevant Biomarkers

Tier	Genomic Alteration	Relevant Therapies (In this cancer type)	Clinical Trials
IA	*KMT2A-MLLT1 fusion* **Prognostic significance:** ELN 2017: Adverse	allogeneic stem cells azacitidine cytarabine cytarabine + daunorubicin cytarabine + daunorubicin + etoposide cytarabine + etoposide + idarubicin cytarabine + fludarabine + idarubicin + filgrastim cytarabine + idarubicin cytarabine + mitoxantrone decitabine gemtuzumab ozogamicin + chemotherapy venetoclax + chemotherapy	16
IIC	*KRAS p.(Q61H) c.183A>C*	None	2

Public data sources included in relevant therapies: FDA[1], NCCN, **EMA**[2], ESMO
Public data sources included in prognostic and diagnostic significance: NCCN, ESMO
Tier Reference: Li et al. *Standards and Guidelines for the Interpretation and Reporting of Sequence Variants in Cancer: A Joint Consensus Recommendation of the Association for Molecular Pathology, American Society of Clinical Oncology, and College of American Pathologists.* J Mol Diagn. 2017 Jan;19(1):4-23.

Variant Details

DNA Sequence Variants

Gene	Amino Acid Change	Coding	Locus	Allele Frequency	Variant Effect
KRAS	p.(Q61H)	c.183A>C	chr12:25380275	5.99%	missense

Gene Fusions (RNA)

Genes	Variant ID	Locus
KMT2A-MLLT1	KMT2A-MLLT1.K8M4	chr11:118353210 - chr19:6230724

FIG. 16.13.5 NGS detected *KMT2A::MLLT1* fusion.

in the rare T-cell cases with long survivors (https://atlasgeneticsoncology.org/haematological/1071/t(11;19)(q23;p13-3)).

Future testing and recommendations

Chromosome analysis, FISH, or NGS can be ordered to monitor the disease progression and treatment efficacy in the future.

Case 16.14 Acute myeloid leukemia with *NUP98::KDM5A* fusion

Clinical indication

A 1-year-old female presented with a differential diagnosis that included acute megakaryoblastic leukemia (AMKL or AML-M7 category of the French-American-British classification). Bone marrow aspirate was sent in for the following laboratory tests.

Test ordered

- Immunophenotypic analysis by flow cytometry
- Chromosome analysis of the bone marrow
- FISH study with pediatric AML panel

Laboratory test performed

Flow cytometry analysis was performed on bone marrow aspirate. Chromosome analysis on bone marrow aspirate was performed using the culture condition optimized for precursor and myeloid cell growth. FISH was performed on bone marrow aspirate with pediatric AML panel including the probe sets (supplied by Oxford Gene Technology, Inc.) covering the following loci: *MECOM* at 3q26.2, *NUP98* at 11p15, *KMT2A* at 11q23, and 7q probes for 7q22 and 7q31. The FISH panel also includes the dual-color dual-fusion probe sets for detecting *DEK::NUP214* fusion derived from t(6;9)(p22;q34), *RUNX1::RUNX1T1* fusion derived from t(8;21)(q21.3;q22), *PML::RARA* fusion derived from t(15;17)(q24;q21), and *CBFB::MYH11* fusion derived from inv(16)(p13;q22) or t(16;16)(p113;q22). *NUP98::KDM5A* quantitative real-time PCR assay was performed as a reflex test based on chromosome and FISH results.

Test results

The flow cytometric findings revealed an aberrant myeloblast population identified at 23% of total nonerythroid cells with abnormal maturing myeloid cells. FISH study with *NUP98* dual-color break-apart probe revealed an abnormal signal pattern (one fusion signal, a separate red signal, and a separate green signal) instead of two intact fusion signals in 29% of nuclei examined (Fig. 16.14.1A and B). The FISH analysis also showed low-level gains of 8q, 16q, and 21q in 1.5%–6% of nuclei examined, suggesting the presence of a low-level subclone. Chromosome analysis revealed a complex abnormal female karyotype with a three-way translocation involving 11q, 12p, a likely 13q; monosomy 13, trisomy 16, add(16p), trisomy 21, and a marker chromosome in 12 out of 20 (60%) of the cells examined (Fig. 16.14.1C). The remaining eight cells appear to be chromosomally normal.

The reflex quantitative real-time PCR assay detected *NUP98::KDM5A* fusion transcript with a quantitative NCN (normalized copy numbers) value at 0.047.

Results with interpretations

The FISH finding revealed a typical *NUP98* break-apart signal pattern, consistent with *NUP98* gene rearrangement. However, this finding cannot define the partner gene/chromosome involving the *NUP98* rearrangement. The RT-PCR results verified the *NUP98* rearrangement at the molecular level and further revealed *NUP98::KDM5A* fusion. Chromosome analysis revealed complex abnormalities including a three-way translocation t(11;12;13)(q24;p13;?q12) involving chromosomes 11q, 12p, and 13q, and other chromosome abnormalities. This three-way translocation cannot be defined with *NUP98::KDM5A*

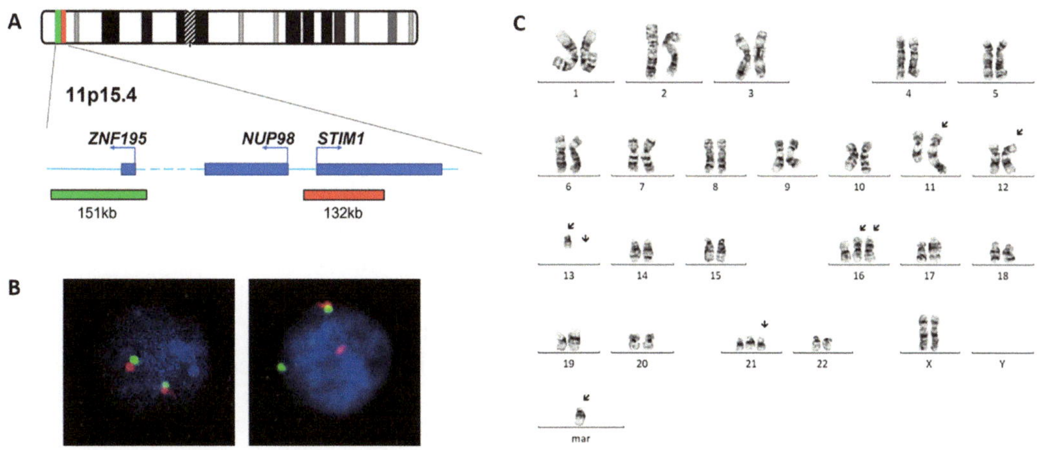

FIG. 16.14.1 Diagram of the *NUP98* break-apart FISH probe set that covers the *5′NUP98* and *3′NUP98* genomic regions (A). A normal interphase nucleus exhibited two fusion signals. An interphase nucleus revealed an abnormal signal pattern: one fusion, one *green*, and one *red* signal (B). G-banded chromosomes demonstrated a complex karyotype with a three-way translocation involving 11q, 12p, a likely 13q, and other abnormalities (C). ISCN: 48,XX,t(11;12;13)(q24;p13;?q12),-13,+16,add(16)(p11.2)x2,+21,+mar[12]/46,XX[8]

gene rearrangement since the *NUP98* locus localizes at 11p15. Due to the cryptic nature of the chromosome morphology changes associated with this translocation, the translocation t(11;12)(p15;p13) that leads to *NUP98::KDM5A* gene fusion cannot be detected by chromosome analysis.

These genetic findings support the diagnosis of AML with *NUP98* rearrangement, which confers a poor prognosis. Studies showed that more than 40 partner genes of *NUP98* have been reported, and *NSD1* at 5q35 and *KMD5A* at 12p13 are the most common partner genes. Abnormal megakaryoblasts are most commonly seen in patients with *NUP98::KMD5A* and *NUP98::JARID1A* fusion [64,65]. Assessment of *NUP98::KDM5A* gene rearrangement was recommended as part of the standard survey in pediatric AML. Hematopoietic stem cell transplant in the first complete remission was also indicated for such patients [68].

Future testing and recommendations

The flow-cytometric assay can be applied to detect measurable residual disease (MRD) in future specimens. *NUP98::KDM5A* quantitative real-time PCR assay can be used for future disease monitoring with the quantitative NCN of 0.047 as the patient-specific baseline.

NUP98 gene rearrangement should be evaluated for AML patients who present a normal karyotype or karyotype with no other AML-related recurrent chromosomal aberrations, particularly for pediatric patients with megakaryoblast. FISH with the *NUP98* break-apart probe can be useful in detection of most of the *NUP98* gene rearrangements. Fusion-specific quantitative RT-PCR can be a sensitive approach for disease monitoring.

The positive finding of *NUP98* rearrangements can classify patients into a high-risk group and warrant appropriate treatment strategies for patients.

Summary of key learning points

- AML is the most common type of acute leukemia in adults.
- AML has many subtypes, and each has different chromosome abnormalities that can be tested for disease diagnosis and treatment decision.
- Chromosomal rearrangements involving *KMT2A* may be missed due to subtle changes near the telomeres of the chromosomes involved.
- FISH, RT-PCR, and NGS can help identify the translocations in case conventional cytogenetics may overlook them.

References

[1] J.D. Khoury, et al., The 5th edition of the World Health Organization classification of haematolymphoid tumours: myeloid and histiocytic/dendritic neoplasms, Leukemia 36 (7) (2022) 1703–1719.

[2] D.A. Arber, et al., International consensus classification of myeloid neoplasms and acute leukemias: integrating morphologic, clinical, and genomic data, Blood 140 (11) (2022) 1200–1228.

[3] S. David, V. Mathews, Mechanisms and management of coagulopathy in acute promyelocytic leukemia, Thromb. Res. 164 (Suppl 1) (2018) S82–S88.

[4] A. Ferrari, et al., Immunophenotype distinction between acute promyelocytic leukaemia and CD15- CD34- HLA-DR- acute myeloid leukaemia with nucleophosmin mutations, Hematol. Oncol. 30 (3) (2012) 109–114.

[5] P. Horna, et al., Diagnostic immunophenotype of acute promyelocytic leukemia before and early during therapy with all-trans retinoic acid, Am. J. Clin. Pathol. 142 (4) (2014) 546–552.

[6] M. Liu, et al., Flow cytometric analysis of CD64 expression pattern and density in the diagnosis of acute promyelocytic leukemia: a multi-center study in Shanghai, China, Oncotarget 8 (46) (2017) 80625–80637.

[7] A. Orfao, et al., The flow cytometric pattern of CD34, CD15 and CD13 expression in acute myeloblastic leukemia is highly characteristic of the presence of PML-RARalpha gene rearrangements, Haematologica 84 (5) (1999) 405–412.

[8] V. Lallemand-Breitenbach, et al., Retinoic acid and arsenic synergize to eradicate leukemic cells in a mouse model of acute promyelocytic leukemia, J. Exp. Med. 189 (7) (1999) 1043–1052.

[9] F. Lo-Coco, et al., Retinoic acid and arsenic trioxide for acute promyelocytic leukemia, N. Engl. J. Med. 369 (2) (2013) 111–121.

[10] U. Platzbecker, et al., Improved outcomes with retinoic acid and arsenic trioxide compared with retinoic acid and chemotherapy in non-high-risk acute promyelocytic leukemia: final results of the randomized Italian-German APL0406 trial, J. Clin. Oncol. 35 (6) (2017) 605–612.

[11] C. Signorelli, M. Ferdico, Use of cefotetan in antibiotic prevention in gynecologic surgical interventions, Clin. Ter. 125 (5) (1988) 345–351.

[12] M.A. Sanz, et al., Management of acute promyelocytic leukemia: updated recommendations from an expert panel of the European LeukemiaNet, Blood 133 (15) (2019) 1630–1643.

[13] S.Y. Han, et al., Secondary cytogenetic abnormalities in core-binding factor AML harboring inv(16) vs t(8;21), Blood Adv. 5 (10) (2021) 2481–2489.

[14] M. Yabe, et al., Systemic mastocytosis associated with t(8,21) acute myeloid leukemia in a child: detection of the D816A mutation of KIT, Pediatr. Blood Cancer 59 (7) (2012) 1313–1316.

[15] R.C. Johnson, et al., Hidden mastocytosis in acute myeloid leukemia with t(8;21)(q22;q22), Am. J. Clin. Pathol. 140 (4) (2013) 525–535.

[16] D.H. Yang, et al., Predictable prognostic factor of CD56 expression in patients with acute myeloid leukemia with t(8:21) after high dose cytarabine or allogeneic hematopoietic stem cell transplantation, Am. J. Hematol. 82 (1) (2007) 1–5.

[17] N. Iriyama, et al., CD56 expression is an independent prognostic factor for relapse in acute myeloid leukemia with t(8;21), Leuk. Res. 37 (9) (2013) 1021–1026.

[18] L. Shang, et al., The immunophenotypic characteristics and flow cytometric scoring system of acute myeloid leukemia with t(8;21) (q22;q22); RUNX1-RUNX1T1, Int. J. Lab. Hematol. 41 (1) (2019) 23–31.

[19] H. Khoury, et al., Acute myelogenous leukemia with t(8;21)—identification of a specific immunophenotype, Leuk. Lymphoma 44 (10) (2003) 1713–1718.

[20] D. Ray, et al., Lineage-inappropriate PAX5 expression in t(8;21) acute myeloid leukemia requires signaling-mediated abrogation of polycomb repression, Blood 122 (5) (2013) 759–769.

[21] R.K. Hills, et al., Addition of gemtuzumab ozogamicin to induction chemotherapy in adult patients with acute myeloid leukaemia: a meta-analysis of individual patient data from randomised controlled trials, Lancet Oncol. 15 (9) (2014) 986–996.

[22] S. Vasu, et al., Ten-year outcome of patients with acute myeloid leukemia not treated with allogeneic transplantation in first complete remission, Blood Adv. 2 (13) (2018) 1645–1650.

[23] C. Ustun, et al., Core-binding factor acute myeloid leukemia with t(8;21): Risk factors and a novel scoring system (I-CBFit), Cancer Med. 7 (9) (2018) 4447–4455.

[24] M.T. Krauth, et al., High number of additional genetic lesions in acute myeloid leukemia with t (8;21)/RUNX1-RUNX1T1: frequency and impact on clinical outcome, Leukemia 28 (7) (2014) 1449–1458.

[25] M. Hamosh, Does infant nutrition affect adiposity and cholesterol levels in the adult? J. Pediatr. Gastroenterol. Nutr. 7 (1) (1988) 10–16.

[26] Y. Wang, et al., In adults with t(8;21)AML, posttransplant RUNX1/RUNX1T1-based MRD monitoring, rather than c-KIT mutations, allows further risk stratification, Blood 124 (12) (2014) 1880–1886.

[27] F.F. Yalniz, et al., Significance of minimal residual disease monitoring by real-time quantitative polymerase chain reaction in core binding factor acute myeloid leukemia for transplantation outcomes, Cancer 126 (10) (2020) 2183–2192.

[28] C. Meyer, et al., The MLL recombinome of acute leukemias in 2017, Leukemia 32 (2) (2018) 273–284.

[29] J.D. Rowley, et al., All patients with the T(11;16)(q23;p13.3) that involves MLL and CBP have treatment-related hematologic disorders, Blood 90 (2) (1997) 535–541.

[30] W. Xie, et al., t(11;16)(q23;p13)/KMT2A-CREBBP in hematologic malignancies: presumptive evidence of myelodysplasia or therapy-related neoplasm? Ann. Hematol. 99 (3) (2020) 487–500.

[31] H. Bolouri, et al., The molecular landscape of pediatric acute myeloid leukemia reveals recurrent structural alterations and age-specific mutational interactions, Nat. Med. 24 (1) (2018) 103–112.

[32] G.C. Issa, et al., Predictors of outcomes in adults with acute myeloid leukemia and KMT2A rearrangements, Blood Cancer J. 11 (9) (2021) 162.

[33] M. Bill, et al., Mutational landscape and clinical outcome of patients with de novo acute myeloid leukemia and rearrangements involving 11q23/KMT2A, Proc. Natl. Acad. Sci. U. S. A. 117 (42) (2020) 26340–26346.

[34] K. Mrozek, et al., Adult patients with de novo acute myeloid leukemia and t(9;11)(p22;q23) have a superior outcome to patients with other translocations involving band 11q23: a cancer and leukemia group B study, Blood 90 (11) (1997) 4532–4538.

[35] B.V. Balgobind, et al., Novel prognostic subgroups in childhood 11q23/MLL-rearranged acute myeloid leukemia: results of an international retrospective study, Blood 114 (12) (2009) 2489–2496.

[36] J.D. de Rooij, et al., Pediatric non-down syndrome acute megakaryoblastic leukemia is characterized by distinct genomic subsets with varying outcomes, Nat. Genet. 49 (3) (2017) 451–456.

[37] Y. Hara, et al., Prognostic impact of specific molecular profiles in pediatric acute megakaryoblastic leukemia in non-down syndrome, Genes Chromosom. Cancer 56 (5) (2017) 394–404.

[38] H. Dohner, et al., Diagnosis and management of AML in adults: 2017 ELN recommendations from an international expert panel, Blood 129 (4) (2017) 424–447.

[39] C. Glass, et al., Global identification of EVI1 target genes in acute myeloid leukemia, PLoS One 8 (6) (2013), e67134.

[40] S. Groschel, et al., A single oncogenic enhancer rearrangement causes concomitant EVI1 and GATA2 deregulation in leukemia, Cell 157 (2) (2014) 369–381.

[41] H. Yamazaki, et al., A remote GATA2 hematopoietic enhancer drives leukemogenesis in inv(3)(q21;q26) by activating EVI1 expression, Cancer Cell 25 (4) (2014) 415–427.

[42] B.C. Medeiros, et al., Immunophenotypic features of acute myeloid leukemia with inv(3)(q21q26.2)/t(3;3)(q21;q26.2), Leuk. Res. 34 (5) (2010) 594–597.

[43] J.M. Raya, et al., Acute myeloid leukemia with inv(3)(q21q26.2) or t(3;3)(q21;q26.2): clinical and biological features and comparison with other acute myeloid leukemias with cytogenetic aberrations involving long arm of chromosome 3, Hematology 20 (8) (2015) 435–441.

[44] C. Haferlach, et al., Three novel cytogenetically cryptic EVI1 rearrangements associated with increased EVI1 expression and poor prognosis identified in 27 acute myeloid leukemia cases, Genes Chromosom. Cancer 51 (12) (2012) 1079–1085.

[45] S. Ottema, et al., Atypical 3q26/MECOM rearrangements genocopy inv(3)/t(3;3) in acute myeloid leukemia, Blood 136 (2) (2020) 224–234.

[46] I. Summerer, et al., Prognosis of MECOM (EVI1)-rearranged MDS and AML patients rather depends on accompanying molecular mutations than on blast count, Leuk. Lymphoma 61 (7) (2020) 1756–1759.

[47] S. Ottema, et al., The leukemic oncogene EVI1 hijacks a MYC super-enhancer by CTCF-facilitated loops, Nat. Commun. 12 (1) (2021) 5679.

[48] Z. Tang, et al., Deciphering the complexities of MECOM rearrangement-driven chromosomal aberrations, Cancer Genet. 233-234 (2019) 21–31.

[49] M. Zhao, et al., Newly designed breakapart FISH probe helps to identify cases with true MECOM rearrangement in myeloid malignancies, Cancer Genet. 262-263 (2022) 23–29.

[50] L.M. Secker-Walker, A. Mehta, B. Bain, Abnormalities of 3q21 and 3q26 in myeloid malignancy: a United Kingdom cancer cytogenetic group study, Br. J. Haematol. 91 (2) (1995) 490–501.

[51] S. Lugthart, et al., Clinical, molecular, and prognostic significance of WHO type inv(3)(q21q26.2)/t(3;3)(q21;q26.2) and various other 3q abnormalities in acute myeloid leukemia, J. Clin. Oncol. 28 (24) (2010) 3890–3898.

[52] C.P. Soupir, et al., Philadelphia chromosome-positive acute myeloid leukemia: a rare aggressive leukemia with clinicopathologic features distinct from chronic myeloid leukemia in myeloid blast crisis, Am. J. Clin. Pathol. 127 (4) (2007) 642–650.

[53] S. Konoplev, et al., Molecular characterization of de novo Philadelphia chromosome-positive acute myeloid leukemia, Leuk. Lymphoma 54 (1) (2013) 138–144.

[54] C. Orsmark-Pietras, et al., Clinical and genomic characterization of patients diagnosed with the provisional entity acute myeloid leukemia with BCR-ABL1, a Swedish population-based study, Genes Chromosom. Cancer 60 (6) (2021) 426–433.

[55] E.P. Nacheva, et al., Deletions of immunoglobulin heavy chain and T cell receptor gene regions are uniquely associated with lymphoid blast transformation of chronic myeloid leukemia, BMC Genomics 11 (2010) 41.

[56] E.P. Nacheva, et al., Does BCR/ABL1 positive acute myeloid leukaemia exist? Br. J. Haematol. 161 (4) (2013) 541–550.

[57] M.L. Slovak, et al., A retrospective study of 69 patients with t(6;9)(p23;q34) AML emphasizes the need for a prospective, multicenter initiative for rare 'poor prognosis' myeloid malignancies, Leukemia 20 (7) (2006) 1295–1297.

[58] J.D. Sandahl, et al., t(6;9)(p22;q34)/DEK-NUP214-rearranged pediatric myeloid leukemia: an international study of 62 patients, Haematologica 99 (5) (2014) 865–872.

[59] K. Tarlock, et al., Acute myeloid leukaemia (AML) with t(6;9)(p23;q34) is associated with poor outcome in childhood AML regardless of FLT3-ITD status: a report from the Children's Oncology Group, Br. J. Haematol. 166 (2) (2014) 254–259.

[60] E. Papaemmanuil, et al., Genomic classification and prognosis in acute myeloid leukemia, N. Engl. J. Med. 374 (23) (2016) 2209–2221.

[61] H.B. Ommen, et al., The kinetics of relapse in DEK-NUP214-positive acute myeloid leukemia patients, Eur. J. Haematol. 95 (5) (2015) 436–441.

[62] H. Fang, et al., Myelodysplastic syndrome with t(6;9)(p22;q34.1)/DEK-NUP214 better classified as acute myeloid leukemia? A multicenter study of 107 cases, Mod. Pathol. 34 (6) (2021) 1143–1152.

[63] K.M. Chisholm, et al., Acute erythroid leukemia is enriched in NUP98 fusions: a report from the Children's Oncology Group, Blood Adv. 4 (23) (2020) 6000–6008.

[64] J.D. de Rooij, et al., NUP98/JARID1A is a novel recurrent abnormality in pediatric acute megakaryoblastic leukemia with a distinct HOX gene expression pattern, Leukemia 27 (12) (2013) 2280–2288.

[65] Y. Hara, et al., Patients aged less than 3 years with acute myeloid leukaemia characterize a molecularly and clinically distinct subgroup, Br. J. Haematol. 188 (4) (2020) 528–539.

[66] N. Niktoreh, et al., Mutated WT1, FLT3-ITD, and NUP98-NSD1 fusion in various combinations define a poor prognostic group in pediatric acute myeloid leukemia, J. Oncol. 2019 (2019) 1609128.

[67] N.L. Michmerhuizen, J.M. Klco, C.G. Mullighan, Mechanistic insights and potential therapeutic approaches for NUP98-rearranged hematologic malignancies, Blood 136 (20) (2020) 2275–2289.

[68] S. Noort, et al., The clinical and biological characteristics of NUP98-KDM5A in pediatric acute myeloid leukemia, Haematologica 106 (2) (2021) 630–634.

[69] V. Bisio, et al., NUP98-fusion transcripts characterize different biological entities within acute myeloid leukemia: a report from the AIEOP-AML group, Leukemia 31 (4) (2017) 974–977.

[70] N.A. McNeer, et al., Genetic mechanisms of primary chemotherapy resistance in pediatric acute myeloid leukemia, Leukemia 33 (8) (2019) 1934–1943.

[71] J.D. de Rooij, et al., Recurrent abnormalities can be used for risk group stratification in pediatric AMKL: a retrospective intergroup study, Blood 127 (26) (2016) 3424–3430.

[72] K. Dohner, et al., Mutant nucleophosmin (NPM1) predicts favorable prognosis in younger adults with acute myeloid leukemia and normal cytogenetics: interaction with other gene mutations, Blood 106 (12) (2005) 3740–3746.

[73] B. Falini, et al., Cytoplasmic nucleophosmin in acute myelogenous leukemia with a normal karyotype, N. Engl. J. Med. 352 (3) (2005) 254–266.

[74] C. Thiede, et al., Prevalence and prognostic impact of NPM1 mutations in 1485 adult patients with acute myeloid leukemia (AML), Blood 107 (10) (2006) 4011–4020.

[75] B. Falini, et al., Acute myeloid leukemia carrying cytoplasmic/mutated nucleophosmin (NPMc+ AML): biologic and clinical features, Blood 109 (3) (2007) 874–885.

[76] G. Nagel, et al., Epidemiological, genetic, and clinical characterization by age of newly diagnosed acute myeloid leukemia based on an academic population-based registry study (AMLSG BiO), Ann. Hematol. 96 (12) (2017) 1993–2003.

[77] G. Cazzaniga, et al., Nucleophosmin mutations in childhood acute myelogenous leukemia with normal karyotype, Blood 106 (4) (2005) 1419–1422.

[78] P. Brown, et al., The incidence and clinical significance of nucleophosmin mutations in childhood AML, Blood 110 (3) (2007) 979–985.

[79] C.G. Mullighan, et al., Pediatric acute myeloid leukemia with NPM1 mutations is characterized by a gene expression profile with dysregulated HOX gene expression distinct from MLL-rearranged leukemias, Leukemia 21 (9) (2007) 2000–2009.

[80] M. Braoudaki, et al., The frequency of NPM1 mutations in childhood acute myeloid leukemia, J. Hematol. Oncol. 3 (2010) 41.

[81] W. Wang, et al., Temporal and spatial control of nucleophosmin by the Ran-Crm1 complex in centrosome duplication, Nat. Cell Biol. 7 (8) (2005) 823–830.

[82] B. Falini, et al., Both carboxy-terminus NES motif and mutated tryptophan(s) are crucial for aberrant nuclear export of nucleophosmin leukemic mutants in NPMc+ AML, Blood 107 (11) (2006) 4514–4523.

[83] B. Falini, et al., Immunohistochemistry predicts nucleophosmin (NPM) mutations in acute myeloid leukemia, Blood 108 (6) (2006) 1999–2005.

[84] L. Federici, B. Falini, Nucleophosmin mutations in acute myeloid leukemia: a tale of protein unfolding and mislocalization, Protein Sci. 22 (5) (2013) 545–556.

[85] J.M. Bennett, et al., Is the association of "cup-like" nuclei with mutation of the NPM1 gene in acute myeloid leukemia clinically useful? Am. J. Clin. Pathol. 134 (4) (2010) 648–652.

[86] S.S. Patel, et al., Clinicopathologic and genetic characterization of nonacute NPM1-mutated myeloid neoplasms, Blood Adv. 3 (9) (2019) 1540–1545.

[87] F. Forghieri, et al., NPM1-mutated myeloid neoplasms with <20% blasts: a really distinct clinicopathologic entity? Int. J. Mol. Sci. 21 (23) (2020).

[88] B. Falini, et al., NPM1-mutated acute myeloid leukemia: from bench to bedside, Blood 136 (15) (2020) 1707–1721.

[89] C. Haferlach, et al., AML with mutated NPM1 carrying a normal or aberrant karyotype show overlapping biologic, pathologic, immunophenotypic, and prognostic features, Blood 114 (14) (2009) 3024–3032.

[90] J. Kronke, et al., Monitoring of minimal residual disease in NPM1-mutated acute myeloid leukemia: a study from the German-Austrian acute myeloid leukemia study group, J. Clin. Oncol. 29 (19) (2011) 2709–2716.

[91] A. Ivey, et al., Assessment of minimal residual disease in standard-risk AML, N. Engl. J. Med. 374 (5) (2016) 422–433.

[92] M. Jongen-Lavrencic, et al., Molecular minimal residual disease in acute myeloid leukemia, N. Engl. J. Med. 378 (13) (2018) 1189–1199.

[93] R.E. Gale, et al., The impact of FLT3 internal tandem duplication mutant level, number, size, and interaction with NPM1 mutations in a large cohort of young adult patients with acute myeloid leukemia, Blood 111 (5) (2008) 2776–2784.

[94] M. Pratcorona, et al., Favorable outcome of patients with acute myeloid leukemia harboring a low-allelic burden FLT3-ITD mutation and concomitant NPM1 mutation: relevance to post-remission therapy, Blood 121 (14) (2013) 2734–2738.

[95] R.F. Schlenk, et al., Differential impact of allelic ratio and insertion site in FLT3-ITD-positive AML with respect to allogeneic transplantation, Blood 124 (23) (2014) 3441–3449.

[96] R.M. Stone, et al., Midostaurin plus chemotherapy for acute myeloid leukemia with a FLT3 mutation, N. Engl. J. Med. 377 (5) (2017) 454–464.

[97] K. Dohner, et al., Impact of NPM1/FLT3-ITD genotypes defined by the 2017 European LeukemiaNet in patients with acute myeloid leukemia, Blood 135 (5) (2020) 371–380.

[98] B. Falini, L. Brunetti, M.P. Martelli, How I diagnose and treat NPM1-mutated AML, Blood 137 (5) (2021) 589–599.

[99] S. Loghavi, et al., Clinical features of de novo acute myeloid leukemia with concurrent DNMT3A, FLT3 and NPM1 mutations, J. Hematol. Oncol. 7 (2014) 74.

[100] M.F. Bezerra, et al., Co-occurrence of DNMT3A, NPM1, FLT3 mutations identifies a subset of acute myeloid leukemia with adverse prognosis, Blood 135 (11) (2020) 870–875.

[101] L.V. Cappelli, et al., Indeterminate and oncogenic potential: CHIP vs CHOP mutations in AML with NPM1 alteration, Leukemia 36 (2) (2022) 394–402.

[102] L. Angenendt, et al., Chromosomal abnormalities and prognosis in NPM1-mutated acute myeloid leukemia: a pooled analysis of individual patient data from nine international cohorts, J. Clin. Oncol. 37 (29) (2019) 2632–2642.

[103] K. Tarlock, et al., CEBPA-bZip mutations are associated with favorable prognosis in de novo AML: a report from the Children's Oncology Group, Blood 138 (13) (2021) 1137–1147.

[104] F. Taube, et al., CEBPA mutations in 4708 patients with acute myeloid leukemia: differential impact of bZIP and TAD mutations on outcome, Blood 139 (1) (2022) 87–103.

[105] S. Wakita, et al., Prognostic impact of CEBPA bZIP domain mutation in acute myeloid leukemia, Blood Adv. 6 (1) (2022) 238–247.

[106] E. Taskesen, et al., Prognostic impact, concurrent genetic mutations, and gene expression features of AML with CEBPA mutations in a cohort of 1182 cytogenetically normal AML patients: further evidence for CEBPA double mutant AML as a distinctive disease entity, Blood 117 (8) (2011) 2469–2475.

[107] B.J. Wouters, et al., Double CEBPA mutations, but not single CEBPA mutations, define a subgroup of acute myeloid leukemia with a distinctive gene expression profile that is uniquely associated with a favorable outcome, Blood 113 (13) (2009) 3088–3091.

[108] A. Dufour, et al., Acute myeloid leukemia with biallelic CEBPA gene mutations and normal karyotype represents a distinct genetic entity associated with a favorable clinical outcome, J. Clin. Oncol. 28 (4) (2010) 570–577.

[109] C.L. Green, et al., Prognostic significance of CEBPA mutations in a large cohort of younger adult patients with acute myeloid leukemia: impact of double CEBPA mutations and the interaction with FLT3 and NPM1 mutations, J. Clin. Oncol. 28 (16) (2010) 2739–2747.

[110] R.C. Lindsley, et al., Acute myeloid leukemia ontogeny is defined by distinct somatic mutations, Blood 125 (9) (2015) 1367–1376.

[111] Y. Gao, et al., Distinct mutation landscapes between acute myeloid leukemia with myelodysplasia-related changes and de novo acute myeloid leukemia, Am. J. Clin. Pathol. 157 (5) (2022) 691–700.

[112] D.A. Arber, et al., Prognostic impact of acute myeloid leukemia classification. Importance of detection of recurring cytogenetic abnormalities and multilineage dysplasia on survival, Am. J. Clin. Pathol. 119 (5) (2003) 672–680.

[113] M. Yanada, et al., Long-term outcomes for unselected patients with acute myeloid leukemia categorized according to the World Health Organization classification: a single-center experience, Eur. J. Haematol. 74 (5) (2005) 418–423.

[114] R. Devillier, et al., Role of ASXL1 and TP53 mutations in the molecular classification and prognosis of acute myeloid leukemias with myelodysplasia-related changes, Oncotarget 6 (10) (2015) 8388–8396.

[115] T.A. Gruber, et al., An Inv(16)(p13.3q24.3)-encoded CBFA2T3-GLIS2 fusion protein defines an aggressive subtype of pediatric acute megakaryoblastic leukemia, Cancer Cell 22 (5) (2012) 683–697.

[116] C. Thiollier, et al., Characterization of novel genomic alterations and therapeutic approaches using acute megakaryoblastic leukemia xenograft models, J. Exp. Med. 209 (11) (2012) 2017–2031.

[117] R. Masetti, et al., CBFA2T3-GLIS2-positive acute myeloid leukaemia. A peculiar paediatric entity, Br. J. Haematol. 184 (3) (2019) 337–347.

[118] A. Zangrando, et al., CD56, HLA-DR, and CD45 recognize a subtype of childhood AML harboring CBFA2T3-GLIS2 fusion transcript, Cytometry A 99 (8) (2021) 844–850.

[119] L. Eidenschink Brodersen, et al., A recurrent immunophenotype at diagnosis independently identifies high-risk pediatric acute myeloid leukemia: a report from Children's Oncology Group, Leukemia 30 (10) (2016) 2077–2080.

[120] T. Haferlach, et al., AML with translocation t(8;16)(p11;p13) demonstrates unique cytomorphological, cytogenetic, molecular and prognostic features, Leukemia 23 (5) (2009) 934–943.

[121] E.A. Coenen, et al., Pediatric acute myeloid leukemia with t(8;16)(p11;p13), a distinct clinical and biological entity: a collaborative study by the International-Berlin-Frankfurt-Munster AML-study group, Blood 122 (15) (2013) 2704–2713.

[122] S. Noort, et al., Prognostic impact of t(16;21)(p11;q22) and t(16;21)(q24;q22) in pediatric AML: a retrospective study by the I-BFM Study Group, Blood 132 (15) (2018) 1584–1592.

[123] A.R. von Bergh, et al., High incidence of t(7;12)(q36;p13) in infant AML but not in infant ALL, with a dismal outcome and ectopic expression of HLXB9, Genes Chromosom. Cancer 45 (8) (2006) 731–739.

[124] A.D.L. Espersen, et al., Acute myeloid leukemia (AML) with t(7;12)(q36;p13) is associated with infancy and trisomy 19: data from Nordic Society for Pediatric Hematology and Oncology (NOPHO-AML) and review of the literature, Genes Chromosom. Cancer 57 (7) (2018) 359–365.

[125] T. Liu, et al., Identification of novel recurrent CPSF6-RARG fusions in acute myeloid leukemia resembling acute promyelocytic leukemia, Blood 131 (16) (2018) 1870–1873.

[126] Y.Z. Qin, X.J. Huang, H.H. Zhu, Identification of a novel CPSF6-RARG fusion transcript in acute myeloid leukemia resembling acute promyelocytic leukemia, Leukemia 32 (10) (2018) 2285–2287.

[127] E. Such, et al., A novel NUP98/RARG gene fusion in acute myeloid leukemia resembling acute promyelocytic leukemia, Blood 117 (1) (2011) 242–245.

[128] J.S. Ha, et al., Identification of a novel PML-RARG fusion in acute promyelocytic leukemia, Leukemia 31 (9) (2017) 1992–1995.

[129] Z. Su, et al., Novel reciprocal fusion genes involving HNRNPC and RARG in acute promyelocytic leukemia lacking RARA rearrangement, Haematologica 105 (7) (2020) e376–e378.

[130] S. Kayser, M.J. Levis, Clinical implications of molecular markers in acute myeloid leukemia, Eur. J. Haematol. 102 (1) (2019) 20–35.

[131] S. Li, et al., Myelodysplastic syndrome/acute myeloid leukemia with t(3;21)(q26.2;q22) is commonly a therapy-related disease associated with poor outcome, Am. J. Clin. Pathol. 138 (1) (2012) 146–152.

[132] C. Glass, et al., The role of EVI1 in myeloid malignancies, Blood Cells Mol. Dis. 53 (1-2) (2014) 67–76.

[133] A.A. Hinai, P.J. Valk, Review: Aberrant EVI1 expression in acute myeloid leukaemia, Br. J. Haematol. 172 (6) (2016) 870–878.

[134] S. Kayser, et al., The impact of therapy-related acute myeloid leukemia (AML) on outcome in 2853 adult patients with newly diagnosed AML, Blood 117 (7) (2011) 2137–2145.

[135] L. Lo Nigro, et al., Prognostic impact of t(9;11) in childhood acute myeloid leukemia (AML), Leukemia 17 (3) (2003) 636.

[136] J.E. Rubnitz, et al., Favorable impact of the t(9;11) in childhood acute myeloid leukemia, J. Clin. Oncol. 20 (9) (2002) 2302–2309.

[137] Y. Ishikawa, et al., Prospective evaluation of prognostic impact of KIT mutations on acute myeloid leukemia with RUNX1-RUNX1T1 and CBFB-MYH11, Blood Adv. 4 (1) (2020) 66–75.

[138] L. Su, et al., Acute myeloid leukemia with CEBPA mutations: current progress and future directions, Front. Oncol. 12 (2022), 806137.

[139] A. Fasan, et al., The role of different genetic subtypes of CEBPA mutated AML, Leukemia 28 (4) (2014) 794–803.

[140] P.A. Ho, et al., Prevalence and prognostic implications of CEBPA mutations in pediatric acute myeloid leukemia (AML): a report from the Children's Oncology Group, Blood 113 (26) (2009) 6558–6566.

[141] M. Breccia, et al., FLT3-ITD confers poor prognosis in patients with acute promyelocytic leukemia treated with AIDA protocols: long-term follow-up analysis, Haematologica 98 (12) (2013) e161–e163.

[142] F. Tripon, et al., Co-occurrence of PML-RARA gene fusion, chromosome 8 trisomy, and FLT3 ITD mutation in a young female patient with de novo acute myeloid leukemia and early death: a CARE case report, Medicine (Baltimore) 99 (14) (2020), e19730.

[143] K. Paulson, et al., Acute promyelocytic leukaemia is characterized by stable incidence and improved survival that is restricted to patients managed in leukaemia referral centres: a pan-Canadian epidemiological study, Br. J. Haematol. 166 (5) (2014) 660–666.

[144] Z.X. Shen, et al., All-trans retinoic acid/As2O3 combination yields a high quality remission and survival in newly diagnosed acute promyelocytic leukemia, Proc. Natl. Acad. Sci. U. S. A. 101 (15) (2004) 5328–5335.

[145] M.J. Kim, et al., Molecular methods for genomic analyses of variant PML-RARA or other RARA-related chromosomal translocations in acute promyelocytic leukemia, Korean J. Hematol. 47 (4) (2012) 307–308.

[146] A. Melnick, J.D. Licht, Deconstructing a disease: RARalpha, its fusion partners, and their roles in the pathogenesis of acute promyelocytic leukemia, Blood 93 (10) (1999) 3167–3215.

[147] C. Callens, et al., Prognostic implication of FLT3 and Ras gene mutations in patients with acute promyelocytic leukemia (APL): a retrospective study from the European APL Group, Leukemia 19 (7) (2005) 1153–1160.

[148] A. Zelent, et al., Translocations of the RARalpha gene in acute promyelocytic leukemia, Oncogene 20 (49) (2001) 7186–7203.

[149] S.E. Langabeer, et al., Molecular profiling: a case of ZBTB16-RARA acute promyelocytic leukemia, Case Rep. Hematol. 2017 (2017) 7657393.

[150] D.A. Pollyea, et al., NCCN guidelines insights: acute myeloid leukemia, version 2.2021, J. Natl. Compr. Cancer Netw. 19 (1) (2021) 16–27.

[151] A. Dash, D.G. Gilliland, Molecular genetics of acute myeloid leukaemia, Best Pract. Res. Clin. Haematol. 14 (1) (2001) 49–64.

[152] G.L. Picharski, et al., The impact of Flt3 gene mutations in acute promyelocytic leukemia: a meta-analysis, Cancers (Basel) 11 (9) (2019).

[153] E.M. Blanco, et al., Cytogenetically cryptic and FISH-negative PML/RARA rearrangement in acute promyelocytic leukemia detected only by PCR: an exceedingly rare phenomenon, Cancer Genet. 207 (1-2) (2014) 48–49.

[154] H. Fan, et al., PML-RARA fusion resulting from a cryptic insertion of RARA gene into PML gene without the reciprocal RARA-PML fusion: clinical, cytogenetic, and molecular characterization and prognosis, Eur. J. Haematol. 93 (4) (2014) 354–358.

[155] K. Karlin, et al., Cytogenetically cryptic PML::RARA fusion in acute promyelocytic leukemia: testing strategies in the modern era, Leuk. Res. Rep. 17 (2022), 100320.

[156] A. Zaccaria, et al., Cryptic translocation of PML/RARA on 17q. A rare event in acute promyelocytic leukemia, Cancer Genet. Cytogenet. 138 (2) (2002) 169–173.

17

Blastic plasmacytoid dendritic cell neoplasm (BPDCN)

Guang Liu
SONORA QUEST LABORATORIES, PHOENIX, AZ, UNITED STATES

Background

Blastic plasmacytoid dendritic cell neoplasm (BPDCN) is a clinically aggressive tumor derived from the precursors of plasmacytoid dendritic cells that involve multiple sites. It is most commonly found in the skin but can also infiltrate the bone marrow, peripheral blood, and lymph nodes, as well as visceral organs. Most often, BPDCN presents with features of both lymphoma and leukemia. However, the clinical presentation of BPDCN is broadly heterogeneous and was most recently categorized as a distinct clinical entity by the World Health Organization (WHO) classification of hematological malignancies in 2022 under histiocytic/dendritic cell neoplasms [1]. The immunophenotype from flow cytometry is essential for the diagnosis of BPDCN. Most BPDCN cells overexpress CD123 and may also be positive for CD4, CD56, CD303, and TCL1. The incidence of BPDCN has been estimated to be 0.04 cases per 100,000 people. The average age at diagnosis is 60–70 years, but this neoplasm can occur at any age, including childhood. There are more men than women who are diagnosed with BPDCN, with approximately 75% of cases occurring in men [2,3]. Although rare, BPDCN has been discussed more frequently in recent years because of a new drug (Elzonris), which has been approved by FDA and could be effective in treating this disease [4]. The diagnosis of BPDCN has been challenging because its clinical features can be heterogeneous and can overlap with other hematologic neoplasms [4]. Studies have looked to identify recurrent genetic abnormalities in BPDCN, and some have shown that 8q24/*MYC* rearrangement is a recurrent cytogenetic abnormality occurring in 10%–15% of BPDCN; the *MYC* gene is often partnered with nonimmunoglobulin chromosomal loci and may play a role in BPDCN pathogenesis [5,6].

Case 17.1 Blastic plasmacytoid dendritic cell neoplasm (BPDCN)

Clinical indication

A 72-year-old male presented with fatigue and weight loss. Flow cytometric analysis detected an abnormal population of cells accounting for approximately 62.4% of analyzed

Cases in Laboratory Genetics and Genomics (LGG) Practice. https://doi.org/10.1016/B978-0-323-99622-8.00032-0

events. This abnormal cell population expressed CD123, CD4 (dim), CD56, HLA-DR, CD38, subset CD117 (dim), and dim variable CD45, while negative for CD34, CD10, CD19, CD20, CD22, cytoplasmic CD3, CD5, CD13, CD33, CD11c, CD11b, CD64, MPO, and TdT. By morphology, the cells were large mononuclear cells with moderate agranular cytoplasm exhibiting occasional vacuoles, fine nuclear chromatin, and prominent nucleoli. The immunophenotype of CD123, CD4, and CD56 with the absence of lineage-specific antigens was highly suggestive of blastic plasmacytoid dendritic cell neoplasm; however, acute undifferentiated leukemia could not be entirely excluded.

Test ordered

- Chromosome analysis of the leukemic blood
- FISH: AML panel and *MYC* gene rearrangement
- NGS Hematology Molecular Profile

Laboratory test performed

Chromosome analysis, FISH, and NGS methods were described in Chapters 1 and 12.

Test results

Firstly, chromosome analysis revealed a 6;8 translocation and an extra chromosome 12 with additional material of unknown origin added to 12q24.1 (Fig. 17.1.1).

Secondly, FISH was performed on interphase nuclei using probes localized to the 5′ and 3′ ends of the *MYC* (8q24) gene region. Two hundred nuclei were examined, and the results were positive for a rearrangement involving *MYC* in 194/200 (97.0%) of the cells scored (Fig. 17.1.2).

FISH was also performed on interphase nuclei using probes localized to the *D5S721* (5p15.2), *EGR1* (5q31), *D7Z1* (7cen), *D7S486* (7q31), *D8Z2* (8cen), *RUNX1T1* (8q22), *ABL1* (9q34.12), *KMT2A* (11q23), *PML* (15q24.1), *CBFB* (16q22.1), *RARA* (17q21.1), *D20S108* (20q12), *RUNX1* (21q22.3), and *BCR* (22q11) gene regions. Two hundred nuclei were examined, and the results were within normal limits for the laboratory's established background rates (Fig. 17.1.3).

Finally, NGS Hematology Myeloid Profile was performed, and the results showed a *KRAS* p.(A146P) mutation (Fig. 17.1.4).

Results with interpretations

Chromosome analysis exhibited a translocation t(6;8)(p21;q24), which might result in the fusion of the *MYC* gene at 8q24 and the *SUPT3H* gene at 6p21. FISH for the *MYC* probe showed *MYC* rearrangement and was concordant with the findings of the chromosome analysis. The t(6;8)(p21;q24)/*MYC::SUPT3H* fusion is a recurrent abnormality seen in BPDCN and is generally associated with a poor prognosis [5,6]. The morphologic and cytometric findings were suggestive of BPDCN for this patient; however, acute undifferentiated

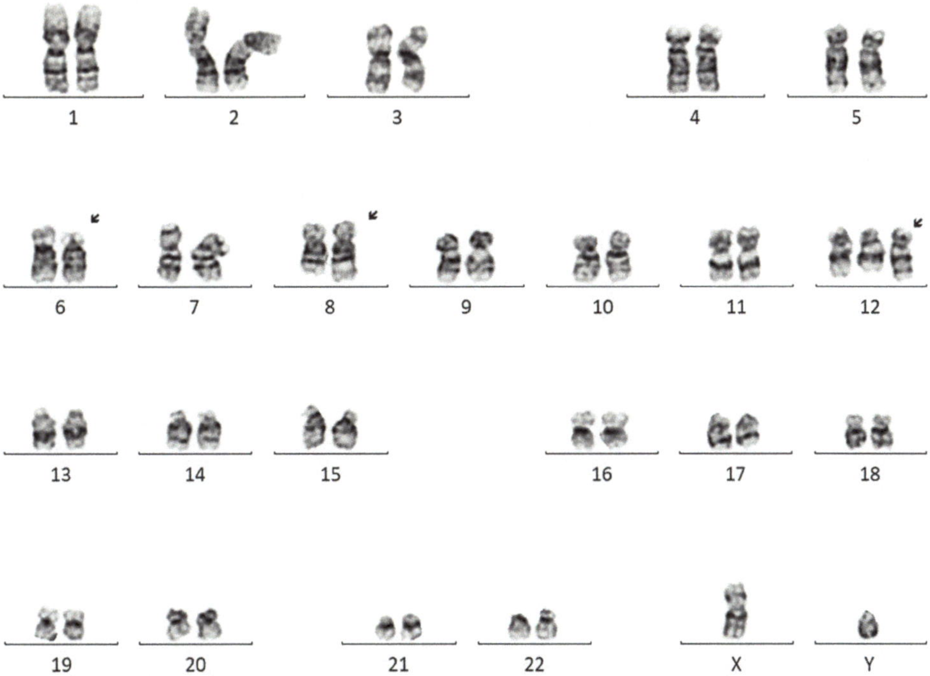

FIG. 17.1.1 The karyotype of leukemic blood from the patient showed a 6;8 translocation and an extra chromosome 12 with additional material added to 12q. ISCN: 47,XY,t(6;8)(p21;q24),+add(12)(q24.1)[20]

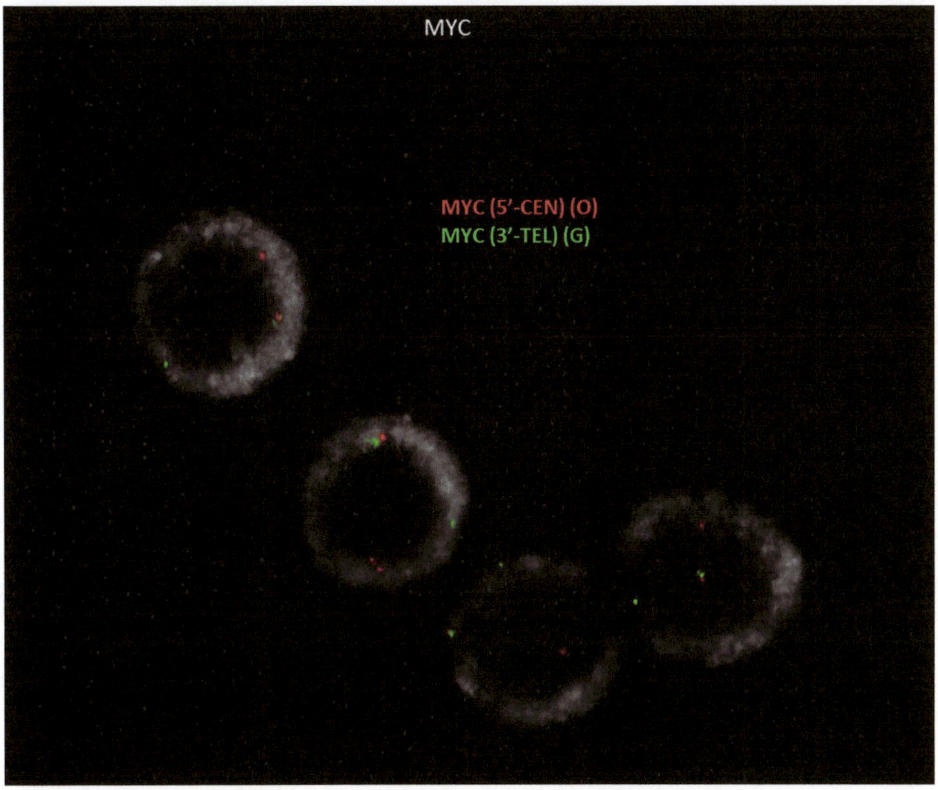

FIG. 17.1.2 FISH with *MYC* break-apart probe was positive for *MYC* rearrangement. ISCN: nuc ish (MYCx2)(5'MYC sep 3'MYCx1)[194/200]

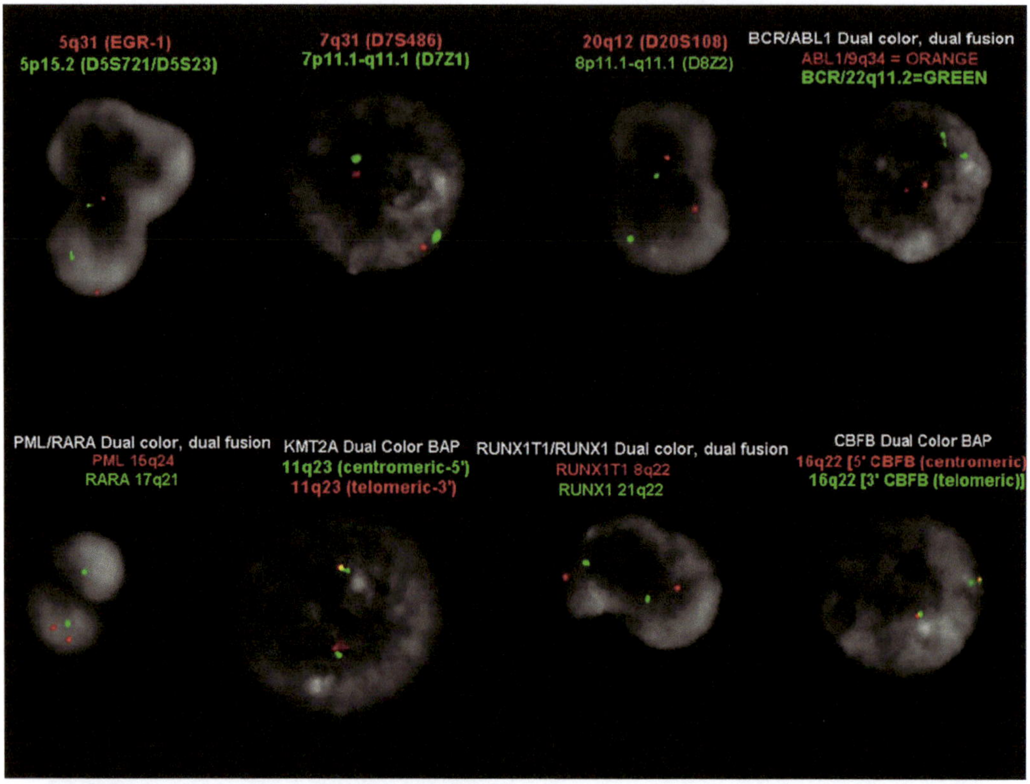

FIG. 17.1.3 FISH with AML panel showed negative results for all probes examined. ISCN: nuc ish(D5S721,EGR1)x2 [200/200],(CEP7,D7S486)x2[200/200],(D8Z2,D20S108)x2[200/200],(RUNX1T1,RUNX1)x2[200/200],(ABL1,BCR)x2 [200/200],(KMT2Ax2)[200/200],(PML,RARA)x2[200/200],(CBFBx2)[200/200]

Relevant Biomarkers

Tier	Genomic Alteration	Relevant Therapies (In this cancer type)	Clinical Trials
IIC	*KRAS p.(A146P) c.436G>C*	None	5

Public data sources included in relevant therapies: FDA1, NCCN, **EMA**2, ESMO
Tier Reference: Li et al. *Standards and Guidelines for the Interpretation and Reporting of Sequence Variants in Cancer: A Joint Consensus Recommendation of the Association for Molecular Pathology, American Society of Clinical Oncology, and College of American Pathologists.* J Mol Diagn. 2017 Jan;19(1):4-23.

Variant Details

DNA Sequence Variants

Gene	Amino Acid Change	Coding	Locus	Allele Frequency	Variant Effect
KRAS	p.(A146P)	c.436G>C	chr12:25378562	20.30%	missense

FIG. 17.1.4 NGS detected a mutation of *KRAS* (p.Ala146Pro).

leukemia cannot be excluded. Therefore, identification of the t(6;8) translocation along with *MYC* rearrangement confirmed the diagnosis of BPDCN and ruled out acute undifferentiated leukemia [6]. *KRAS* mutation detected by NGS has also been reported in BPDCN [7]. The prognosis of BPDCN is generally poor. The translocation t(6;8)(p21;q24) may indicate an even more aggressive disease process [8]. The new FDA-approved drug, tagraxofusp-erzs (Elzonris), has shown to be safe and effective in the treatment of BPDCN [9].

Future testing and recommendations

FISH and chromosome analysis on leukemic blood can be ordered to monitor disease progression and treatment efficacy in the future.

Summary of key learning points

- BPDCN is a clinically aggressive heterogeneous hematologic malignancy that can be easily misdiagnosed as other forms of leukemia or lymphoma.
- 8q24/*MYC* rearrangement is a recurrent cytogenetic abnormality in BPDCN. Identification of 8q24/*MYC* rearrangement is important for diagnosis of BPDCN.
- The new FDA-approved drug, tagraxofusp-erzs, has shown to be safe and effective in the treatment of BPDCN.

References

[1] J.D. Khoury, et al., The 5th edition of the World Health Organization classification of haematolymphoid tumours: myeloid and histiocytic/dendritic neoplasms, Leukemia 36 (7) (2022) 1703–1719.

[2] G.S. Guru Murthy, N. Pemmaraju, E. Atallah, Epidemiology and survival of blastic plasmacytoid dendritic cell neoplasm, Leuk. Res. 73 (2018) 21–23.

[3] Y. Gonzaga, M. Fontes, Blastic plasmacytoid dendritic-cell neoplasm, N. Engl. J. Med. 383 (22) (2020) 2158.

[4] K. Sweet, Blastic plasmacytoid dendritic cell neoplasm: diagnosis, manifestations, and treatment, Curr. Opin. Hematol. 27 (2) (2020) 103–107.

[5] P.C. Boddu, et al., 8q24/MYC rearrangement is a recurrent cytogenetic abnormality in blastic plasmacytoid dendritic cell neoplasms, Leuk. Res. 66 (2018) 73–78.

[6] Y. Nakamura, et al., Identification of SUPT3H as a novel 8q24/MYC partner in blastic plasmacytoid dendritic cell neoplasm with t(6;8)(p21;q24) translocation, Blood Cancer J. 5 (2015), e301.

[7] A. Stenzinger, et al., Targeted ultra-deep sequencing reveals recurrent and mutually exclusive mutations of cancer genes in blastic plasmacytoid dendritic cell neoplasm, Oncotarget 5 (15) (2014) 6404–6413.

[8] L. Sumarriva Lezama, et al., An analysis of blastic plasmacytoid dendritic cell neoplasm with translocations involving the MYC locus identifies t(6;8)(p21;q24) as a recurrent cytogenetic abnormality, Histopathology 73 (5) (2018) 767–776.

[9] D. Hammond, N. Pemmaraju, Tagraxofusp for blastic plasmacytoid dendritic cell neoplasm, Hematol. Oncol. Clin. North Am. 34 (3) (2020) 565–574.

18

Acute leukemias of ambiguous lineage

Xia Li

SONORA QUEST LABORATORIES, PHOENIX, AZ, UNITED STATES

Background

Acute leukemias of ambiguous lineage are heterogeneous groups of hematopoietic malignancies. In these leukemias, the blasts show markers of multiple developmental lineages, which cannot be classified as acute myeloid or lymphoblastic leukemias. This group of diseases accounts for 2%–5% of all acute leukemias depending on the diagnostic criteria used [1]. The fifth World Health Organization (WHO) Classification of Haematolymphoid Tumours (2022) classified this group of diseases as acute leukemia of mixed or ambiguous lineage including acute leukemia of ambiguous lineage (ALAL) and mixed-phenotype acute leukemia (MPAL). Both ALAL and MPAL have overlapping clinical and immunophenotypic features and share common molecular pathogenic mechanisms [2].

ALAL with defining genetic abnormalities includes MPAL with *BCR::ABL1* fusion, *KMT2A, ZNF384,* or *BCL11B* rearrangements. ALAL, immunophenotypically defined includes MPAL with B/myeloid, T/myeloid, rare types, ALAL, not otherwise specified NOS, and acute undifferentiated leukemia (AUL) [2].

Because of the presence of adverse cytogenetics abnormalities such as *KMT2A* and *BCR::ABL1* rearrangements in these cases, the prognosis is in general worse than either acute myeloid leukemia (AML) or acute lymphoid leukemia (ALL) [3,4]. As for the treatment, there are not enough trial data to guide therapy. The treatment generally relies on ALL-like regimens followed by consolidation chemotherapy or hematopoietic stem cell transplant (HSCT) [4,5]. A study by Orgel et al. discussed the challenges and uncertainty of the optimal chemotherapy for treating MPAL and the role of HSCT for pediatric patients. The results of the Children's Oncology Group (COG) MPAL cohort and a literature review suggest that ALL chemotherapy without HSCT may be the preferred initial therapy. AML chemotherapy and HSCT are the best practices for those with treatment failure as addressed by minimal residual disease [6].

In most clinical diagnostics laboratories, FISH, karyotyping, RT-PCR, and NGS are the common technologies to identify copy number variations, mutations, or translocations (*KMT2A, BCR::ABL1* fusion) seen in MPAL [7–9]. A study by Quesada et al. examined 14 cases of MPAL and showed that seven were B-cell/myeloid MPALs immunophenotype, six were

Cases in Laboratory Genetics and Genomics (LGG) Practice. https://doi.org/10.1016/B978-0-323-99622-8.00019-8

T-cell/myeloid immunophenotype, and one was B-cell/T-cell immunophenotype. A total of 25 distinct mutations were identified in 15 different genes in 9/14 (64%) patients. *FLT3*-ITD was the only recurrent mutation in two patients [10]. In this chapter, a few cases will be illustrated to demonstrate how these genetic aberrations are detected by different assays and what limitations each technology may have when evaluating the concordance of the test results.

Case 18.1 Mixed phenotype acute (B/myeloid) leukemia (MPAL) with a complex karyotype

Clinical indication

A 42-year-old male with a past medical history of erosive gastritis presented with epistaxis and pancytopenia. Additionally, he just received a new diagnosis of acute mixed phenotype leukemia. He had an allogeneic bone marrow transplant and was now presenting with body aches, fevers, shortness of breath, and cough. Peripheral blood smear showed 60% blast concerning relapse and no concern for Auer rods.

Test ordered

- Chromosome analysis of the bone marrow at diagnosis and relapse
- FISH: AML panel
- FISH: ALL hyperdiploidy panel
- FISH: *ETV6::RUNX1* rearrangement and *RARA* break-apart probe
- Flow MRD (send out)
- NGS Hematology Molecular Profile

Laboratory test performed

Chromosome analysis, FISH, and NGS methods were described in Chapters 1 and 12.

Test results

Chromosome analysis was performed on bone marrow collected at diagnosis. Of the 21 cells examined, three exhibited an isochromosome for 17q. The remaining 18 cells appear to be chromosomally normal (Fig. 18.1.1).

Chromosome analysis was also performed on bone marrow collected at relapse. Of the 21 cells examined, 14 were abnormal with multiple structural chromosomal aberrations including unbalanced translocations involving 1p&3q, 1q&2q, possible insertion of part of chromosome 3 onto 1p, a balanced translocation involving 1p&8q, add(8q), del(12p), and del(20q). These cells were consistent with the patient's gender (Recipient Karyotype). The remaining seven cells were all donor-derived and appeared to be chromosomally normal (Fig. 18.1.2).

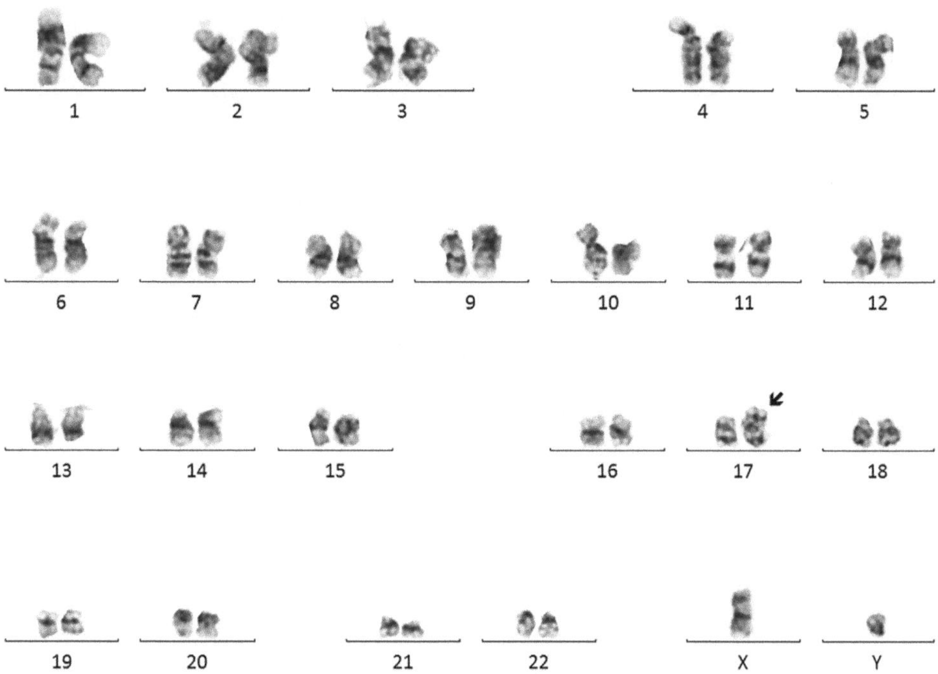

FIG. 18.1.1 Chromosome analysis on the bone marrow at diagnosis showed an isochromosome for 17q. ISCN: 46,XY,i(17)(q10)[3]/46,XY[18]

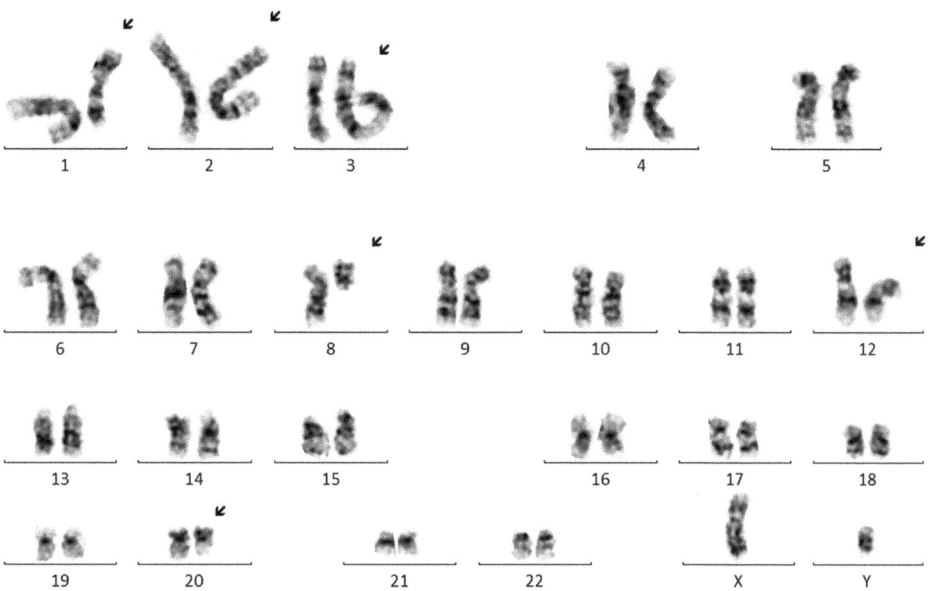

FIG. 18.1.2 Chromosome analysis on the bone marrow at relapse showed a complex karyotype with multiple structural chromosome abnormalities. ISCN: 46,XY,der(1)t(1;3)(p13;q21)t(1;2)(q25;q33),der(2)t(1;2),der(3)t(1;3)ins(1;?)(p36.1;?) t(1,8)(p36.1;q12),add(8)(q12),del(12)(p13),del(20)(q11.2q13.1)[14]//46,XX[7]

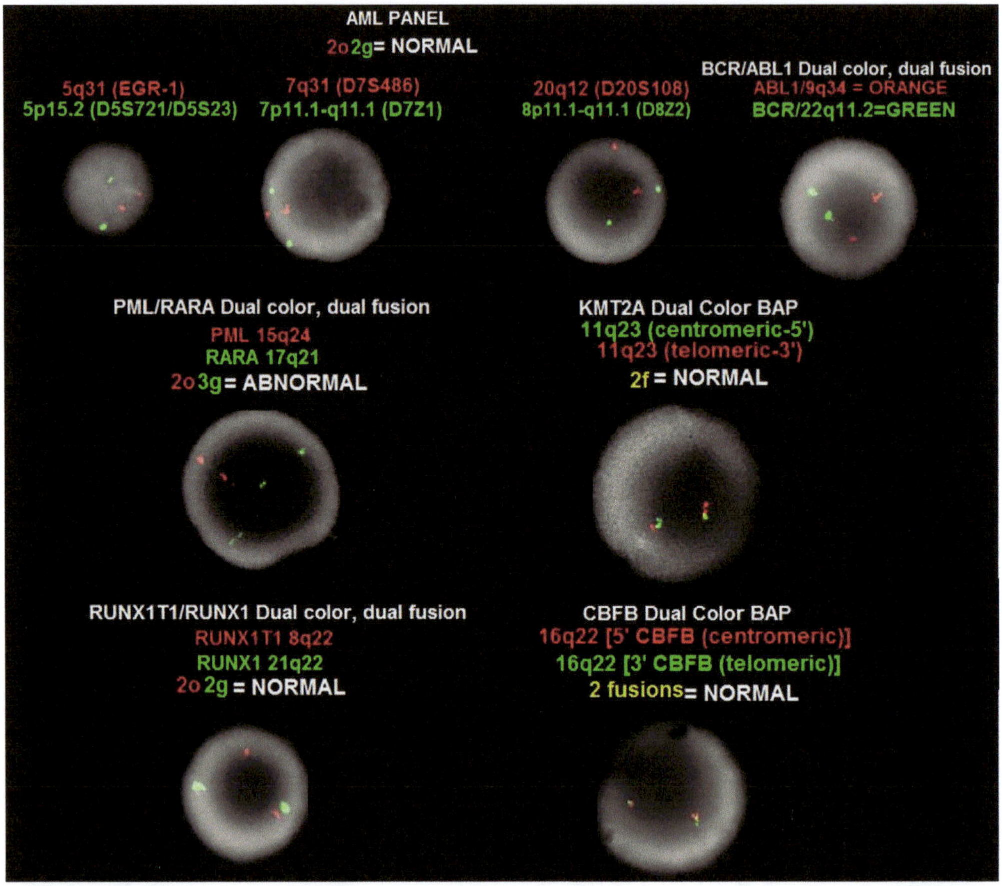

FIG. 18.1.3 FISH for the AML panel showed a gain of *RARA* in 20/200 (10%) of cells scored. ISCN: nuc ish(D5S721,EGR1)x2[200/200],(CEP7,D7S486)x2[200/200],(D8Z2,D20S108)x2[199/200],(RUNX1T1,RUNX1)x2[200/200],(ABL1,BCR)x2[200/200],(KMT2Ax2)[200/200],(PMLx2,RARAx3)[20/200],(CBFBx2)[200/200]

FISH for the AML panel was performed on interphase nuclei using probes localized to the *D5S721* (5p15.2), *EGR1* (5q31), *D7Z1* (7cen), *D7S486* (7q31), *D8Z2* (8cen), *RUNX1T1* (8q22), *ABL1* (9q34.12), *KMT2A* (11q23), *PML* (15q24.1), *CBFB* (16q22.1), *RARA* (17q21.1), *D20S108* (20q12), *RUNX1* (21q22.3), and *BCR* (22q11) gene regions. Two hundred nuclei were examined, and the results demonstrated a gain of *RARA* (17q21.1) in 20/200 (10.0%) of cells scored (Fig. 18.1.3).

FISH for the ALL hyperdiploidy panel, RARA break-apart probe, *ETV6::RUNX1* rearrangement probes was also performed on interphase nuclei using probes localized to the chromosome 4, 10, and 17 centromeric regions, *RARA* (17q21), *ETV6* (12p13), and *RUNX1* (21q22.3) gene region. Two hundred nuclei were examined, and the results demonstrated a gain of *RARA* in 37/200 (18.5%) of the cells scored (Fig. 18.1.4).

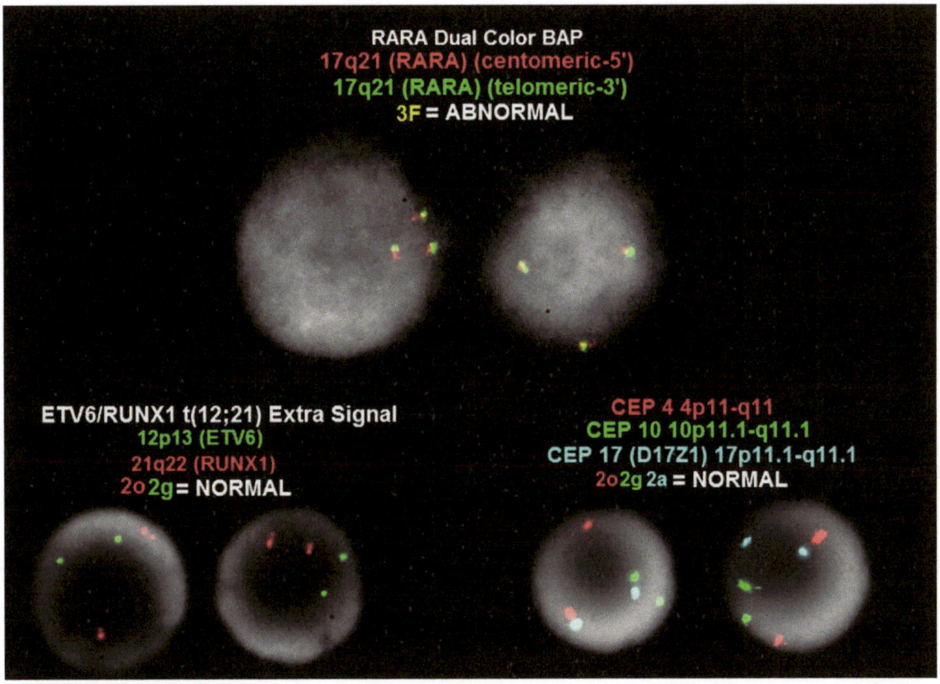

FIG. 18.1.4 FISH for ALL hyperdiploidy panels including *RARA* break-apart probe and *ETV6::RUNX1* rearrangement probes showed gain of *RARA* in 18.5% of cells examined. ISCN: nuc ish(CEP4,CEP10,CEP17)x2[200/200],(ETV6,RUNX1) x2[200/200],(RARA)x3[37/200]

Flow MRD was performed in another lab; the results showed abnormal CD34-positive blasts representing approximately 8% of the white cells and expressing several markers, consistent with persistent/recurrent acute leukemia (B/myeloid) (data not shown).

Finally, the NGS hematology profile was performed and detected a mutation of *RUNX1* p.(R166fs)c.494_495insG (Fig. 18.1.5).

Results with interpretations

The karyotype from the bone marrow collected at diagnosis identified an isochromosome for 17q. A gain of 17q was also seen from the concurrent FISH testing. i(17q) has been seen in myeloid disorders, although it was rarely reported in mixed acute leukemia.

The karyotype from the bone marrow collected at relapse showed complex rearrangements involving chromosomes 1, 2, 3, and 8, an additional material on 8q, a deletion of 12p, and a deletion of 20q. These abnormalities were all from the recipient cells, indicating engraftment failure, and were consistent with relapse of the patient's known acute leukemia.

FISH for AML, ALL panel, *RARA* break-apart probe, and *ETV6::RUNX1* rearrangement probes showed gain of *RARA* only from the diagnostic bone marrow sample, concordant with the findings from the karyotype.

Clinically Significant Biomarkers ▒ Indicated ■ Contraindicated

Genomic Alteration	Relevant Therapies (In this cancer type)	Clinical Trials
RUNX1 p.(R166fs) c.494_495insG Tier: IA **Prognostic significance:** Poor **Diagnostic significance:** None	None	0

Sources included in relevant therapies: FDA[1], NCCN
Sources included in prognostic and diagnostic significance: NCCN

Variant Details

DNA Sequence Variants

Gene	Amino Acid Change	Coding	Locus	Allele Frequency	Variant Effect
RUNX1	p.(R166fs)	c.494_495insG	chr21:36252867	39.21%	frameshift Insertion

FIG. 18.1.5 NGS detected a mutation of *RUNX1* p.(R166fs)c.494_495insG.

Finally, NGS detected a mutation of *RUNX1* p.(R166fs)c.494_495insG. This is a frameshift insertion at 39.21% allele frequency and is associated with a poor prognosis based on the NCCN guidelines [11].

Future testing and recommendations

The patient was on the status post stem cell transplant and was treated with inotuzumab. Chromosome analysis, FISH, and NGS on bone marrow can be ordered to monitor disease progression and treatment efficacy in the future.

Case 18.2 Mixed phenotype acute (B/myeloid) leukemia (MPAL) with *FLT3*-ITD and other mutations

Clinical indication

A 67-year-old woman initially presented with pancytopenia. She reported bilateral leg aches, shortness of breath, and some tightness around the lower ribs. The patient denied experiencing any nausea, vomiting, or diarrhea. She underwent a bone marrow biopsy for evaluation of the abnormal myeloid population and blasts. Cytometric and morphologic analysis of bone marrow revealed a small abnormal myeloid blast population. Eighty-seven percent (87%) of the cells were in the analyzed lymphocyte region. The diagnosis was bi-phenotypic leukemia.

Test ordered

- Chromosome analysis of the bone marrow
- FISH: AML panel

- FISH: ALL hyperdiploidy panel and *ETV6::RUNX1* rearrangement
- NGS Hematology Molecular Profile

Laboratory test performed

Chromosome analysis, FISH, and NGS methods were described in Chapters 1 and 12.

Test results

Chromosome analysis identified two distinct abnormal clones. Clone 1 with six cells exhibited an unbalanced translocation involving 1q and 9p, resulting in del(9p) and dup(1q) (Fig. 18.2.1). Clone 2 with five cells revealed trisomy 6 (Fig. 18.2.2).

FISH for the AML panel was performed on interphase nuclei using probes localized to the *D5S721* (5p15.2), *EGR1* (5q31), *D7Z1* (7cen), *D7S486* (7q31), *D8Z2* (8cen), *RUNX1T1* (8q22), *ABL1* (9q34.12), *KMT2A* (11q23), *PML* (15q24.1), *CBFB* (16q22.1), *RARA* (17q21.1), *D20S108* (20q12), *RUNX1* (21q22.3), and *BCR* (22q11) gene regions. Two hundred nuclei were examined for each probe, and the results were within normal limits for the laboratory's established background rates (Fig. 18.2.3).

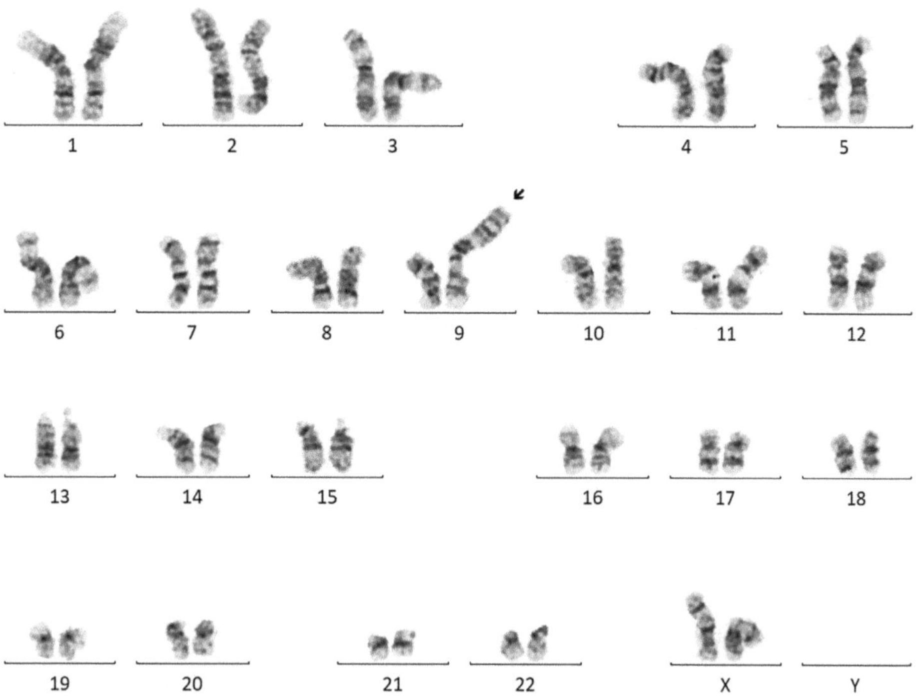

FIG. 18.2.1 Chromosome analysis identified abnormal clone 1 with an unbalanced translocation involving 1q and 9p resulting in del(9p). ISCN: 46,XX,der(9)t(1;9)(q12;p24)[6]/47,XX,+6[5]/46,XX[9]

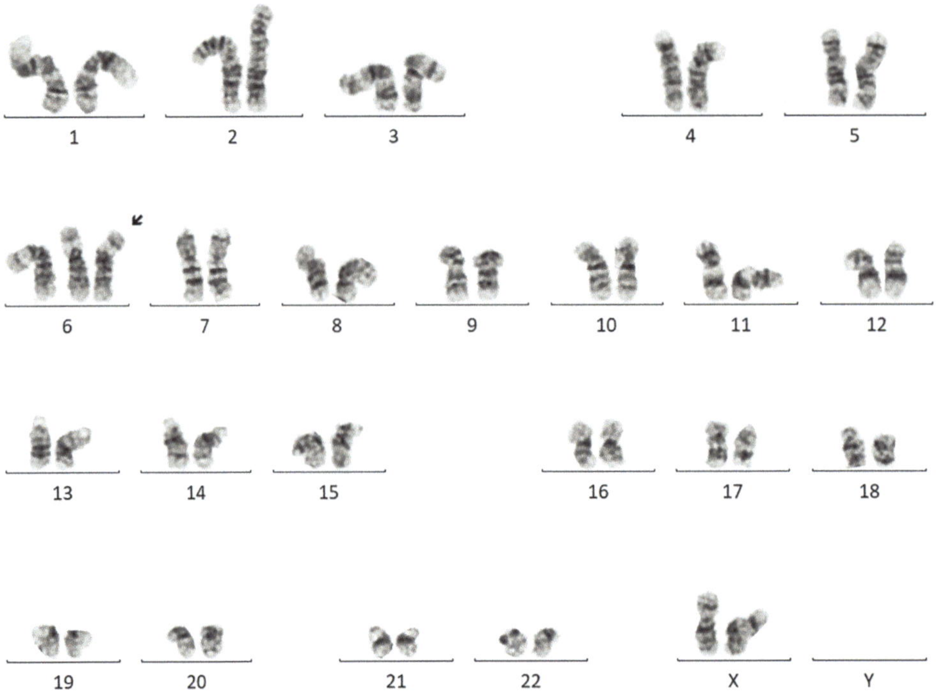

FIG. 18.2.2 Chromosome analysis identified abnormal clone 2 with trisomy 6.

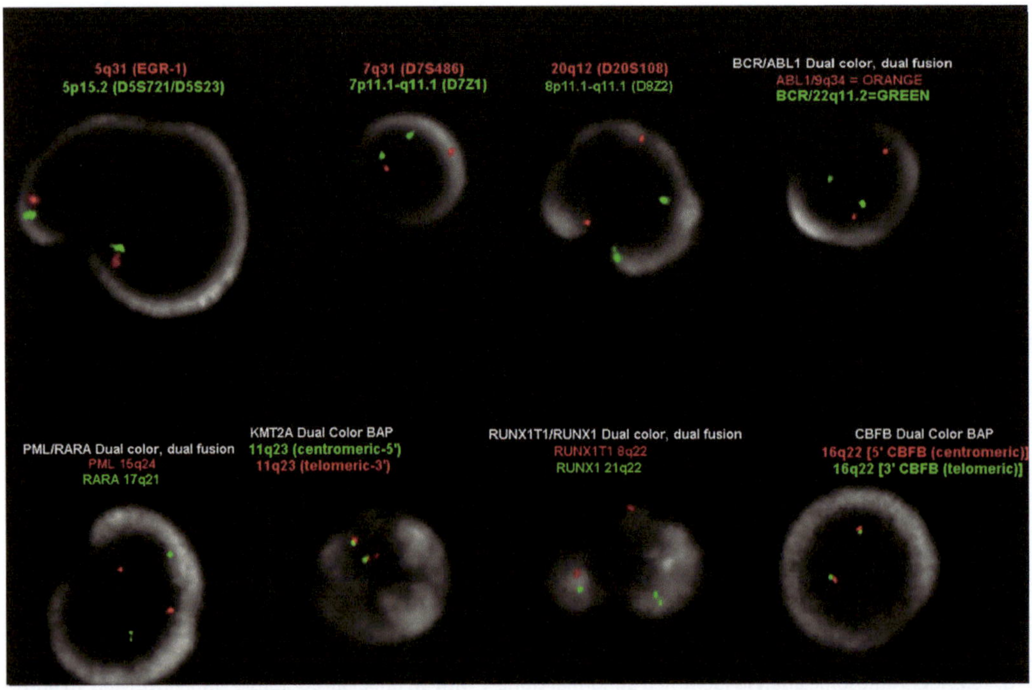

FIG. 18.2.3 FISH for the AML panel showed negative results for all probes examined. ISCN: nuc ish(D5S721,EGR1)x2[200/200],(CEP7,D7S486)x2[200/200],(D8Z2,D20S108)x2[200/200],(RUNX1T1,RUNX1)x2[200/200],(ABL1,BCR)x2[200/200],(KMT2Ax2)[200/200],(PML,RARA)x2[200/200],(CBFBx2)[200/200]

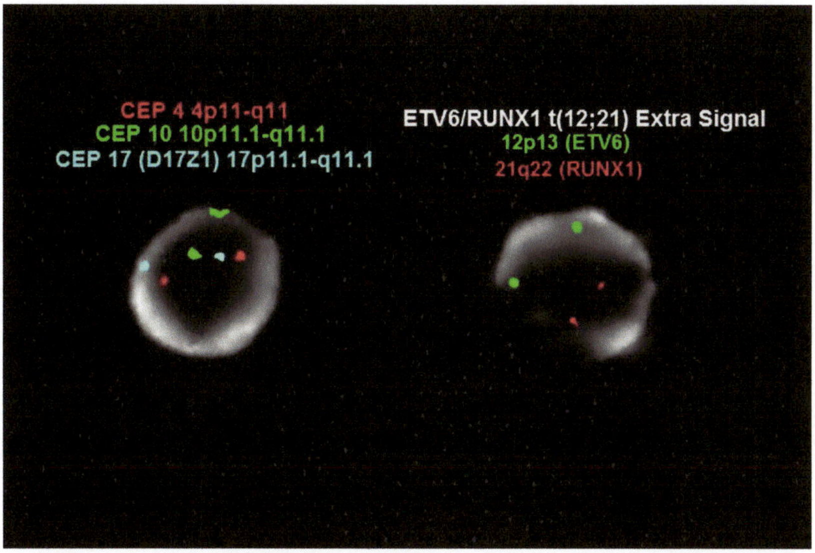

FIG. 18.2.4 FISH for the ALL hyperdiploidy panel and probes for *ETV6::RUNX1* rearrangement revealed negative results for all probes examined. ISCN: nuc ish(CEP4,CEP10,CEP17)x2[199/200],(ETV6,RUNX1)x2[197/200]

FISH for the ALL hyperdiploidy panel and *ETV6::RUNX1* rearrangement probes was also performed on interphase nuclei using probes localized to the chromosome 4, 10, and 17 centromeric regions, *ETV6* (12p13) and *RUNX1* (21q22.3) gene region. Two hundred nuclei were examined for each probe, and the results were within normal limits for the laboratory's established background rates (Fig. 18.2.4).

Finally, the NGS hematology profile was performed and detected several mutations including *FLT3*-ITD, two *RUNX1* mutations, *NRAS*, *PHF6*, and *PRPF8* mutations (Fig. 18.2.5).

Results with interpretations

The chromosome finding is genetically most compatible with MPAL in the context of the flow cytometry report. Two unrelated clones were detected in 11 of the 20 metaphase cells analyzed. Six cells (clone 1) showed an unbalanced translocation between chromosomes 1 and 9 leading to a gain of 1q and a deletion of 9p. Five cells (clone 2) showed trisomy 6. The fact that two distinct clones were present along with the mutations detected by NGS confirmed the diagnosis of mixed phenotype leukemia. Some studies showed that *FLT3*-positive MPAL can be treated successfully using midostaurin [12]. Clinicopathologic correlation is recommended.

Future testing and recommendations

Chromosome analysis and NGS on bone marrow can be ordered to monitor disease progression and treatment efficacy in the future.

Relevant Biomarkers

Tier	Genomic Alteration	Relevant Therapies (In this cancer type)	Clinical Trials
IA	*FLT3 ITD mutation*	**gilteritinib** [1, 2] **midostaurin + chemotherapy** [1, 2] azacitidine cytarabine + daunorubicin cytarabine + daunorubicin + etoposide cytarabine + etoposide + idarubicin cytarabine + fludarabine + idarubicin + filgrastim cytarabine + idarubicin cytarabine + mitoxantrone decitabine gemtuzumab ozogamicin + chemotherapy sorafenib sorafenib + chemotherapy venetoclax + chemotherapy	27
	Prognostic significance: ELN 2017: Intermediate to Adverse		
IA	*RUNX1 p.(S172Afs*4) c.514delA, RUNX1 p.(R162K) c.485G>A*	allogeneic stem cells azacitidine cytarabine cytarabine + daunorubicin cytarabine + daunorubicin + etoposide cytarabine + etoposide + idarubicin cytarabine + fludarabine + idarubicin + filgrastim cytarabine + idarubicin cytarabine + mitoxantrone decitabine gemtuzumab ozogamicin + chemotherapy midostaurin + chemotherapy venetoclax + chemotherapy	5
	Prognostic significance: ELN 2017: Adverse		
IIC	*NRAS p.(Q61K) c.181C>A*	None	2
IIC	*PHF6 p.(L32Tfs*4) c.93_94insA*	None	1

Public data sources included in relevant therapies: **FDA**1, NCCN, **EMA**2, ESMO
Public data sources included in prognostic and diagnostic significance: NCCN, ESMO
Tier Reference: Li et al. *Standards and Guidelines for the Interpretation and Reporting of Sequence Variants in Cancer: A Joint Consensus Recommendation of the Association for Molecular Pathology, American Society of Clinical Oncology, and College of American Pathologists.* J Mol Diagn. 2017 Jan;19(1):4-23.

Prevalent cancer biomarkers without relevant evidence based on included data sources
*PRPF8 p.(V1314Sfs*7) c.3939_3939delGinsCA*

Variant Details

DNA Sequence Variants

Gene	Amino Acid Change	Coding	Locus	Allele Frequency	Variant Effect
NRAS	p.(Q61K)	c.181C>A	chr1:115256530	9.60%	missense
FLT3	p.(E598_Y599insSEEVVQVTGSSDNEYFYVDFREYE)	c.1794_1795insTCTGAGGAGGTGGTACAGGTGACCGGCTCCTCAGATAATGAGTACTTCTACGTTGATTTCAGAGAATATGAA	chr13:28608261	2.00%	nonframeshift Insertion
PRPF8	p.(V1314Sfs*7)	c.3939_3939delGinsCA	chr17:1565283	45.50%	frameshift Block Substitution
RUNX1	p.(S172Afs*4)	c.514delA	chr21:36231869	30.08%	frameshift Deletion
RUNX1	p.(R162K)	c.485G>A	chr21:36252877	19.50%	missense
PHF6	p.(L32Tfs*4)	c.93_94insA	chrX:133511739	32.78%	frameshift Insertion

FIG. 18.2.5 NGS identified *FLT3*-ITD, *RUNX1*, and other mutations.

Case 18.3 Mixed phenotype acute (B/myeloid) leukemia (MPAL) with *RUNX1* mutation

Clinical indication

A 65-year-old male presented with abdominal and pelvic pain. He had a past medical history of mixed lineage acute leukemia which was diagnosed about 2 years ago. Upon recent surveillance, he was found to have a rising *BCR::ABL1* fusion on peripheral blood. This prompted a bone marrow biopsy, the results of which were consistent with relapsed disease. He underwent reinduction chemotherapy with Decitabine and Venetoclax with the addition of Potaninb.

Test ordered

- Chromosome analysis of the bone marrow
- FISH: AML panel
- AML MRD (send out)
- *BCR::ABL1* quantitative real-time PCR
- NGS Hematology Molecular Profile

Laboratory test performed

Chromosome analysis, FISH, and NGS methods were described in Chapters 1 and 12.

Test results

Chromosome analysis on bone marrow was performed. Of the 20 cells examined, 18 exhibited a balanced translocation involving the long arms of chromosomes 2 and 14. The remaining two cells appear to be chromosomally normal (Fig. 18.3.1). While not associated with any disorder, this is a patient-specific abnormality that can be monitored.

FISH for the AML panel was performed on interphase nuclei using probes localized to the *D5S721* (5p15.2), *EGR1* (5q31), *D7Z1* (7cen), *D7S486* (7q31), *D8Z2* (8cen), *RUNX1T1* (8q22), *ABL1* (9q34.12), *KMT2A* (11q23), *PML* (15q24.1), *CBFB* (16q22.1), *RARA* (17q21.1), *D20S108* (20q12), *RUNX1* (21q22.3), and *BCR* (22q11) gene regions. Two hundred nuclei were examined for each probe, and the results were within normal limits for the laboratory's established background rates (Fig. 18.3.2).

AML MRD results were positive for an abnormal myeloid blast population (9.3%) (data not shown).

Then *BCR::ABL1* quantitative real-time PCR designed for detecting p210 and p190 was performed. The results were positive for p210 (Fig. 18.3.3).

Finally, the NGS hematology profile was performed and detected the *RUNX1* frameshift mutation that is associated with many relevant therapies recommended by the NCCN guidelines (Fig. 18.3.4).

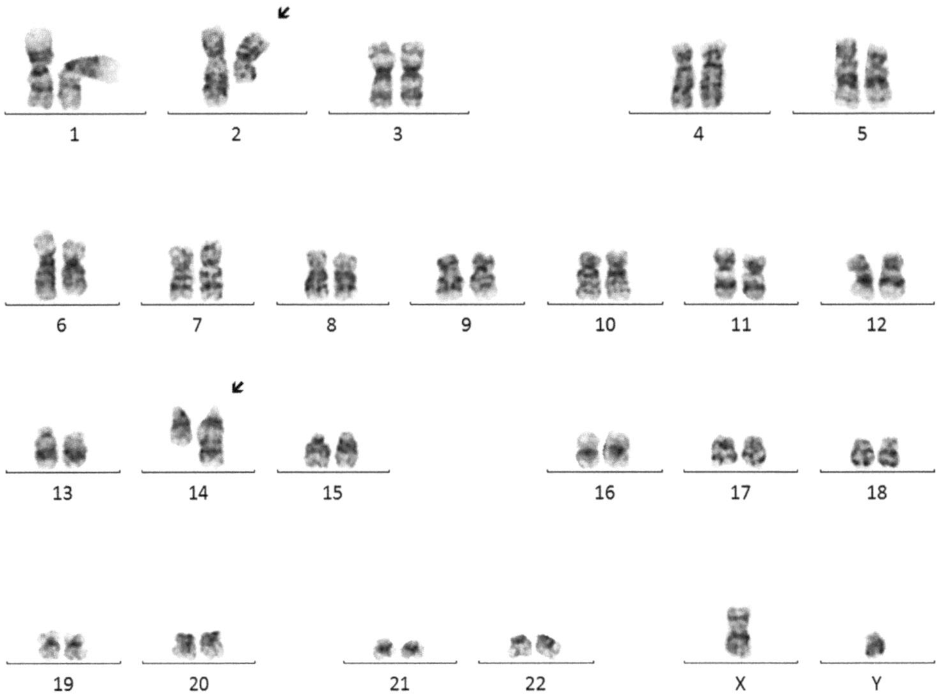

FIG. 18.3.1 Chromosome analysis identified a balanced translocation involving 2q and 14q. ISCN: 46,XY,t(2;14)(q23;q32)[7]/46,XY[8]

Results with interpretations

Although chromosome analysis with t(2;14) did not show an association with MPAL, NGS did show a *RUNX1 p.(A251Mfs*4) c.749_750insGA* mutation with several treatment options for the patient. According to European LeukemiaNet, this mutation is associated with an adverse prognosis. Clinicopathologic correlation is recommended.

Future testing and recommendations

Chromosome analysis, q-PCR, and NGS on bone marrow can be ordered to monitor disease progression and treatment efficacy in the future.

Summary of key learning points

- ALAL is a heterogeneous group of hematopoietic malignancies.
- This group of diseases has an adverse prognosis.
- The treatment relies on ALL-like regimens followed by consolidation chemotherapy or hematopoietic stem cell transplant (HSCT).
- FISH, karyotyping, RT-PCR, and NGS are the common technologies used to identify copy number variations, mutations, or translocations (*KMT2A, BCR::ABL1* fusion) seen in MPAL.

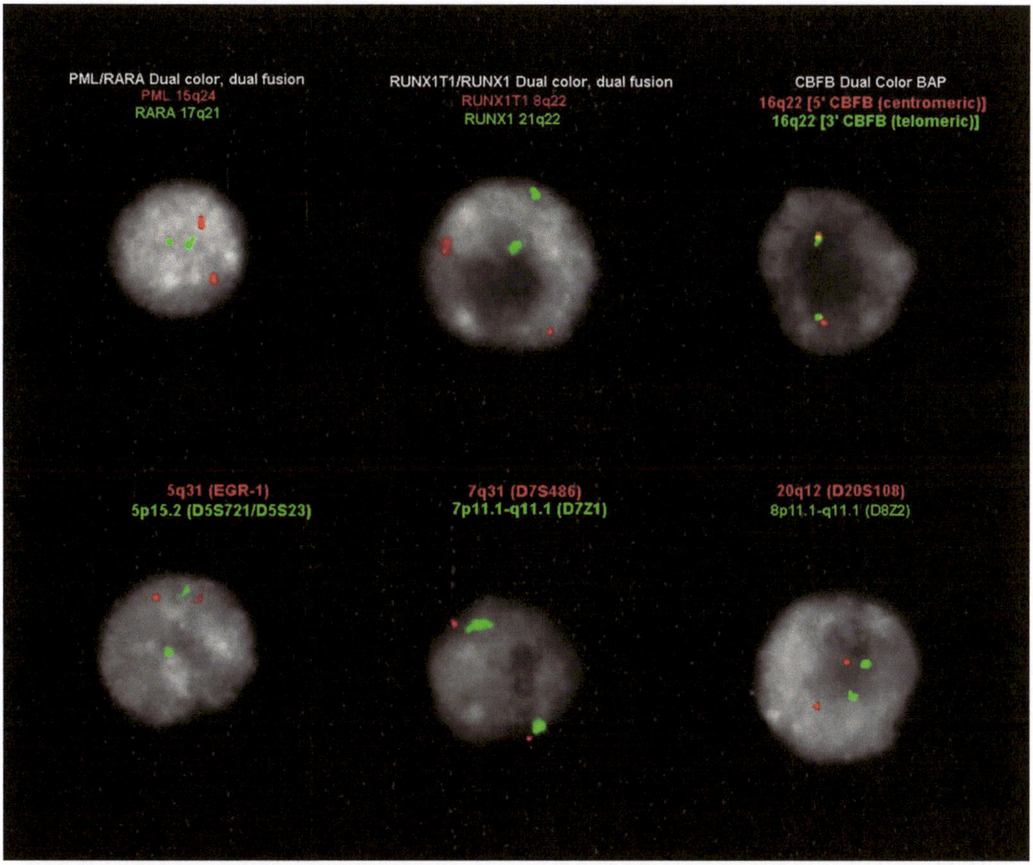

FIG. 18.3.2 FISH for the AML panel revealed negative results for all probes examined. ISCN: nuc ish(D5S721,EGR1)x2[200/200],(CEP7,D7S486)x2[200/200],(D8Z2,D20S108)x2[200/200],(RUNX1T1,RUNX1)x2[200/200],(ABL1,BCR)x2[200/200],(KMT2Ax2)[200/200],(PML,RARA)x2[200/200],(CBFBx2)[200/200]

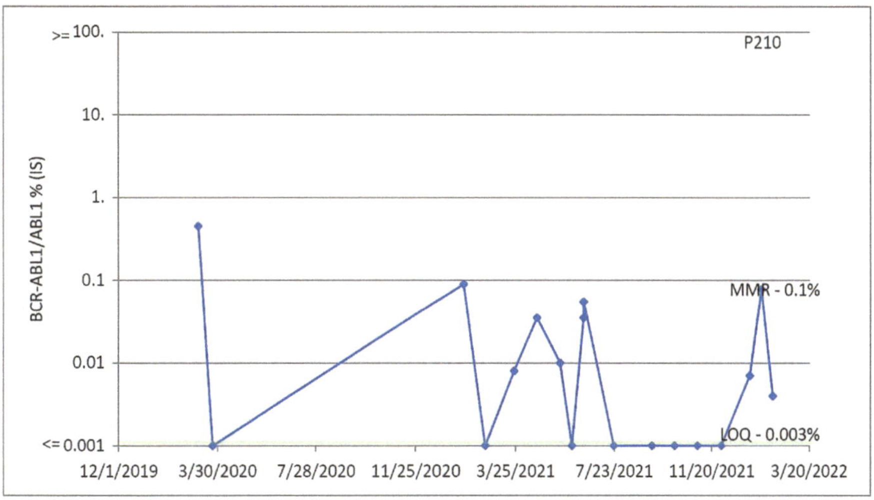

FIG. 18.3.3 Trend and chart for p210 by qPCR showed positive results for *BCR::ABL1* rearrangement.

Relevant Biomarkers

Tier	Genomic Alteration	Relevant Therapies (In this cancer type)	Clinical Trials
IA	*RUNX1 p.(A251Mfs*4) c.749_750insGA* **Prognostic significance:** ELN 2017: Adverse **Diagnostic significance:** None	allogeneic stem cells azacitidine cytarabine cytarabine + daunorubicin cytarabine + daunorubicin + etoposide cytarabine + etoposide + idarubicin cytarabine + fludarabine + idarubicin + filgrastim cytarabine + idarubicin cytarabine + mitoxantrone decitabine gemtuzumab ozogamicin + chemotherapy venetoclax + chemotherapy	4

Public data sources included in relevant therapies: FDA1, NCCN, **EMA**2, ESMO
Public data sources included in prognostic and diagnostic significance: NCCN, ESMO
Tier Reference: Li et al. *Standards and Guidelines for the Interpretation and Reporting of Sequence Variants in Cancer: A Joint Consensus Recommendation of the Association for Molecular Pathology, American Society of Clinical Oncology, and College of American Pathologists.* J Mol Diagn. 2017 Jan;19(1):4-23.

Variant Details

DNA Sequence Variants

Gene	Amino Acid Change	Coding	Locus	Allele Frequency	Variant Effect
RUNX1	p.(A251Mfs*4)	c.749_750insGA	chr21:36206762	4.18%	frameshift Insertion

FIG. 18.3.4 NGS results showed *RUNX1* p.(A251Mfs*4) c.749_750insGA mutation.

References

[1] M. Cernan, T. Szotkowski, Z. Pikalova, Mixed-phenotype acute leukemia: state-of-the-art of the diagnosis, classification and treatment, Biomed. Pap. Med. Fac. Univ. Palacky Olomouc Czech Repub. 161 (3) (2017) 234–241.

[2] J.D. Khoury, et al., The 5th edition of the World Health Organization classification of haematolymphoid tumours: myeloid and histiocytic/dendritic neoplasms, Leukemia 36 (7) (2022) 1703–1719.

[3] D.A. Arber, et al., The 2016 revision to the World Health Organization classification of myeloid neoplasms and acute leukemia, Blood 127 (20) (2016) 2391–2405.

[4] M. Khan, R. Siddiqi, K. Naqvi, An update on classification, genetics, and clinical approach to mixed phenotype acute leukemia (MPAL), Ann. Hematol. 97 (6) (2018) 945–953.

[5] O. Wolach, R.M. Stone, How I treat mixed-phenotype acute leukemia, Blood 125 (16) (2015) 2477–2485.

[6] E. Orgel, et al., Mixed-phenotype acute leukemia: a cohort and consensus research strategy from the Children's Oncology Group Acute Leukemia of Ambiguous Lineage Task Force, Cancer 126 (3) (2020) 593–601.

[7] Q. Li, et al., FISH improves risk stratification in acute leukemia by identifying KMT2A abnormal copy number and rearrangements, Sci. Rep. 12 (1) (2022) 9585.

[8] A. Pallavajjala, et al., Genomic characterization of chromosome translocations in patients with T/myeloid mixed-phenotype acute leukemia, Leuk. Lymphoma 59 (5) (2018) 1231–1238.

[9] J. Suttorp, et al., Optical genome mapping as a diagnostic tool in pediatric acute myeloid leukemia, Cancers (Basel) 14 (9) (2022).

[10] A.E. Quesada, et al., Mixed phenotype acute leukemia contains heterogeneous genetic mutations by next-generation sequencing, Oncotarget 9 (9) (2018) 8441–8449.

[11] M.M. Li, et al., Standards and guidelines for the interpretation and reporting of sequence variants in cancer: a joint consensus recommendation of the association for molecular pathology, American Society of Clinical Oncology, and College of American Pathologists, J. Mol. Diagn. 19 (1) (2017) 4–23.

[12] Z. Tremblay, et al., Use of midostaurin in mixed phenotype acute leukemia with FLT3 mutation: a case series, Eur. J. Haematol. 108 (2) (2022) 163–165.

19

Precursor lymphoid neoplasms

Xia Li and Guang Liu
SONORA QUEST LABORATORIES, PHOENIX, AZ, UNITED STATES

Background

Precursor lymphoid neoplasm consists mainly of B-lymphoblastic leukemia/lymphoma (B-ALL) and T-lymphoblastic leukemia/lymphoma (T-ALL). They are a group of genetically heterogeneous lymphoid neoplasms derived from B- and T-lymphoid progenitors. These are malignant disorders of immature B or T-cells that occur typically in children, and 75% of patients are under the age of 6. T-ALL accounts for about 15% of childhood ALL cases and is more common in adolescents than in younger children. Each year, there are approximately 6000 new cases of ALL diagnosed in the United States. Over 80% of these cases represent B-ALL forms and are leukemic, whereas T-ALL presents with a mediastinal mass with or without leukemic involvement [1].

To establish the diagnosis of these disorders, morphologic, immunophenotypic, and genetic features are to be evaluated so that the differentiation between normal progenitors and other hematopoietic and nonhematopoietic neoplasms can be determined [2]. According to the revised fourth edition of the WHO Classification (2017) of Tumors of Hematopoietic and Lymphoid Tissues, there are 10 subgroups of B-ALL including B-lymphoblastic leukemia/lymphoma, not otherwise specified, B-lymphoblastic leukemia/lymphoma with recurrent genetic abnormalities (9 subtypes). These nine subtypes include B-ALL with hyperdiploidy, B-ALL with hypodiploidy, B-ALL with iAMP21, B-ALL with *BCR::ABL1* fusion, B-ALL with *BCR::ABL1*-like features, B-ALL with *KMT2A* rearrangement, B-ALL with *ETV6::RUNX1* fusion, B-ALL *TCF3::PBX1*, and B-ALL with *IGH::IL3* fusion [1,3].

In the fifth edition of the WHO Classification (2022) of Tumors of Hematopoietic and Lymphoid Tissues, two new entities were added to this group: B-lymphoblastic leukemia/lymphoma with *ETV6::RUNX1*-like features and B-lymphoblastic leukemia/lymphoma with *TCF3::HLF* fusion [4]. For T-ALL, T-lymphoblastic leukemia/lymphoma and early T-precursor lymphoblastic leukemia/lymphoma were in both the fourth revised and the fifth editions [1,4].

In the genetic diagnostic lab, karyotyping, FISH, PCR, and next-generation sequencing (NGS) are used to identify chromosome abnormalities including rearrangements of *BCR::ABL1*, *KMT2A*, *ETV6::RUNX1*, *TCF3::PBX1*, *IGH::IL*, *TCF::HLF*, hyperdiploidy, hypodiploidy, iAML21, T-cell receptor gene rearrangement, etc. Gene mutations associated with precursor lymphoid neoplasms are usually identified by NGS [5]. The information from these

Cases in Laboratory Genetics and Genomics (LGG) Practice. https://doi.org/10.1016/B978-0-323-99622-8.00002-2

tests will help patients with targeted therapy, predictive prognosis, and risk stratification. In this chapter, 14 cases will be used to illustrate how these tests will benefit patients, including a large pediatric population.

Case 19.1 B-lymphoblastic leukemia (B-ALL, Ph+) with T315I resistance mutation

A 70-year-old female felt very fatigued and weak. The bone marrow biopsy report showed 90% lymphoblasts. Immunohistochemistry (IHC) stained positive for CD34 and Terminal deoxynucleotidyl transferase (TdT), an assay specifically created for the detection of acute lymphoblastic leukemia (ALL). Flow cytometry identified 80% of lymphoblasts as positive for CD10, CD19, CD22, CD34, CD79a, and TdT, consistent with acute leukemia, B-ALL. The patient also went through chemotherapy and had induced neutropenia and pancytopenia. She was treated with HyperCVAD plus dasatinib initially. After the first relapse, she was given prednisone and dasatinib. On the second relapse, she was found to have a T315I mutation in the *ABL1* kinase domain, which is a resistance mutation to the first-line therapy. Then she started with ponatinib, an inhibitor targeted to this T315I mutation.

Test ordered

- Chromosome analysis of the bone marrow
- FISH: *BCR::ABL1* probes
- *BCR::ABL1* quantitative real-time PCR
- NGS Hematology Molecular Profile
- *ABL1* kinase domain mutation analysis

Laboratory test performed

Chromosome analysis, FISH, q-PCR, and NGS methods were described in Chapters 1 and 12.

ABL1 kinase domain mutation analysis is a test designed for the detection of mutations in imatinib-treated CML patients with acquired resistance. The early detection of mutations should provide clinical benefit by allowing for early intervention. Candidates for the *BCR::ABL1*1 kinase domain mutation analysis include those who fail to respond to Gleevec (imatinib) and other TKIs therapy or are in an accelerated phase/blast crisis. Most of the labs use RT-PCR and Sanger sequencing method for mutation detection. The analysis includes the detection of all mutations recommended by guidelines, including the common T315I, Y253H, E255K/V, F359V/C/I, F317L/V/I/C, T315A, and V299L. Alternatively, NGS is used to detect these mutations with a higher sensitivity than Sanger sequencing [6]. In the following patient, this assay was performed in another lab.

Test results

Firstly, a karyotype revealed a complex structural rearrangement that involved chromosomes 9 and 17 leading to the loss of chromosomes 9p and 17p. On the same chromosome 9, at the q arm, there was a translocation involving chromosomes 9q and 22q resulting in a derivative chromosome 22 (Philadelphia chromosome), which is associated with a poor prognosis of B-ALL. In addition, there was monosomy 7 (Fig. 19.1.1).

Secondly, FISH was performed on interphase nuclei using probes localized to the *ABL1* (9q34.12) and *BCR* (22q11) gene regions. Two hundred nuclei were examined, and the results were positive for a rearrangement involving *BCR::ABL1* in 181/200 (85.5%) of the cells scored (Fig. 19.1.2).

Then *BCR::ABL1* quantitative real-time PCR designed for detecting p210 and p190 showed negative results (see the trend table and chart for p210 in Fig. 19.1.3). The discordant results were discussed in "Results with Interpretations" section.

Next, NGS Hematology Molecular Profile was conducted, and the results showed an atypical transcript e13a3 (Fig. 19.1.4).

Finally, RT-PCR and Sanger sequencing were conducted in another lab showing T315I mutation by *ABL1* kinase domain mutation analysis (data not shown).

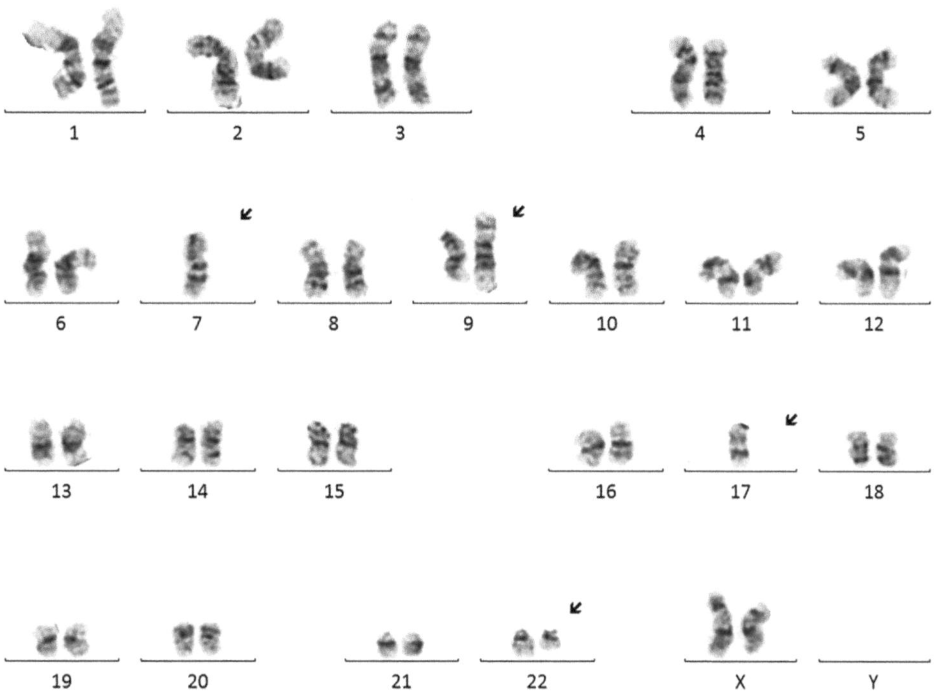

FIG. 19.1.1 The karyotype of bone marrow from the patient showed complex rearrangements involving chromosomes 9&22, chromosomes 9&17, and monosomy 7. ISCN: 44,XX,-7,der(9;17)(q10;q10)t(9;22)(q34.2;q11.2),der(22)t(9;22)[19]/46,XX[3]

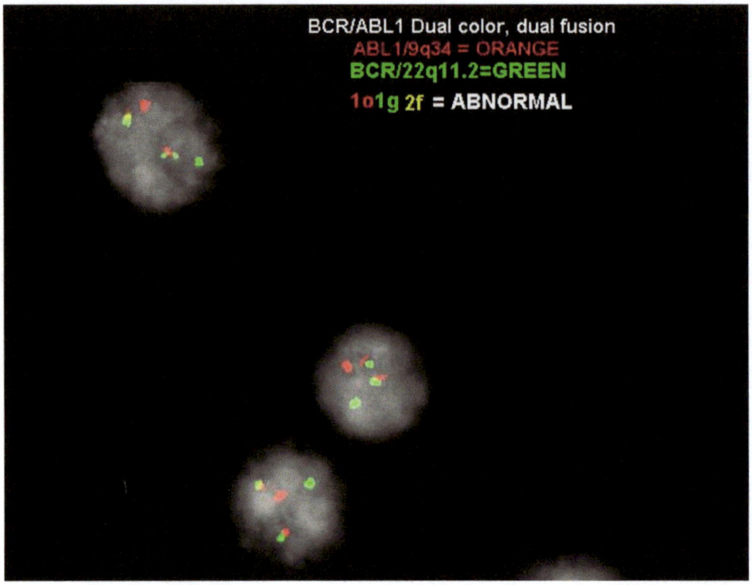

FIG. 19.1.2 FISH with *BCR::ABL1* dual-color dual-fusion probes showed a positive *BCR::ABL1* rearrangement. ISCN: nuc ish(*ABL1*,*BCR*)x3(*ABL1* con *BCR*x2)[171/200]

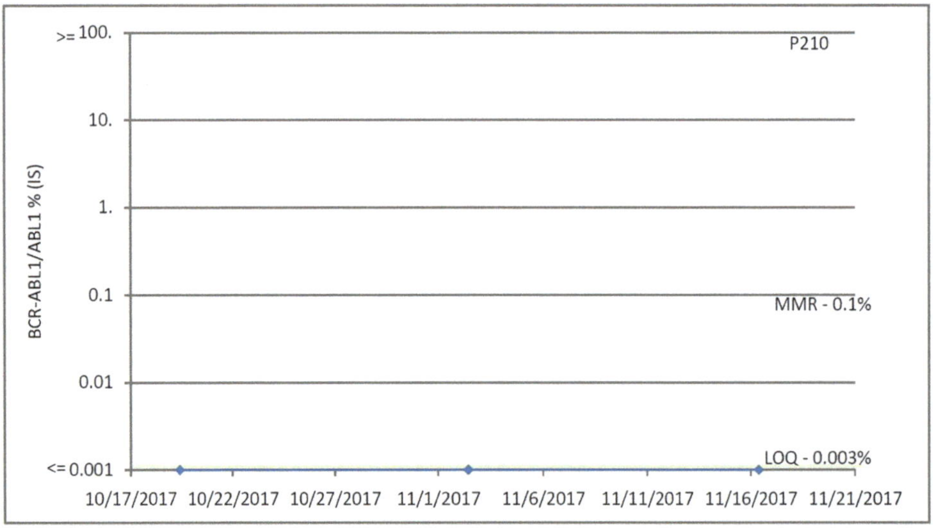

FIG. 19.1.3 qPCR results showed the trend table and chart of p210 negative results due to failure to detect the e13a3 (a rare transcript for *BCR::ABL1*).

Results with interpretations

Among the tests performed on this patient, chromosome analysis, FISH, and NGS were concordant, all of which detected t(9;22) or *BCR::ABL1* rearrangement. The patient has an atypical transcript e13a3 identified by NGS. The qPCR was designed to detect p210

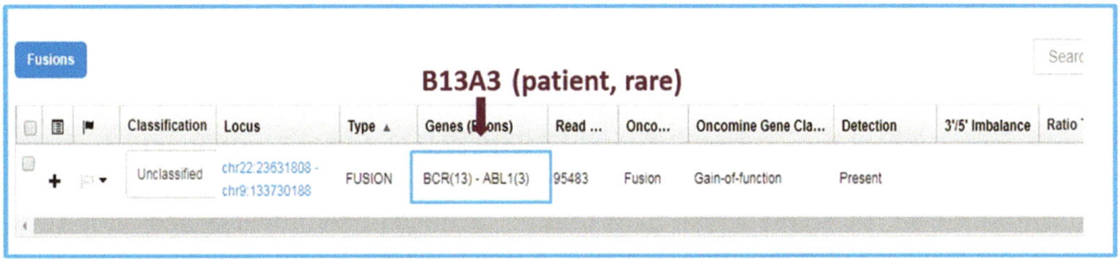

FIG. 19.1.4 NGS showed the fusion of *BCR::ABL1* with an atypical transcript e13a3.

and p190 transcripts, which explained why *BCR::ABL1* rearrangement was not detected by qPCR. The patient has had several relapses after the initial diagnosis and did not respond well to the therapy with imatinib. Therefore, the *ABL1* kinase domain mutation analysis was performed in another lab, and the T315I mutation was identified. This is a common mutation that accounts for ~20% clinical resistance to TKIs. Ponatinib is a novel, orally active, multitargeted TKI. It binds to *BCR::ABL1* with high potency, to inhibit the whole spectrum of mutants conferring resistance against other TKIs, including the T315I mutant that is resistant to all current therapies [6–8].

Future testing and recommendations

Since the patient does not have p210 and p190 transcripts, chromosome analysis, FISH, or NGS can be ordered to monitor disease progression. Otherwise, a send-out to the lab that performs qPCR detecting atypical transcripts is recommended.

Case 19.2 B-lymphoblastic leukemia (B-ALL, Ph+) with a complex karyotype

Clinical indication

A 57-year-old male had Ph+ B-lymphoblastic leukemia on chemotherapy. He was in remission for a few months, but then his blood counts began to decrease, requiring transfusions. The patient was placed on chemotherapy and allopurinol.

Test ordered

- Chromosome analysis of the bone marrow
- FISH: ALL panel
- *BCR::ABL1* quantitative real-time PCR

Laboratory test performed

Chromosome analysis, FISH, and RT-PCR methods were described previously in Chapters 1 and 12.

Test results

Chromosome analysis identified three abnormal clones (two related; one unrelated). Clone 1 with six cells showed t(9;22)(q34;q11.2)/*BCR::ABL1* together with gains of the X chromosome and chromosomes 7, 8, and 17 (Fig. 19.2.1). Clone 2 with four cells showed t(9;22) and a deletion 3p (Fig. 19.2.2). Clone 3 with two cells showed a gain of the Y chromosome as the sole aberration, a finding of unclear significance, possibly representing low-level sex chromosome mosaicism (Fig. 19.2.3).

FISH for ALL panel was performed on interphase nuclei using probes localized to the chromosome 4, 10, and 17 centromeric regions; *ABL1* (9q34.12); *BCR* (22q11); *KMT2A* (11q23); *ETV6* (12p13); *RUNX1* (21q22.3) gene regions. Two hundred nuclei were examined, and the results were positive for a typical *BCR::ABL1* rearrangement (2F1O1G) in 164/200 (84.0%) of nuclei scored, and trisomy 17 in 132/200 (66.0%) of the cells scored (Fig. 19.2.4). One additional fusion gene was observed in 7/200 (3.5%) of nuclei scored (data not shown).

Then the *BCR::ABL1* quantitative real-time PCR designed for detecting p210 and p190 showed positive results for p190, concordant with those of karyotyping and FISH (Fig. 19.2.5).

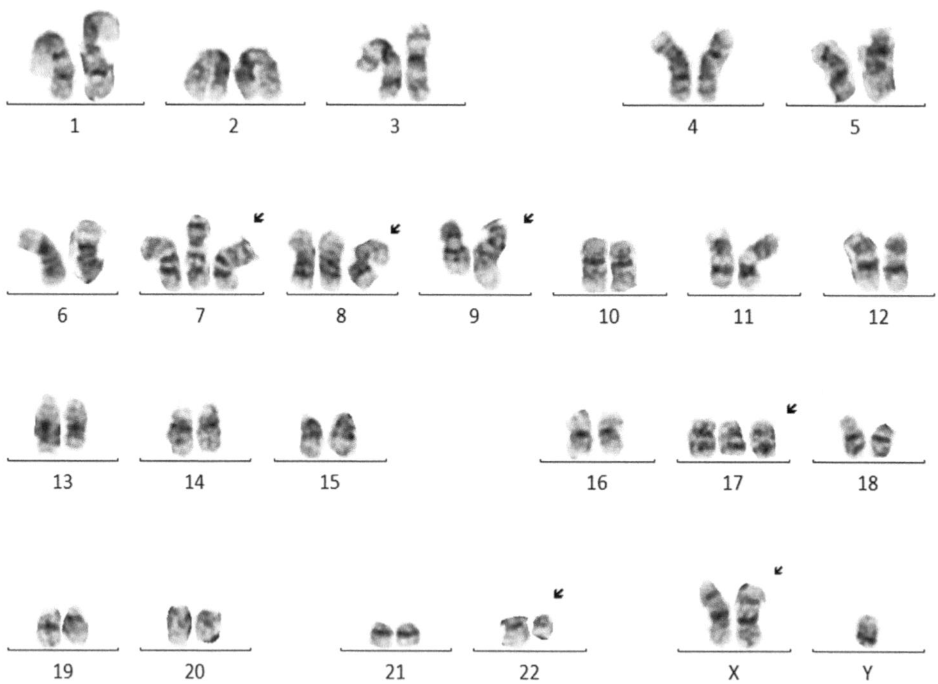

FIG. 19.2.1 Chromosome analysis showed abnormal clone 1 with a Philadelphia chromosome and gains of chromosomes X, 7, 8, and 17. ISCN: 50,XY,+X,+7,+8,t(9;22)(q34;q11.2),+17[6]/46,XY,del(3)(p21),t(9;22)[4]/47,XY,+Y[2]/46,XY[8]

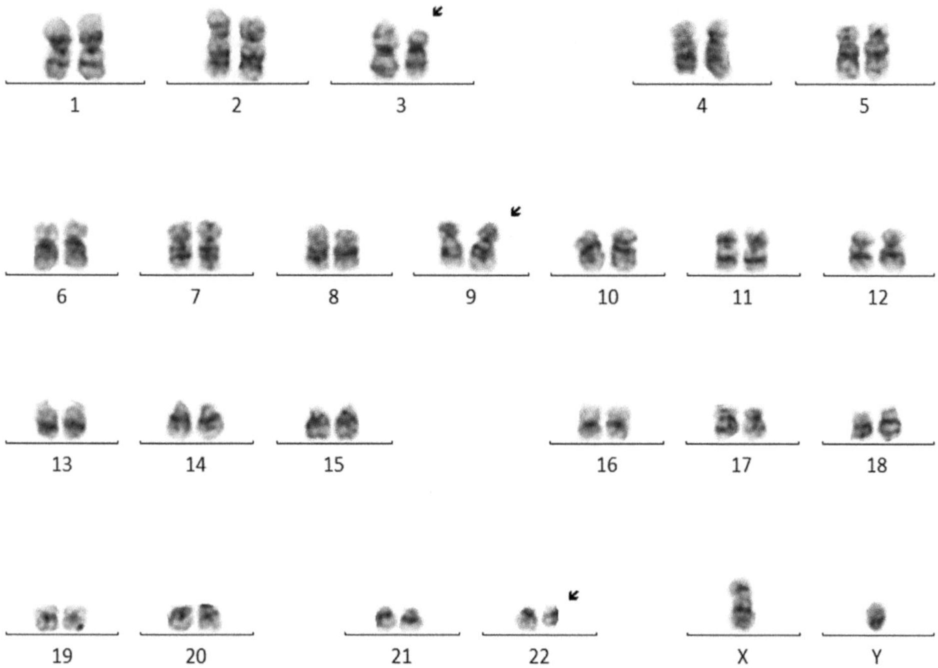

FIG. 19.2.2 Chromosome analysis showed abnormal clone 2 with a Philadelphia chromosome and del(3p).

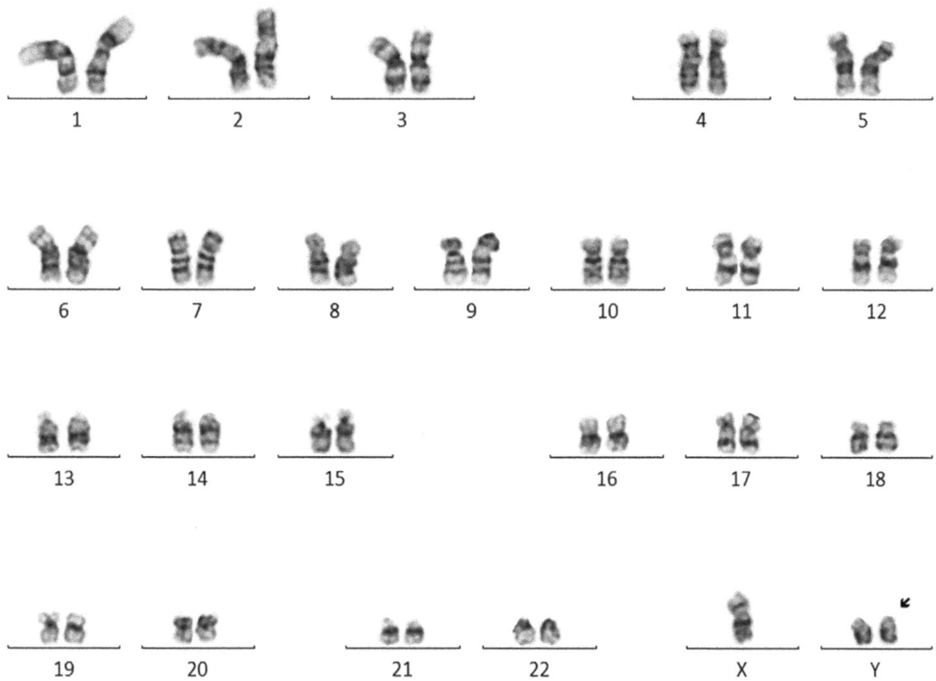

FIG. 19.2.3 Chromosome analysis showed abnormal clone 3 with a gain of the Y chromosome.

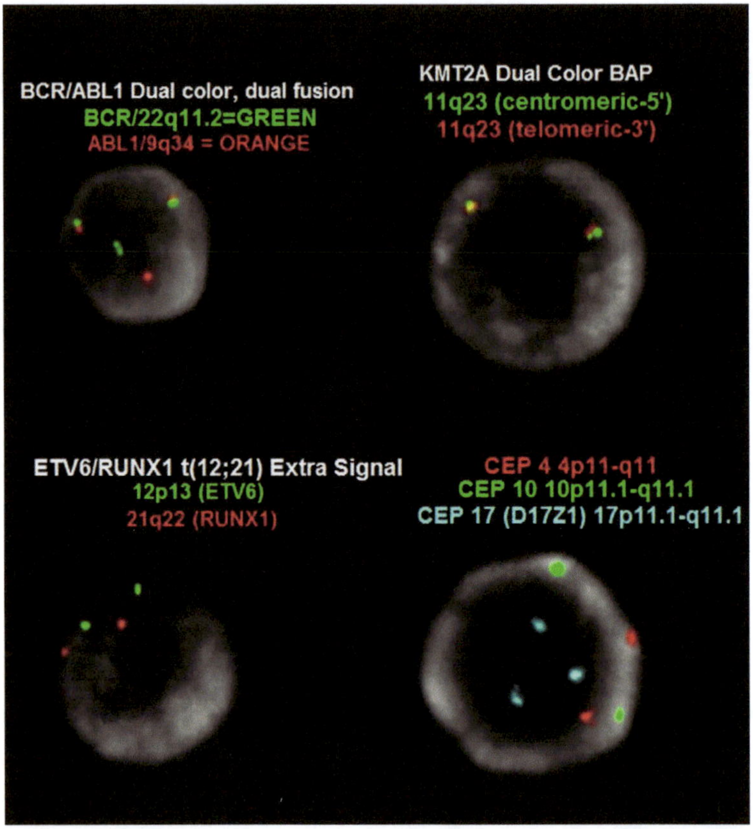

FIG. 19.2.4 FISH for ALL panel showed positive results for *BCR::ABL1* rearrangement in 87.5% of the cells scored and trisomy 17 in 66.0% of the cells scored. ISCN: nuc ish(CEP4x2,CEP10x2,D17Z1x3)[132/200],(ABL1,BCR)x3(ABL1 con BCRx2)[168/200]/(ABL1,BCR)x4(ABL1 con BCRx3)[7/200],(KMT2A)x2[200/200],(ETV6,RUNX1)x2[199/200]

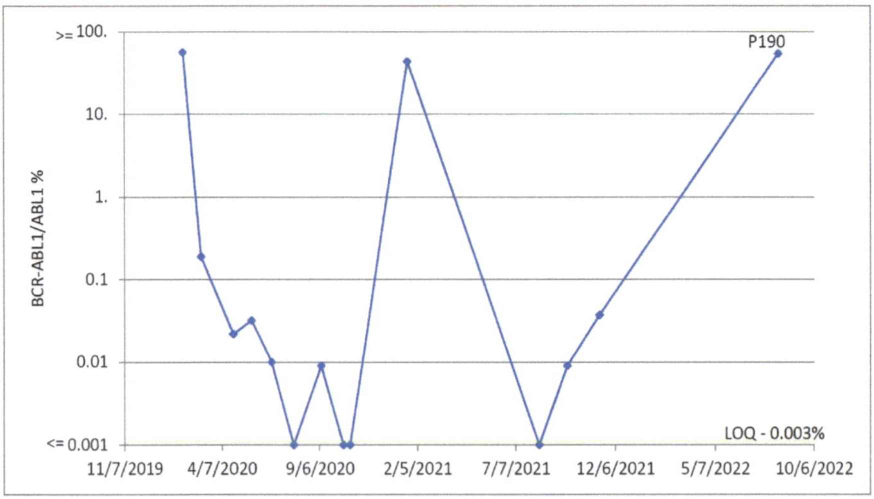

FIG. 19.2.5 qPCR results showed the trend table and chart of p190 positive results.

Results with interpretations

The karyotyping findings support recurrent B-lymphoblastic leukemia/lymphoma (B-ALL) in the context of the flow cytometry report and previous history. The t(9;22) with trisomy for the X chromosome and chromosomes 7 and 17 were detected previously supporting recurrent disease. Trisomy 8 in clone 1 and the deletion of 3p in clone 2 were novel abnormalities suggesting clonal evolution. The presence of the t(9;22) in B-ALL is associated with an unfavorable prognosis. However, the incorporation of tyrosine kinase inhibitors in the treatment has proven beneficial for these patients.

The FISH findings of the typical *BCR::ABL1* rearrangement and trisomy 17 have been reported previously, along with positive results for p190 from *BCR::ABL1* quantitative real-time PCR, all of which are consistent with recurrent B-lymphoblastic leukemia.

Future testing and recommendations

Regular follow-ups and clinicopathologic correlation are recommended. Chromosome, FISH, and q-PCR can be ordered for disease monitoring and treatment efficacy in the future.

Case 19.3 Relapsed B-lymphoblastic leukemia (B-ALL, Ph−) with a complex karyotype

Clinical indication

The patient was a 75-year-old with relapsed B-ALL. He had pancytopenia and reported feeling unwell but denied fever, chills, shortness of breath, chest pain, or abdominal pain.

Test ordered

- Chromosome analysis of the bone marrow
- FISH: ALL panel
- *BCR::ABL1* quantitative real-time PCR

Laboratory test performed

Chromosome analysis and FISH methods were described previously in Chapters 1 and 12.

Test results

Of the 20 cells examined, 16 were abnormal and 3 different clones were observed. Clone 1 with 11 cells exhibited multiple numerical and structural chromosome abnormalities, including loss of the Y chromosome, and gains of chromosomes 1, 6, 8, 10, 18, 19, 21, and add (21p) in a haploid background (Fig. 19.3.1). Clone 2 with three cells was a double clone from clone 1 (Fig. 19.3.2). Clone 3 with two cells had an additional Y chromosome (Fig. 19.3.3). The remaining four cells appeared to be chromosomally normal.

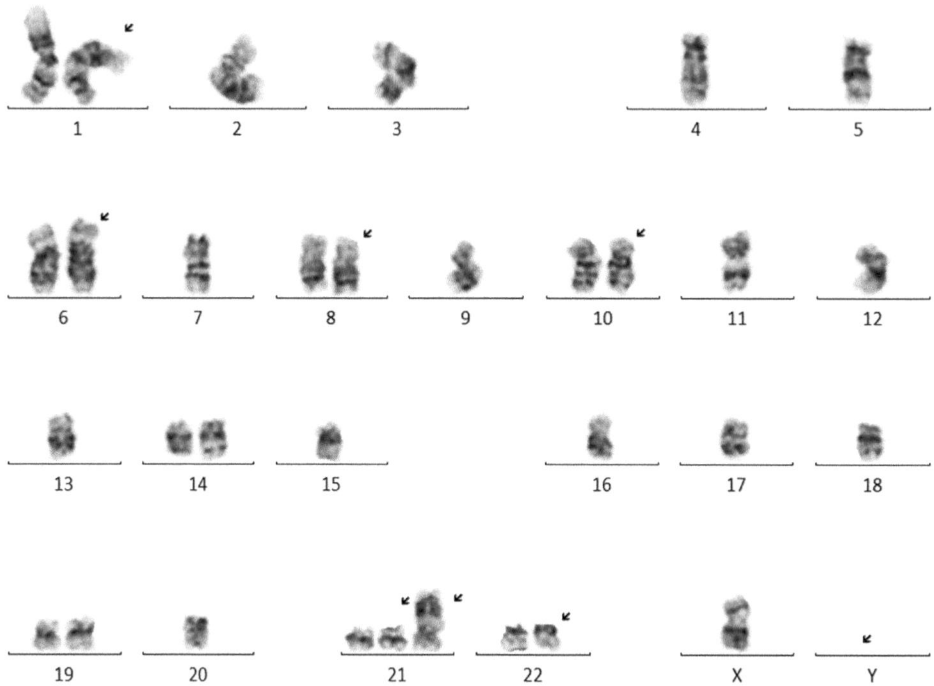

FIG. 19.3.1 Chromosome analysis showed abnormal clone 1 with gains of chromosomes 1, 6, 8, 10, 18, 19, and 21, loss of the Y chromosome, and add(21p) in a hypodiploid background. ISCN: 31-32<n>,X,-Y,+1,+6,+8,+10,+18,+19,+21,+21,add(21)(p11.2) [cp11]/ 64-65,idemx2[cp3]/47,XY,+Y[2]/46,XY[4]

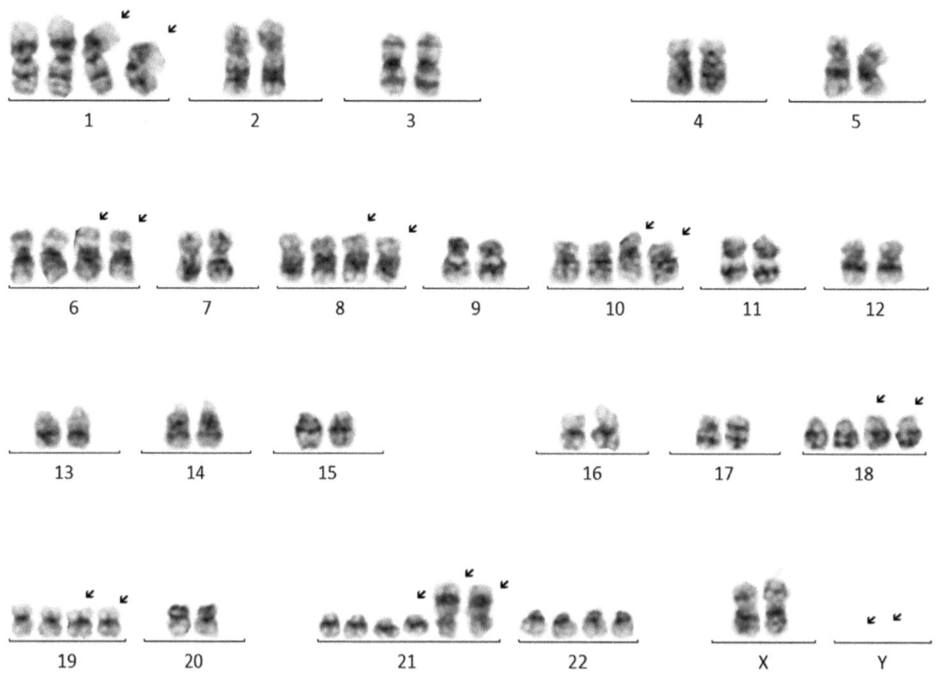

FIG. 19.3.2 Chromosome analysis showed abnormal clone 2, which is a double clone of clone 1.

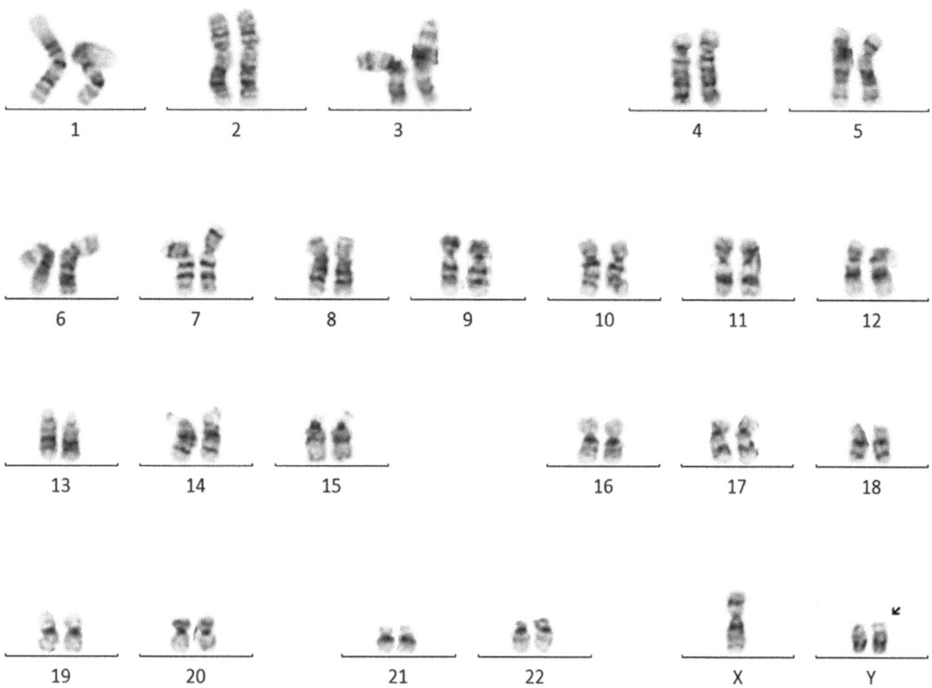

FIG. 19.3.3 Chromosome analysis showed abnormal clone 3 with a gain of the Y chromosome.

FISH for ALL panel was performed on interphase nuclei using probes localized to the chromosome 4, 10, and 17 centromeric regions, *ABL1* (9q34.12), *BCR* (22q11), *KMT2A* (11q23), *ETV6* (12p13), and *RUNX1* (21q22.3) gene regions. Two hundred nuclei were examined, and the results demonstrated multiple abnormalities (Fig. 19.3.4).

Loss of chromosomes 4 and 17 was observed in 26/200 (13.0%) of the cells scored.

Loss of 9q34 (*ABL1*) was observed in 59/200 (29.5%) of the cells scored.

Gain of 9q34 (*ABL1*) was observed in 38/200 (19.0%) of the cells scored.

Gain of chromosome 10 was observed in 43/200 (21.0%) of the cells scored.

Loss of 11q (*KMT2A*) was observed in 45/200 (22.5%) of the cells scored.

Loss of 12p (*ETV6*) was observed in 60/200 (30.0%) of the cells scored.

Gain of 21q (*RUNX1*) was observed in 50/200 (25.0%) of the the cells scored.

Gain of 22q (*BCR*) was observed in 38/200 (19.0%) of the cells scored.

BCR::ABL1 quantitative real-time PCR showed negative results (data not shown).

Results with interpretations

Chromosome analysis identified a near-haploid karyotype with gains of multiple chromosomes and other abnormalities. FISH results showed multiple gains or losses of multiple chromosome regions. The losses of chromosomes 4 and 17, along with multiple gene region losses, indicate a near-haploid component, which is considered a poor risk.

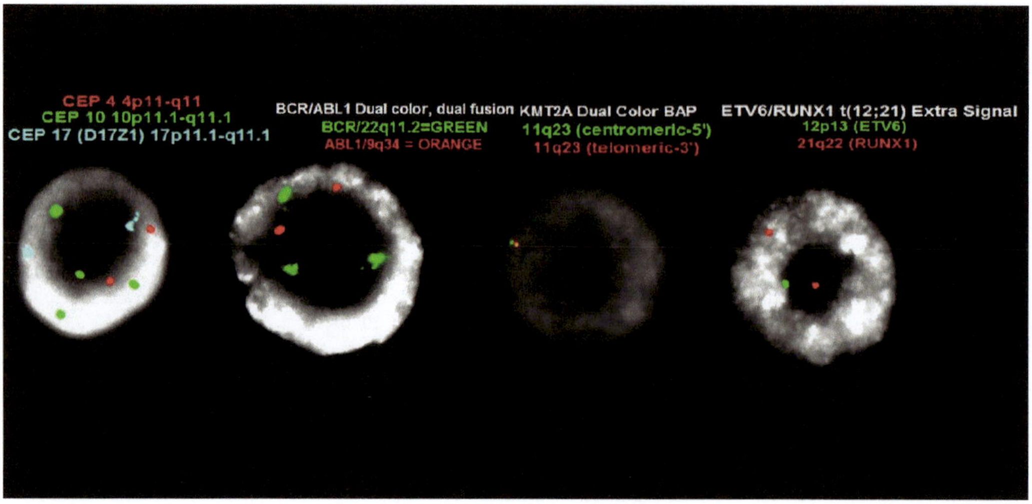

FIG. 19.3.4 FISH for ALL panels showed multiple gains and losses of multiple chromosomes without rearrangement of *BCR::ABL1*, *ETV6::RUNX1*, and *KMT2A*. ISCN: nuc ish(CEP4x2,CEP10x3-4,CEP17x2)[43/200]/(CEP4x1,CEP10x2,CEP17x1)[26/200],(ABL1x2,BCRx3-4)[38/200]/(ABL1x1,BCRx2)[59/200],(KMT2Ax1)[45/200],(ETV6x2,RUNX1x3-4)[50/200]/(ETV6x1,RUNX1x2)[60/200]

Future testing and recommendations

Regular follow-ups and clinicopathologic correlation are recommended. Chromosome analysis and FISH can be ordered for disease monitoring and treatment efficacy in the future.

Case 19.4 B-lymphoblastic leukemia (B-ALL) with t(12;21)(p13;q22)/*ETV6::RUNX1* fusion

Clinical indication

A 13-year-old female with no significant past medical history presented with an enlarged lymph node, chest pain, and shortness of breath. She denied any history of fevers, weight loss, or sore throat. Blood work showed a normal WBC count with lymphocytosis and a blast. Mild anemia was noted with normal platelet count. Chest X-ray was done with no abnormal findings. Leukemia was suspected.

Test ordered

- Chromosome analysis of the bone marrow
- FISH: ALL and Ph-like ALL panels
- *BCR::ABL1* quantitative real-time PCR

Laboratory test performed

Chromosome analysis, FISH, and q-PCR methods were described previously in Chapters 1 and 12.

Test results

Of the 25 cells examined, 16 exhibited an abnormal chromosome complement with 2 related clones; clonal evolution was evident. Clone 1 with 11 cells showed a del(6q), a del(12p), and a der(18) derived from an 18;21 translocation (Fig. 19.4.1). Clone 2 with five cells showed a der(1) derived from a 1;21 translocation in addition to the del(12p) (Fig. 19.4.2). The remaining nine cells appeared to be chromosomally normal.

FISH ALL panel was performed on interphase nuclei using probes localized to the chromosome 4, 10, and 17 centromeric regions; *ABL1* (9q34.12); *BCR* (22q11); *KMT2A* (11q23); *ETV6* (12p13); *RUNX1* (21q22.3) gene regions. Two hundred nuclei were examined, and the results demonstrated rearrangement involving *ETV6::RUNX1* or t(12;21), deletion of 12p13 (*ETV6*), and gains of 21q22 (*RUNX1*) in 164/200 (82.0%) nuclei scored (Fig. 19.4.3). Metaphase FISH confirmed the *ETV6::RUNX1* fusion and gain of 21q22 (*RUNX1*) (Fig. 19.4.4).

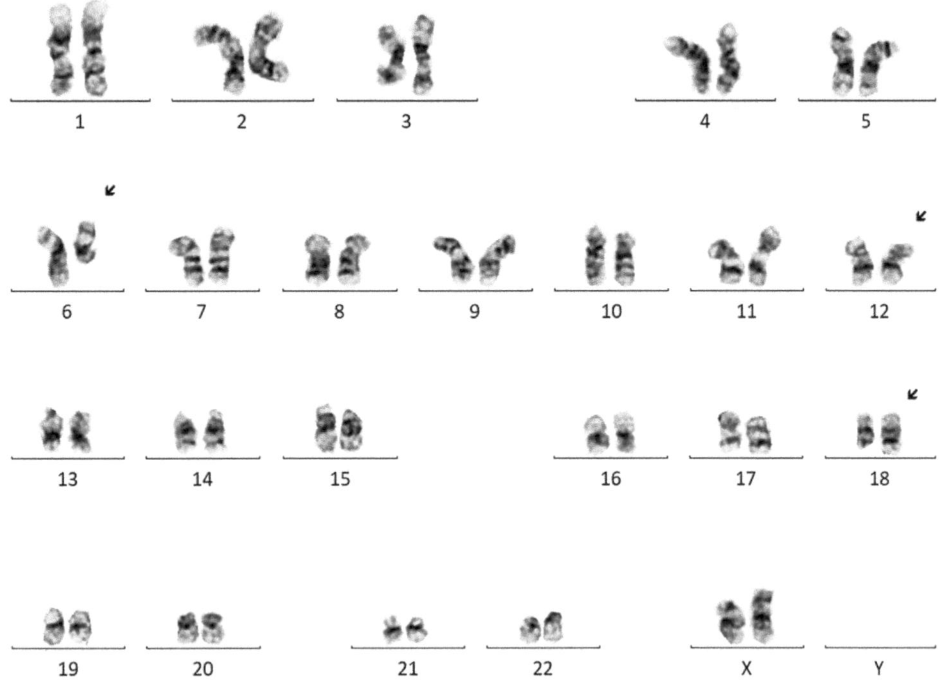

FIG. 19.4.1 Chromosome analysis exhibited an abnormal clone 1 with del(6q), del(12p), and an unbalanced translocation involving 18q and 21q. ISCN: 46,XX,del(6)(q16),del(12)(p12p13),der(18)t(18;21)(q22;q21)[11]/ 46,XX,der (1)t(1;21)(q41;q21),del(12)(p12p13)[5]/46,XX[9]

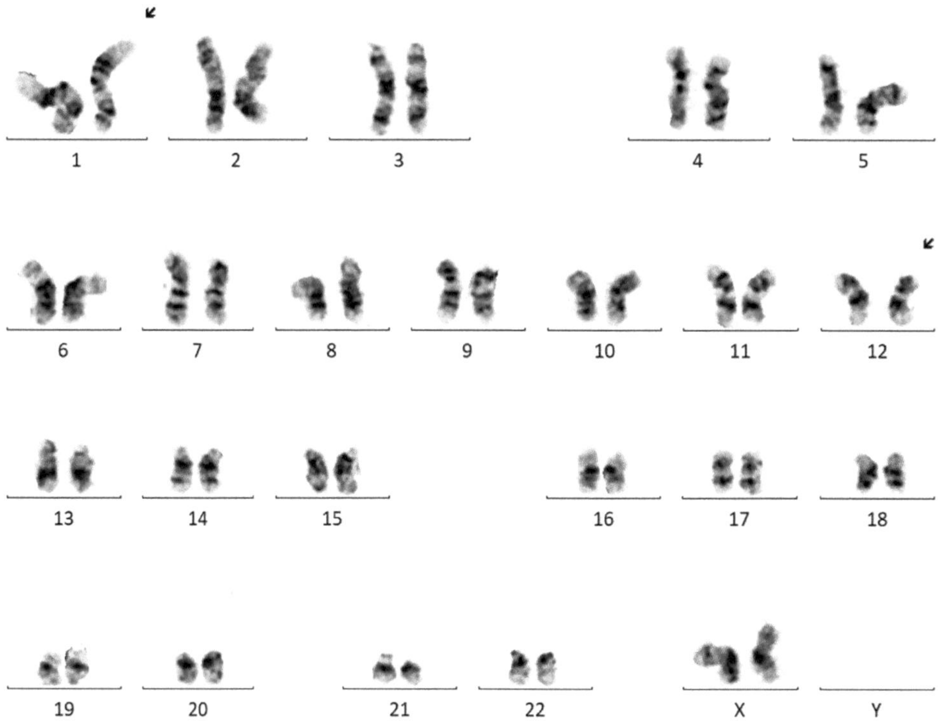

FIG. 19.4.2 Chromosome analysis exhibited an abnormal clone 2 with del(12p) and an unbalanced translocation involving 1q and 21q.

FISH Ph-like ALL panel was also performed on interphase nuclei using probes localized to the *CRLF2* (Xp22.33/Yp11.32), *ABL2* (1q25.2), *CSF1R* (5q32), *PDGFRB* (5q33.2), *JAK2* (9p24.1), *ABL1* (9q34.13), and *EPOR* (19p13.2) gene regions. Two hundred nuclei were examined, and the results were within normal limits for the laboratory's established background rates (Fig. 19.4.3).

BCR::ABL1 quantitative real-time PCR showed negative results for the fusion (data not shown).

Results with interpretations

Chromosome analysis revealed a complex karyotype with two related abnormal clones. The del(12p)(*ETV6*) was also seen in the concurrent FISH studies. The "cryptic" *ETV6::RUNX1* rearrangement by FISH was not appreciated by the karyotype due to the low chromosome banding resolution. *ETV6::RUNX1* rearrangement is a recurrent abnormality found in 25% of childhood B acute lymphoblastic leukemia and is generally associated with a good prognosis.

FISH ALL panel detected t(12;21) and a deletion of *ETV6*. This combination of abnormalities is associated with a favorable prognosis in acute lymphoblastic leukemia.

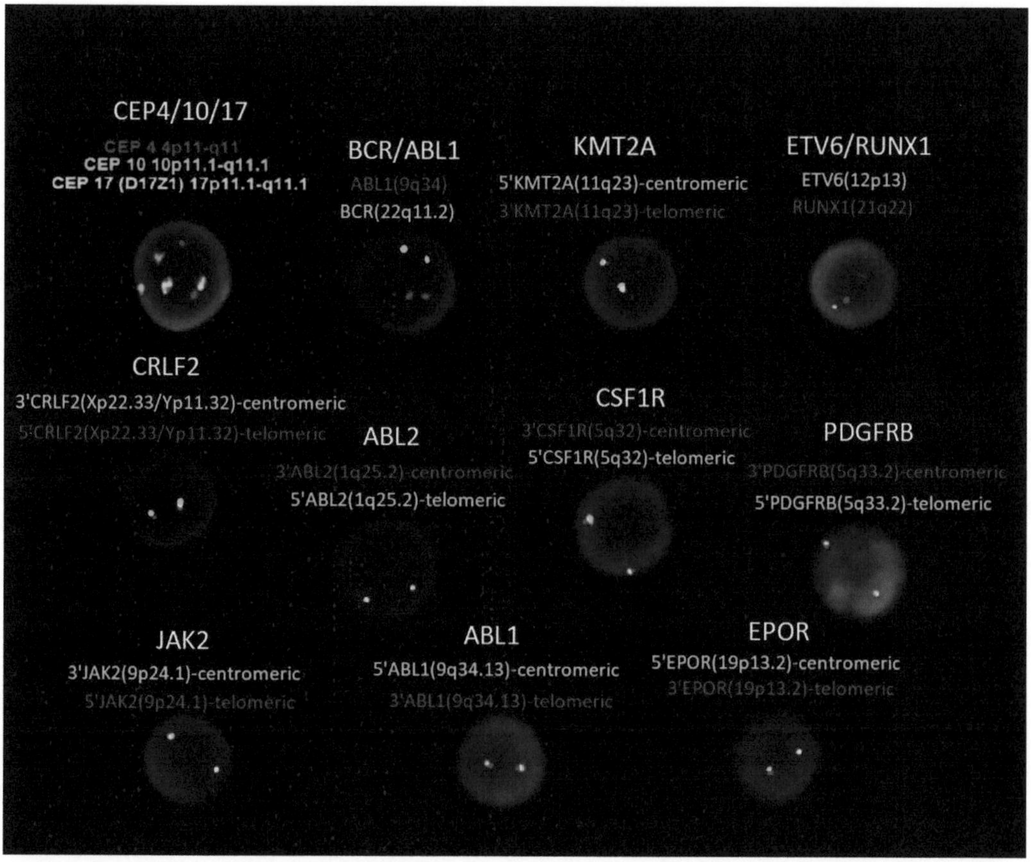

FIG. 19.4.3 FISH for ALL panel showed positive results for t(12:21) or *ETV6::RUNX1* rearrangement with deletion of 12p13 (*ETV6*) and gain of 21q22 (*RUNX1*) observed. FISH for Ph-like ALL panel was negative for all probes examined. ISCN for ALL panel: nuc ish(CEP4,CEP10,D17Z1)x2[200/200],(ABL1,BCR)x2[200/200],(KMT2A)x2[200/200],(ETV6x1, RUNX1x4)(ETV6 con RUNX1x1)[164/200]; ISCN for Ph-like ALL panel: nuc ish(CRLF2x2)[200/200],(ABL2x2)[200/200], (CSF1Rx2)[200/200],(PDGFRBx2)[200/200],(JAK2x2)[200/200],(ABL1x2)[200/200],(EPORx2)[200/200]

Future testing recommendations

Regular follow-ups and clinicopathologic correlation are recommended. Chromosome analysis and FISH can be ordered for disease monitoring and treatment efficacy in the future.

Case 19.5 Ph-like B-lymphoblastic leukemia (Ph-like ALL) with *CRLF2* rearrangement

Clinical indication

The patient was an 86-year-old female with a past medical history of hypertension, hyperlipidemia, and congestive heart failure who had initially presented with complaints of generalized fatigue and weakness. The patient was subsequently admitted to the hospital

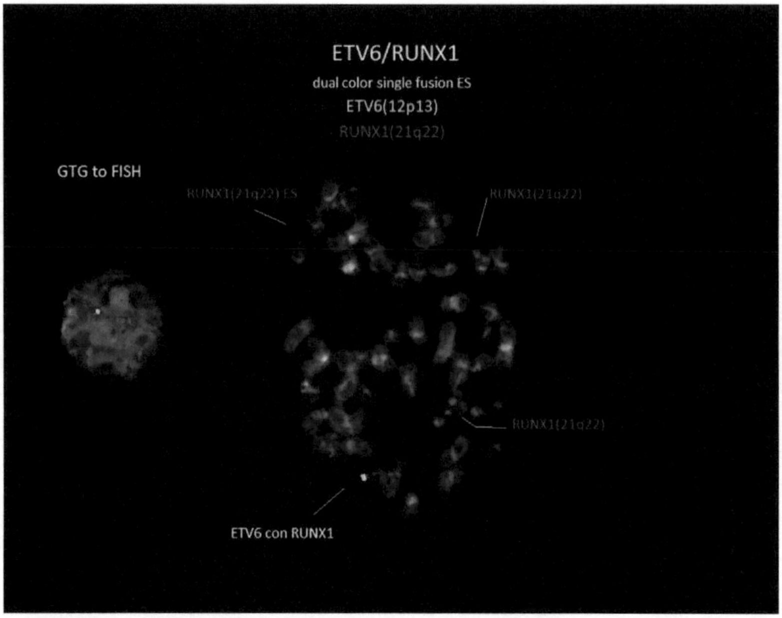

FIG. 19.4.4 Metaphase FISH confirmed the *ETV6::RUNX1* fusion and gain of 21q22 (*RUNX1*).

due to concern for acute leukemia as the patient had an elevated WBC count. The patient denied noticing any bleeding from anywhere, as well as any nausea, vomiting, diarrhea, abdominal pain, dysuria, cough, or shortness of breath. She was found to be in tumor lysis with 88% blasts on peripheral blood and had an acute kidney injury.

Test ordered

- Chromosome analysis of the bone marrow
- FISH: AML panel
- FISH: ALL panel
- FISH: Ph-like ALL panel
- NGS Hematology Molecular Profile
- *BCR::ABL1* quantitative real-time PCR

Laboratory test performed

Chromosome analysis, FISH, q-PCR, and NGS methods were described previously in Chapters 1 and 12.

Test results

Chromosome analysis was performed initially. All 20 cells examined exhibited monosomy 18 and a complex derivative chromosome 20 (der20) derived from t(18;20) and additional material of unknown origin attached to 20p12. Clonal evolution was evident. Clone 1 (15 cells) showed trisomy 22 in addition to monosomy 18 and der(20) (Fig. 19.5.1). Clone 2 (five cells) revealed trisomy X in addition to monosomy 18 and der(20) (Fig. 19.5.2).

FISH for AML panel was performed on interphase nuclei using probes localized to the *D5S721* (5p15.2), *EGR1* (5q31), *D7Z1* (7cen), *D7S486* (7q31), *D8Z2* (8cen), *RUNX1T1* (8q22), *ABL1* (9q34.12), *KMT2A* (11q23), *PML* (15q24.1), *CBFB* (16q22.1), *RARA* (17q21.1), *D20S108* (20q12), *RUNX1* (21q22.3), *and BCR* (22q11) gene regions. Two hundred nuclei were examined, and the results demonstrate the following abnormalities (Fig. 19.5.3).

Deletion of 20q12 was observed in 187/200(93.5%) of the cells scored.

Gain of *RUNX1* (21q22) was observed in 9/200 (4.5%) of the cells scored.

Gain of *BCR* (22q11.2) was observed in 182/200 (91.0%) of the cells scored.

FISH for ALL panel was performed on interphase nuclei using probes localized to the chromosome 4, 10, and 17 centromeric regions; *ABL1* (9q34.12), *BCR* (22q11), *KMT2A* (11q23), *ETV6* (12p13), and *RUNX1* (21q22.3) gene regions. Two hundred nuclei were examined, and the results demonstrated a gain of 21q in 12/200 (6.0%) of the cells scored and a gain of 22q in 172/200 (86.0%) of the cells scored (Fig. 19.5.4).

FISH Ph-like ALL panel was performed on interphase nuclei using probes localized to the *CRLF2* (Xp22.33/Yp11.32), *ABL2* (1q25.2), *CSF1R* (5q32), *PDGFRB* (5q33.2), *JAK2*

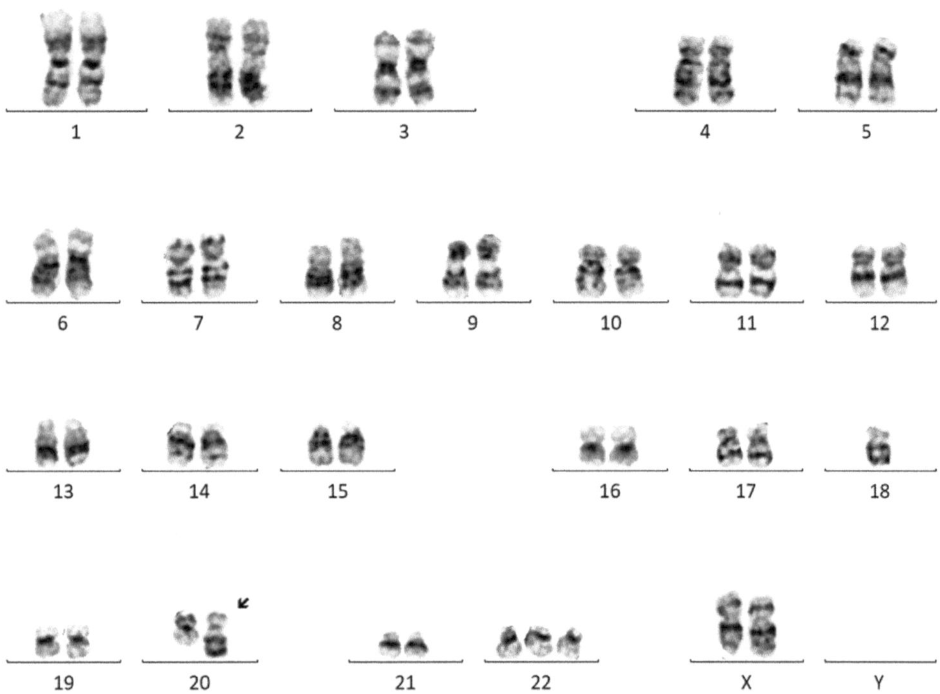

FIG. 19.5.1 Chromosome analysis revealed abnormal clone 1 with a complex rearrangement involving chromosomes 18 and 20. ISCN: 46,XX,-18,der(20)add(20)(p12)t(18;20)(q11.2;q11.2),+22[15]/46,idem,+X,-22[5]

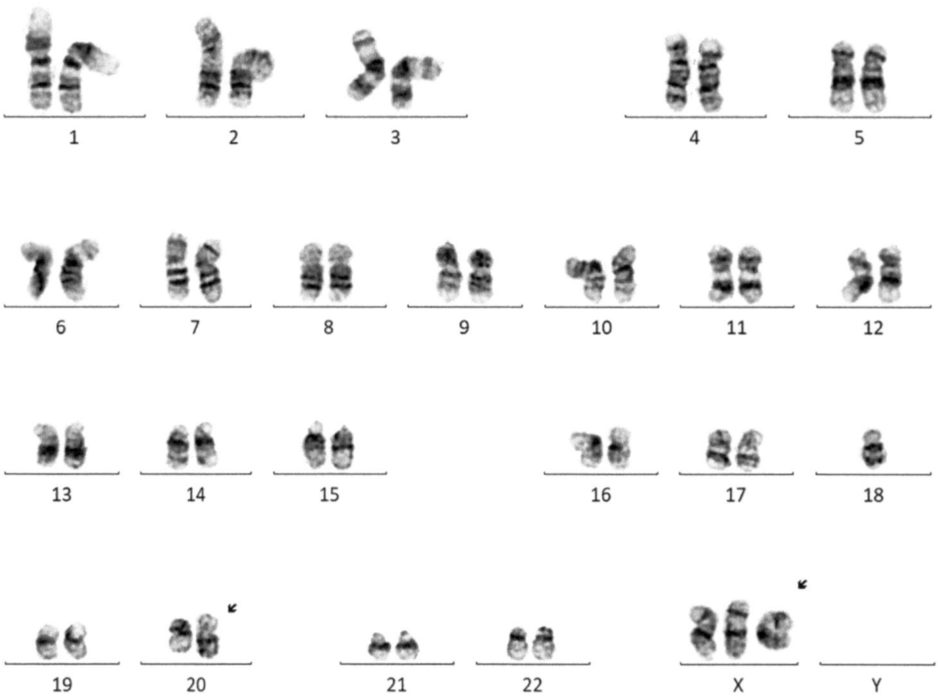

FIG. 19.5.2 Chromosome analysis revealed abnormal clone 2 with a complex rearrangement involving chromosomes 18 and 20, loss of chromosome 22, and gain of the X chromosome.

(9p24.1), *ABL1* (9q34.13), and *EPOR* (19p13.2) gene regions. Two hundred nuclei were examined, and the results demonstrated a *CRLF2* rearrangement in 183/200 (91.5%) of the cells scored (Fig. 19.5.4).

NGS Hematology Molecular Profile identified an *NRAS p.(Q61H) c.183A>C* mutation (Fig. 19.5.5).

BCR::ABL1 quantitative real-time PCR results were negative for both p210 and p190 (Fig. 19.5.6, see trend table and chart of p210).

Results with interpretations

Chromosome analysis detected two abnormal clones including trisomy 22. FISH studies were positive for deletion of 20q, gains of *RUNX1* (21q22) and *BCR* (22q11) with no evidence for *BCR::ABL1* rearrangement, concordant with the negative *BCR::ABL1* quantitative PCR results. FISH also detected a *CRLF2* rearrangement. NGS results showed NRAS mutation. These results are consistent with *BCR::ABL1*-like B-lymphoblastic leukemia (Ph-like B-ALL). *CRLF2* rearrangement accounts for about half of the Ph-like B-ALL cases and are associated with poor prognosis.

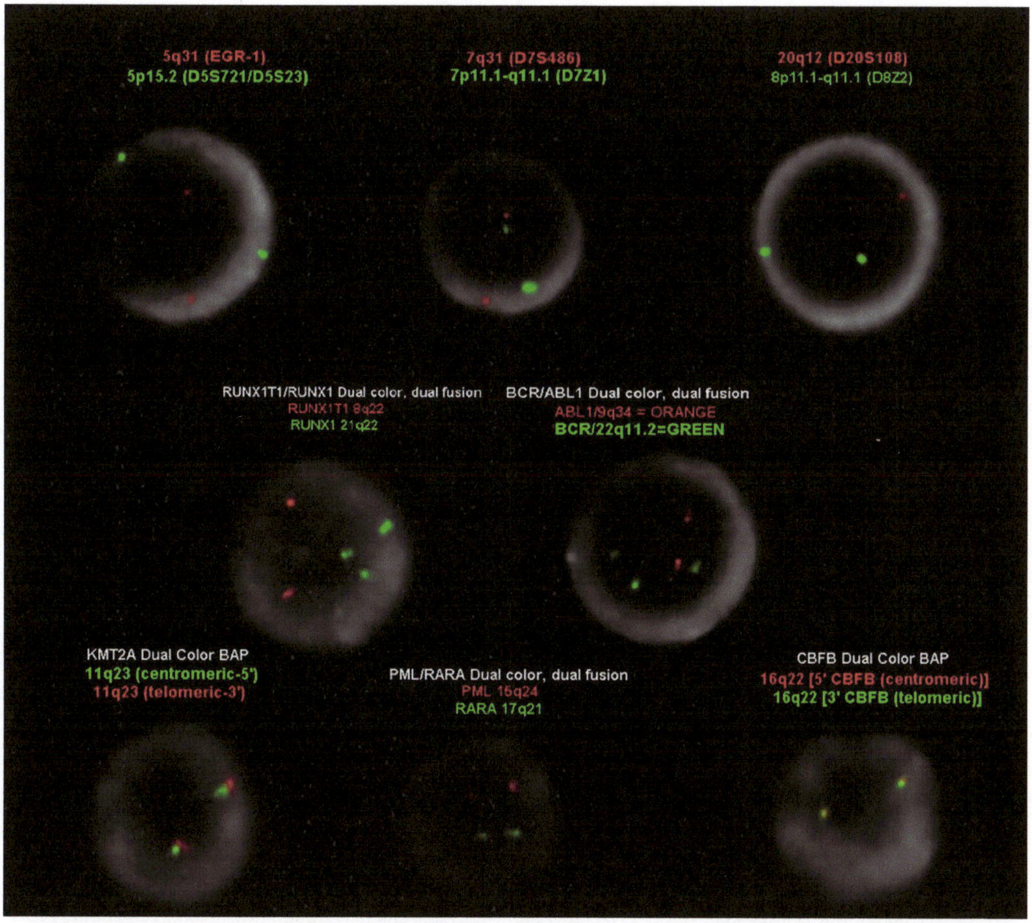

FIG. 19.5.3 FISH for the AML panel showed a deletion of 20q12, a gain of *RUNX1* (21q22), and a gain of *BCR* (22q11.2). ISCN: nuc ish(D5S721,EGR1)x2[200/200],(CEP7,D7S486)x2[200/200],(D8Z2x2,D20S108x1)[187/200],(RUNX1T1x2, RUNX1x3)[9/200],(ABL1x2,BCRx3)[182/200],(KMT2Ax2)[200/200], (PML,RARA)x2[200/200],(CBFBx2)[200/200]

Future testing recommendations

Regular follow-ups and clinicopathologic correlation are recommended. Chromosome analysis, FISH, and NGS can be ordered for disease monitoring and treatment efficacy in the future.

Case 19.6 T-lymphoblastic leukemia (T-ALL) with t(10;11)(p12;q21)/*PICALM::MLLT10* fusion

Clinical indication

A 32-year-old man presented with complaints of fatigue, night sweats, headache, and facial droop. According to bone marrow flow cytometry, 16% of dim CD45

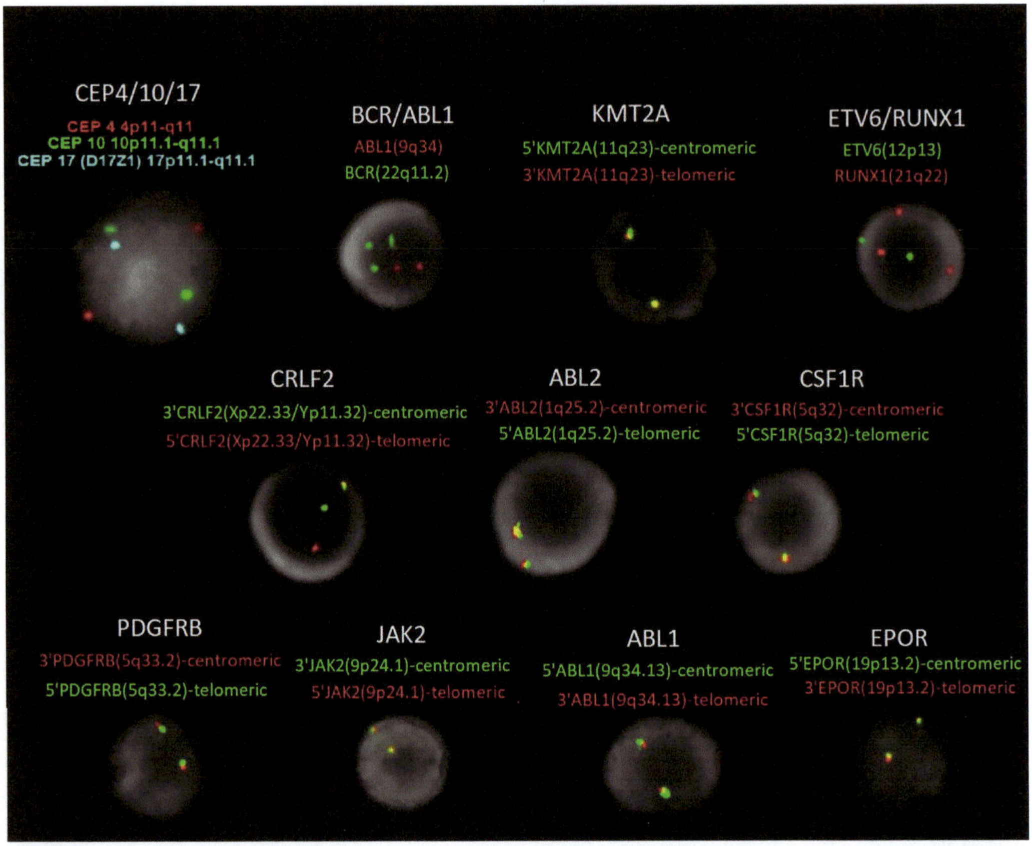

FIG. 19.5.4 FISH for ALL panel was positive for a gain of *RUNX1* (21q22) (6.0%), and a gain of *BCR* (22q11) (86.0%), concordant with the results from the AML panel. FISH for Ph-like ALL panel revealed *CRLF* rearrangements (91.5%). ISCN for ALL panel: nuc ish(CEP4,CEP10,CEP17)x2[200/200],(ABL1x2,BCRx3)[172/200], (KMT2Ax2)[200/200],(ETV6x2, RUNX1x3)[12/200]; ISCN for Ph-like ALL panel: nuc ish(CRLF2x2)(3′CRLF2 sep 5′CRLF2 x1)[159/200]/ (CRLF2x3)(3′CRLF2 sep 5′CRLF2 x1)[24/200], (ABL2x2)[200/200],(CSF1Rx2)[200/200],(PDGFRBx2)[200/200], (JAK2x2)[197/200],(ABL1x2)[199/200],(EPORx2)[199/200]

Relevant Biomarkers

Tier	Genomic Alteration	Relevant Therapies (In this cancer type)	Clinical Trials
IA	*NRAS p.(Q61H) c.183A>C* **Prognostic significance:** NCCN: Poor	None	4

Public data sources included in relevant therapies: FDA1, NCCN, EMA2, ESMO
Public data sources included in prognostic and diagnostic significance: NCCN, ESMO
Tier Reference: Li et al. *Standards and Guidelines for the Interpretation and Reporting of Sequence Variants in Cancer: A Joint Consensus Recommendation of the Association for Molecular Pathology, American Society of Clinical Oncology, and College of American Pathologists.* J Mol Diagn. 2017 Jan;19(1):4-23.

Variant Details

DNA Sequence Variants

Gene	Amino Acid Change	Coding	Locus	Allele Frequency	Variant Effect
NRAS	p.(Q61H)	c.183A>C	chr1:115256528	4.02%	missense

FIG. 19.5.5 NGS Hematology Molecular Profile detected an *NRAS* p.(Q61H) mutation.

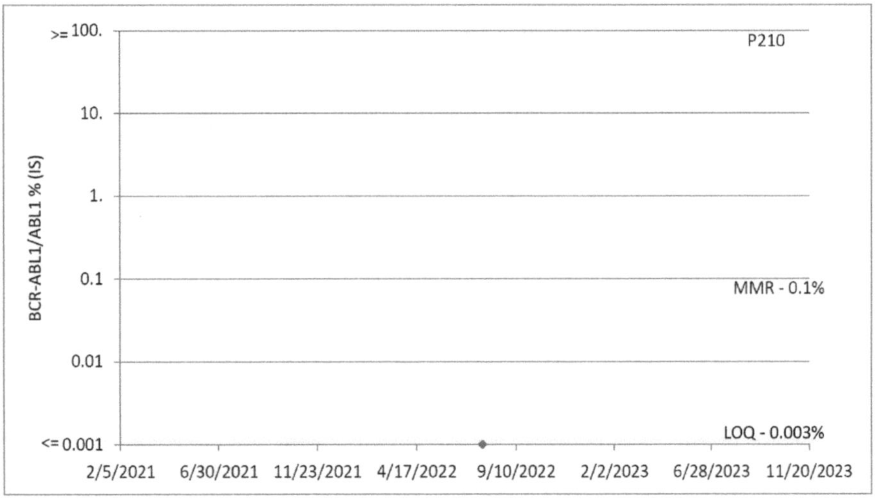

FIG. 19.5.6 qPCR results showed the trend table and chart of p210 negative results.

T-lymphoblasts were positive for CD2 (dim partial), CD3 (dim partial) CD5 (dim), CD7, CD10, and CD34 (subset), and negative for CD13, CD19, CD20, CD33, and CD56, which was most consistent with T-lymphoblastic leukemia. According to cerebrospinal fluid flow cytometry, there was an increased population (64%) of dim CD45 T-lymphoblasts that were positive for cytoplasmic CD3, CD5, CD7, and TdT. A subset of the blasts were positive for CD4 (dim), CD2, surface CD3, CD10, CD34, and CD38. The blasts were negative for the myeloid and B-cell markers that were evaluated. This was most consistent with T-lymphoblastic leukemia involving the central nervous system (CNS).

Test ordered

- Chromosome analysis of the bone marrow
- FISH: ALL panel; *PDGFRA*, *PDGFRB*, and *FGFR1* break-apart probes
- NGS Hematology Molecular Profile

Laboratory test performed

Chromosome analysis, FISH, and NGS methods were described in Chapters 1 and 12.

Test results

Chromosome analysis was performed initially on bone marrow. Of the 20 metaphase cells analyzed, 3 exhibited a structural abnormality involving chromosome 9, and a translocation between chromosomes 10 and 11 with breakpoints at band 10p12 and 11q21 (Fig. 19.6.1). The remaining 17 cells were chromosomally normal.

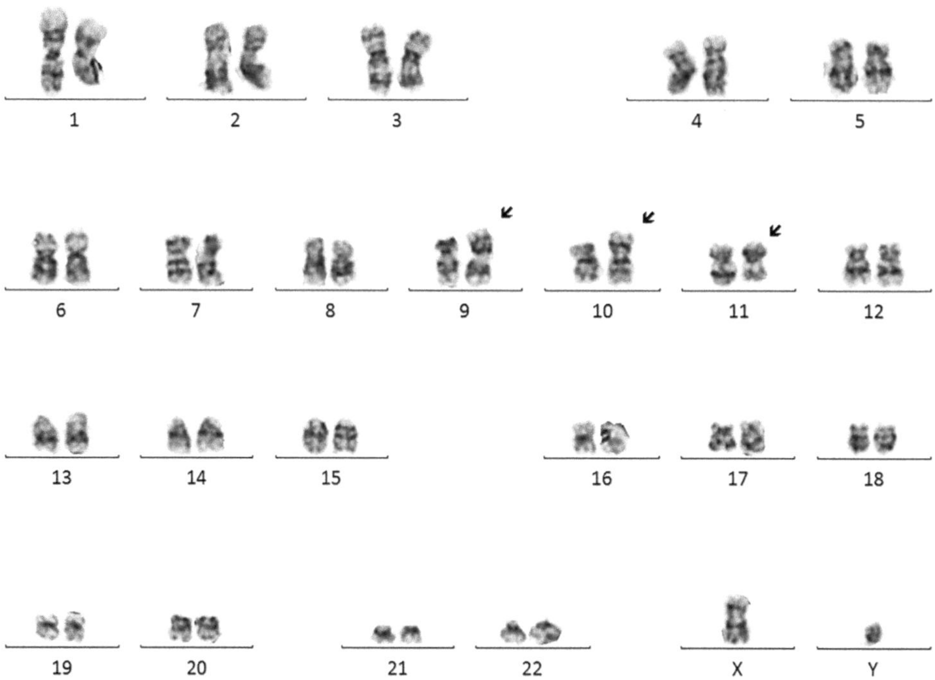

FIG. 19.6.1 The karyotype showed a structural abnormality involving chromosome 9 and a translocation between 10p12 and 11q21. ISCN: 46,XY,der(9)t(9;9)(p12;q21),t(10;11)(p12;q21)[3]/46,XY[17]

Then FISH was also performed on interphase nuclei using probes localized to the chromosome 4, 10, and 17 centromeric regions, *ABL1* (9q34.12), *BCR* (22q11), *KMT2A* (11q23), *ETV6* (12p13), *RUNX1* (21q22.3), *SCFD2/FIP1L1/LNX1/PDGFRA* (4q12), the 3′ and 5′ ends of the *PDGFRB* (5q33.1), and the 5′ and 3′ ends of the *FGFR1* (8p12.23-p11.22) gene regions (supplied by Abbott Molecular, Inc.). Two hundred nuclei were examined for all probes, and the results demonstrated a gain of *ABL1* in 7/200 (3.5%) of the cells scored. FISH was negative for the rest of probes examined (Fig. 19.6.2 and Fig. 19.6.3).

Finally, NGS Hematology Molecular Profile was conducted on bone marrow, and the results showed a *PICALM::MLLT10* fusion (Fig. 19.6.4).

Results with interpretations

Chromosome analysis exhibited a translocation t(10;11)(p12;q21). NGS showed an *MLLT10::PICALM* fusion, which was concordant with the cytogenetic finding. FISH was negative for this because the specific probes are not available in our lab. The *PICALM::MLLT10* fusion (also called *CALM::AF10* fusion) is found in 10% of T-lymphoblastic leukemia/lymphoma (TALL/LBL) cases and is generally associated with an adverse prognosis [1,9,10]. Patients with initial CNS involvement have a worse prognosis than patients without CNS disease.

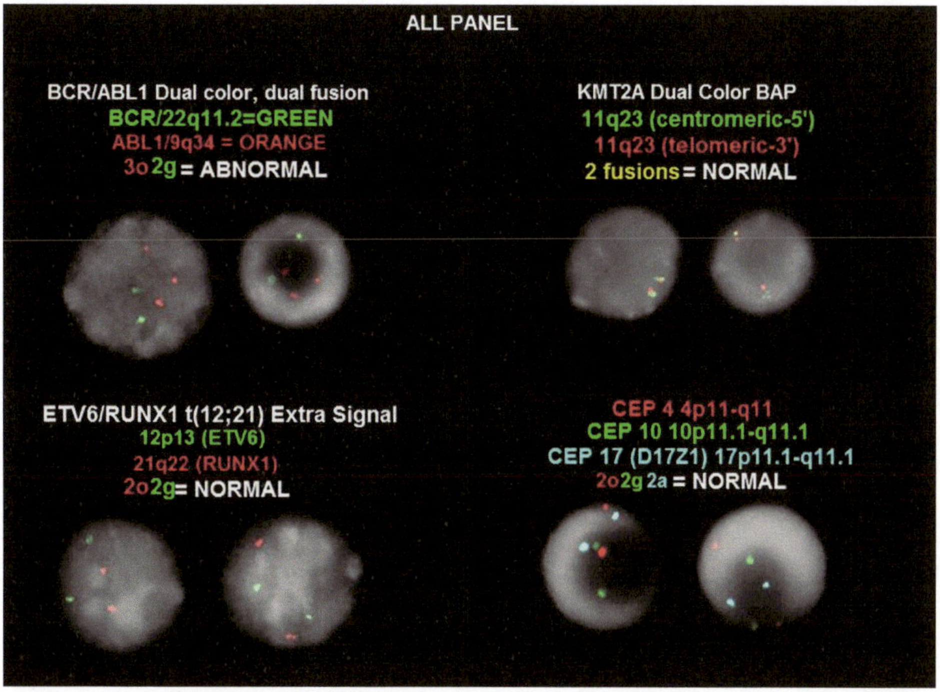

FIG. 19.6.2 FISH with ALL panel showed positive results for a gain of *ABL1*. ISCN: nuc ish(CEP4,CEP10,CEP17)x2 [200/200],(ABL1x3,BCRx2)[7/200],(KMT2Ax2)[200/200],(ETV6,RUNX1)x2[200/200]

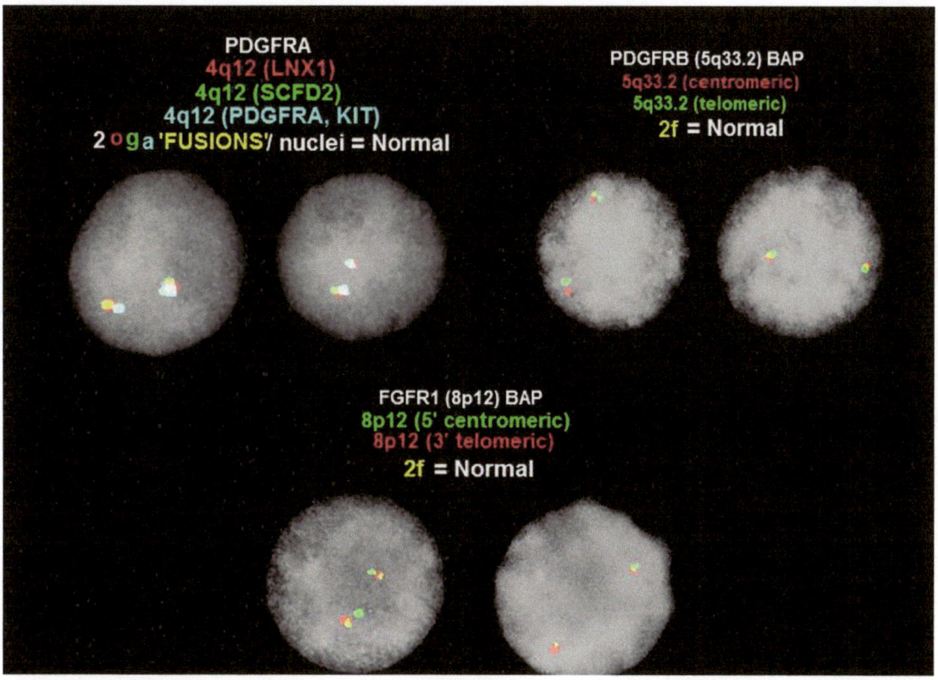

FIG. 19.6.3 FISH was negative for rearrangements of *PDGFRA*, *PDGFRB*, and *FGFR1*. ISCN: nuc ish(SCFD2,FIP1L1,LNX1, PDGFRAx2)[200/200],(SCFD2,FIP1L1,LNX1,PDGFRAx2)[200/200], (FGFR1x2)[200/200]

Clinically Significant Biomarkers — Indicated ■ Contraindicated

Genomic Alteration	Relevant Therapies (In this cancer type)	Clinical Trials
PICALM-MLLT10 fusion Tier: IIC **Prognostic significance:** None **Diagnostic significance:** None	None	1

Sources included in relevant therapies: FDA1, NCCN
Sources included in prognostic and diagnostic significance: NCCN

Variant Details

Gene Fusions (RNA)

Genes	Variant ID	Locus
PICALM-MLLT10	PICALM-MLLT10.P19M4	chr11:85685751 - chr10:21875223

FIG. 19.6.4 NGS showed a fusion of *PICALM:: MLLT10*, concordant with the cytogenetic finding.

However, the introduction of early and effective CNS-directed therapy might no longer portend a poor prognosis for CNS leukemia [11]. Clinicopathologic correlation is advised.

Future testing and recommendations

Chromosome analysis and NGS on bone marrow can be ordered to monitor disease progression and treatment efficacy in the future.

Case 19.7 T-lymphoblastic leukemia (T-ALL) with t(11;18)(p15;q12)/*NUP98::SETBP1* fusion

Clinical indication

A 13-year-old male presented with circulating lymphoblasts with dim CD3 positivity. FISH for *PML::RARA* was normal. The bone marrow biopsy showed a normocellular marrow (95%) with sheets of lymphoblasts replacing most of the marrow parenchyma. Peripheral blood smear showed normocytic normochromic anemia, neutropenia, monocytopenia, and numerous circulating lymphoblasts (42%). A flow cytometry analysis was consistent with T-lymphoblastic leukemia with 98% T-lymphoblasts. The immunophenotype was suggestive of near-early T-cell precursor lymphoblastic leukemia.

Test ordered

- Chromosome analysis of the bone marrow
- FISH: AML panel
- FISH: ALL hyperdiploidy panel
- FISH: *ETV6::RUNX1*

- FISH: *TRA/D* Gene Rearrangement
- NGS Hematology Molecular Profile
- B-cell Gene rearrangement (send out)
- T-cell Receptor Gamma Gene Rearrangement (send out)

Laboratory test performed

Chromosome analysis, FISH, and NGS methods were described in Chapters 1 and 12.

Test results

Chromosome analysis was performed initially. Of the 22 cells examined, 16 exhibited a balanced translocation involving the short arm of chromosome 11 and the long arm of chromosome 18. t(11;18)(p15;q12) involves *NUP98::SETBP1* fusion and is a recurrent chromosome rearrangement seen in T-ALL (Fig. 19.7.1). The remaining six cells appear to be chromosomally normal.

FISH for AML/ALL hyperploidy panels and *ETV6::RUNX1* rearrangement were performed on interphase nuclei using probes localized to the centromeres of chromosomes 4, 10, and 17, *D5S721* (5p15.2), *EGR1* (5q31), *D7Z1* (7cen), *D7S486* (7q31), *D8Z2* (8cen), *RUNX1T1* (8q22), *ABL1* (9q34.12), *KMT2A* (11q23), *PML* (15q24.1), *CBFB* (16q22.1), *RARA* (17q21.1), *D20S108* (20q12), *RUNX1* (21q22.3), *BCR* (22q11), *ETV6* (12p13), and *RUNX1*

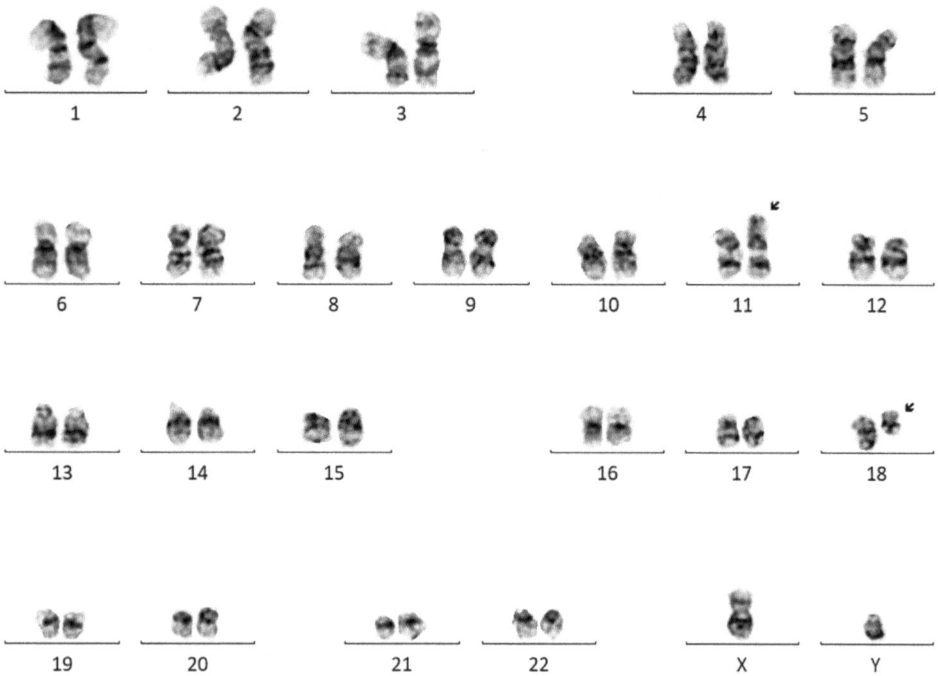

FIG. 19.7.1 The karyotype showed a balanced translocation of t(11;18). ISCN: 46,XY,t(11;18)(p15;q12)[16]/46,XY[6]

(21q22.3) gene regions. Two hundred nuclei were examined, and the results were within normal limits for the laboratory's established background rates (Fig. 19.7.2).

FISH For *TRA/D* rearrangement was performed on interphase nuclei using probes localized to the 5′ and 3′ ends of the T-cell receptor alpha/delta (*TRA/D*; 14q11.2) gene region. Two hundred nuclei were examined, and the results were within normal limits for the laboratory's established background rates (Fig. 19.7.3).

NGS detected a *NUP98::SETBP1* Fusion (also seen in the karyotype) and *PHF6 p.(C20Lfs*21) c. c.59_60delGTinsTAAGAGTAGGCCCGCACCAGGGACC* mutation (Fig. 19.7.4).

The B-cell gene rearrangement test failed due to an insufficient amount of DNA/RNA from the specimen. T-cell receptor gamma gene rearrangement was positive by PCR (these were external tests, data not shown).

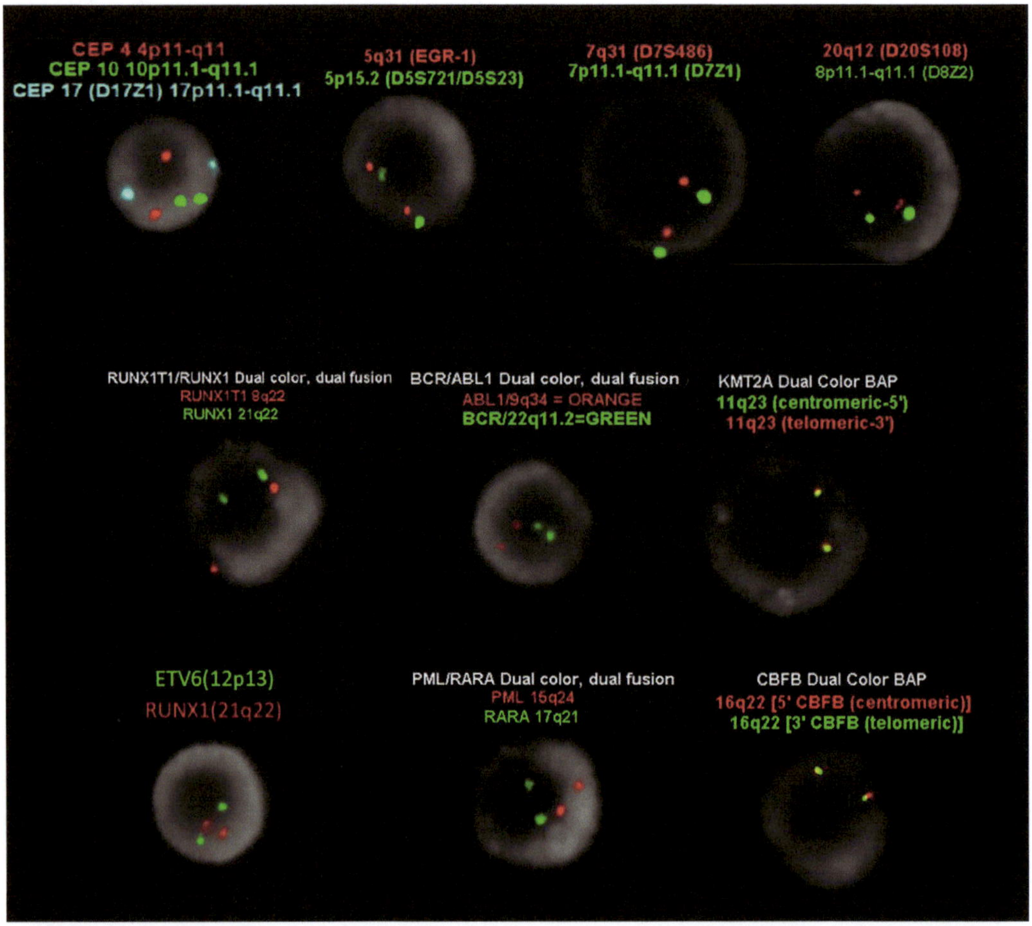

FIG. 19.7.2 FISH for AML/ALL hyperdiplody panels and *ETV6::RUNX1* rearrangement probes revealed normal results for all probes examined. ISCN: nuc ish(D5S721,EGR1)x2[200/200],(CEP7,D7S486)x2[200/200],(D8Z2,D20S108)x2[200/200],(RUNX1T1,RUNX1)x2[200/200],(ABL1,BCR)x2[200/200],(KMT2Ax2)[200/200],(PML,RARA)x2[200/200],(CBFBx2)[200/200],(CEP4,CEP10,CEP17)x2[200/200], ETV6,RUNX1)x2[200/200]

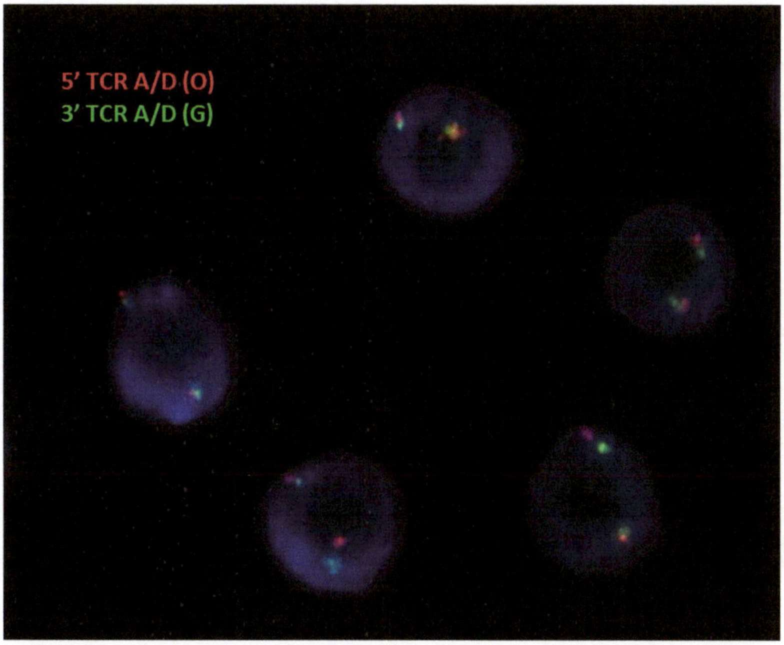

FIG. 19.7.3 FISH for *TRA/D* rearrangement showed normal results for all probes examined. ISCN: nuc ish(*TRA/D*x2)[200/200]

Relevant Biomarkers

Tier	Genomic Alteration	Relevant Therapies (In this cancer type)	Clinical Trials
IIC	*PHF6 p.(C20Lfs*21) c.59_60delGTinsTAAGAGTAGGC CCGCACCAGGGACC*	None	1

Public data sources included in relevant therapies: FDA1, NCCN, **EMA**2, ESMO
Tier Reference: Li et al. *Standards and Guidelines for the Interpretation and Reporting of Sequence Variants in Cancer: A Joint Consensus Recommendation of the Association for Molecular Pathology, American Society of Clinical Oncology, and College of American Pathologists.* J Mol Diagn. 2017 Jan;19(1):4-23.

Prevalent cancer biomarkers without relevant evidence based on included data sources
NUP98-SETBP1 fusion

Variant Details

DNA Sequence Variants

Gene	Amino Acid Change	Coding	Locus	Allele Frequency	Variant Effect
PHF6	p.(C20Lfs*21)	c.59_60delGTinsTAAGAGTA GGCCCGCACCAGGGACC	chrX:133511706	94.39%	frameshift Block Substitution

Gene Fusions (RNA)

Genes	Variant ID	Locus
NUP98-SETBP1	NUP98-SETBP1.N12S6	chr11:3765739 - chr18:42643044

FIG. 19.7.4 NGS detected a fusion of *NUP98::SETBP1* and a frameshift mutation of *PHF6*. The fusion was also seen from the karyotype.

Results with interpretations

Chromosome analysis detected a balanced translocation of t(11;18). t(11;18)(p15;q12) [*NUP98::SETBP1*] was a recurrent chromosome translocation seen in T-ALL (https://atlasgeneticsoncology.org/haematological/1466/t(11;18)(p15;q12)) [12]. All FISH testing in this patient was negative.

NGS identified a *NUP98::SETBP1* fusion, concordant with the cytogenetic finding. NGS also revealed a frameshift mutation of *PHF6*. The T-cell receptor gamma gene rearrangement was diagnosed as positive by PCR, concordant with the diagnosis of T-ALL for this patient.

Future testing recommendations

Chromosome analysis, NGS, and PCR for T-cell receptor gamma gene rearrangement should be ordered for follow-up appointments. Clinicopathologic correlation is recommended.

Case 19.8 T-lymphoblastic leukemia (T-ALL) with t(1;14)(p32;q11.2)/*TRA::TAL1* fusion

Clinical indication

A 22-year-old male presented with a large mediastinal mass. The patient denied fever, chills, diarrhea, constipation, mouth lesions, dysuria, chest pain, and dyspnea. Cytometric and morphologic analysis of peripheral blood revealed an abnormal population of T cells. T-cell lymphoblastic leukemia was suspected.

Test ordered

- Chromosome analysis of the bone marrow
- FISH: ALL panel
- FISH: *TRA/D* rearrangement
- NGS Hematology Molecular Profile Complete (send out)
- T-Cell MRD (send out)

Laboratory test performed

Chromosome analysis and FISH methods were described previously in Chapters 1 and 12.

Test results

Chromosome analysis was performed initially. Of the 20 cells analyzed, 6 exhibited a translocation between the short arm of chromosome 1 and the long arm of chromosome 14. The remaining 14 cells were chromosomally normal (Fig. 19.8.1).

FISH for ALL panel was performed on interphase nuclei using probes localized to the chromosome 4, 10, and 17 centromeric regions; *ABL1* (9q34.12), *BCR* (22q11), *KMT2A*

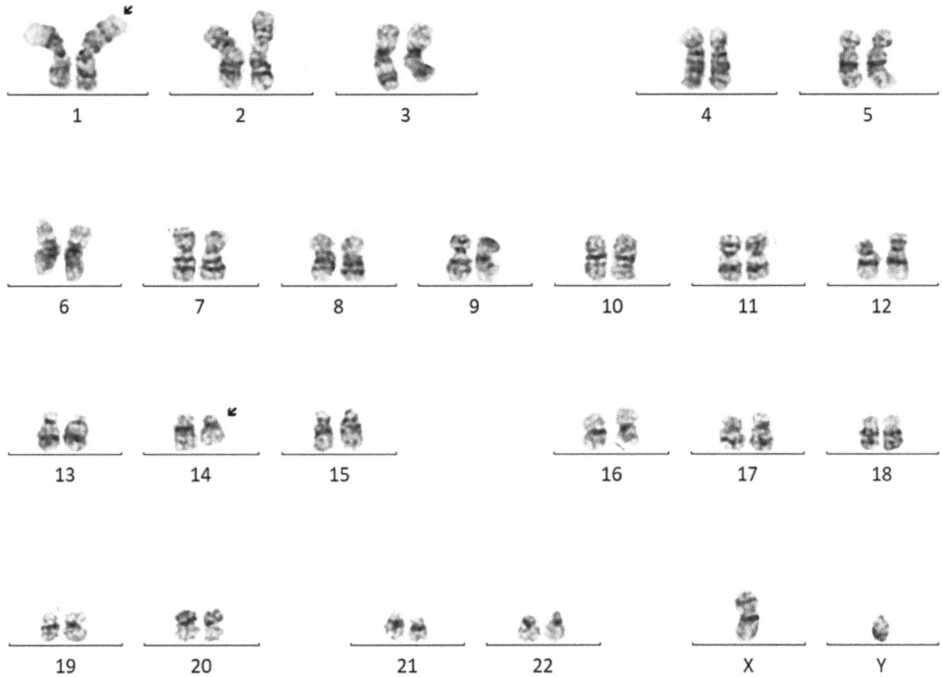

FIG. 19.8.1 Chromosome analysis identified a balanced translocation involving 1p and 14q. ISCN: 46,XY,t(1;14)(p32;q11.2)[6]/46,XY[14]

(11q23), *ETV6* (12p13), and *RUNX1* (21q22.3) gene regions. Two hundred nuclei were examined, and the results were within normal limits for the laboratory's established background rates (Fig. 19.8.2).

FISH was also performed on interphase nuclei using probes localized to the 5′ and 3′ ends of the T-cell receptor alpha/delta (*TRA/D*; 14q11.2) gene region. Two hundred nuclei were examined, and the results were positive for a rearrangement involving *TRA/D* in 98/200 (49.0%) of the cells scored (Fig. 19.8.3).

NGS and T-cell TCRG MRD results from other labs came back negative (data not shown).

Results with interpretations

This translocation t(1;14) identified by chromosome analysis appeared to involve the 1p32 and 14q11.2 regions, which resulted in a fusion between the *TRA* (14q11.2) and *TAL1* (1p32) genes. The translocation t(1;14)(p32;q11.2) *TRA::TAL1* has been reported in T-cell ALL [13–15]. FISH with *TRA/D* probes was positive for *TRA/D* rearrangement, concordant with the cytogenetic findings.

Future testing and recommendations

Chromosome analysis and FISH for *TRA/D* should be tested during the regular follow-ups to monitor the disease progression. Clinicopathologic correlation is recommended.

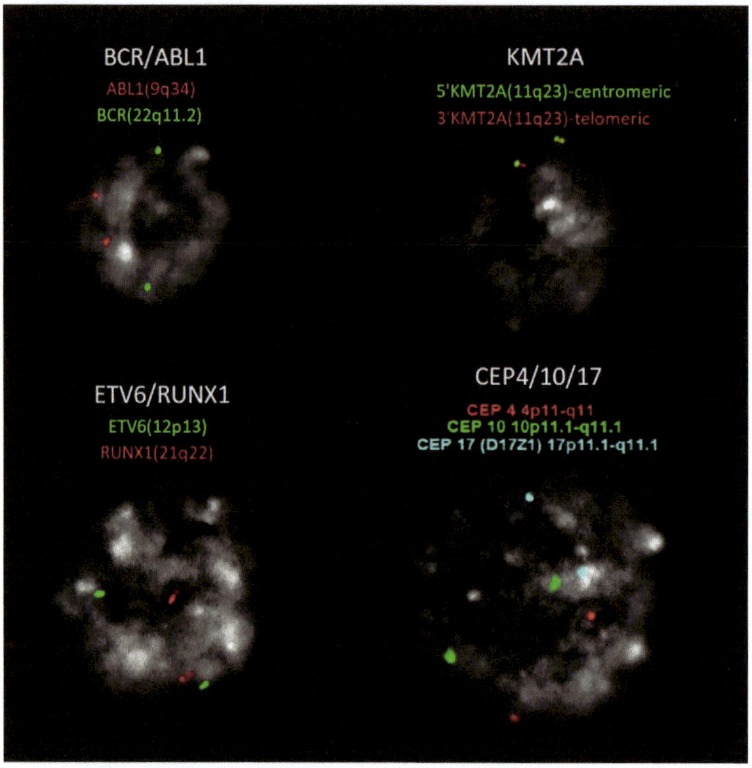

FIG. 19.8.2 FISH for ALL panel showed normal results for all probes examined. ISCN: nuc ish(CEP4,CEP10,CEP17)x2[200/200],(ABL1,BCR)x2[200/200], (KMT2Ax2)[200/200],(ETV6,RUNX1)x2[200/200]

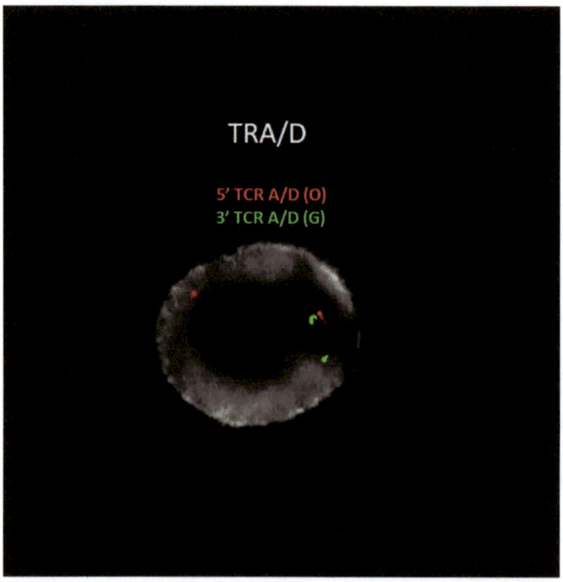

FIG. 19.8.3 FISH for *TRA/D* gene rearrangement was positive for the *TRA/D* rearrangement. ISCN: nuc ish(TRA/ Dx2)(5'TRA/ D sep 3'TRA/ Dx1)[98/200]

Case 19.9 T-lymphoblastic leukemia (T-ALL) with t(11;14)(p13;q11.2)/*LMO2::TRD* fusion

Clinical indication

An 8-year-old female presented with lymphadenopathy, significantly elevated WBC, intraparenchymal brain hemorrhages, respiratory failure, tumor lysis syndrome with acute kidney injury (AKI). The current diagnosis of leukemia concerns for T-ALL.

Test ordered

- Chromosome analysis of the bone marrow
- FISH: *BCR::ABL1*, *KMT2A* rearrangements and *CDKN2A*
- FISH: *TRA/D* rearrangement

Laboratory test performed

Chromosome analysis and FISH methods were described previously in Chapters 1 and 12.

Test results

Chromosome analysis was performed initially. Of the 20 cells examined, 15 exhibited a balanced translocation involving 11p13 and 14q11.2, which resulted in *LMO2::TRD* fusion. In addition, chromosome analysis also detected an unbalanced translocation involving 17q and 20q resulting in del(20q) (Fig. 19.9.1). The remaining five cells appear to be chromosomally normal.

FISH was performed on interphase nuclei using probes localized to *ABL1* (9q34.12), *BCR* (22q11), *CDKN2A* (9p21), *CEP9* (9cen), the 5′ and 3′ ends of the *KMT2A* (11q23), the 5′ and 3′ ends of the T-cell receptor alpha/delta (*TRA/D*; 14q11.2) gene region. Two hundred nuclei were examined for each probe, and the results demonstrated a rearrangement involving *TRA/D* in 191/200 (95.5%) of the cells scored (Fig. 19.9.2).

Results with interpretations

Chromosome analysis identified t(11;14), resulting in *LMO2::TRD* fusion. The concurrent FISH identified *TRA/D* rearrangement, concordant with the cytogenetic findings. These results are consistent with a diagnosis of T-ALL. *LMO2::TRD* fusion is a recurrent chromosome abnormality seen in 5%–10% of pediatric T-ALL [16,17]. Although the prognostic significance of this translocation has not been clearly determined yet, it was accompanied by an early relapse after hematopoietic stem cell transplantation in the patient reported [17].

Future testing recommendations

Chromosome analysis and FISH for *TRA/D* should be ordered during regular follow-ups to monitor disease progression. Clinicopathologic correlation is recommended.

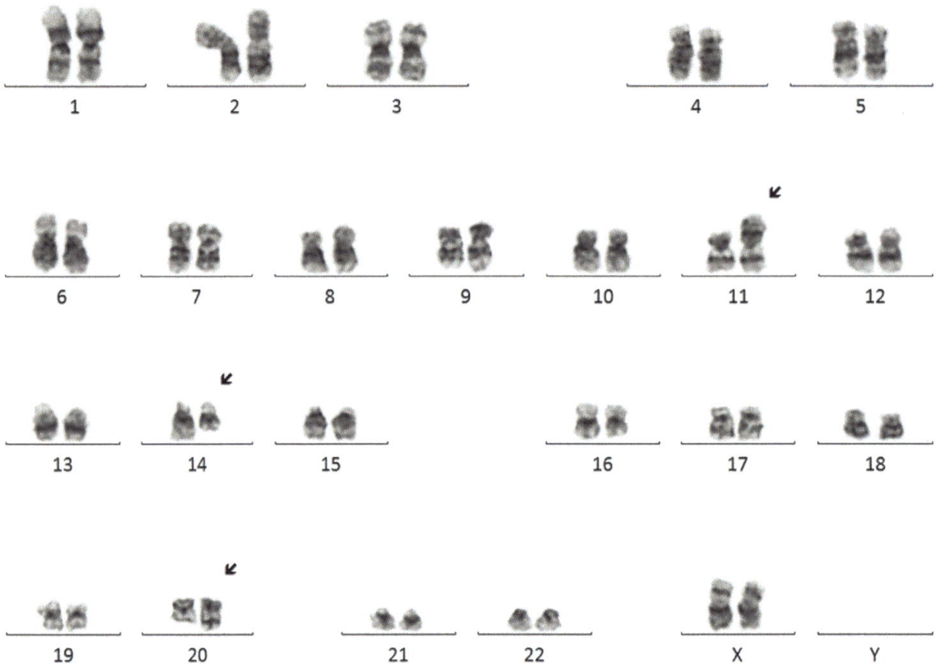

FIG. 19.9.1 Chromosome analysis revealed an abnormal karyotype with t(11;14)(p13;q11.2)/*LMO2::TRD* fusion and an unbalanced translocation involving 17q and 20q. ISCN: 46,XX,t(11;14)(p13;q11.2),der(20)t(17;20)(q12;q13.2)[15]/46,XX[5]

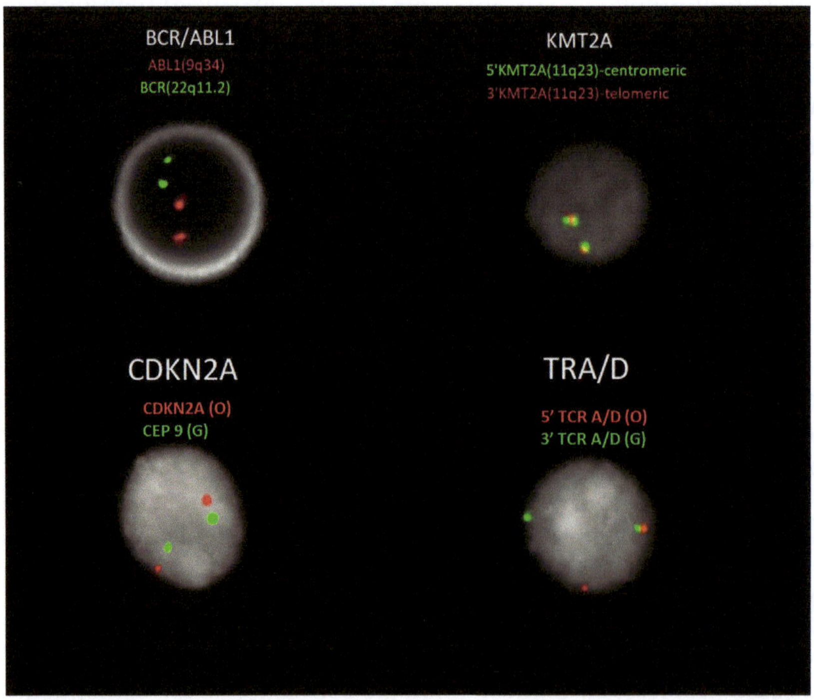

FIG. 19.9.2 FISH for *BCR::ABL1*, *KMT2A*, *CDKN2A*, and *TRA/D* results showed a positive rearrangement for *TRA/D* in 95.5% of cells examined. ISCN: nuc ish (ABL1,BCR)x2[200/200],(KMT2Ax2)[200/200],CDKN2A,CEP9)x2[200/200],(TRA/Dx2)(5'TRA/D sep 3'TRA/Dx1)[191/200]

Case 19.10 Ph-like B-lymphoblastic leukemia (Ph-like ALL) with *CRLF2* rearrangement and t(2;8)(p12;q24)/*IGK::MYC* fusion

Clinical indication

A 27-year-old male presented with pancytopenia, and macrocytic anemia. He had a medical history of Down syndrome. Flow cytometry showed 76% blasts that are CD34(−), CD20(−), CD19(+), CD10(+), and TDT(+). B-ALL or Ph-like ALL was suspected.

Test ordered

- Chromosome analysis of the bone marrow
- FISH: ALL panel
- FISH: Ph-like ALL panel
- FISH: *MYC* Gene Rearrangement

Laboratory test performed

Chromosome analysis and FISH methods were described previously in Chapters 1 and 12.

Test results

Chromosome analysis was performed initially. Of the 21 cells examined, all had trisomy 21, which was consistent with the patient's known history of Down syndrome. Eight of these also had other abnormalities, and clonal evolution was evident. Clone 1 with five cells revealed t(2;8), del(12p), del(16q), and trisomy 21 (Fig. 19.10.1). Clone 2 with three cells had add(18q) and trisomy 21 (Fig. 19.10.2). Clone 3 with 13 cells exhibited only trisomy 21 (Fig. 19.10.3).

FISH for ALL panel was performed on interphase nuclei using probes localized to the chromosome 4, 10, and 17 centromeric regions; *ABL1* (9q34.12); *BCR* (22q11); *KMT2A* (11q23); *ETV6* (12p13); *RUNX1* (21q22.3) gene regions. Two hundred nuclei were examined, and the results were positive for a gain of *RUNX1* in 200/200 (100.0%) of the cells scored, which is consistent with the patient's known history of Down syndrome and a loss of one copy of *ETV6* in 166/200 (83.0%) of the cells scored (Fig. 19.10.4).

FISH for Ph-like ALL panel was performed on interphase nuclei using probes localized to the *CRLF2* (Xp22.33/Yp11.32), *ABL2* (1q25.2), *CSF1R* (5q32), *PDGFRB* (5q33.2), *JAK2* (9p24.1), *ABL1* (9q34.13), and *EPOR* (19p13.2) gene regions. Two hundred nuclei were examined, and the results demonstrated a *CRLF2* rearrangement in 177/200 (88.5%) of the cells scored (Fig. 19.10.4).

FISH for *MYC* rearrangement was performed on interphase nuclei using probes localized to the 5′ and 3′ ends of the *MYC* (8q24) gene region. Two hundred nuclei were examined, the results were positive for a rearrangement involving *MYC* in 186/200 (93.0%) of the cells scored (Fig. 19.10.5).

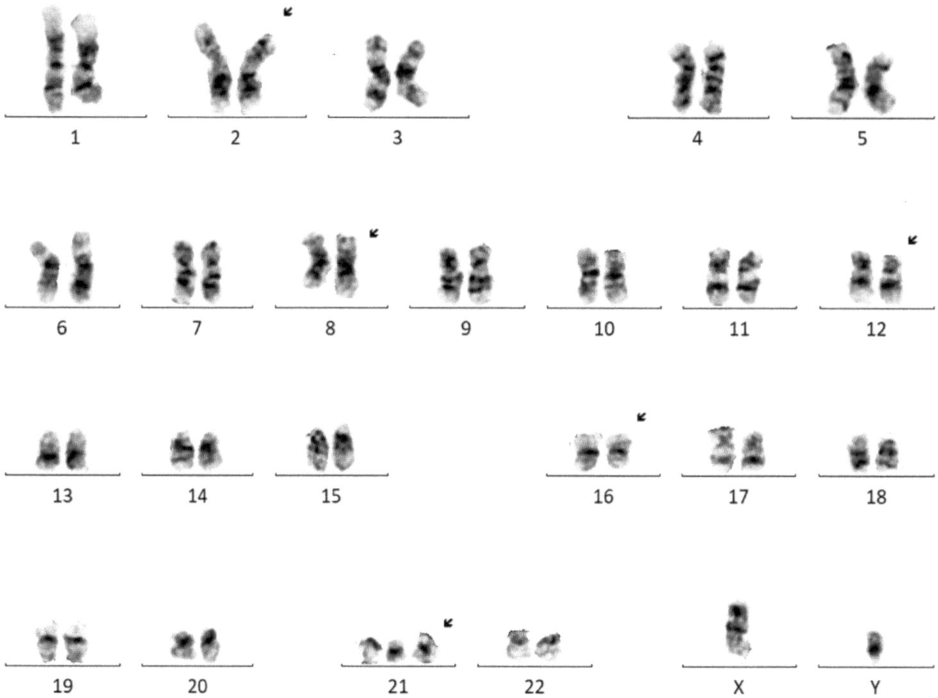

FIG. 19.10.1 Chromosome analysis exhibited an abnormal clone 1 with t(2;8), del(12p), del(16q) and trisomy 21. ISCN: 47,XY,t(2;8)(p21;?q24.2),del(12)(p12),del(16)(q13q22),+21c[cp5]/ 47,idem,add(18)(q21)[cp3]/47,XY,+21c[13]

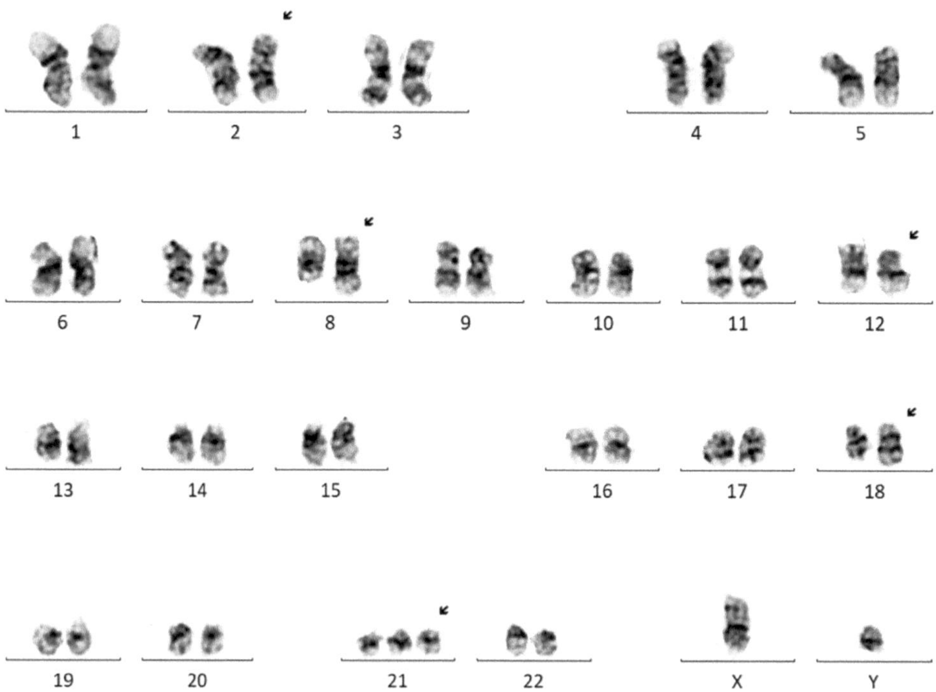

FIG. 19.10.2 Chromosome analysis exhibited an abnormal clone 2 with add(18q) and trisomy 21 in addition to the abnormalities seen in clone 1.

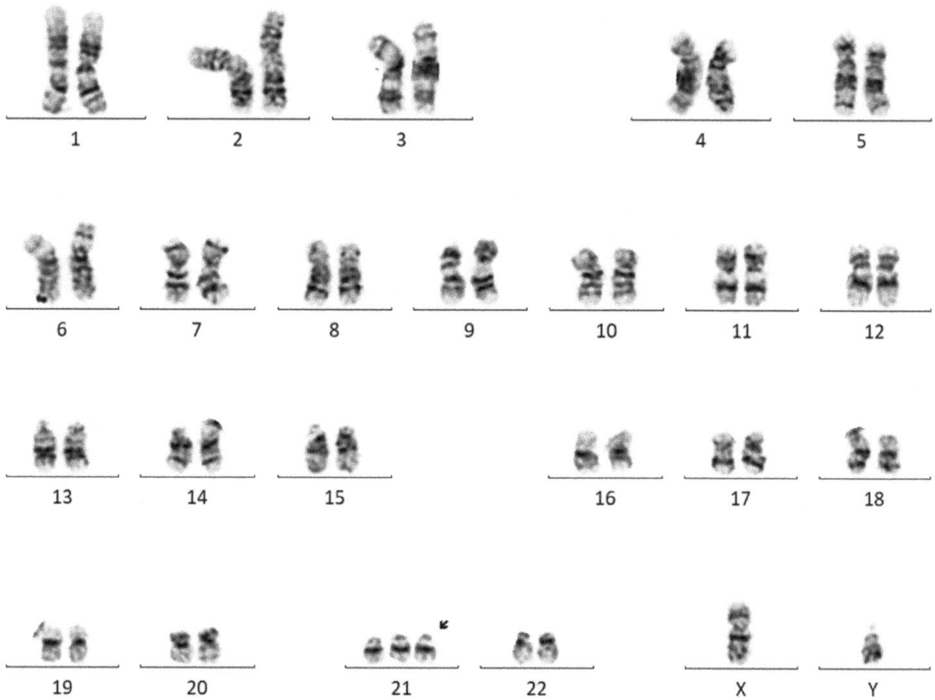

FIG. 19.10.3 Chromosome analysis exhibited an abnormal clone 3 with trisomy 21.

Results with interpretations

Chromosome analysis revealed trisomy 21 in all cells examined. Additional abnormalities were also observed including t(2;8), del(12p), del(16q) and add(18q). del(12p), the gain of 21q, and *MYC* rearrangement were also seen from the concurrent FISH testing. del(12p) has been seen in B-ALL.

MYC rearrangement has been described both in B-cell acute lymphoblastic leukemia (ALL) and in non-Hodgkin lymphomas (NHL), especially in Burkitt lymphoma, and double-hit diffuse large B-cell lymphomas (DLBCL) (https://atlasgeneticsoncology.org). *MYC* rearrangement without other recurrent genetic abnormalities is rare in lymphoblastic leukemia/lymphoma (B-ALL/LBL), with most cases reported in pediatric patients. Three adult cases were reported in B-ALL with *IGH::MYC* fusion [18]. Since this patient had a *CRLF2* rearrangement, the diagnosis for Ph-like ALL is more accurate, although the treatment is very similar to those of B-ALL [19].

Future testing recommendations

Chromosome analysis and FISH for MYC rearrangement should be ordered during regular follow-ups to monitor disease progression. Clinicopathologic correlation is recommended.

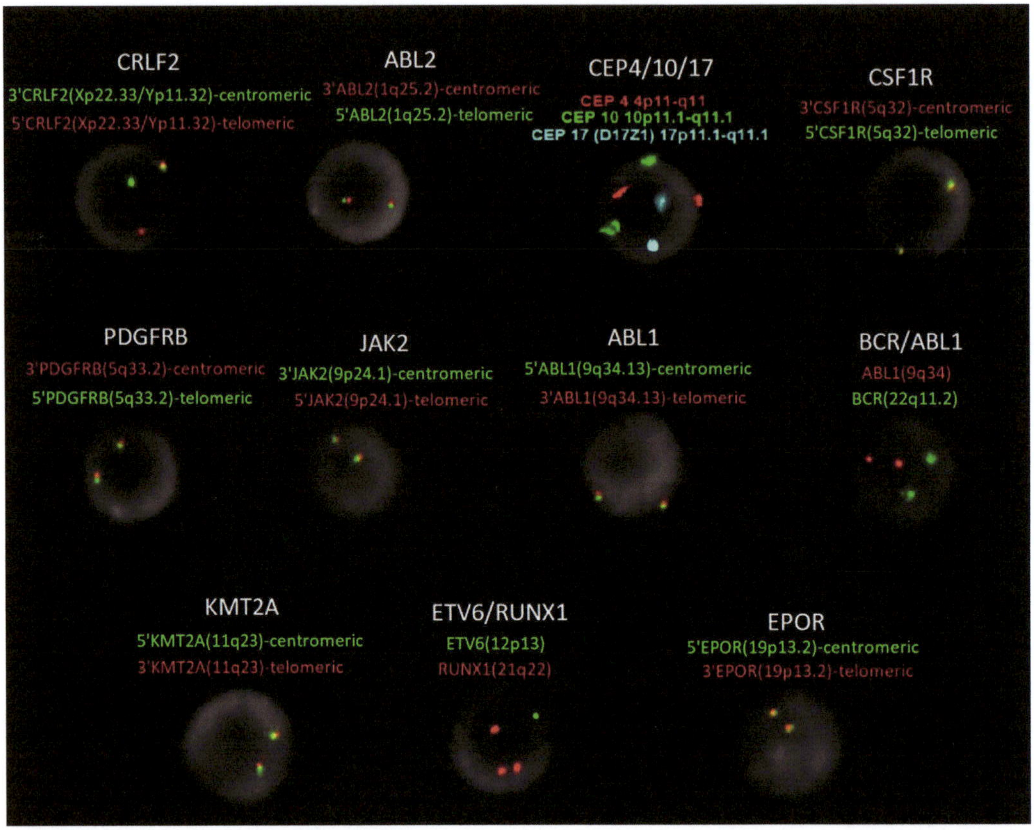

FIG. 19.10.4 FISH for ALL and Ph-like ALL panels was positive for deletion 12p13 and gain of 21q (*RUNX1*), which was consistent with the patient's known history of Down syndrome. FISH also revealed positive results for *CRLF2* rearrangement. ISCN for ALL panel: nuc ish(ABL1,BCR)x2[200/200],(KMT2A)x2[200/200],(ETV6x1,RUNX1x3)[166/200]/(ETV6x2,RUNX1x3)[34/200]; ISCN for Ph-like ALL panel: nuc ish(CRLF2x2)(3'CRLF2 sep 5'CRLF2 x1)[177/200],(ABL2x2)[200/200],(CSF1Rx2)[200/200],(PDGFRBx2)[200/200],(JAK2x2)[200/200],(ABL1x2)[200/200],(EPORx2)[200/200]

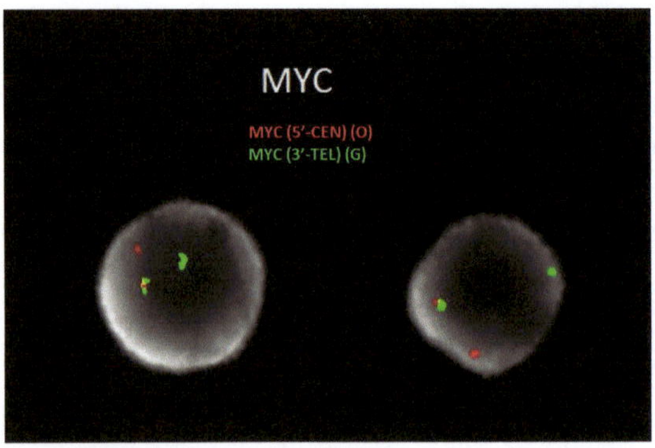

FIG. 19.10.5 FISH for *MYC* gene rearrangement showed positive results *MYC* rearrangement. ISCN: nuc ish (MYCx2)(5'MYC sep 3'MYCx1)[186/200]

Case 19.11 B-lymphoblastic leukemia (B-ALL) with t(1;19)(q23;p13.3)/*TCF3::PBX1* fusion

Clinical indication

A 47-year-old female presented with a relapsed Philadelphia negative B-cell ALL. She was undergoing re-induction chemotherapy at the time and was admitted for fever and sepsis. Her hospital course was complicated by severe hyperbilirubinemia and transaminitis.

Test ordered

- Chromosome analysis of the bone marrow
- FISH: ALL panel

Laboratory test performed

Chromosome analysis and FISH assays were described previously (see Chapter 1 and 12).

Test results

Eleven of the 20 metaphase cells analyzed showed a complex karyotype with t(1;19)(q23;p13.3)/*TCF3::PBX1*, a deletion of 6q, an apparently balanced translocation between 7q and 12p, and an isochromosome 9q resulting in a gain of 9q and a loss of 9p (Fig. 19.11.1). The remaining nine cells appear to be chromosomally normal.

FISH for ALL panel was performed on interphase nuclei using probes localized to the chromosome 4, 10, and 17 centromeric regions; *ABL1* (9q34.12), *BCR* (22q11), *KMT2A* (11q23), *ETV6* (12p13), and *RUNX1* (21q22.3) gene regions. Two hundred nuclei were examined, the results demonstrated previous reported gain of *ABL1* (9q34) in 180/200 (90.0%) of the cells scored. This abnormality was associated with an isochromosome 9 involving two long arms of chromosome 9 and consistent with persistent B-lymphoblastic leukemia (Fig. 19.11.2).

Results with interpretations

The chromosome analysis revealed a complex karyotype and was consistent with persistent B-lymphoblastic leukemia (B-ALL) with t(1;19)(q23;p13.3). A gain of 9q was also demonstrated by the FISH ALL panel testing, concordant with the cytogenetic findings of iso(9q). The t(1;19)(q23;p13.3)/*TCF3::PBX1* is a recurrent chromosomal abnormality in B-ALL and is typically associated with CD10 positive, CD34 negative B-lymphoblasts. Some studies suggest that chromosomal abnormalities involving 9p may have a significant negative impact on survival in adult B-precursor acute lymphoblastic leukemia [20]. Therefore, the combination of del(9p) and t(1;19) indicates an adverse prognosis. Clinicopathologic correlation is advised.

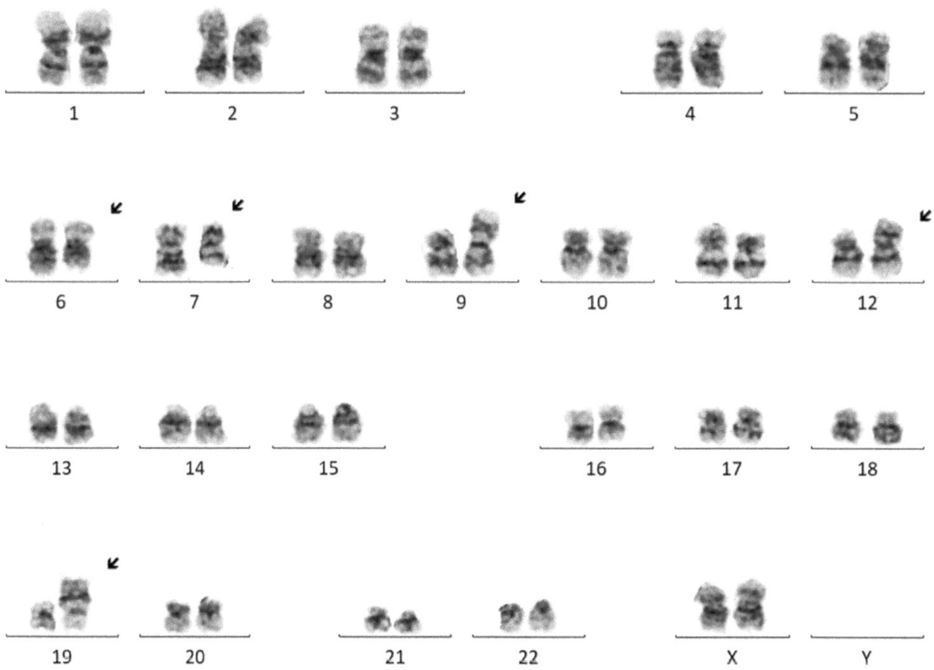

FIG. 19.11.1 Chromosome analysis showed an abnormal clone with del(6q, t(7;12)(q22;p13), i(9q), and der(19)t(1;19)(q23;p13). ISCN: 46,XX,del(6)(q15q23),t(7;12)(q22;p13),i(9)(q10),der(19)t(1;19)(q23;p13)[11]/ 46,XX[9]

Future testing and recommendations

Chromosome analysis and FISH can be ordered to monitor disease progression and treatment efficacy in the future.

Case 19.12 B-lymphoblastic leukemia (B-ALL) with iAMP21

Clinical indication

An 11-year-old male presented with body pain/weakness, paleness, fatigue, decreased oral intake, and a low-grade fever. Examination of the bone marrow showed extensive involvement by B-lymphoblastic leukemia/lymphoma (B-ALL). The B-lymphoid blasts were estimated to represent roughly 90% of the marrow elements (confirmed by immunohistochemistry).

Test ordered

- Chromosome analysis of the bone marrow
- FISH: ALL panel, *PDGFRB*, *ABL1*, *ABL2*, and *CRLF2* rearrangements, and *TP53* deletion

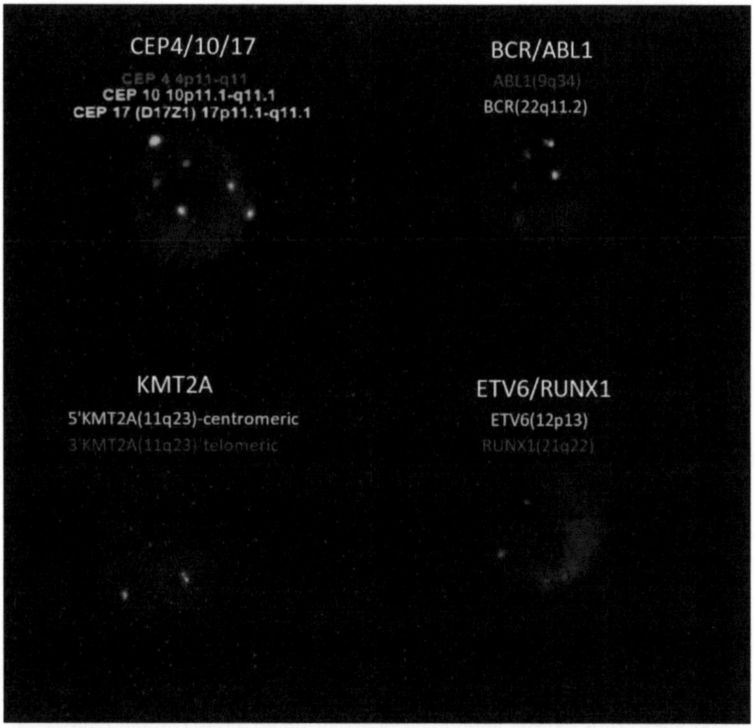

FIG. 19.11.2 FISH for the ALL panel revealed a gain of 9q34 (*ABL1*) in 90.0% of the the cells scored. ISCN: nuc ish(CEP4, CEP10,CEP17)x2[200/200],(ABL1x3,BCRx2)[180/200],(KMT2Ax2)[200/200],(ETV6,RUNX1)x2[200/200].

Laboratory test performed

Chromosome analysis and FISH assays were described previously (see Chapters 1 and 12).

Test results

Of the 23 cells examined, 15 were abnormal, and clonal evolution was evident. All abnormal cells exhibited a ring chromosome 21 with an insertion of unknown material and amplification of *RUNX1* on the ring chromosome 21 (Fig. 19.12.1); seven of these also had a balanced translocation involving Xp and 13q (Fig. 19.12.2); four of these showed add(3p), add(6p), add(6q), and a whole-arm translocation involving 8q and 17q in addition to the ring chromosome 21 (Fig. 19.12.3). The remaining eight cells appear to be chromosomally normal.

FISH for ALL panel, break-apart probes with *ABL1*, *ABL2*, and *CRLF2*, *PDGFRB*, and probes for *TP53* was performed on interphase nuclei using probes localized to the chromosome 4, 10, and 17 centromeric regions; *ABL1* (9q34.12), *BCR* (22q11), *KMT2A*

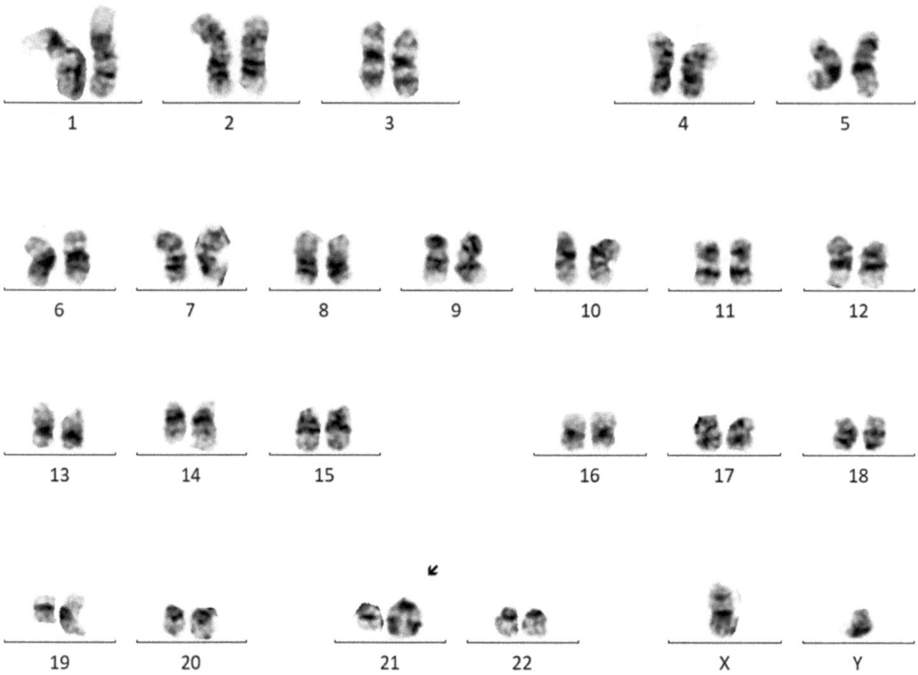

FIG. 19.12.1 Chromosome analysis showed abnormal clone 1 including a ring chromosome 21 with an insertion of unknown material and amplification of *RUNX1*, confirmed by FISH. ISCN: 46,XY,der(21)ins(21;?)(q22;?)r(21)(p13q22).ish der(21)(RUNX1 amp)[4]/ 46,idem,t(X;13)(p22.3;q31)[7]/ 45,idem,add(3)(p25),add(6)(p23),add(6)(q23),der(8;17)(q10;q10)[4]/ 46,XY[8]

(11q23), *ETV6* (12p13), *RUNX1* (21q22.3), and probes localized to the 5′ and 3′ ends of the *ABL1* (9q34.12), *ABL2* (1q25.2), *CRLF2* (Xp22.33/Yp11.32), *PDGFRB* (5q33), as well as probes localized to *Tp53* (17p13.1) and *CEP17* (17cen) gene regions. Two hundred nuclei were examined, and the results demonstrated *RUNX1* amplification in 191/200 (95.5%) of the the cells scored, and *TP53* deletion in 36/200 (18%) of the the cells scored (Fig. 19.12.4). Metaphase FISH confirmed iAMP21 (Fig. 19.12.5). This combination of abnormalities is associated with a poor prognosis in B-cell ALL.

Results with interpretations

Chromosome analysis revealed a complex karyotype with clonal evolution observed. FISH testing confirmed iAMP21 and *Tp53* deletion. Intrachromosomal amplification of chromosome 21 (iAMP21) is a rare high-risk chromosomal abnormality reported in approximately 2%–5% of pediatric patients with B-cell precursor ALL (BCP-ALL) [21,22]. This abnormality has been associated with a poor prognosis in patients treated by a standard

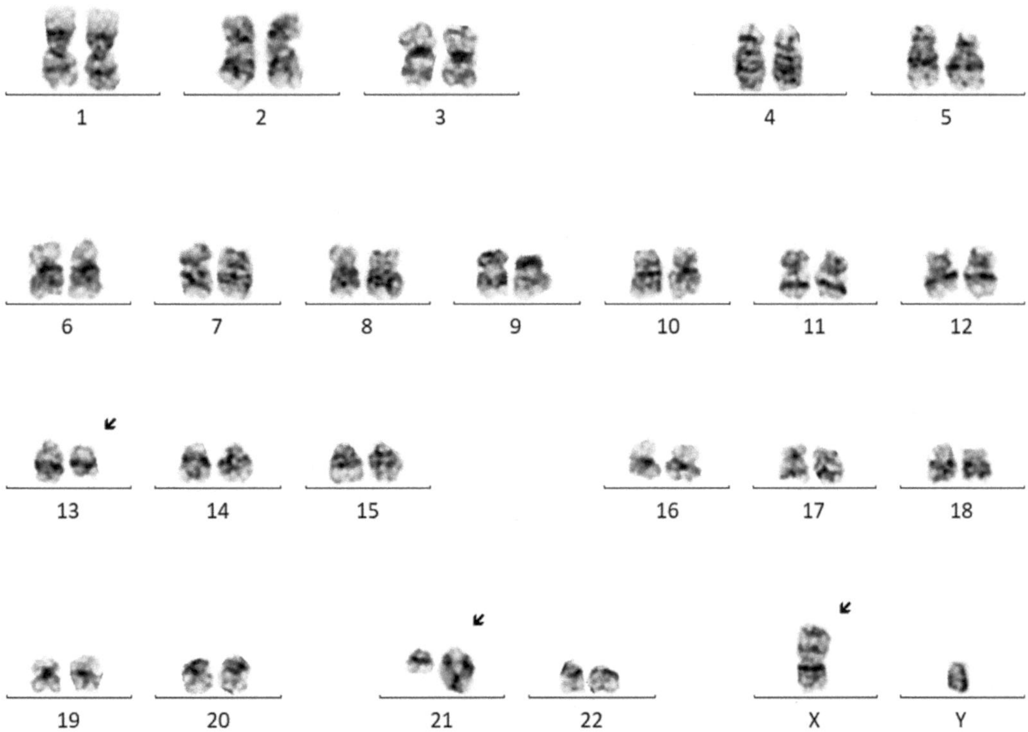

FIG. 19.12.2 Chromosome analysis showed abnormal clone 2 with t(X;13) and ring chromosome 21.

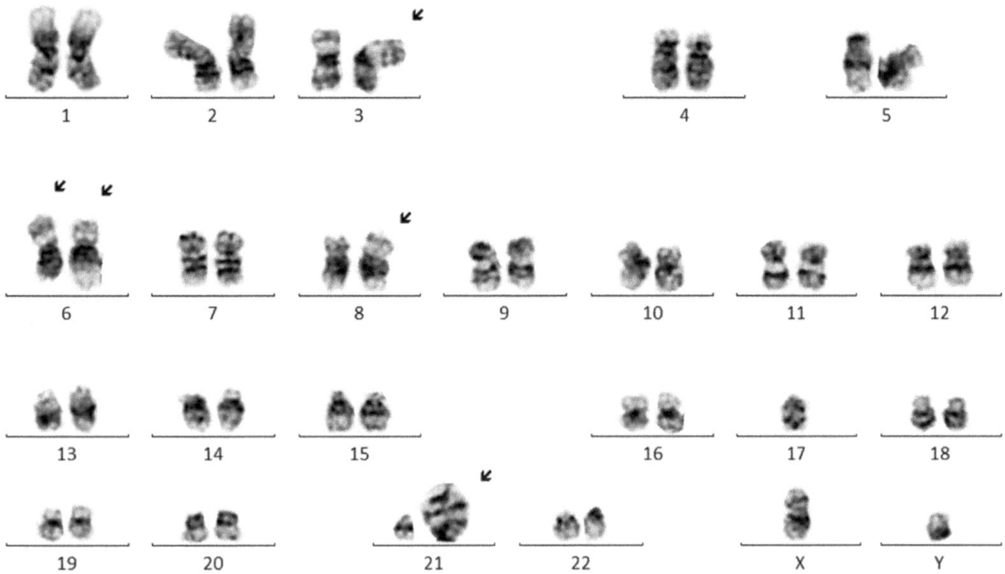

FIG. 19.12.3 Chromosome analysis showed abnormal clone 3 with add(3p), add(6p), add(6q), and a whole-arm translocation involving 8q and 17q in addition to the ring chromosome 21.

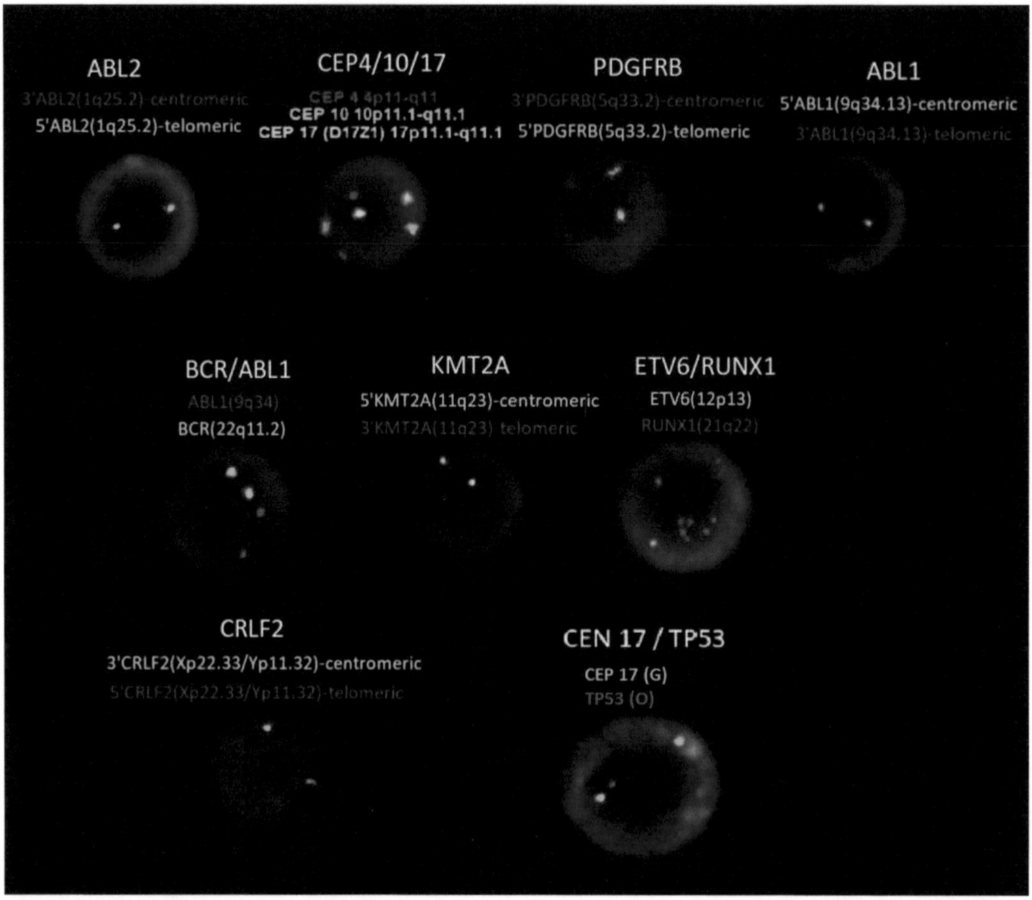

FIG. 19.12.4 FISH for the ALL panel plus break-apart probes with *PDGFRB*, *ABL1*, *ABL2*, and *CRLF2*, and probes for *TP53* revealed *RUNX1* amplification in 191/200 (95.50%) of the the cells scored and *TP53* deletion in 36/200 (18%) of the the cells scored. ISCN: nuc ish(CEP4,CEP10,CEP17)x2[200/200],(ABL1,BCR)x2[200/200],(KMT2Ax2)[200/200],(ETV6x2, RUNX1 amp)[191/200],(PDGFRBx2)[200/200],(ABL1x2)[200/200],(ABL2x2)[200/200],(CRLF2x2)[200/200],(TP53x1, D17Z1x2)[36/200]

protocol [23,24]. In conclusion, the degree of karyotypic complexity along with iAMP21 observed indicates a more aggressive disease process. Clinicopathologic correlation is advised.

Future testing and recommendations

Chromosome analysis and FISH can be ordered to monitor disease progression and treatment efficacy in the future.

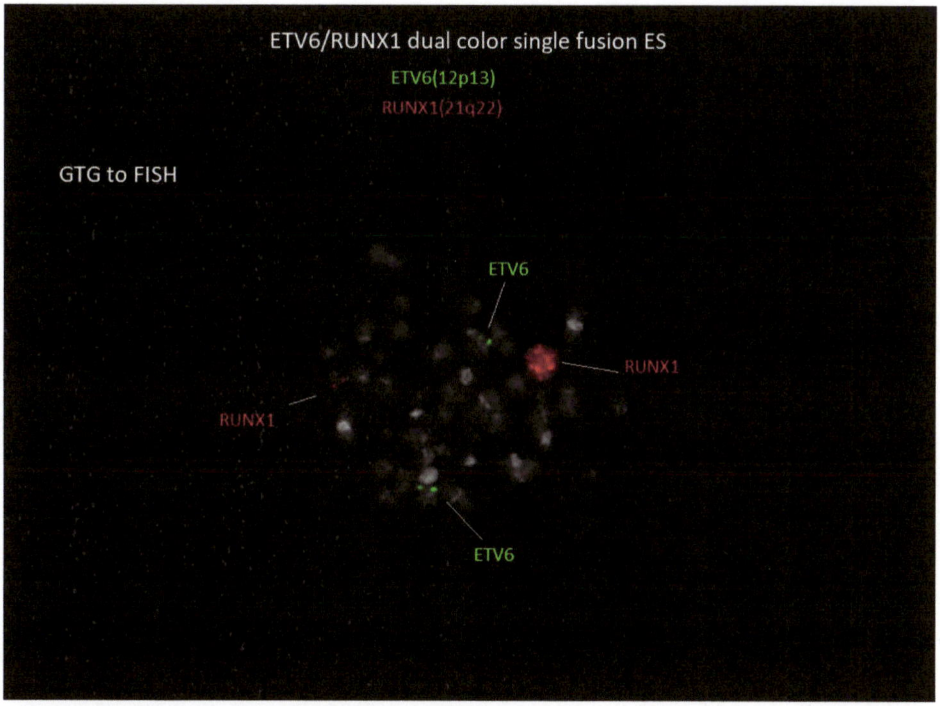

FIG. 19.12.5 Metaphase FISH confirmed *RUNX1* amplification.

Case 19.13 Ph-like B-cell lymphoblastic leukemia (Ph-like ALL) with *IGH* and *CRLF2* rearrangements

Clinical indication

A 20-year-old male presented with dyspnea, generalized weakness, night sweats, and weight loss. He was being investigated for acute leukemia. His CBC showed marked leukocytosis with circulating blasts. Bone marrow report showed a new diagnosis of B-cell ALL with high risk for tumor lysis syndrome.

Test ordered

- Chromosome analysis of the bone marrow
- FISH: ALL panel
- FISH: Ph-like ALL panel
- FISH: *IGH* Gene Rearrangement

Laboratory test performed

Chromosome analysis and FISH assays were described previously (see Chapters 1 and 12).

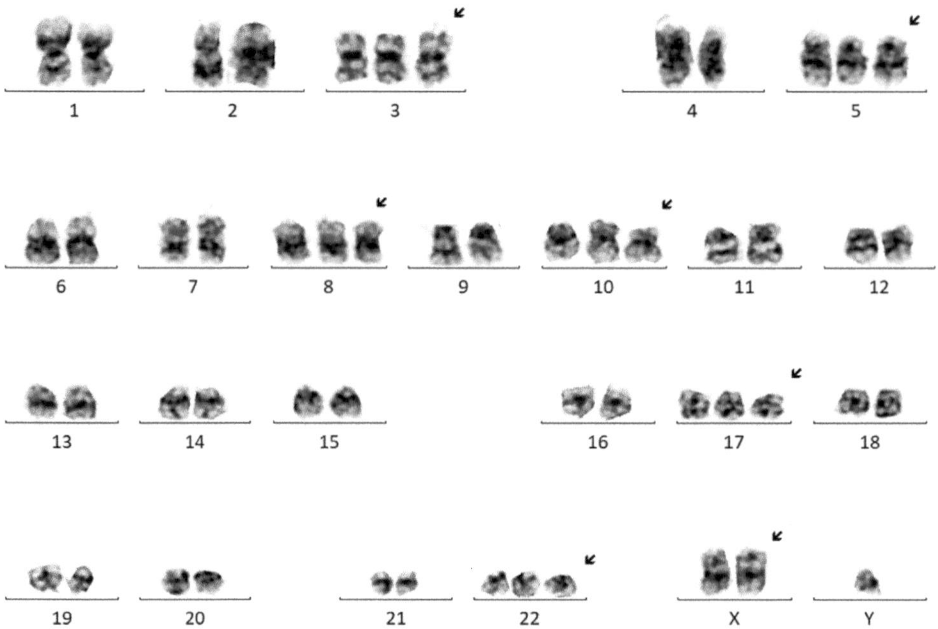

FIG. 19.13.1 Chromosome analysis revealed a hyperdiploid karyotype with gains of chromosomes X, 3, 5, 8, 10, 17, and 22 in all 20 cells examined. ISCN: 53,XY,+X,+3,+5,+8,+10,+17,+22[20]

Test results

Of the 20 cells examined, all exhibited a hyperdiploid karyotype with gains of the X chromosome, and chromosomes 3, 5, 8, 10, 17, and 22 (Fig. 19.13.1).

FISH for ALL panel was performed on interphase nuclei using probes localized to the chromosome 4, 10, and 17 centromeric regions; *ABL1* (9q34.12), *BCR* (22q11), *KMT2A* (11q23), *ETV6* (12p13), and *RUNX1* (21q22.3) gene regions. Two hundred nuclei were examined, and the results demonstrated trisomy 10 and 17 in 196/200 (98.0%) of the cells scored and an extra copy of BCR (22q11.2) in 194/200 (97.0%) of the cells scored (Fig. 19.13.2).

FISH for Ph-like ALL was performed on interphase nuclei using probes localized to the *CRLF2* (Xp22.33/Yp11.32), *ABL2* (1q25.2), *CSF1R* (5q32), *PDGFRB* (5q33.2), *JAK2* (9p24.1), *ABL1* (9q34.13), and *EPOR* (19p13.2) gene regions. Two hundred nuclei were examined, and the results demonstrated a *CRLF2* rearrangement in 196/200 (98.0%) of the cells scored and a gain of 5q (*CSF1R*, *PDGFRB*) in 194/200 (97.0%) of the cells scored (Fig. 19.13.3).

When analyzing the results, we noticed that the patient is of Hispanic descent. According to published studies, it is likely that this patient has an *IGH::CRLF2* rearrangement based on FISH results with *CRLF2* probe [25–27]. Therefore, a FISH for *IGH* rearrangement was performed on interphase nuclei using probes localized to the 5′ and 3′ ends of the *IGH*

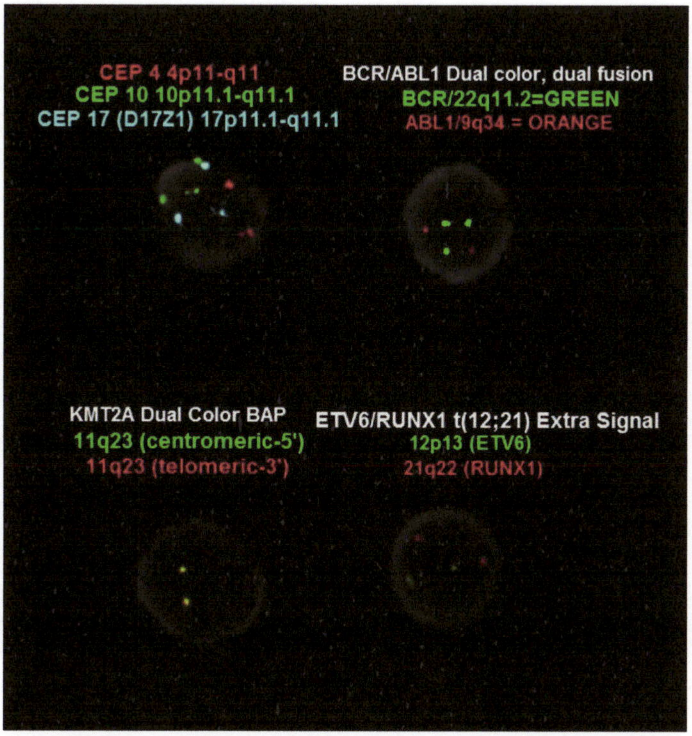

FIG. 19.13.2 FISH for ALL panel showed trisomy 10 and 17 in 98% and gain of BCR in 97% of the the cells scored. ISCN: nuc ish(CEP4x2,CEP10x3,CEP17x3)[196/200],(ABL1x2,BCRx3)[194/200],(KMT2Ax2)[200/200],(ETV6,RUNX1)x2[199/200]

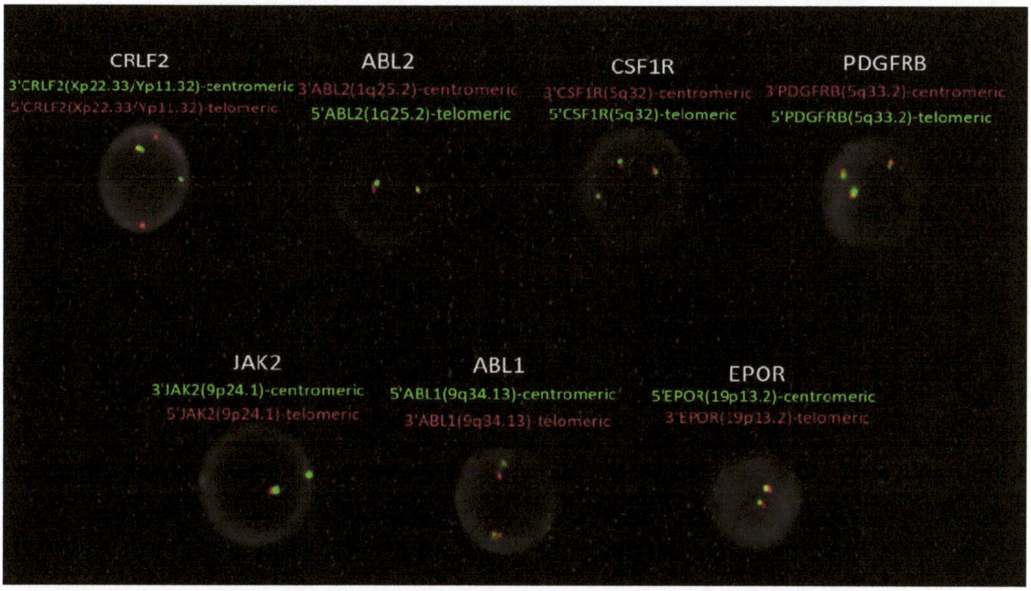

FIG. 19.13.3 FISH for the Ph-like ALL panel revealed *CRLF2* rearrangement in 98% and a gain of 5q in 97% of the the cells scored. ISCN: nuc ish(5'CRLF2x3,3'CRLF2x2)(3'CRLF2 con 5'CRLF2 x1)[196/200],(ABL2x2)[200/200],(CSF1Rx3)[194/200],(PDGFRBx3)[194/200],(JAK2x2)[200/200],(ABL1x2)[200/200],(EPORx2)[200/200]

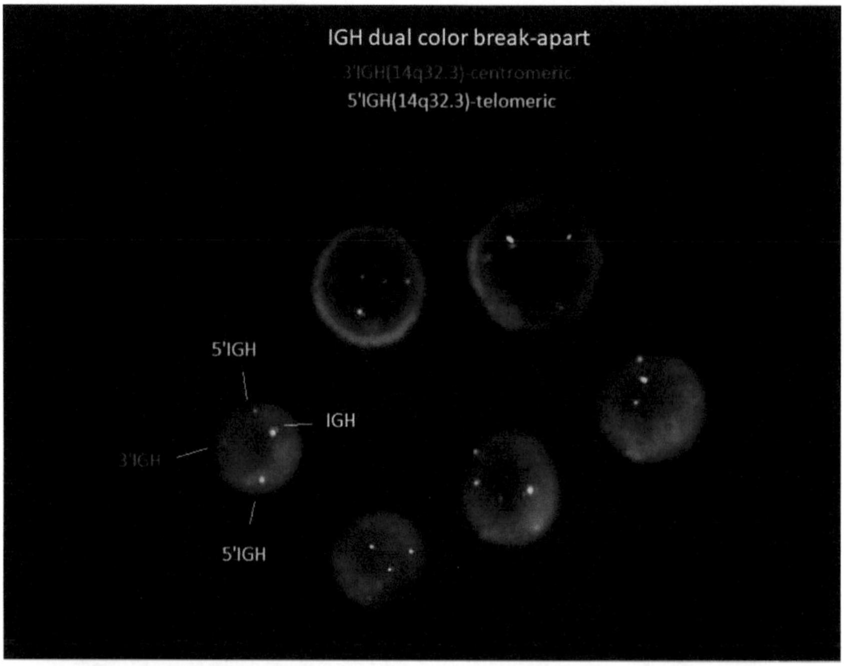

FIG. 19.13.4 FISH for the *IGH* break-apart probe showed positive results for *IGH* rearrangement. ISCN: nuc ish(3′IGHx2,5′IGHx3)(3′IGH con 5′IGHx1)[196/200]

(14q32.33) gene region. Two hundred nuclei were examined, and the results demonstrated an *IGH* rearrangement, including an additional 5′IGH signal, in 196/200 (98.0%) of the the cells scored (Fig. 19.13.4).

In order to support the presence of an *IGH::CRLF2* rearrangement, metaphase FISH with *IGH* and *CRLF2* dual-color break-apart probes on two separate slides were performed. The results demonstrated the expected fusion. For metaphase FISH with the *CRLF2* break-apart probes, the 5′-*CRLF2* signals (red) were on the two derivative X chromosomes, the 3′-*CRLF2* signal (green) was on the derivative 14, and the *CRLF* intact signal (yellow) was on the Y chromosome (Fig. 19.13.5). For the metaphase FISH with the *IGH* break-apart probes, the 5′-*IGH* signals (green) were on the two derivative X chromosomes, the 3′-*IGH* signal (red) was on the derivative chromosome 14; the intact *IGH* signal (yellow) was on the normal chromosome 14 (Fig. 19.13.6).

Results with interpretations

Chromosome analysis revealed a hyperdiploid karyotype with gains of multiple chromosomes. FISH identified *CRLF2* and *IGH* rearrangement. This finding is diagnostic of B-lymphoblastic leukemia with *BCR::ABL1*-like features and is associated with high-risk clinical features. Clinicopathologic correlation is advised.

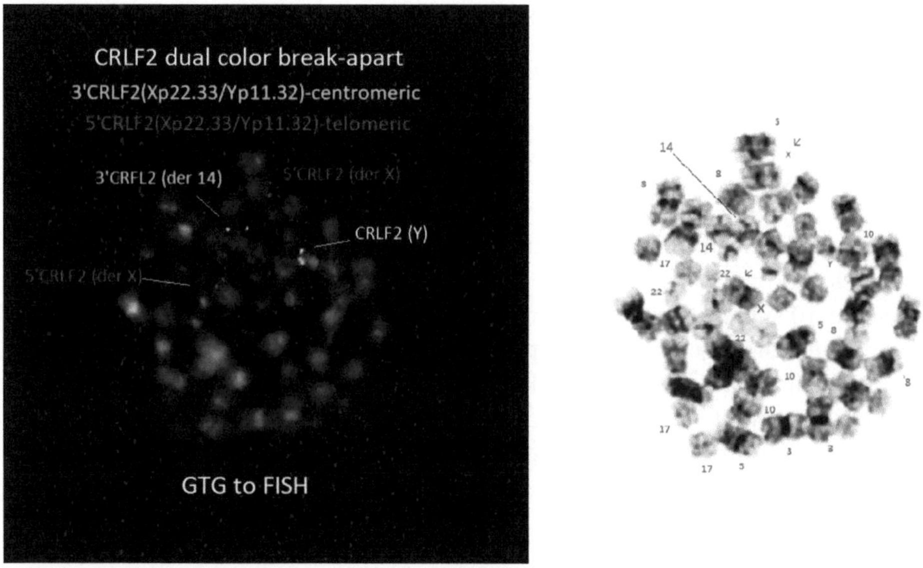

FIG. 19.13.5 Metaphase FISH for the *CRLF2* break-apart probes showed positive results for the fusion. Note that the 3′-*CRLF2 (green)* was on derivative chromosome 14.

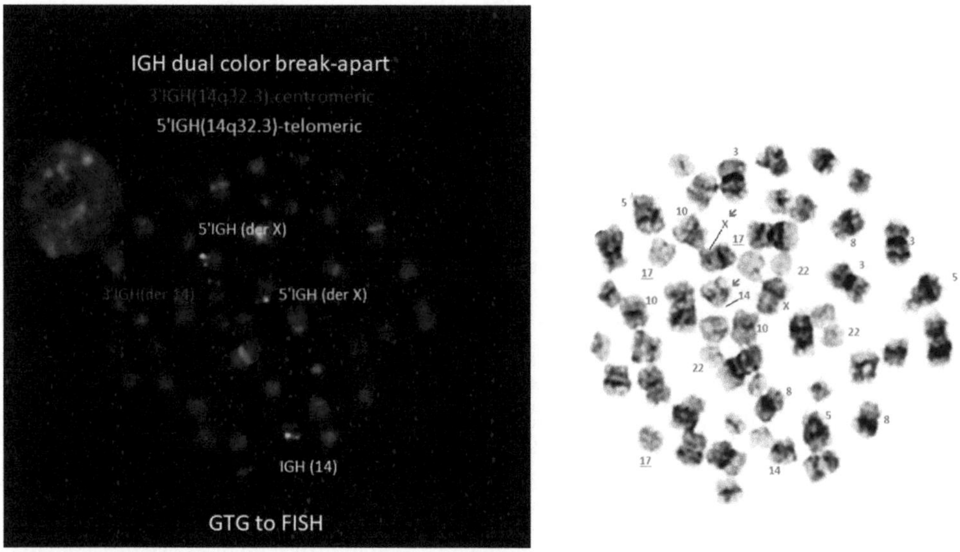

FIG. 19.13.6 Metaphase FISH for the *IGH* break-apart probes showed positive results for the fusion. Note that the two 5′-*IGH* signals *(green)* were on the two derivative X chromosomes.

Future testing and recommendations

Chromosome analysis and FISH can be ordered to monitor disease progression and treatment efficacy in the future.

Summary of the key learning points

- B-ALL is commonly seen in children under the age of 6. The associated diseases are usually aggressive.
- There are over 10 subtypes of precursor lymphoid neoplasms in B-ALL, and most of them have gene rearrangements or amplifications.
- Genetic testing including karyotyping, FISH, RT-PCR, or NGS is crucial to patients when seeking treatment options, predicative prognosis, and risk stratification.
- Regular follow-ups to monitor disease progression are important for patients due to the risk of relapse in disease.

References

[1] C. Wenzinger, E. Williams, A.A. Gru, Updates in the pathology of precursor lymphoid neoplasms in the revised fourth edition of the WHO classification of tumors of hematopoietic and lymphoid tissues, Curr. Hematol. Malig. Rep. 13 (4) (2018) 275–288.

[2] M. Onciu, Acute lymphoblastic leukemia, Hematol. Oncol. Clin. North Am. 23 (4) (2009) 655–674.

[3] D.A. Arber, et al., The 2016 revision to the World Health Organization classification of myeloid neoplasms and acute leukemia, Blood 127 (20) (2016) 2391–2405.

[4] R. Alaggio, et al., The 5th edition of the World Health Organization classification of haematolymphoid tumours: lymphoid neoplasms, Leukemia 36 (7) (2022) 1720–1748.

[5] N. Coccaro, et al., Next-generation sequencing in acute lymphoblastic leukemia, Int. J. Mol. Sci. 20 (12) (2019).

[6] S. Soverini, et al., Next-generation sequencing for BCR-ABL1 kinase domain mutation testing in patients with chronic myeloid leukemia: a position paper, J. Hematol. Oncol. 12 (1) (2019) 131.

[7] R. DeBoer, et al., Clinical impact of ABL1 kinase domain mutations and IKZF1 deletion in adults under age 60 with Philadelphia chromosome-positive (Ph+) acute lymphoblastic leukemia (ALL): molecular analysis of CALGB (Alliance) 10001 and 9665, Leuk. Lymphoma 57 (10) (2016) 2298–2306.

[8] C. Baer, et al., Detection of ABL1 kinase domain mutations in therapy-naive BCR-ABL1-positive acute lymphoblastic leukemia, Haematologica 107 (2) (2022) 562–563.

[9] N.M. Savage, et al., Acute leukemia with PICALM-MLLT10 fusion gene: diagnostic and treatment struggle, Cancer Genet. Cytogenet. 202 (2) (2010) 129–132.

[10] P. Lopez-Nieva, et al., Detection of novel fusion-transcripts by RNA-Seq in T-cell lymphoblastic lymphoma, Sci. Rep. 9 (1) (2019) 5179.

[11] T.H. Jaing, et al., Clinical significance of central nervous system involvement at diagnosis of childhood T-cell acute lymphoblastic leukemia, Pediatr. Blood Cancer 45 (2) (2005) 135–138.

[12] I. Panagopoulos, et al., Fusion of NUP98 and the SET binding protein 1 (SETBP1) gene in a paediatric acute T cell lymphoblastic leukaemia with t(11;18)(p15;q12), Br. J. Haematol. 136 (2) (2007) 294–296.

[13] O. Bernard, et al., Two site-specific deletions and t(1;14) translocation restricted to human T-cell acute leukemias disrupt the 5' part of the tal-1 gene, Oncogene 6 (8) (1991) 1477–1488.

[14] O. Bernard, et al., Characterization of translocation t(1;14)(p32;q11) in a T and in a B acute leukemia, Leukemia 7 (10) (1993) 1509–1513.

[15] O. Bernard, et al., Molecular analysis of T-cell receptor transcripts in a human T-cell leukemia bearing a t(1;14) and an inv(7); cell surface expression of a TCR-beta chain in the absence of alpha chain, Leukemia 7 (10) (1993) 1645–1653.

[16] K. Rack, et al., Optimizing the diagnostic workflow for acute lymphoblastic leukemia by optical genome mapping, Am. J. Hematol. 97 (5) (2022) 548–561.

[17] M. Safavi, A.S. Sharari, T cell acute lymphoblastic leukemia with t(11;14)(p13;q11) and trisomy 8, Blood Res. 56 (3) (2021) 128.

[18] Y. Li, et al., B lymphoblastic leukemia/lymphoma with Burkitt-like morphology and IGH/MYC rearrangement: report of 3 cases in adult patients, Am. J. Surg. Pathol. 42 (2) (2018) 269–276.

[19] C.H. Pui, et al., Philadelphia chromosome-like acute lymphoblastic leukemia, Clin. Lymphoma Myeloma Leuk. 17 (8) (2017) 464–470.

[20] H. Nahi, et al., An investigation into whether deletions in 9p reflect prognosis in adult precursor B-cell acute lymphoblastic leukemia: a multi-center study of 381 patients, Haematologica 93 (11) (2008) 1734–1738.

[21] A.V. Moorman, The clinical relevance of chromosomal and genomic abnormalities in B-cell precursor acute lymphoblastic leukaemia, Blood Rev. 26 (3) (2012) 123–135.

[22] S.K. Ma, et al., Characterization of additional genetic events in childhood acute lymphoblastic leukemia with TEL/AML1 gene fusion: a molecular cytogenetics study, Leukemia 15 (9) (2001) 1442–1447.

[23] D.R. Garcia, et al., Intrachromosomal amplification of chromosome 21 (iAMP21) detected by ETV6/RUNX1 FISH screening in childhood acute lymphoblastic leukemia: a case report, Rev. Bras. Hematol. Hemoter. 35 (5) (2013) 369–371.

[24] A.V. Moorman, et al., Prognosis of children with acute lymphoblastic leukemia (ALL) and intrachromosomal amplification of chromosome 21 (iAMP21), Blood 109 (6) (2007) 2327–2330.

[25] T. Herold, N. Gokbuget, Philadelphia-like acute lymphoblastic leukemia in adults, Curr. Oncol. Rep. 19 (5) (2017) 31.

[26] T. Herold, et al., Adults with Philadelphia chromosome-like acute lymphoblastic leukemia frequently have IGH-CRLF2 and JAK2 mutations, persistence of minimal residual disease and poor prognosis, Haematologica 102 (1) (2017) 130–138.

[27] M.D.R. Juarez-Velazquez, et al., High occurrence of CRLF2 abnormalities in Mexican children with B-cell acute lymphoblastic leukemia, Cytokine 155 (2022) 155896.

20

Mature B-cell neoplasms

Xia Li
SONORA QUEST LABORATORIES, PHOENIX, AZ, UNITED STATES

Background

Mature B-cell neoplasms comprise over 90% of lymphoid neoplasms worldwide, and there are 4% of new cases each year. They are more common in developing countries. In this group of disorders, there are over 50 different types according to the fifth edition of the WHO Classification of Hematolymphoid Tumors [1]. The most common types are follicular lymphoma (FL) and diffuse large B-cell lymphoma (DLBCL), which make up 50% of non-Hodgkin's lymphomas. The other common types include B-cell chronic lymphocytic leukemia/small lymphocytic lymphoma (CLL/SLL) and mantle cell lymphoma (MCL). The specific B-cell neoplasm frequency varies in different parts of the world. FL is more common in the United States and Western Europe, while Burkitt's lymphoma is endemic to Africa (https://wiki.clinicalflow.com). Treatment options are available for many of the disorders [2]. Phosphoinositide 3-kinase (PI3K) inhibitors are used to treat FL and other relapsed or refractory indolent non-Hodgkin lymphomas (iNHL) [3]. Bruton's tyrosine kinase (BTK) inhibitors are used to treat SLL, Waldenstrom macroglobulinemia (WM), and marginal zone lymphoma (MZL) [4,5].

Biomarker testing is becoming a more important tool in the diagnosis and prognosis of mature B-cell lymphoma, in addition to the pathology evaluation of tissue. For example, karyotyping, FISH, CMA, and NGS can identify del(6q), del(13q), del(11q) (ATM), del(17p) (TP53), and trisomy 12 in CLL/SLL. These abnormalities provide valuable prognostic information [6]. T(14;18) is commonly seen in FL and DLBCL [7–9], and t(11;14)(q23; q32) is a hallmark of MCL [10,11]. *C-MYC*, *BCL2*, and *BCL6* rearrangements were used to distinguish double or triple-hit lymphomas [12]. However, *BCL6* was no longer used as a criterion for triple-hit lymphoma [1]. There are numerous mutations in mature B-cell lymphomas identified by next-generation sequencing (NGS) or whole-genome sequencing (WGS) such as *CCND3*, *E2A/ID3*, *SF3B1*, *NOTCH1*, *MYD88*, *ATM*, *MLL2*, *EZH2*, *CREBBP*, *BRAF*, *TP53*, and *BIRC3* [13]. With the high number of mutations found in mature B-cell neoplasms, the opportunity exists for personalized therapy. In this chapter, a few cases will be described to illustrate the importance of genetic testing for the identification of mutations and translocations to help physicians with diagnosis, prognosis, and treatment options.

Cases in Laboratory Genetics and Genomics (LGG) Practice. https://doi.org/10.1016/B978-0-323-99622-8.00018-6

Case 20.1 Atypical B-cell chronic lymphocytic leukemia (CLL)

Clinical indication

A 45-year-old male presented with hyperleukocytosis, bilateral palpable lymphadenopathy, and shortness of breath. He was initially treated with a short course of steroids, but then developed anemia, thrombocytopenia, and pseudo-hyperkalemia.

Test ordered

- Chromosome analysis of the bone marrow
- FISH: low-grade lymphoma, CLL panel and *MYC* gene rearrangement
- NGS (send-out)

Laboratory test performed

Chromosome analysis, FISH, and NGS methods were described previously in Chapters 1 and 12.

Test results

Chromosome analysis showed a complex karyotype with two rearrangements involving chromosome 14 *(IGH)* [t(2;14) and t(14;19)] and trisomy 12 (Fig. 20.1.1). NGS was performed in another lab, and the results were negative for DNA mutations and RNA fusions (data not shown).

FISH for low-grade lymphoma panel was performed on interphase nuclei using probes localized to *IGH* (14q32.3) and *CCND1* (11q13.3), plus the 5′ and 3′ ends of *BCL6* (3q27), *IGH* (14q32.3), *BCL2* (18q21.3), and *MALT1* (18q21.3) gene regions. Two hundred nuclei were examined, and the results demonstrated multiple abnormalities (Fig. 20.1.2).

Gain of *BCL6* was observed in 26/200 (13.0%) of the cells scored.

Gain of *CCND1* was observed in 15/200 (7.5%) of the cells scored.

Gain of *IGH* was observed in 18/200 (9.0%) of the cells scored.

IGH rearrangement was observed in 166/200 (83.0%) of the cells scored.

BCL2 rearrangements was observed in 5/200 (2.5%) of the cells scored.

Gain of *MALT1* was observed in 17/200 (8.5%) of the cells scored.

FISH was also performed on interphase nuclei using probes localized to the *D12Z3* (12 CEN), 13q14.3, 13q34, *IGH::BCL2* dual-color dual-fusion probes, and probes localized to the 5′ and 3′ ends of the *MYC* (8q24), *ATM* (11q22.3), and *TP53* (17p13.1) gene regions. Two hundred nuclei were examined for each probe, and the results demonstrated multiple abnormalities (Fig. 20.1.2).

Gain of chromosome 12 was observed in 164/200 (82.0%) of the cells scored.

Gain of chromosome 13 was observed in 20/200 (10%) of the cells scored.

Gain of *IGH* was observed in 173/200 (86.5%) of the cells scored.

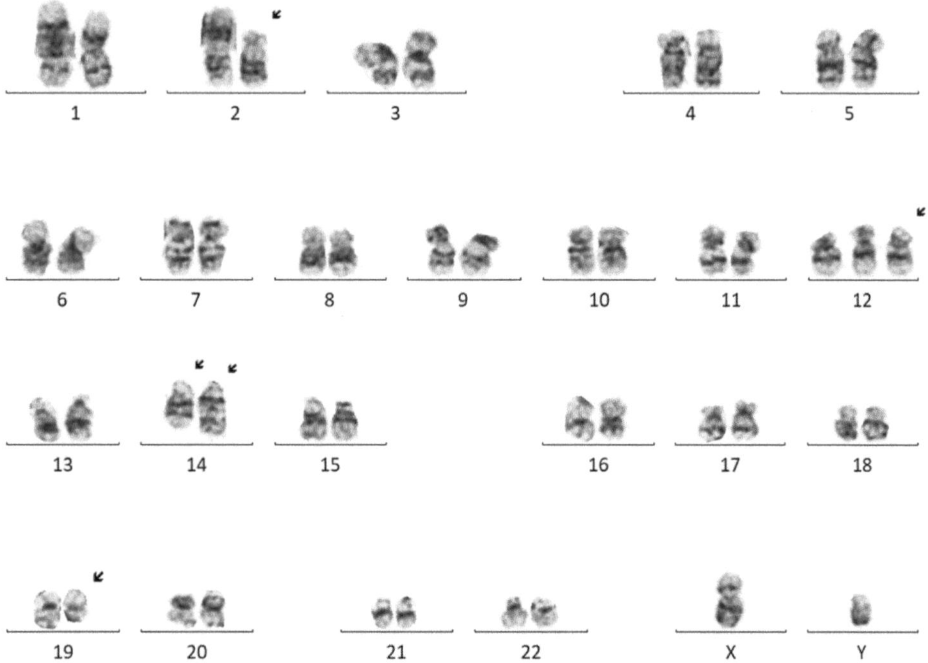

FIG. 20.1.1 Chromosome analysis of the bone marrow showed one abnormal clone from a complex karyotype with t(2;14), t(14;19), and trisomy 12. ISCN: 47,XY,t(2;14)(p13;q32),+12,t(14;19)(q32;q13.3)[cp7]/46,XY[14]

Gain of *BCL2* was observed in 30/200 (15%) of the cells scored.

Gain of *MYC* was observed in 31/200 (15.5%) of the cells scored.

Gains of *ATM* and 17p13.1 were observed in 30/200 (15.0%) of the cells scored.

No rearrangement of *IGH::CCND1* and *IGH::BCL2* was observed. No *MYC* rearrangement was detected.

Results with interpretations

Of the 21 cells examined, 7 exhibited balanced translocations involving 2p&14q and 14q&19q, and trisomy 12. The remaining 14 cells appear to be chromosomally normal.

FISH with probes for low-grade lymphoma, *C-MYC,* chromosomes 12, 13, *ATM* (11q23), *TP53* (17p13.1), *IGH::CCND1,* and *IGH::BCL2* was performed, and the results showed rearrangements of *BCL2* and *IGH* and gains of many chromosomes. However, the fusion partner of *IGH* was not *CCND1* or *BCL2.* Trisomy 12 and *IGH* rearrangements were seen from the FISH testing, concordant with the results of karyotyping. t(2;14) and t(14;19) were reported in atypical CLL. The degree of karyotypic complexity along with multiple abnormalities from FISH indicates a more aggressive disease process.

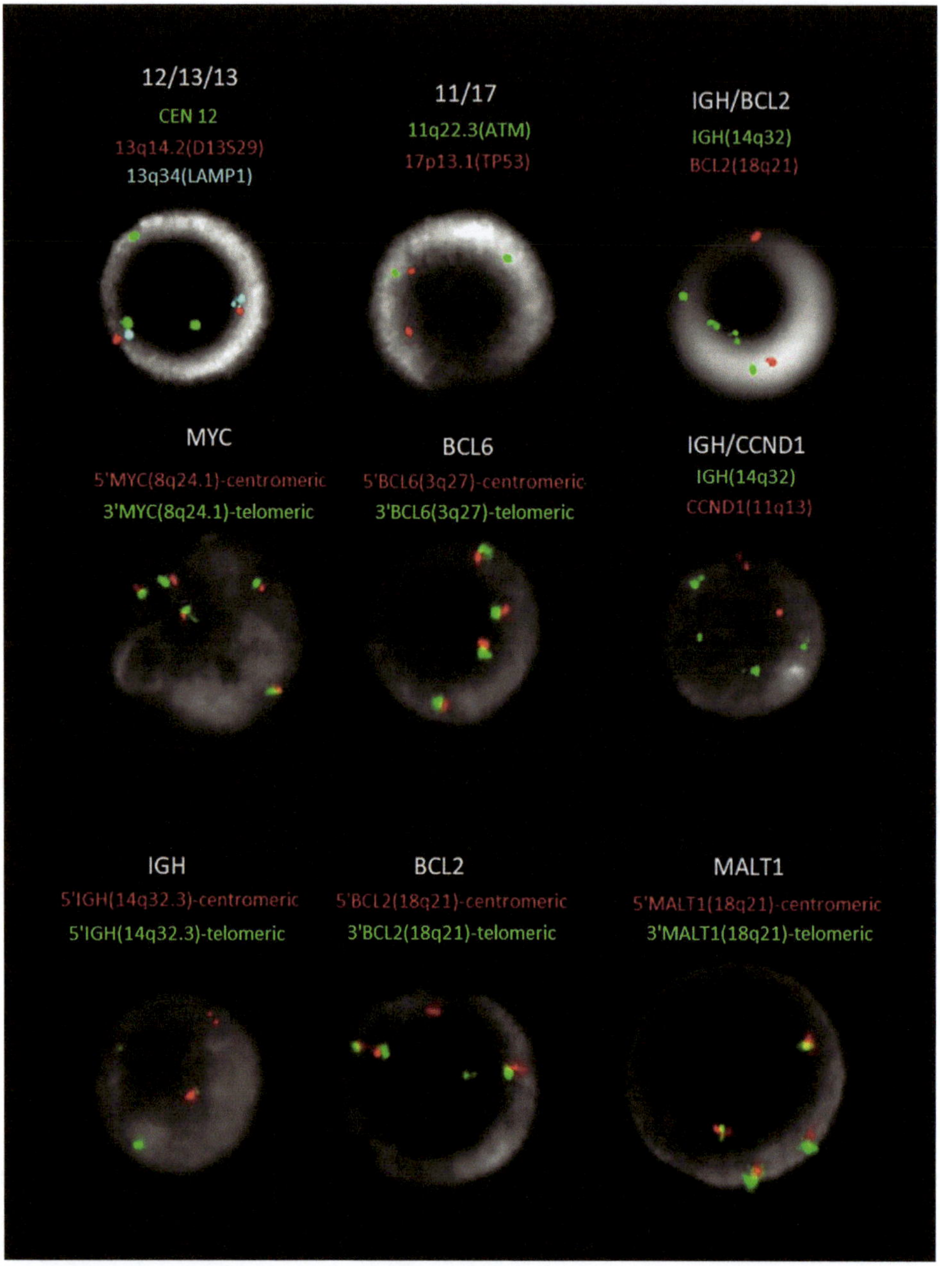

FIG. 20.1.2 FISH for *MYC*, low-grade lymphoma panel, chromosome 12, del(13q), *ATM*, and *TP53* were performed, and the results revealed rearrangements involving *IGH* and *BCL2*, and multiple gains of many chromosomes. ISCN: nuc ish(MYCx4-5)[31/200],(BCL6x4)[26/200],(IGHx2)(3′IGH sep 5′IGHx2)[166/200]/(IGHx4)(3′IGH sep 5′IGHx4)[18/200],(MALT1x4)[17/200],(BCL2x4)(5′BCL2 sep 3′BCL2x1)[5/200]/(BCL2x4)[18/200],(CCND1x2,IGHx3-4)[173/200]/(CCND1x4,IGHx4-8)[15/200],(D13S319,LAMP1)x4[20/200],(D12Z3x3)[144/200]/(D12Z3x6)[20/200],(IGHx4,BCL2x2)[143/200]/(IGHx8,BCL2x4)[30/200],(ATM,TP53)x4[30/200]

Future testing and recommendations

Regular follow up and clinicopathologic correlation are recommended. Chromosome analysis and FISH can be ordered for monitoring disease progression and treatment efficacy in the future.

Case 20.2 Mantle cell lymphoma (MCL)

Clinical indication

A 75-year-old male presented with pancytopenia, abdominal pain, and bruising. A CT scan of his abdomen showed splenomegaly and possible infarct. His bone marrow biopsy was consistent with mantle cell lymphoma with blastoid features.

Test ordered

- Chromosome analysis of the bone marrow
- FISH: CLL panel
- NGS (send-out)
- B-cell MRD (send-out)

Laboratory test performed

Chromosome analysis and FISH methods were described previously in Chapters 1 and 12. NGS was performed externally.

Test results

Of the 20 cells examined, 16 were abnormal and clonal evolution was evident. The stemline with three cells exhibited balanced translocations involving 11q&14q and Xp&22p, and add(7q) (Fig. 20.2.1); The sideline 1 with eight cells showed a balanced translocation involving 2p&8q, an unbalanced translocation involving 7q&22p with an unknown material inserted to 7q21 in addition to the abnormalities in the stemline (Fig. 20.2.2); The sideline 2 with five cells revealed an unbalanced three-way translocation among 1q&14q&11q, and an unbalanced translocation involving 7q&22p with an unknown material inserted to 7q21 in addition to the abnormalities in the stemline (Fig. 20.2.3). The remaining four cells appear to be chromosomally normal.

FISH for CLL panel was performed on interphase nuclei using probes localized to the *CCND1* (11q13), *ATM* (11q22.3), *D12Z3* (12cen), *D13S319* (13q14.2), *LAMP1* (13q34), *IGH* (14q32.3), and *TP53* (17p13.1) gene regions. Two hundred nuclei were examined, and the results demonstrated an *IGH::CCND1* rearrangement in 178/200 (89.0%) of the cells scored. This abnormality is associated with mantle cell lymphoma (Fig. 20.2.4).

Results from NGS and B-cell MRD were all negative (data not shown).

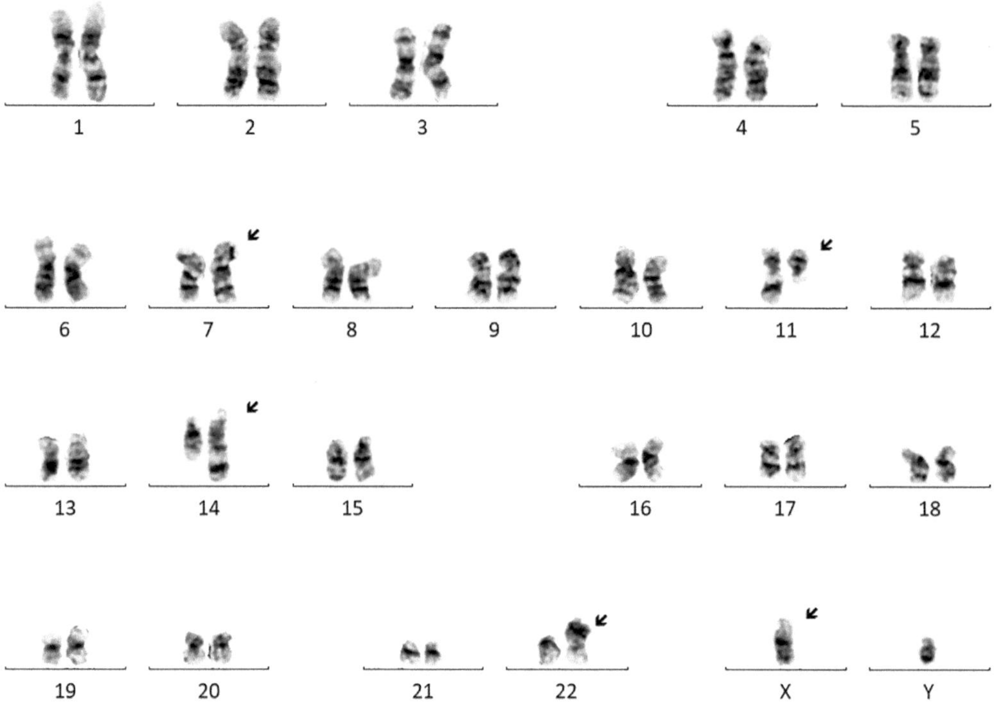

FIG. 20.2.1 Chromosome analysis of the bone marrow showed clone 1 from a complex karyotype with t(X;22), add(7q), and t(11;14). ISCN: 46,Y,t(X;22)(p11.2;p11.2),add(7)(q22),t(11;14)(q13;q32)[3]/ 46,sl,t(2;8)(p12;q24),der(22)t(7;22)(q11.2;p11.2)ins(7;?)(q21;?)[8]/ 46,sdl,der(1)t(1;14;11)(q42;q32;q13),-der(14)t(11;14),+der(14)t(1;14;11),der(22) t(7;22)(q11.2;p11.2)ins(7;?)(q21;?)[5]/46,XY[4]

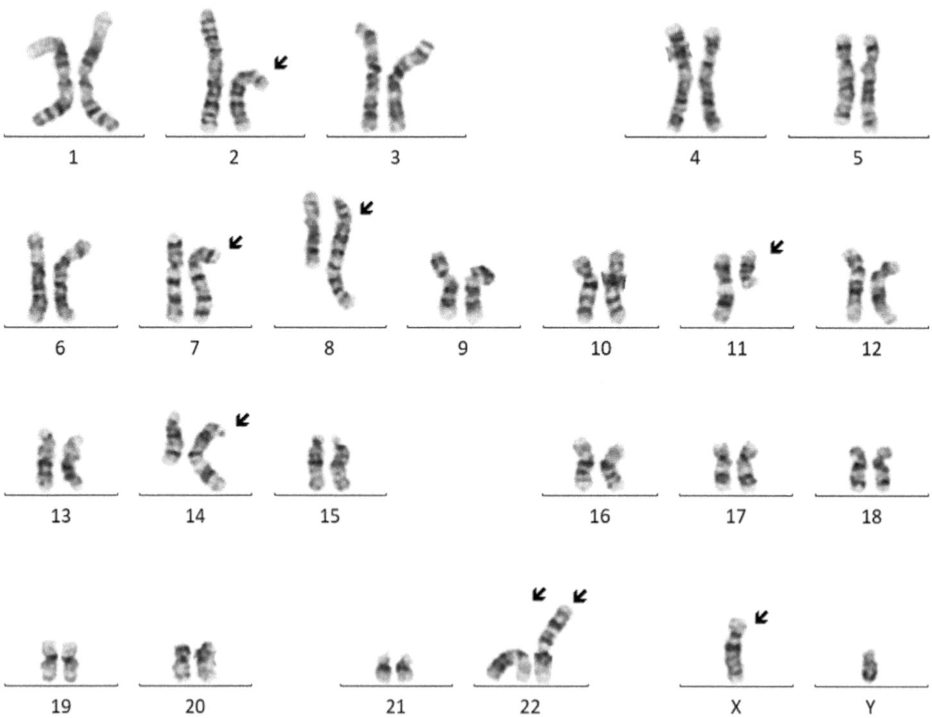

FIG. 20.2.2 Chromosome analysis of the bone marrow showed clone 2 from a complex karyotype with t(2;8), der(22)t(7;22)ins(7;?), t(X;22), add(7q), and t(11;14).

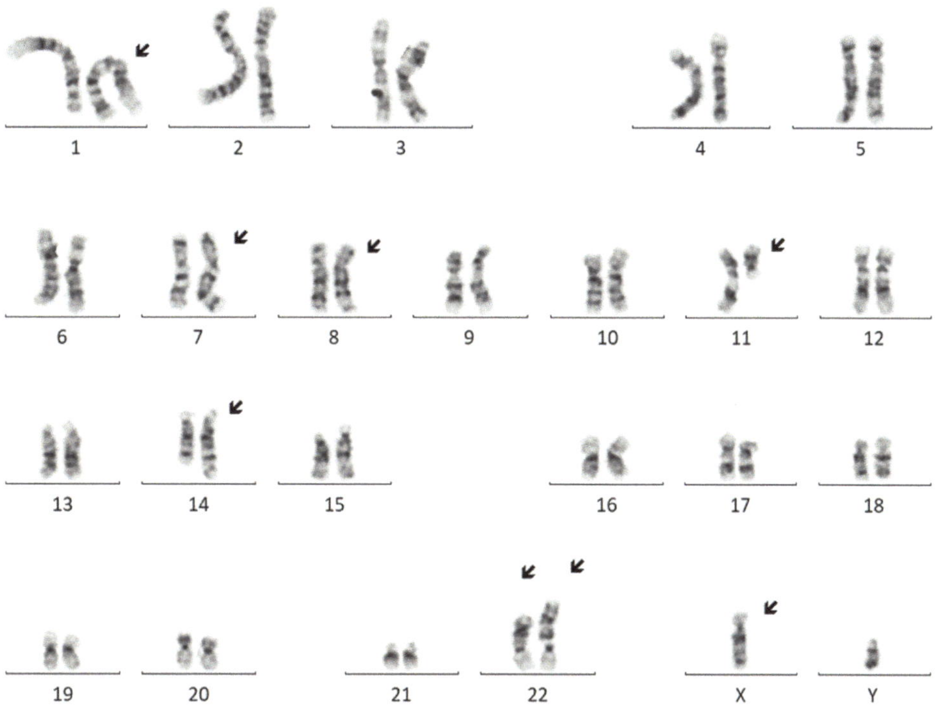

FIG. 20.2.3 Chromosome analysis of the bone marrow showed clone 3 from a complex karyotype with der(1)t(1;14;11), −der(14)t(11;14), +der(14)t(1;14;11), der(22)t(7;22)ins(7;?), t(X;22), and add(7q).

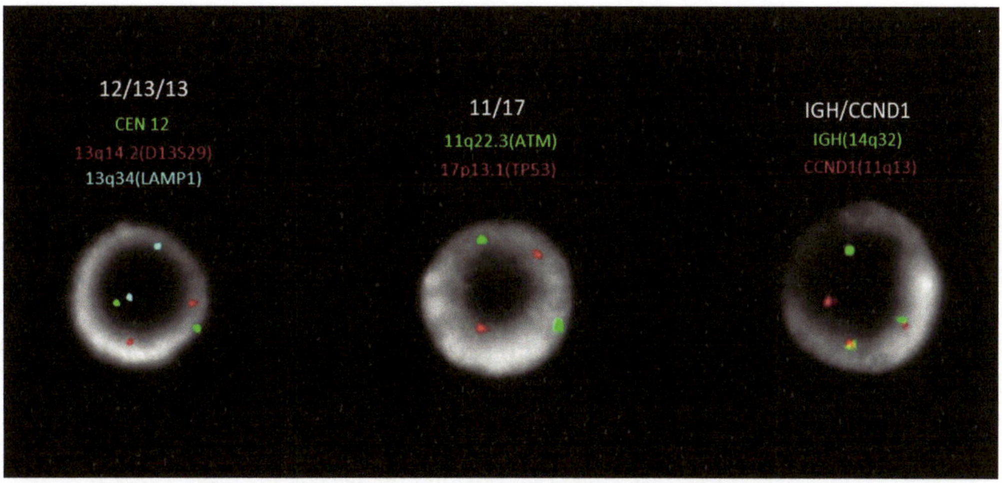

FIG. 20.2.4 FISH for the CLL panel was performed, and the results showed *IGH::CCND1* rearrangement in 89% of the cells scored. ISCN: nuc ish(CCND1,IGH)x3(CCND1 con IGHx2)[178/200],(ATM,TP53)x2[200/200],(D12Z3,D13S319,LAMP1)x2[198/200]

Results with interpretations

The *IGH::CCND1* rearrangement was seen from both the karyotyping and the concurrent FISH testing. These results are consistent with a diagnosis of mantle cell lymphoma. The degree of karyotypic complexity indicates a more aggressive disease process.

Future testing and recommendations

Regular follow up and clinicopathologic correlation are recommended. Chromosome analysis and FISH can be ordered for monitoring disease progression and treatment efficacy in the future.

Case 20.3 Small B-cell lymphoma/follicular lymphoma

Clinical indication

A 72-year-old male presented with atrial fibrillation, shallow breathing, difficulty swallowing, dry cough, face erythema with throat and head swelling, excessive postnasal drip, and slight weight loss. B-cell lymphoma was suspected.

Test ordered

- Chromosome analysis of the bone marrow
- FISH: High-grade lymphoma panel

Laboratory test performed

Chromosome analysis and FISH methods were described previously in Chapters 1 and 12.

Test results

Of the 20 cells examined, 10 exhibited balanced translocations involving 2q&3q, 14q&18q, and trisomy 8. t(14;18)(q32;q21) resulted in fusion of *IGH::BCL2* (Fig. 20.3.1). The remaining 10 cells appear to be chromosomally normal.

FISH was performed on interphase nuclei using probes localized to *IGH* (14q32.3) and *CCND1* (11q13.3), plus the 5′ and 3′ ends of *BCL6* (3q27), *IGH* (14q32.3), *BCL2* (18q21.3), and *MALT1* (18q21.3) gene regions. Two hundred nuclei were examined, and the results demonstrated multiple abnormalities (Fig. 20.3.2).

A *BCL6* rearrangement was observed in 166/200 (83.0%) of the cells scored.

An *IGH* rearrangement was observed in 150/200 (75.0%) of the cells scored.

A *BCL2* rearrangement was observed in 150/200 (75.0%) of the cells scored.

FISH was negative for *IGH::CCND1* rearrangement; however, three copies of *IGH* observed in 165/200 (82.5%) represent an *IGH* rearrangement not involving *CCND1* but *BCL2*.

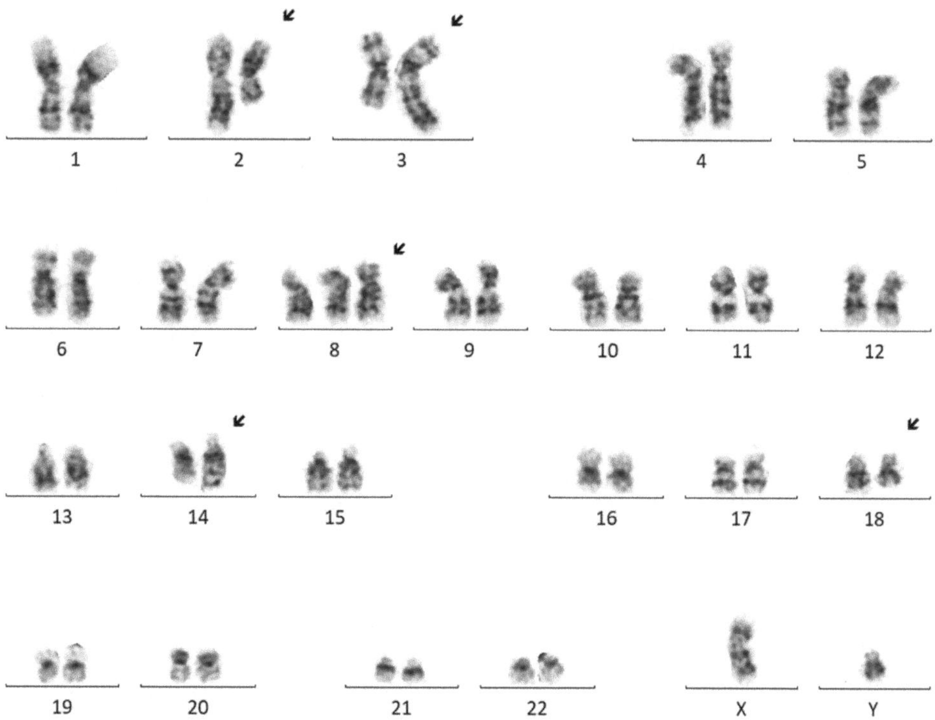

FIG. 20.3.1 Chromosome analysis of the bone marrow showed a complex karyotype with t(2;3), trisomy 8, and t(14;18). ISCN: 47,XY,t(2;3)(q22;q27),+8,t(14;18)(q21;q32)[10]/46,XY[10]

Results with interpretations

Chromosome analysis revealed balanced translocations involving 2q&3q, 14q&18q, and trisomy 8. *BCL2*, *BCL6*, and *IGH* rearrangements were also seen from the concurrent FISH testing.

These results are consistent with a diagnosis of B-cell lymphoma. t(14;18)/*IGH::BCL2* was commonly seen in follicular lymphoma. The degree of karyotypic complexity indicates a more aggressive disease process.

Future testing and recommendations

Regular follow up and clinicopathologic correlation are recommended. Chromosome analysis and FISH can be ordered for monitoring disease progression and treatment efficacy in the future.

Case 20.4 Double-hit lymphoma

Clinical indication

A 73-year-old gentleman with a known history of diffuse large B-cell lymphoma was admitted for planned chemotherapy. He had hypertension, osteoarthritis, GI bleeding,

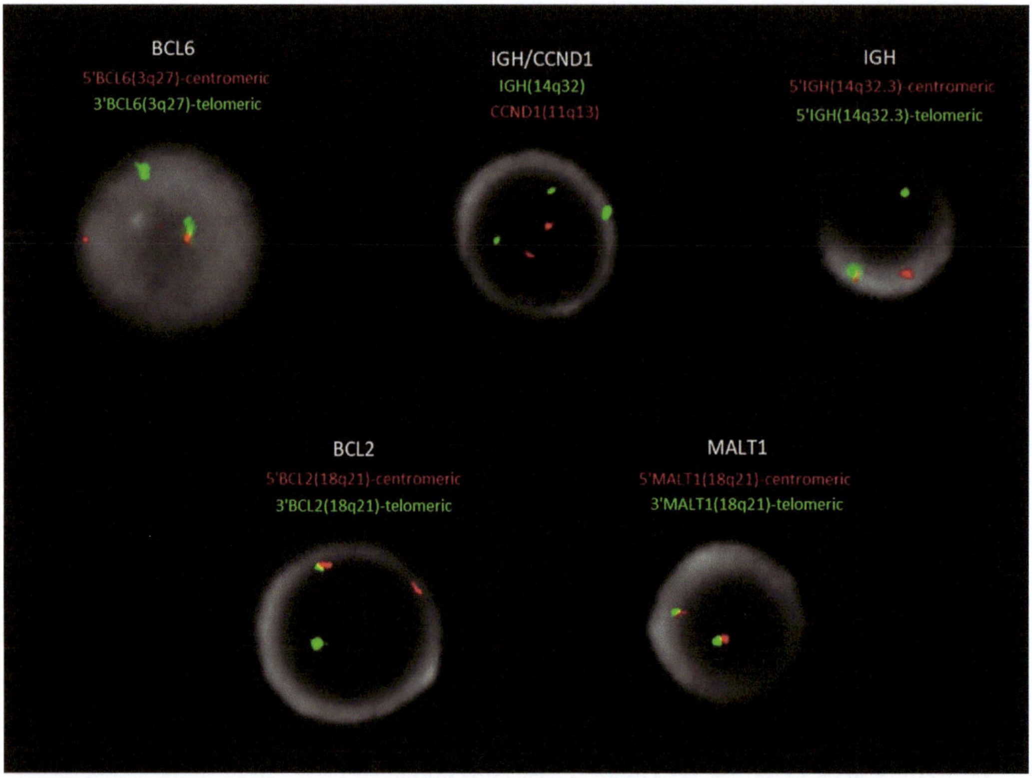

FIG. 20.3.2 FISH for low-grade lymphoma panel was performed, and the results showed *BCL2*, *BCL6*, and *IGH* rearrangements. ISCN: nuc ish (BCL6x2)(3′BCL6 sep 5′BCL6x1)[166/200],(IGHx2)(3′IGH sep 5′IGHx1)[150/200],(MALT1x2)[200/200],(BCL2x2)(5′BCL2 sep 3′BCL2x1)[150/200],(CCND1x2,IGHx3)[165/200]

peptic ulcer disease, and acute renal failure requiring hemodialysis, but was no longer dialysis-dependent.

Test ordered

- Chromosome analysis of the bone marrow
- FISH: High-grade lymphoma panel and *IGH::CCND1* rearrangement

Laboratory test performed

Chromosome analysis and FISH methods were described previously in Chapters 1 and 12.

Test results

Of the 22 cells examined, 15 exhibited trisomy 5, balanced translocations involving 8q&22q and 14q&18q, and trisomy 12 (Fig. 20.4.1). The remaining seven cells appear to be chromosomally normal.

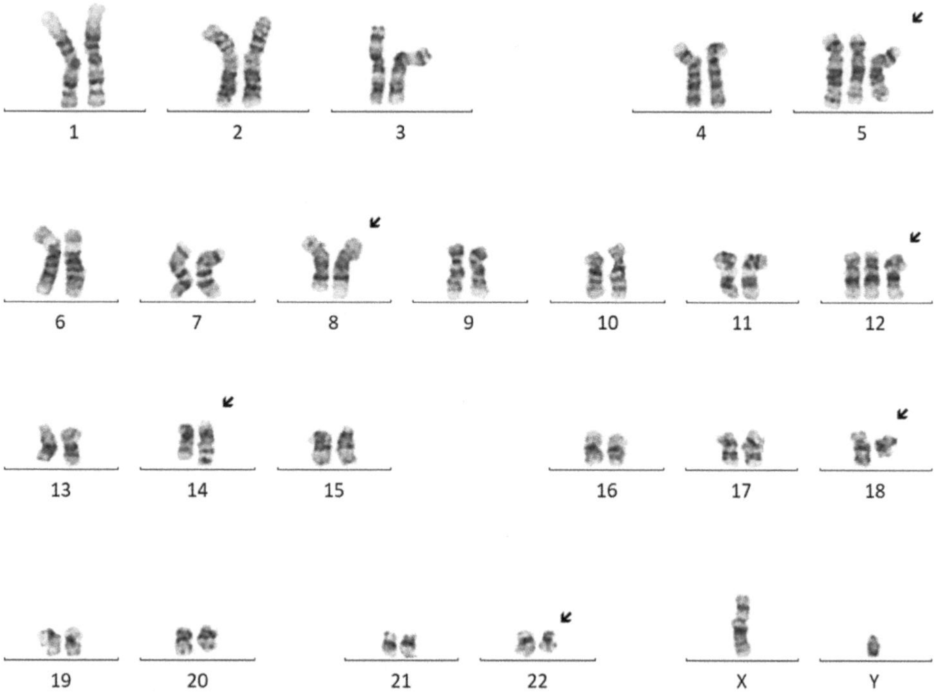

FIG. 20.4.1 Chromosome analysis of the bone marrow showed a complex karyotype with trisomy 5 and 12, t(8;22) and t(14;18). ISCN: 48,XY,+5,t(8;22)(q24.1;q11.2),+12,t(14;18)(q32;q21)[15]/46,XY[7]

FISH for high-grade lymphoma was performed on interphase nuclei using probes localized to the 5′ and 3′ ends of *BCL6* (3q27), *MYC* (8q24.21), *CCND1*(11q13), and *IGH*(14-q32.33), and *BCL2* (18q21.3) gene regions. Two hundred nuclei were examined, and the results demonstrated *MYC* rearrangements in 102/200 (51.0%) of the cells scored, *BCL2* rearrangements in 78/200 (39.0%) of the cells scored, and three copies of *IGH* in 88/200 (44.0%) of the cells scored (Fig. 20.4.2).

Results with interpretations

Chromosome analysis identified balanced translocations involving *MYC::IGL* [t(8;22)] and *BCL2::IGH* [t(14;18)]; these results were also confirmed by the concurrent FISH testing.

FISH for *IGH::CCND1* was negative for rearrangement. Three *IGH* signals are concordant with *BCL2::IGH* rearrangement. These results are consistent with a diagnosis of double-hit lymphoma. The degree of karyotypic complexity indicates a more aggressive disease process.

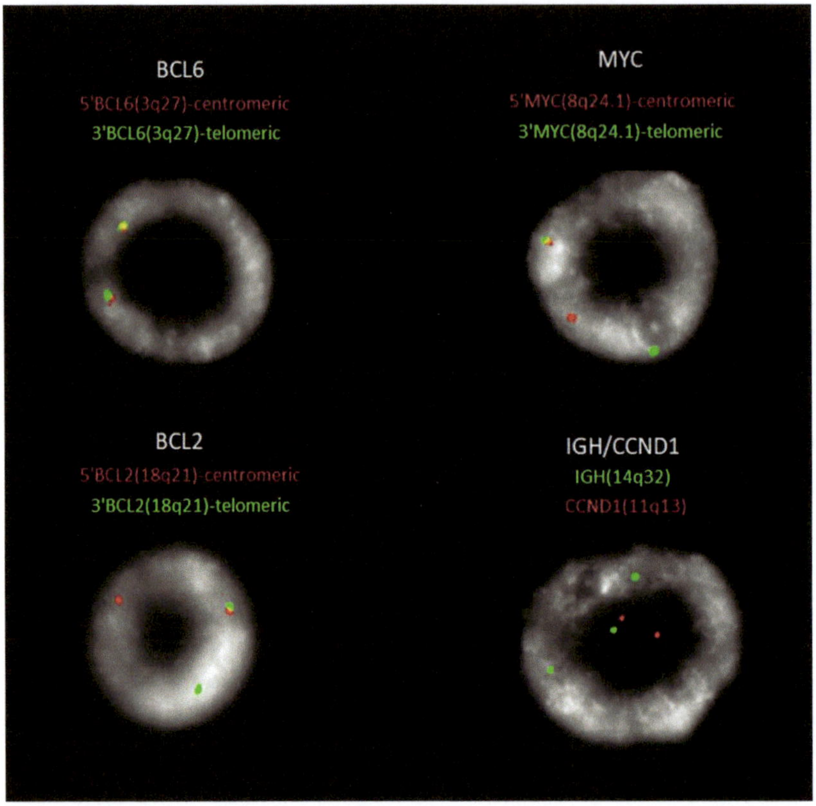

FIG. 20.4.2 FISH for high-grade lymphoma panel and *IGH::CCND1* rearrangement probes was performed, and the results showed rearrangements involving *BCL2* and *MYC*. FISH was negative for *IGH::CCND1* rearrangement. Three copies of *IGH* were seen due to *BCL2::IGH* rearrangement. ISCN: nuc ish (BCL6x2)[199/100],(MYCx2)(5'MYC sep 3'MYCx1)[102/100], (CCND1x2,IGHx3)[88/200],(BCL2x2)(3'BCL2 sep 5'BCL2x1)[78/100]

Future testing and recommendations

Regular follow up and clinicopathologic correlation are recommended. Chromosome and FISH can be ordered for monitoring disease progression and treatment efficacy in the future.

Case 20.5 Double-hit lymphoma with *BCL6* rearrangement

Clinical indication

The patient is a 70-year-old woman with a history of diabetes mellitus type 2, hypertension, and double-hit lymphoma who was recently treated with chemotherapy with post-DA-R-EPOCH cycle 1 chemotherapy. She had right pleural effusion and uncontrolled hyperglycemia.

Test ordered

- Chromosome analysis of the bone marrow
- FISH: High-grade lymphoma panel

Laboratory test performed

Chromosome analysis and FISH methods were described previously in Chapters 1 and 12.

Test results

Of the 20 cells examined, 19 exhibited multiple numerical and structural chromosome abnormalities, including t(3;8), gains of 2 copies of the derivative chromosome 8, add(9p), t(14;18), trisomy 19, del(20q), and a marker chromosome (Fig. 20.5.1). The remaining cell appears to be chromosomally normal.

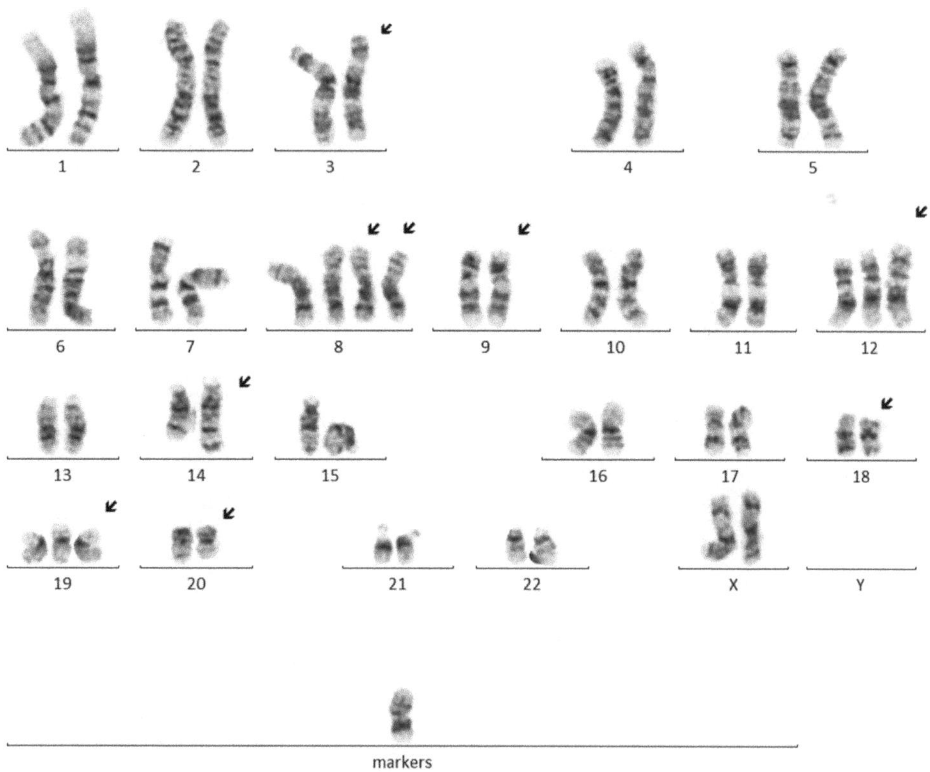

FIG. 20.5.1 Chromosome analysis of the bone marrow showed a complex karyotype with r(3;8), trisomy 8, 12, and 19, add(9p), t(14;18), and del(20q). ISCN: 51,XX,t(3;8)(q27;q24),+8,+der(8)t(3;8),add(9)(p24), +12,t(14;18)(q32;q21),+19, del(20)(q11.2q13.1),+mar[cp19]/46,XX[1]

FISH was performed on interphase nuclei using probes localized to the 5′ and 3′ ends of *BCL6* (3q27), *MYC* (8q24.21), and *BCL2* (18q21.3) gene regions. Two hundred nuclei were examined, and the results demonstrated multiple abnormalities (Fig. 20.5.2).

MYC rearrangements were observed in 169/200 (84.5%) of the cells scored.

BCL2 rearrangements were observed in 157/200 (78.5%) of the cells scored.

BCL6 rearrangements were observed in 170/200 (85.0%) of the cells scored.

Results with interpretations

Chromosome analysis revealed a complex karyotype including t(3;8)/*BCL6::MYC*, t(14;18)/*IGH::BCL2*, and gains of chromosomes 8, 12 and 19, add(9p), and del(20q). A marker chromosome was also present.

Rearrangements involving *BCL2*, *BCL6*, and *MYC* were also seen from the concurrent FISH testing. These results are consistent with a diagnosis of double-hit lymphoma. According to the fifth edition of the WHO classification of hematolymphoid tumors—lymphoid neoplasms, *BCL6* is no longer considered part of this category [1]. The degree of karyotypic complexity indicates a more aggressive disease process.

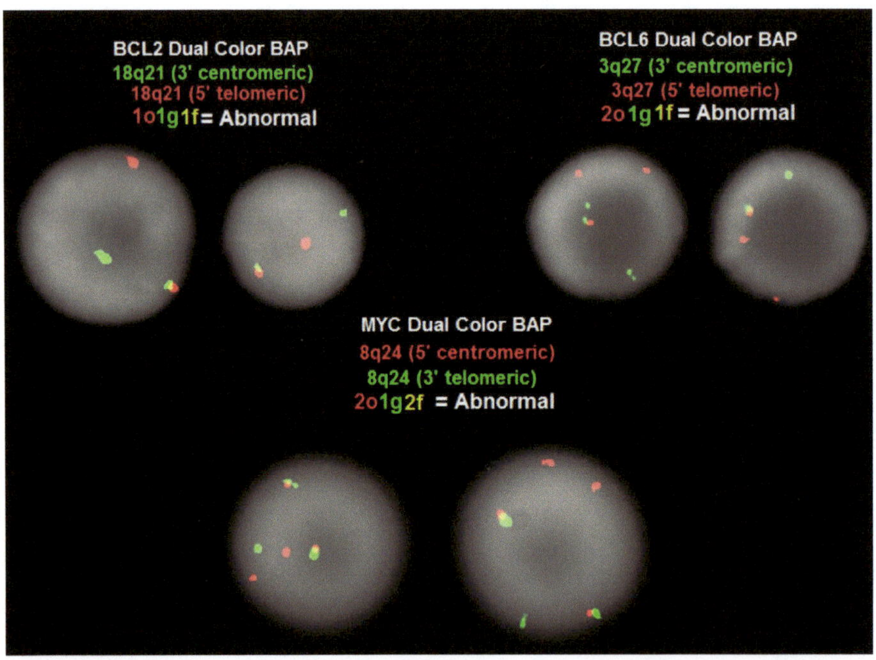

FIG. 20.5.2 FISH for a high-grade lymphoma panel was performed, and the results showed rearrangements involving *BCL2*, *BCL6*, and *MYC*. ISCN: nuc ish (3′BCL6x2,5′BCL6x3)(3′BCL6 con 5′BCL6x1)[170/200],(5′MYCx3,3′MYCx4)(5′MYC con 3′MYCx2)[169/200],(BCL2x2)(3′BCL2 sep 5′BCL2x1)[157/200]

Future testing and recommendations

Regular follow up and clinicopathologic correlation are recommended. Chromosome analysis and FISH can be ordered for monitoring disease progression and treatment efficacy in the future.

Case 20.6 Plasma cell neoplasm

Clinical indication

A 46-year-old female presented with normocytic normochromic anemia. Imaging studies (CT and MRI) showed diffuse bony lesions suspicious of malignancy. The bone marrow biopsy was performed to assess the monoclonal gammopathy. The presence of kappa-restricted plasma cells was in keeping with the diagnosis of plasma cell neoplasm. Plasma cells were at 42% and atypical.

Test ordered

- Chromosome analysis of the bone marrow
- FISH: Plasma cell neoplasm panel

Laboratory test performed

Chromosome analysis and FISH methods were described previously in Chapters 1 and 12.

Test results

Of the 21 cells examined, 1 exhibited multiple numerical and structural chromosome abnormalities, including losses of the X chromosome and chromosome 17, gains of chromosomes 3, 5, 9, 11, 13, 15, 19, and 21, a whole arm translocation involving 1q and 13q, add(6q), add(9p), and add(18p) (Fig. 20.6.1). The remaining 20 cells appear to be chromosomally normal.

FISH for plasma cell neoplasm panel was performed on interphase nuclei using probes localized to the *CDKN2C (*1p32.3), *CKS1B* (1q21.2), *IGH* (14q32.33), *D13S319* (13q14.3), *LAMP1* (13q34), *CEP17* (D17Z1), and *TP53* (17p13.1) gene regions. One hundred nuclei were examined, and the results were positive for deletion of 1p32.3 in 100/100 (100.0%) and trisomy 13 in 100/100 (100.0%) of the cells scored (Fig. 20.6.2).

Results with interpretations

Chromosome analysis identified gains of many chromosomes and other structural abnormalities, including a whole-arm translocation involving 1q&13q, add(6q), add(9p), and add(18p). del(1p) and trisomy 13 were also seen from the concurrent FISH testing. Gains of the odd number of chromosomes are usually associated with a favorable prognosis in

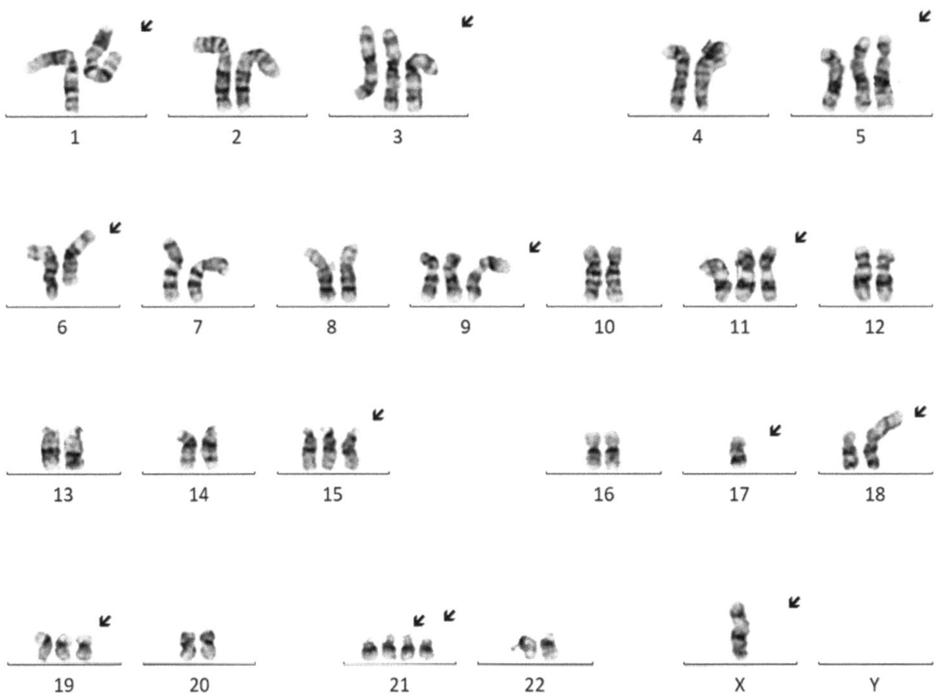

FIG. 20.6.1 Chromosome analysis of the bone marrow showed a complex karyotype with gains of many odd numbers of chromosomes and other structural chromosome abnormalities. ISCN: 52,X,-X,der(1;13)(q10;q10),+3,+5,add(6)(q23),+add(9)(p13),+11,+13,+15,-7,add(18)(p11.3), +19,+21,+21[1]/46,XX[20]

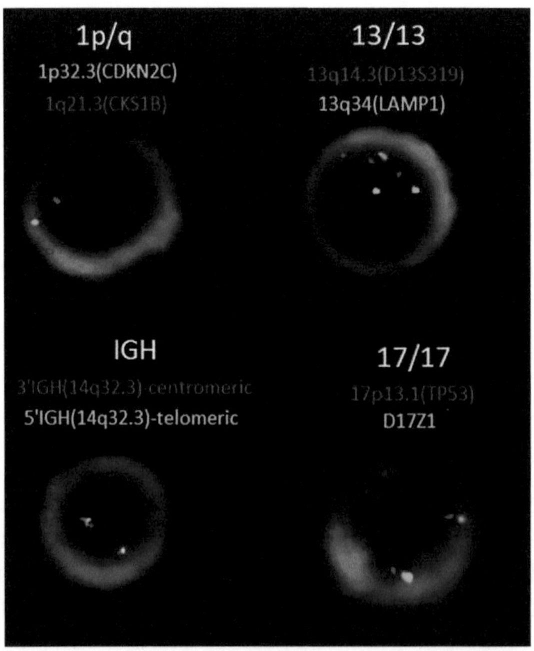

FIG. 20.6.2 FISH for plasma cell neoplasm panel was performed, and the results showed a deletion of 1p and trisomy 13. ISCN: nuc ish(CDKN2Cx1,CKS1Bx2)[100/100],(D13S319,LAMP1)x3[100/100],(IGHx2)[100/100],(TP53,D17Z1)x2[98/100]

plasma cell neoplasm; however, the degree of karyotypic complexity may indicate a more aggressive disease process.

Future testing and recommendations

Regular follow up and clinicopathologic correlation are recommended. Chromosome analysis and FISH can be ordered for monitoring disease progression and treatment efficacy in the future.

Case 20.7 ALK-positive large B-cell lymphoma

Clinical indication

A 60-year-old male presented with a diagnosis of diffuse large B-cell lymphoma. He was treated with a CHOP regimen and was in remission. CHOP regimen is a type of chemotherapy combination that is used to treat non-Hodgkin lymphoma and is being studied in the treatment of other types of cancer. It includes the drugs cyclophosphamide, doxorubicin hydrochloride (hydroxydaunorubicin), vincristine sulfate (Oncovin), and prednisone. He presented due to a new onset of anemia and suspicion of relapse. He complained of some night sweats, fatigue, and tiredness. A bone marrow biopsy was collected.

Test ordered

- Chromosome analysis of the bone marrow
- FISH: *ALK* gene rearrangement
- NGS Hematology Molecular Profile

Laboratory test performed

Chromosome analysis, FISH, and NGS methods were described previously in Chapters 1 and 12.

Test results

Firstly, chromosome analysis was performed. Of the 20 cells examined, 5 exhibited a very complex karyotype with multiple numerical and structural chromosome abnormalities including t(2;17)(p23;q23) in a tetraploid background. Additional abnormalities include losses of the Y chromosome and chromosomes 1, 10, 11, 12, 13, 17, and 18, a gain of chromosome 22, a whole-arm translocation involving Xq&7q; add(1q), add(4p), add(4q), add(6q), add(7p), add(9p), add(10p), add(14q), dup(2p), del(2q), del(8p), der(2)t(2;17), a pseudo dicentric chromosome involving the short arms of two copies of the chromosome 3, dup(5q), an isochromosome 5 for the short arm, unbalanced translocations involving 1q&14q, 7p&15q, 1q&18q, 8q&18p, a balanced translocation involving 19q&21p, and a marker chromosome (Fig. 20.7.1). The remaining 15 cells appear to be chromosomally normal.

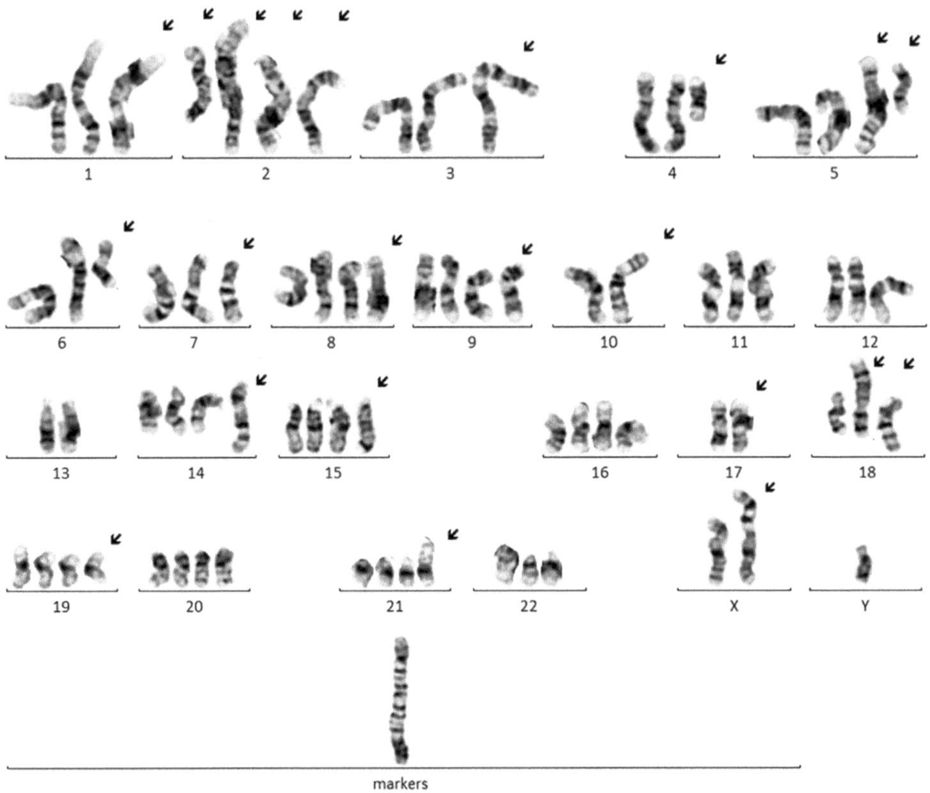

FIG. 20.7.1 Chromosome analysis of the bone marrow showed a complex karyotype with t(2;17) and many additional numerical and structural chromosome abnormalities. ISCN: 77~80<4N>,XY,-Y,der(X;7)(q10;q10),-1,add(1)(q21),dup(2)(p23p21),del(2)(q31), t(2;17)(p23;q23),der(2)t(2;17),psu dic(3;3)(p26;p12),add(4)(p11),add(4)(q21),dup(5)(q33q35),i(5)(p10),add(6)(q12),add(7)(p12),del(8)(p12),add(9)(p21),-10,-10,add(10) (p11.1),-11,-12,-13,add(14)(q32),der(14)t(1;14)(q21;q22),der(15)t(7;15)(p12;q22), -17,-17,-18,der(18)t(1;18)(q25;q21),der(18)t(8;18)(q11.2;p11.3),t(19;21)(q13.2;p13),+22, +mar[cp5]/46,XY[15]

FISH was performed on interphase nuclei using probes localized to the 5′ and 3′ ends of the *ALK* (2p23) gene region. Two hundred nuclei were examined, and the results were positive for a rearrangement involving *ALK* in 25/200 (12.5%) of the cells scored (Fig. 20.7.2).

NGS was performed, and the results were negative for DNA mutations and RNA fusions (data now shown).

Results with interpretations

Chromosome analysis revealed a highly complex karyotype with t(2;17). Along with the concurrent FISH results showing positive rearrangement for *ALK* break-apart probe, this is consistent with a diagnosis of a relapsed anaplastic lymphoma kinase-positive large B-cell lymphoma (*ALK*+ LBCL). *ALK*+ LBCL is a rare hematopoietic neoplasm that accounts for less than 1% of diffuse large B-cell lymphomas (DLBCL). The majority of

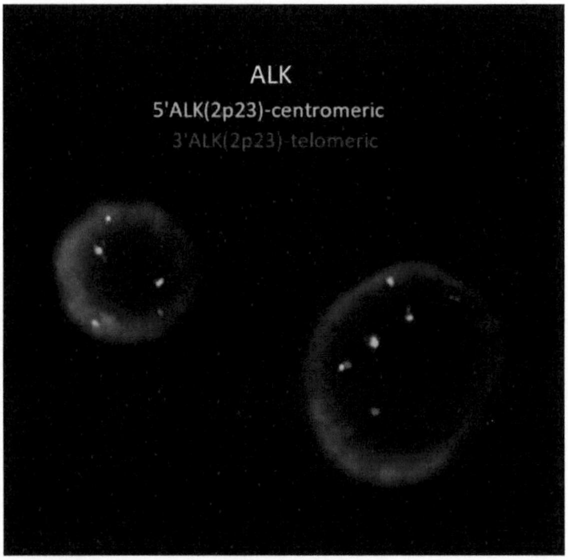

FIG. 20.7.2 FISH for *ALK* break-apart probe was performed, and the results showed *ALK* rearrangement in 12.5% of the cells scored. ISCN: nuc ish(ALKx4)(5'ALK sep 3'ALKx2)[25/200]

ALK+ LBCL harbor the t(2;17)(p23;q23) with a fusion of the clathrin gene (*CLTC*) on chromosome 17q23 with the *ALK* gene on chromosome 2p23 [14]. The degree of karyotypic complexity indicates a more aggressive disease process.

Future testing and recommendations

Regular follow up and clinicopathologic correlation are recommended. Chromosome analysis and FISH can be ordered for monitoring disease progression and treatment efficacy in the future.

Case 20.8 Burkitt lymphoma (BL)

Clinical indication

A 3-year-old boy presented with a mild cough, low-grade fever, and cervical lymphadenopathy; his appetite decreased, and he appeared tired. CBC showed leukocytosis, mild anemia, and abnormal circulating cells. Peripheral blood flow cytometry revealed a population of lambda light-chain-restricted CD10+ B-cells. The patient has undergone a cervical lymph node in addition to bone marrow sampling.

Test ordered

- Chromosome analysis of the bone marrow
- FISH: *MYC, BCL2, BCL6, ABL1, ABL2,* and *PDGFRB* gene rearrangement

- FISH: *IGH::MYC* rearrangement
- FISH: ALL panel

Laboratory test performed

Chromosome analysis, FISH, and NGS methods were described previously in Chapters 1 and 12.

Test results

Chromosome analysis was performed and identified a complex karyotype. Of the 21 cells examined, 17 were abnormal and clonal evolution was evident. In clone 1 (four cells), all abnormal cells exhibited del(3q) and a balanced translocation involving 8q&14q (Fig. 20.8.1). Clone 2 to clone 5 had additional abnormalities besides those seen in clone 1, and these abnormalities are associated with jumping translocations involving chromosome 1q except for clone 3.

In detail, these additional changes are listed below:

Clone 2 with five cells showed a dup(1q) (Fig. 20.8.2).
Clone 3 with three cells had trisomy 1 with add(1p) (Fig. 20.8.3).

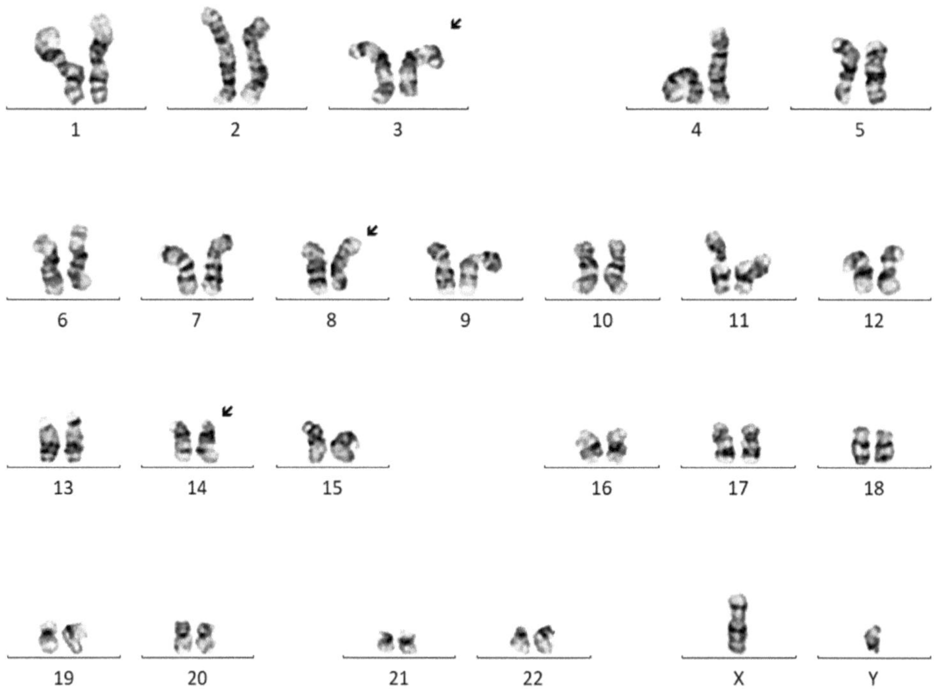

FIG. 20.8.1 Chromosome analysis of the bone marrow showed clone 1 with t(8;14) and del(3q). ISCN: 46,XY,del(3)(q25q26.2),t(8;14)(q24.2;q32)[4]/ 46,idem,dup(1)(q21q32)[5]/ 47,idem,+add(1)(p13)[3]/ 46,idem,der(6)t(1;6)(q21;q27)[3]/ 46,idem,der(13)t(1;13)(q21;q34)[2]/ 46,XY[4]

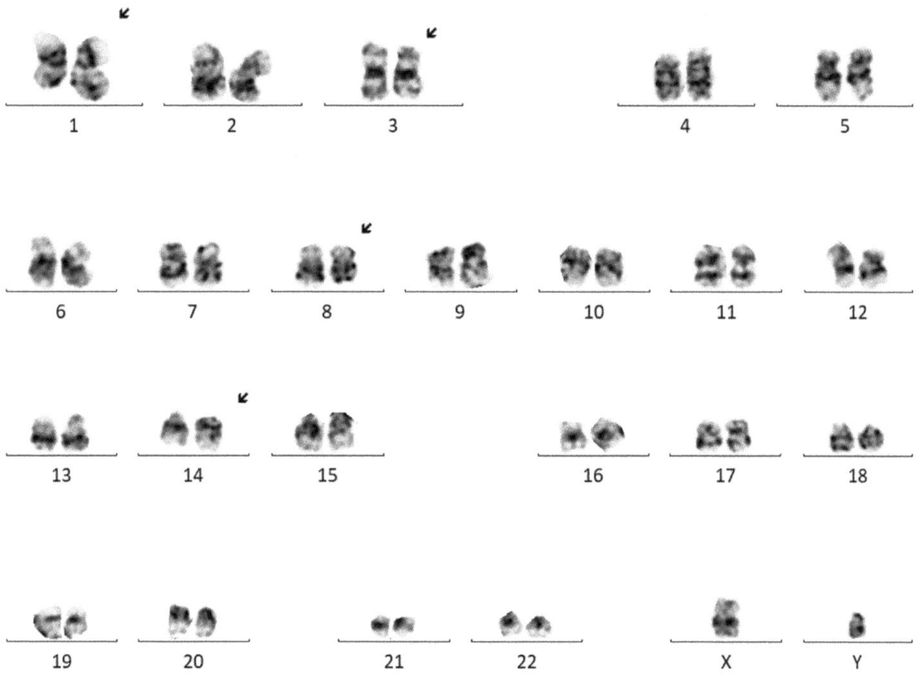

FIG. 20.8.2 Chromosome analysis of the bone marrow showed clone 2 with t(8;14), del(3q), and dup(1q).

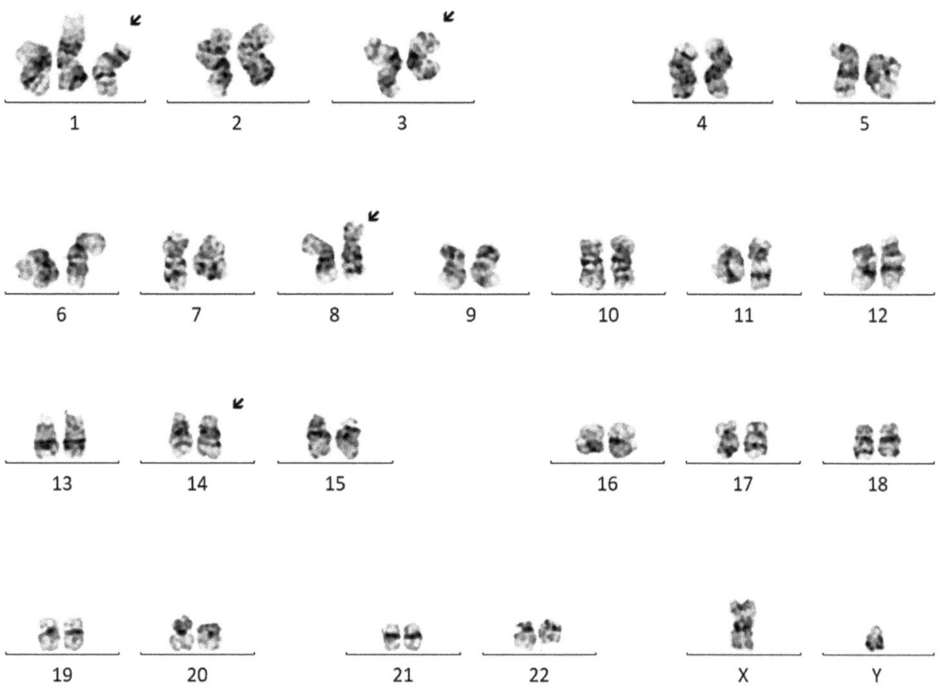

FIG. 20.8.3 Chromosome analysis of the bone marrow showed clone 3 with t(8;14), del(3q), and gain of chromosome 1 with add(1p).

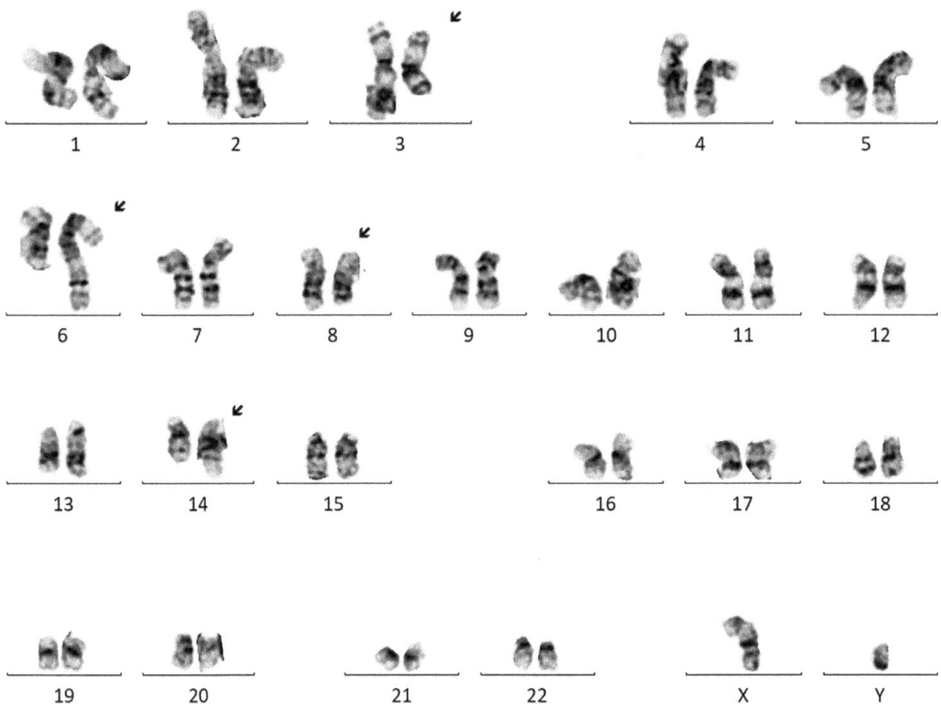

FIG. 20.8.4 Chromosome analysis of the bone marrow showed clone 4 with t(8;14), del(3q), and an unbalanced translocation involving 1q&6q.

Clone 4 with three cells revealed an unbalanced translocation involving 1q&6q (Fig. 20.8.4).

Clone 5 with two cells had an unbalanced translocation involving 1q&13q (Fig. 20.8.5).

The remaining four cells appear to be chromosomally normal.

FISH was performed on interphase nuclei using probes localized to the 5′ and 3′ ends of the *MYC* (8q24), *BCL2* (18q21.3), *BCL6* (3q27), *ABL1* (9q34.12), and *PDGFRB* (5q33.1) gene region. Two hundred nuclei were examined, and the results were positive for a rearrangement involving *MYC* in 181/200 (90.5%) of the cells scored (Fig. 20.8.6).

FISH was also performed on interphase nuclei using probes localized to the *IGH* (14q32.33), *MYC* (8q24), and *D8Z2* (8cen) gene regions. Two hundred nuclei were examined, and the results were positive for *IGH::MYC* rearrangement in 191/200 (95.5%) of the cells scored (Fig. 20.8.7).

Finally, FISH for ALL panel was performed on interphase nuclei using probes localized to the chromosome 4, 10, and 17 centromeric regions; *ABL1* (9q34.12), *ABL2* (1q25), *BCR* (22q11), *KMT2A* (11q23), *ETV6* (12p13), and *RUNX1* (21q22.3) gene regions. Two hundred nuclei were examined, and the results were positive for a gain of *ABL2* in 131/200 (65.5%) of the cells scored (Fig. 20.8.8).

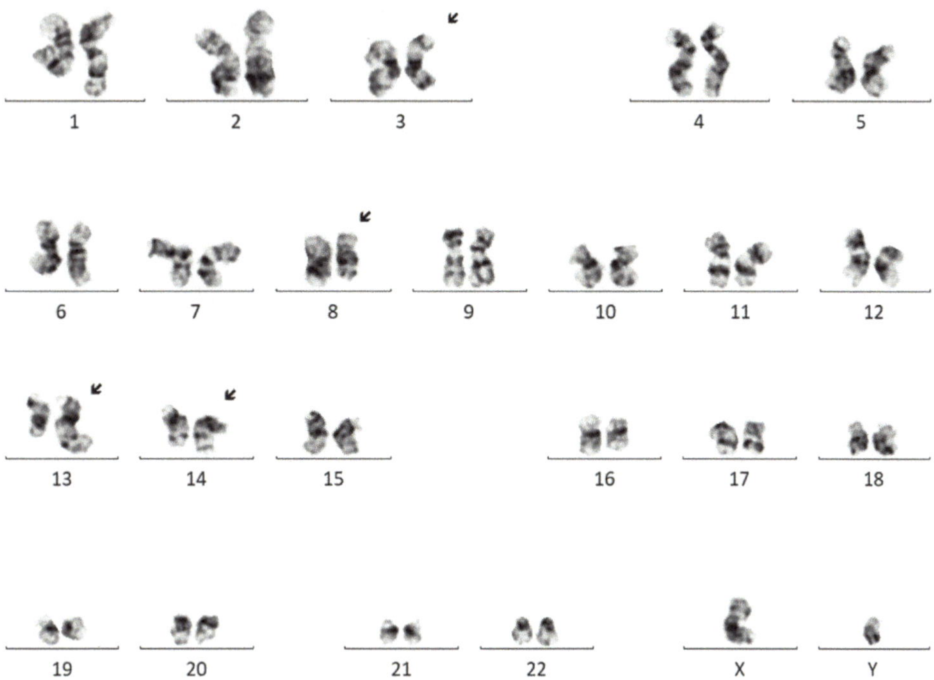

FIG. 20.8.5 Chromosome analysis of the bone marrow showed clone 5 with t(8;14), del(3q), and an unbalanced translocation involving 1q&13q.

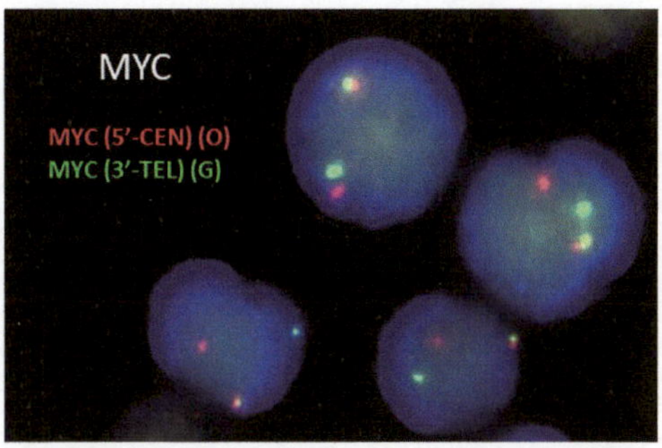

FIG. 20.8.6 FISH with *MYC* break-apart probe showed *MYC* rearrangement in 181/200 (90.5%) of the cells scored. ISCN: nuc ish (MYCx2)(5'MYC sep 3'MYCx1)[181/200]

Results with interpretations

Chromosome analysis revealed a complex karyotype with clonal evolution observed. *MYC::IGH* rearrangement was also seen from the concurrent FISH testing. These results are consistent with a diagnosis of Burkitt lymphoma. The degree of karyotypic complexity indicates a more aggressive disease process.

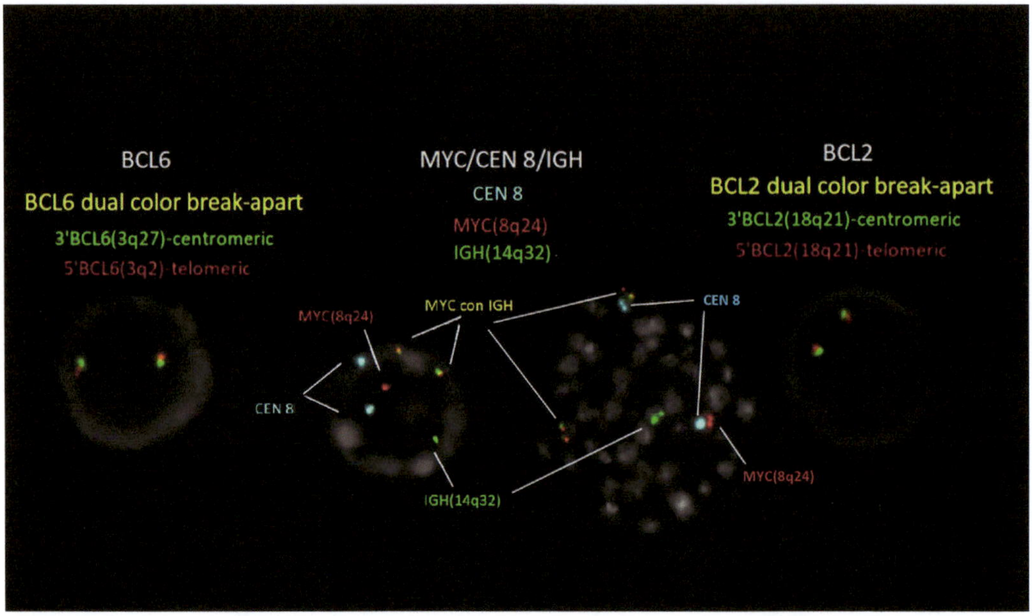

FIG. 20.8.7 FISH with *MYC::IGH*, *BCL2*, and *BCL6* gene rearrangement probes showed *MYC::IGH* rearrangement in 191/200 (95.5%) of the cells scored. ISCN: nuc ish(*MYC*,*IGH*)x3(MYC con IGHx2)[191/200],(BCL2x2)[200/200],(BCL6x2)[200/200]

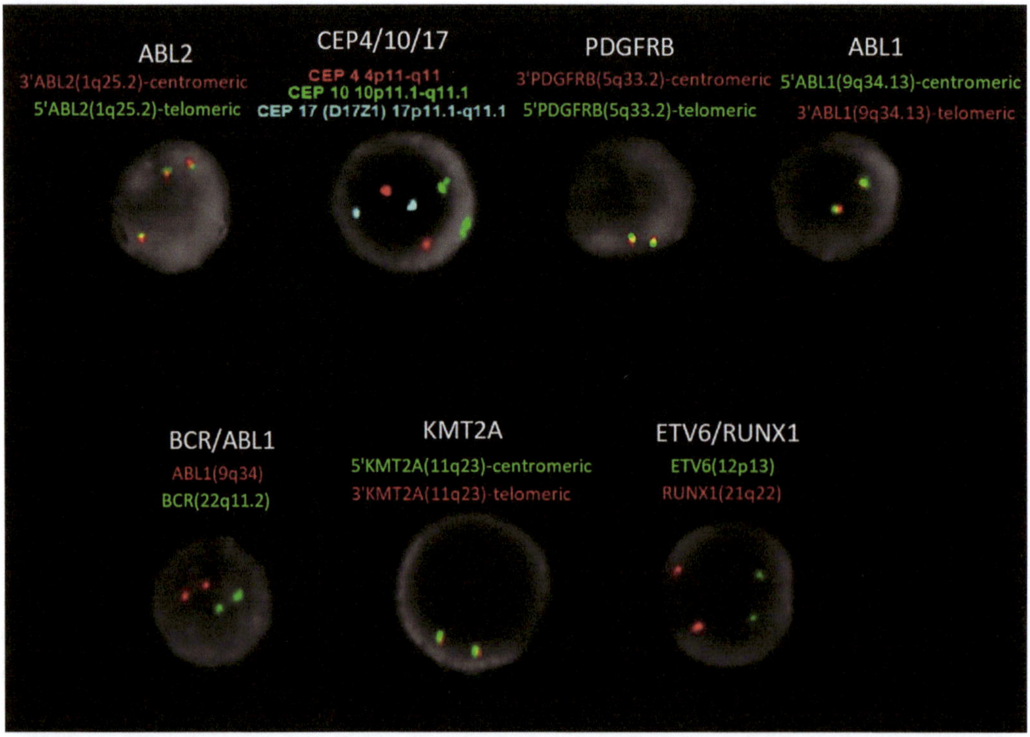

FIG. 20.8.8 FISH with ALL panel, *ABL1*, *ABL2*, and *PDFGRB* probes showed gain of *ABL2* in 131/200 (65.5%) of the cells scored, and negative results for the rest of the probes. ISCN: nuc ish(CEP4,CEP10,CEP17)x2[200/200],(ABL1,BCR)x2[200/200], (KMT2Ax2)[200/200],(ETV6,RUNX1)x2[200/200]

Future testing and recommendations

Regular follow up and clinicopathologic correlation are recommended. Chromosome analysis and FISH can be ordered for monitoring disease progression and treatment efficacy in the future.

Case 20.9 High-grade B-cell lymphoma with t(8;22)(q24;q11)/*IGL::MYC* fusion and *JAK2* rearrangement

Clinical indication

A 70-year-old male presented with acute renal failure with a serum creatinine of 2.08. The hospital course has been complicated by progressive thrombocytopenia for which hematology was consulted. Bone marrow biopsy results showed aggressive B-cell lymphoma.

Test ordered

- Chromosome analysis of the bone marrow
- FISH: High-grade lymphoma panel including *MYC*, *BCL2*, and *BCL6* rearrangements
- FISH: *IGH::MYC* rearrangement
- FISH: *JAK2* rearrangement

Laboratory test performed

Chromosome analysis and FISH methods were described previously in Chapters 1 and 12.

Test results

Of the 20 cells examined, 18 were abnormal, and clonal evolution was evident.

Clone 1 (stemline) with three cells showed XXY and t(8;9)(p22;p24)/*PCM1::JAK2* (Fig. 20.9.1); additional abnormalities seen in the following clones.

Clone 2 (sideline 1) with two cells exhibited t(8;9) and t(8;22) (Fig. 20.9.2).

Clone 3 (sideline 2) with eight cells revealed dup(1q), and der(9)t(8;9)(Fig. 20.9.3).

Clone 4 (sideline 3) with three cells showed trisomy 1 and an isochromosome 1 for the long arm (Fig. 20.9.4).

Clone 5 (sideline 4) with two cells had losses of multiple chromosomes and a double clone of sideline 1 (Fig. 20.9.5).

The remaining two cells appear to be 47,XXY (Fig. 20.9.6).

All cells analyzed showed XXY. It is unclear whether this is an acquired or constitutional abnormality. A peripheral blood sample for constitutional chromosome analysis is recommended to clarify this finding.

FISH was performed on interphase nuclei using probes localized to the 5′ and 3′ ends of *BCL6* (3q27), *MYC* (8q24.21), *BCL2* (18q21.3), and *JAK2* (9p24) gene regions. Two hundred nuclei were examined, and the results demonstrated an *MYC* rearrangement in 54/200

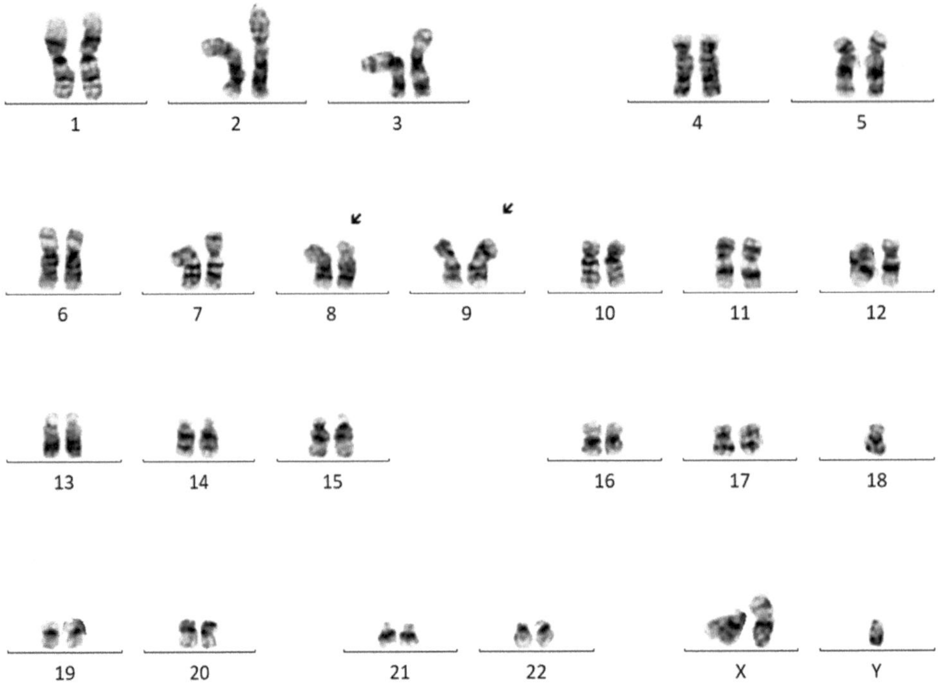

FIG. 20.9.1 Chromosome analysis revealed a complex karyotype with abnormal clone 1 including t(8;9) and XXY sex chromosome complement. ISCN: 47,XXY?c,t(8;9)(p21;p24)[3]/ 47,sl,-t(8;9),+der(8)t(8;9)(p21;p24)t(8;22)(q24.2;q11.2), +der(9)t(8;9),der(22)t(8;22)[2]/ 47,sdl1,dup(1)(q11q42),der(9)t(8;9)[8]/ 47~48,sdl2,-dup(1),+1,+i(1)(q10)[cp3]/ 75, sdlx2,-X,-2,-3,-4,-5,-8,-9,-11,-12,-13,-13,-14,-15,-16,-16,-18,-18,-19,-21[2]/ 47,XXY?c[2]

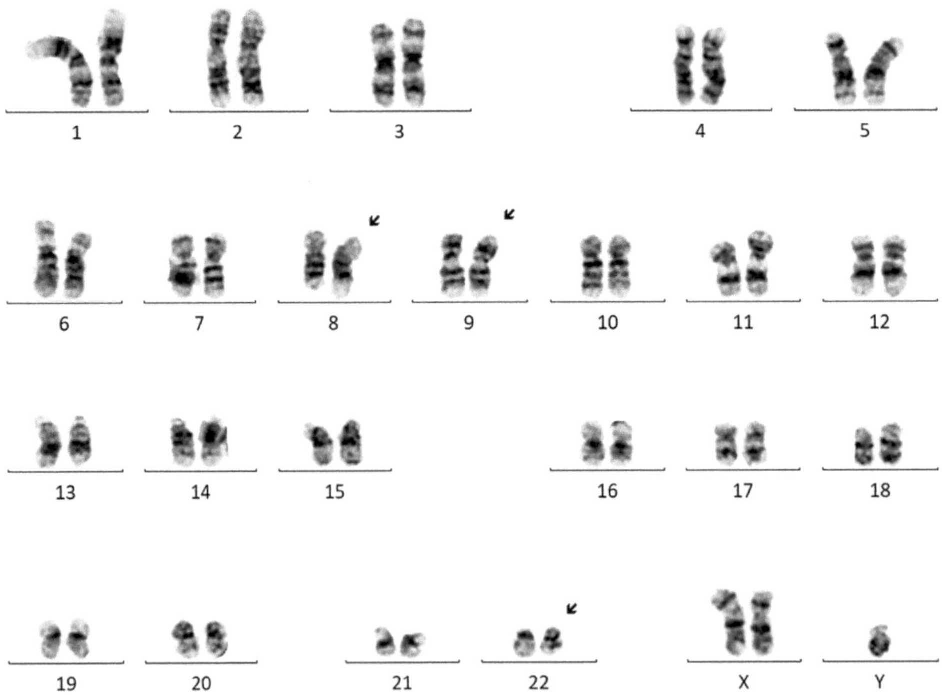

FIG. 20.9.2 Chromosome analysis revealed a complex karyotype with abnormal clone 2 with der(8)t(8;22) in addition to the abnormalities seen in clone 1.

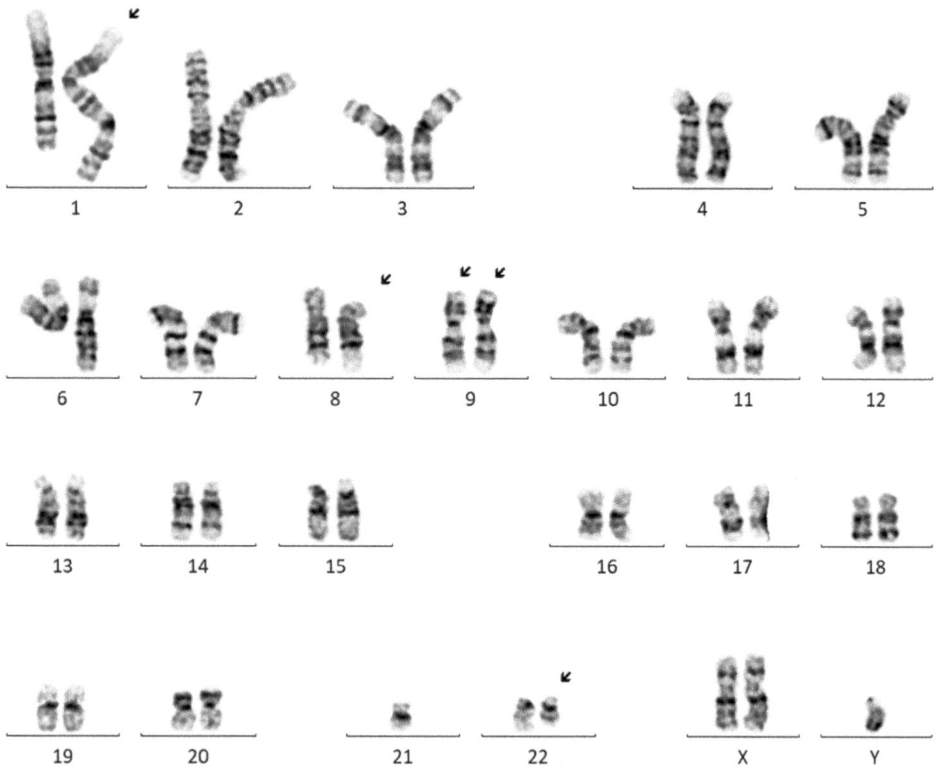

FIG. 20.9.3 Chromosome analysis revealed a complex karyotype with abnormal clone 3 including dup(1q), and der(9) in addition to the abnormalities seen in clone 2.

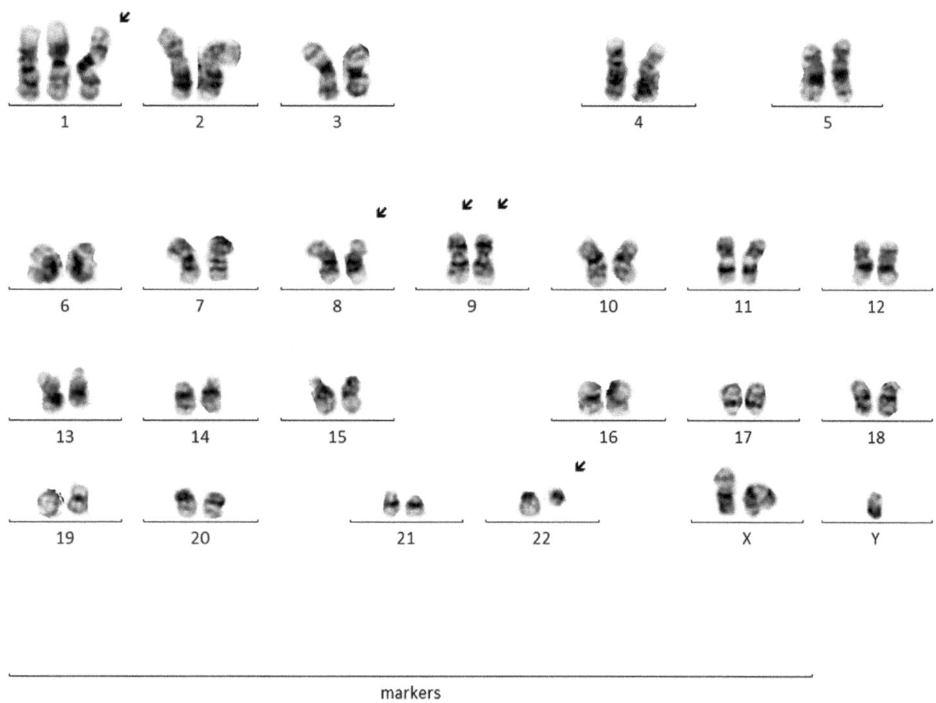

FIG. 20.9.4 Chromosome analysis showed a complex karyotype with abnormal clone 4 including trisomy 1, an isochromosome 1q in addition to the abnormalities seen in clone 2.

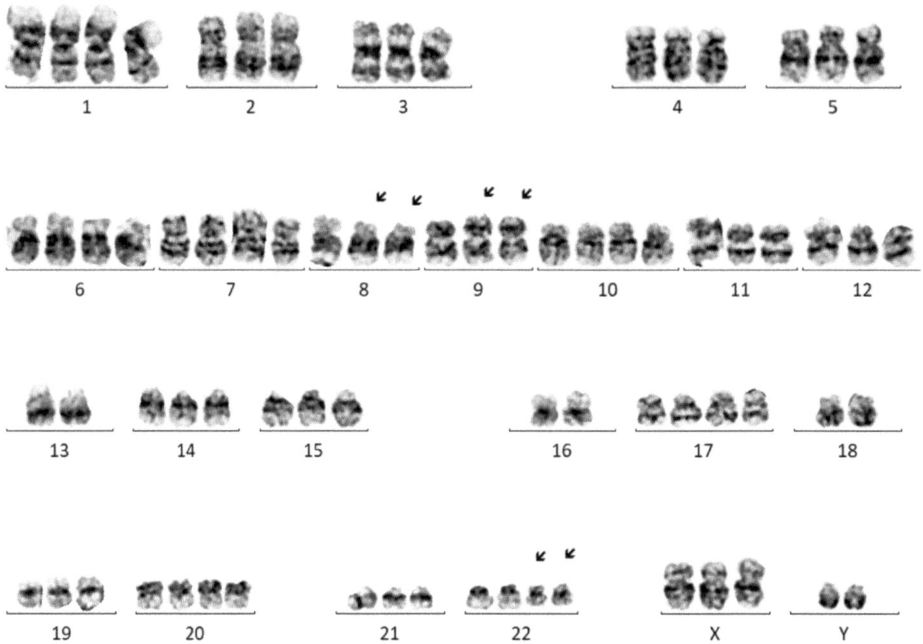

FIG. 20.9.5 Chromosome analysis revealed a complex karyotype with abnormal clone 5, which was a double clone of clone 2 with losses of multiple chromosomes.

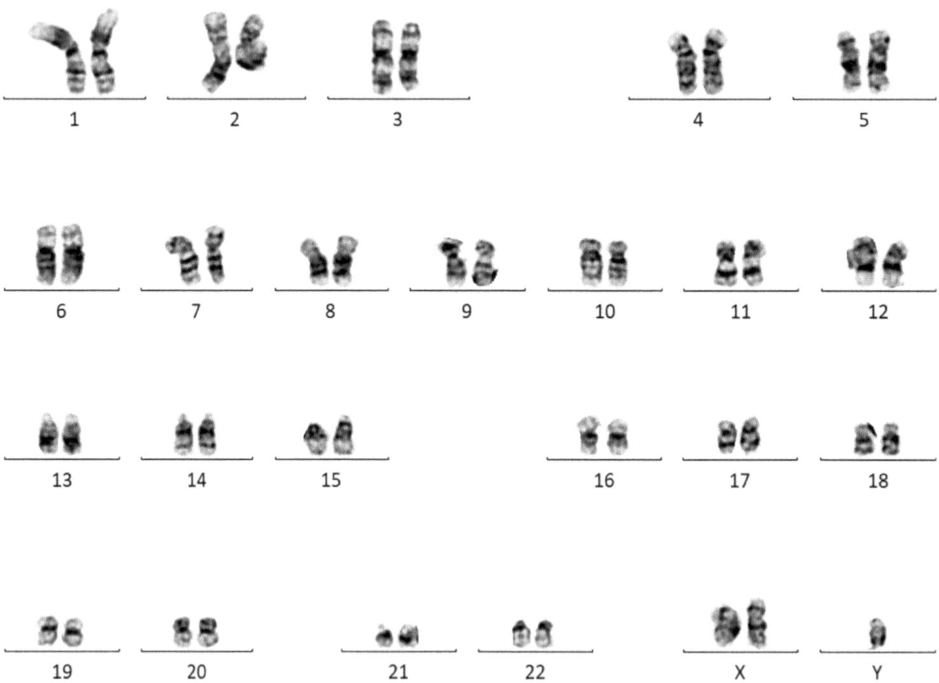

FIG. 20.9.6 Chromosome analysis revealed a complex karyotype with abnormal clone 6 including XXY sex chromosome complement.

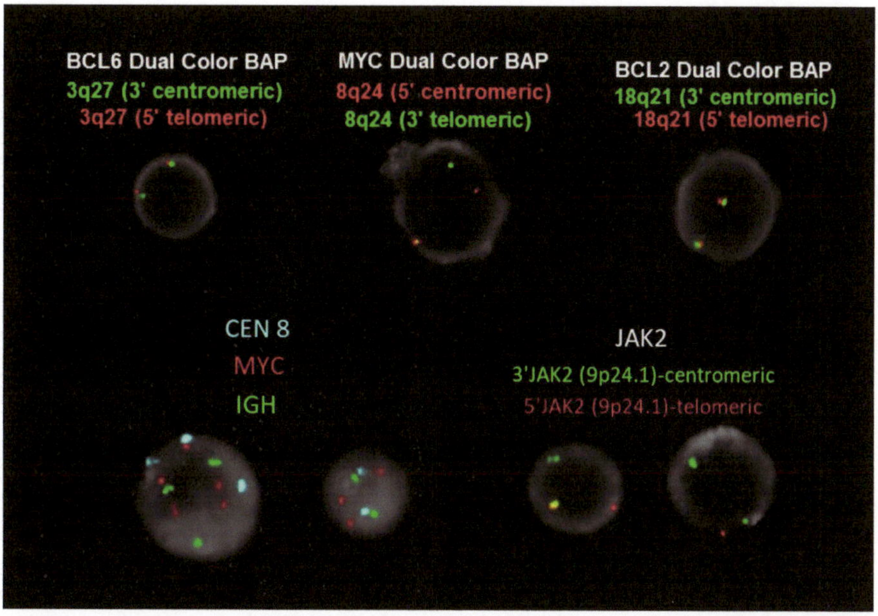

FIG. 20.9.7 FISH for high-grade lymphoma detected *MYC* rearrangement in 27% of the cells scored. *IGH::MYC* probes showed negative results for rearrangement with gains of *D8Z2 (aqua)*, *MYC (orange)*, and *IGH (green)* detected. ISCN for high-grade lymphoma: nuc ish (BCL6x2)[200/200],(MYCx2)(5′MYC sep 3′MYCx1)[42/200]/(MYCx3)(5′MYC sep 3′MYCx2)[12/200],(BCL2x2)[200/200]; ISCN for *IGH::MYC* translocation: nuc ish (D8Z2x2,MYCx3,IGHx2)[42/200]/(D8Z2x3,MYCx5,IGHx3)[16/200]

(27.0%) of the the cells scored and *JAK2* rearrangement in 120/200 (60%) of the the cells scored (Fig. 20.9.7).

FISH was also performed on interphase nuclei using probes localized to the *IGH* (14q32.33), *MYC* (8q24), and *D8Z2* (8cen) gene regions (Abbott, Vysis probes). Two hundred nuclei were examined, and the results demonstrated copy number gains of multiple gene regions with no evidence for an *IGH::MYC* rearrangement. Three copies of *MYC* were detected in 42/200 (21.0%) of the the cells scored. Five copies of *MYC* and three copies of *IGH* and *D8Z2* were detected in 16/200 (8.0%) of the the cells scored (Fig. 20.9.7).

Results with interpretations

Chromosome analysis detected a very complex karyotype with six related abnormal clones. t(8;22)(q24;q11) *IGL::MYC* is commonly seen in B-cell lymphoma including B-ALL, Burkitt lymphoma, and double-hit diffuse large B-cell lymphoma. This fusion results in the constitutive expression of *MYC* and induces proliferation even in the absence of growth factors (https://atlasgeneticsoncology.org/chromosome-explorer/8?cookies=1). In this case, *MYC* rearrangement was also seen from the concurrent FISH testing. These results are consistent with a diagnosis of high-grade B-cell lymphoma.

Interestingly, t(8;9) involving *PCM1::JAK2* was confirmed by FISH, indicating that the patient also has myeloid neoplasms with eosinophilia [15]. These eosinophilia cells were observed in this case. *JAK2* fusion detected exhibits a more aggressive course and variable sensitivity to current TK inhibitors, and in most cases, long-term disease-free survival may only be achievable with allogeneic hematopoietic stem cell transplantation [15]. Overall, the degree of karyotypic complexity indicates a more aggressive disease process. Clinicopathologic correlation is advised.

Future testing and recommendations

Regular follow up and clinicopathologic correlation are recommended. Chromosome analysis and FISH can be ordered for monitoring disease progression and treatment efficacy in the future.

Summary of key learning points

- Mature B-cell neoplasms have many different subtypes of diseases.
- Classification of the disorder is complex. Genetic testing and pathology evaluations from tissue, bone marrow, and blood are both important aspects of an accurate classification.
- FISH, karyotyping, and NGS can identify mutations, gain/loss of chromosomes, and gene fusions that will assist physicians in obtaining valuable information on diagnosis and prognosis.

References

[1] R. Alaggio, et al., The 5th edition of the World Health Organization classification of haematolymphoid tumours: lymphoid neoplasms, Leukemia 36 (7) (2022) 1720–1748.

[2] M. Lumish, et al., How we treat mature B-cell neoplasms (indolent B-cell lymphomas), J. Hematol. Oncol. 14 (1) (2021) 5.

[3] A. Forero-Torres, et al., Parsaclisib, a potent and highly selective PI3Kdelta inhibitor, in patients with relapsed or refractory B-cell malignancies, Blood 133 (16) (2019) 1742–1752.

[4] A. Noy, et al., Targeting Bruton tyrosine kinase with ibrutinib in relapsed/refractory marginal zone lymphoma, Blood 129 (16) (2017) 2224–2232.

[5] M.A. Dimopoulos, et al., Phase 3 trial of ibrutinib plus rituximab in Waldenstrom's macroglobulinemia, N. Engl. J. Med. 378 (25) (2018) 2399–2410.

[6] A.M. Dubuc, et al., FISHing in the dark: how the combination of FISH and conventional karyotyping improves the diagnostic yield in CpG-stimulated chronic lymphocytic leukemia, Am. J. Hematol. 91 (10) (2016) 978–983.

[7] C.S. Rabkin, et al., t(14;18) translocations and risk of follicular lymphoma, J. Natl. Cancer Inst. Monogr. 39 (2008) 48–51.

[8] H.W. Zhang, et al., Clinical impact of t(14;18) in diffuse large B-cell lymphoma, Chin. J. Cancer Res. 23 (2) (2011) 160–164.

[9] H.W. Zhang, et al., Clinical characteristics and prognosis of concurrent positive t(14; 18) and myc gene rearrangement in diffuse large B cell lymphoma, Zhonghua Zhong Liu Za Zhi 38 (3) (2016) 206–210.

[10] J.Y. Li, et al., Detection of translocation t(11;14)(q13;q32) in mantle cell lymphoma by fluorescence in situ hybridization, Am. J. Pathol. 154 (5) (1999) 1449–1452.

[11] E.D. Remstein, et al., Diagnostic utility of fluorescence in situ hybridization in mantle-cell lymphoma, Br. J. Haematol. 110 (4) (2000) 856–862.

[12] H. Kim, H.J. Kim, S.H. Kim, Diagnostic approach for double-hit and triple-hit lymphoma based on immunophenotypic and cytogenetic characteristics of bone marrow specimens, Ann. Lab. Med. 40 (5) (2020) 361–369.

[13] J.P. Vaque, et al., B-cell lymphoma mutations: improving diagnostics and enabling targeted therapies, Haematologica 99 (2) (2014) 222–231.

[14] J. Corean, K.D. Li, A rare case of ALK-positive large B-cell lymphoma with CD33 expression, Case Rep. Hematol. 2018 (2018) 5320590.

[15] A. Reiter, J. Gotlib, Myeloid neoplasms with eosinophilia, Blood 129 (6) (2017) 704–714.

21

Mature T-cell neoplasms

Xia Li[a], Dongbin Xu[b], and Hanyin Cheng[b]
[a]SONORA QUEST LABORATORIES, PHOENIX, AZ, UNITED STATES [b]HEMATOLOGICS, INC., SEATTLE, WA, UNITED STATES

Background

Mature T-lymphomas are one of a group of aggressive (fast-growing) non-Hodgkin lymphomas that begin in mature T-lymphocytes (T cells that have matured in the thymus gland and moved to other lymphatic sites in the body, including lymph nodes, bone marrow, and spleen). These lymphomas are also called peripheral T-cell lymphomas (cancer.org). T-cell lymphoproliferative disorders are a diverse group of lymphoid neoplasms that are a clonal expansion of the mature T-lymphocytes in bone marrow, blood, or other tissues. Because natural killer cells (NK) are closely related and share some phenotypic features, these are categorized together (wiki.clinicalworkflow.com). T and NK neoplasms are less common, and they make up approximately 12% of the lymphoproliferative group. Incidence varies by geography and is more prevalent in Asia. This may be related to the HTLV-1 virus, which is endemic in Japan.

According to the WHO classification, mature T-cell lymphoma and NK neoplasms can be divided into four categories: 1. Leukemic or disseminated type includes T-cell prolymphocytic leukemia (T-PLL), T-cell granular lymphocytic leukemia (T-LGLL), aggressive NK leukemia (ANKL), and adult T-cell lymphoma/leukemia (ATLL), Sezary syndrome (SS); 2. Extranodal type includes extranodal NK/T cell lymphoma (nasal type), Enteropathy-type T-cell lymphoma, Hepatosplenic T-cell lymphoma (HTCL), and subcutaneous panniculitis-like T-cell lymphoma; 3. Cutaneous type contains blastic NK lymphoma, mycosis fungoides/Sezary syndrome, and primary cutaneous anaplastic large cell lymphoma; 4. Nodal type includes peripheral T-cell lymphoma, unspecified, angioimmunoblastic T-cell lymphoma (AITL), and anaplastic large cell lymphoma [1–4].

In recent years, genetic testing has played an important role in the diagnosis and prognosis of mature T-cell lymphoma [5]. One study from 51 cases of T-PLL by cytogenetic and molecular genetic characterization revealed *TRA/D* rearrangements (86%); deletions were detected for *ATM* (69%) and *TP53* (31%), whereas i(8q10) was observed in 61%, and mutations in *ATM*, *TP53*, *JAK1*, and *JAK3* were also observed [6]. Many mature T-cell neoplasms have T-cell receptor (TCR) gene rearrangements including *TRA/D* (14q11), *TCRB* (7q34), and *TCRG* (7p14) [7]. Anaplastic large cell lymphoma (ALCL) such as anaplastic lymphoma kinase (*ALK*)-positive (*ALK*+ ALCL) is an aggressive CD30-positive T-cell lymphoma that exhibits a chromosomal translocation involving the *ALK* gene and the

Cases in Laboratory Genetics and Genomics (LGG) Practice. https://doi.org/10.1016/B978-0-323-99622-8.00008-3

expression of ALK protein [8,9]. FISH, karyotyping, and NGS are the best technologies to identify chromosome translocations involving *ALK*. The prognosis of *ALK*+ ALCL is remarkably better than other T-cell lymphomas [9]. In this chapter, a few cases will be shown to illustrate the benefit of genetic testing for patients with mature T-cell neoplasms.

Case 21.1 T-cell prolymphocytic leukemia (T-PLL)

Clinical indication

A 71-year-old female presented with a history of T-PLL. She had mild lymphocytosis and a rapid rise in her lymphocyte count, as well as worsening lymphadenopathy and transaminitis. A punch biopsy demonstrated a population of atypical lymphoid cells positive for CD3, CD8, CD7, CD5, and CD43 compatible with involvement by T-PLL. PCR confirmed a monoclonal pattern for two sets of T-cell receptor genes.

Test ordered

- Chromosome analysis of the bone marrow
- FISH for *TRA/D* gene rearrangement

Laboratory test performed

Chromosome analysis and FISH methods were described previously in Chapters 1 and 12.

Test results

Of the 20 cells examined, 12 were abnormal, and clonal evolution was evident. All abnormal cells exhibited t(X;14) and trisomy 8 (Fig. 21.1.1); seven cells also showed del(1p), t(1;17), add(5p), and add(17q) (Fig. 21.1.2); three cells showed additional aberrations including add(1q), add(3q), del(11p), del(11q), and add(17p). Monosomy 21 was not a clonal change (Fig. 21.1.3). The remaining eight cells appear to be chromosomally normal.

FISH was performed on interphase nuclei using probes localized to the 5′ and 3′ ends of the T-cell receptor alpha/delta (*TRA/D*; 14q11.2) gene region. Two hundred nuclei were examined, and results demonstrated a rearrangement involving *TRA/D* in 75/200 (37.5%) of the cells scored (Fig. 21.1.4). A PCR send-out test confirmed a monoclonal pattern for two sets of T-cell receptor genes.

Results with interpretations

These results are consistent with a diagnosis of T-cell prolymphocytic leukemia (T-PLL). T-PLL is a rare and aggressive postthymic lymphoid neoplasm characterized by recurrent chromosome rearrangements that lead to activation of the *TCL1A* (14q32.1) or the *MTCP1* (Xq28) genes. In this case, chromosome analysis identified t(X;14)(q28;q11.2), which is

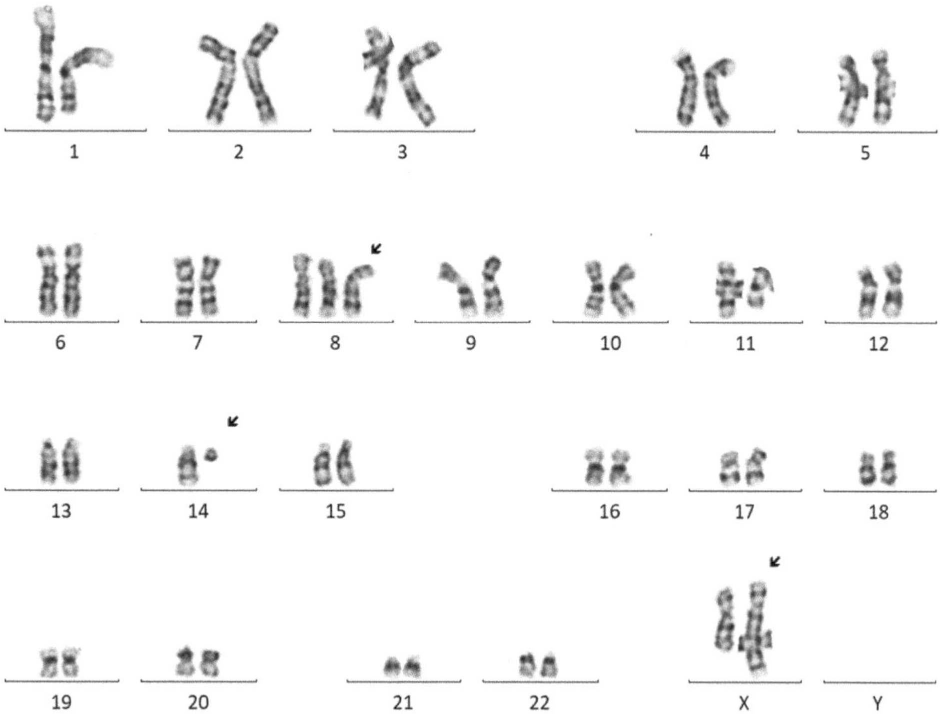

FIG. 21.1.1 Chromosome analysis of the bone marrow showed clone 1 from a complex karyotype with t(X;14) and trisomy 8. ISCN: 46,X,t(X;14)(q28;q11.2),+8[2]/ 46,idem,der(1)del(1)(p32p36.1)t(1;17)(q23q23),add(5)(p14),add(17)(q23)[7]/ 46,idem,add(1)(q32),add(3)(q27),der(11)del(11)(p13p15)del(11)(q21q23),add(17)(p11.2)[3]/ 46,XX [8]

found to occur in approximately 20% of T-PLL cases and leads to overexpression of the *MTCP1* gene by relocation to the T-cell receptor alpha/delta (*TRA*/*D*) located at 14q11.2 locus. Approximately 80% of T-PLL cases, however, are characterized by the inv(14)(q11.2q32.1) and variants, which lead to the activation of the *TCL1A* (14q32.1) gene by relocation to the *TRA*/*D* or *TRB* gene loci. The additional abnormalities in cases with *MTCP1* or *TCL1A*-related abnormalities are similar and include a gain of 8q usually in the form of i(8q), as well as deletions 6q, 9p, 11q, and 13q (https://atlasgeneticsoncology.org/) [10].

Rearrangement with *TRA*/*D* was also seen from the concurrent FISH testing. These results confirmed the diagnosis of T-PLL. The degree of karyotypic complexity indicates a more aggressive disease process.

Future testing and recommendations

Regular follow up and clinicopathologic correlation are recommended. Chromosome analysis and FISH can be ordered for monitoring disease progression and treatment efficacy in the future.

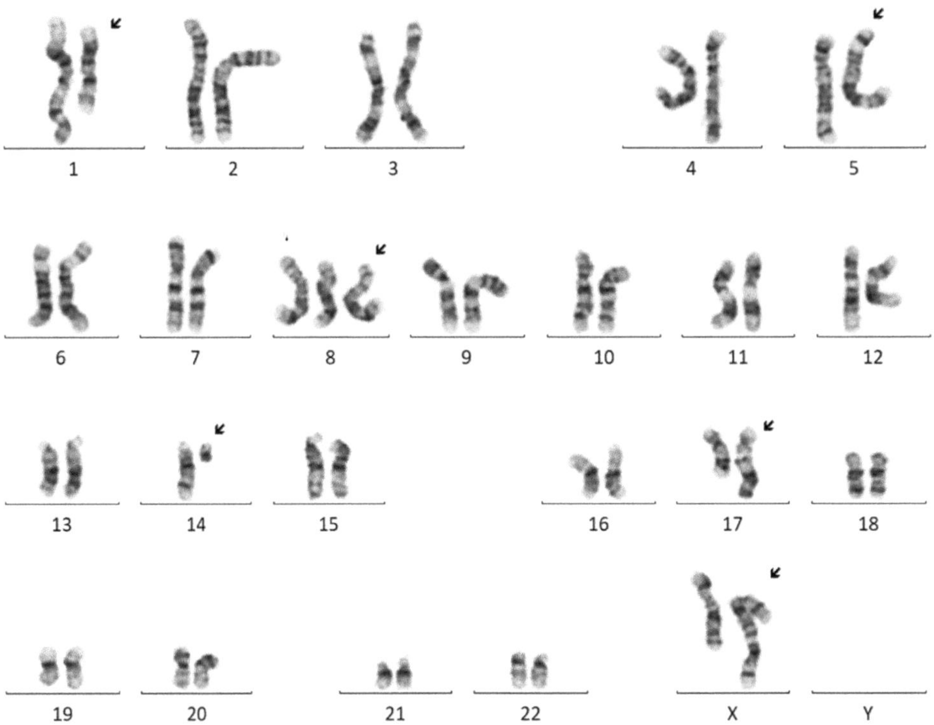

FIG. 21.1.2 Chromosome analysis of the bone marrow showed clone 2 from a complex karyotype with t(X;14), trisomy 8, and many other numerical and structural chromosome abnormalities including an abnormal chromosome 1 with t(1;17), add(5p), and add(17q).

Case 21.2 Mycosis fungoides/Sezary syndrome (MF/SS)

Clinical indication

A 75-year-old with a history of T-cell lymphoma on chemo, mycosis fungoides, suspected sleep apnea, and hyperlipidemia presented with symptoms of shortness of breath and fatigue for the last 4 days. Associated symptoms include dry cough, chills, decreased appetite, decreased activity, and bilateral lower extremity swelling.

Test ordered

- Chromosome analysis of the bone marrow

Laboratory test performed

The chromosome analysis method was described previously in Chapter 1.

Test results

From chromosome analysis, two related complex clones were detected in 10 of 20 metaphase cells analyzed. Clone 1 with six cells showed multiple numerical and structural

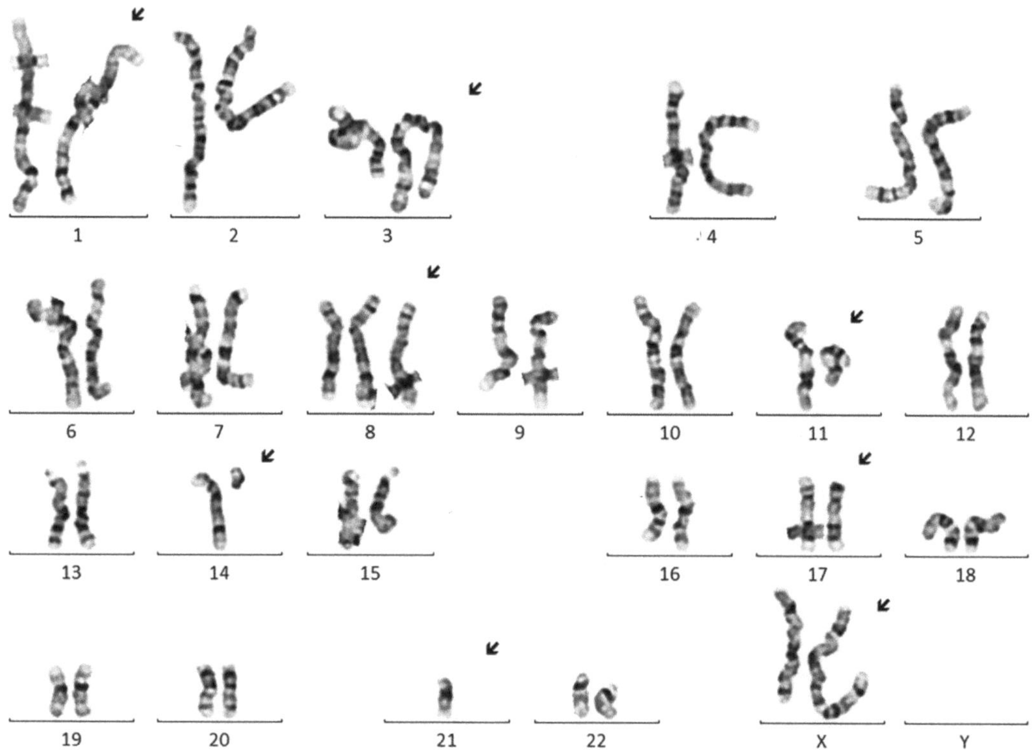

FIG. 21.1.3 Chromosome analysis of the bone marrow showed clone 3 from a complex karyotype with t(X;14), trisomy 8, and many other numerical and structural chromosome abnormalities add(1q), add(3q), del(11p), del(11q), add(17p).

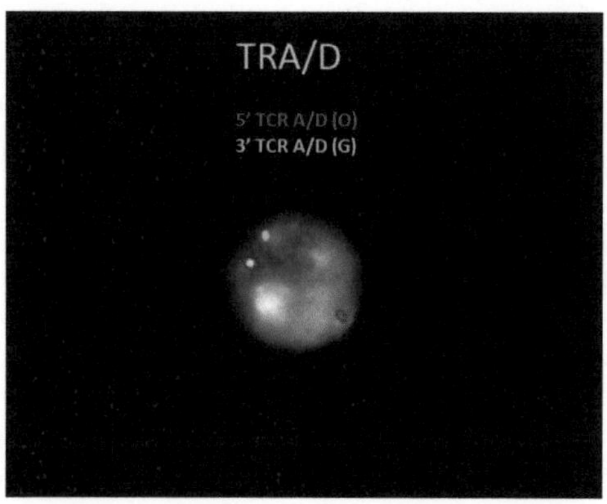

FIG. 21.1.4 FISH for *TRA/D* gene rearrangement showed positive results in 37.5% of cells scored. ISCN: nuc ish (TRA/Dx2)(5'TRA/D sep 3'TRA/ Dx1)[75/200]

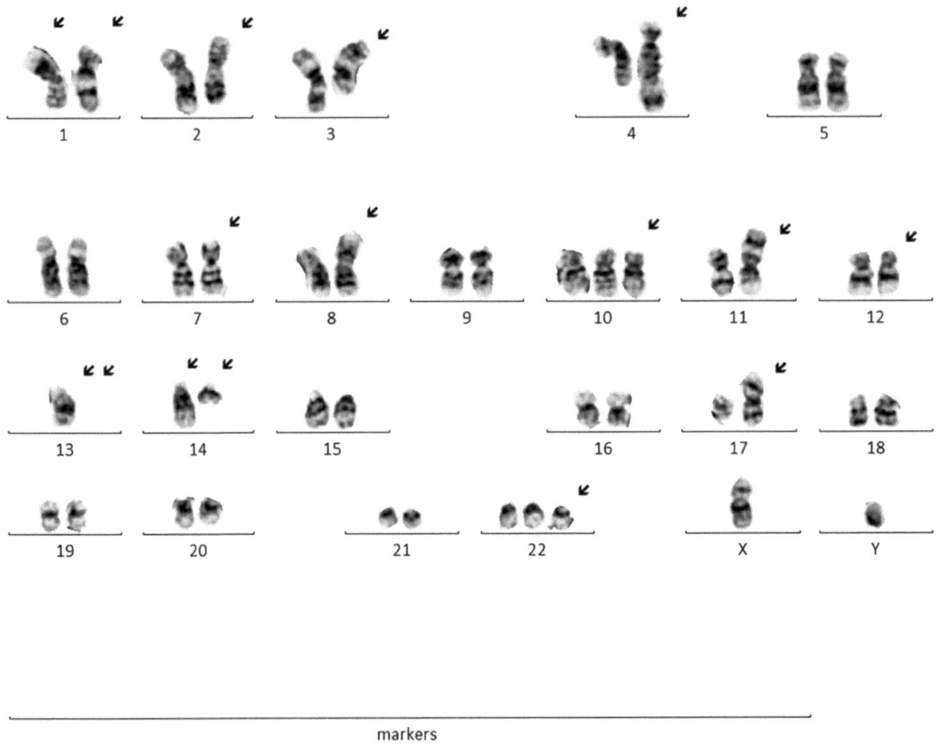

FIG. 21.2.1 Chromosome analysis of the bone marrow showed clone 1 from a complex karyotype with t(14;14)(q11.2;q32.1)/*TRA-D::TCL1A* and many other chromosome abnormalities. ISCN: 46~47,XY,der(1)del(1)(p36.1)add(1)(q21),der(1)add(1)(p36.1) inv(1)(p35q21),del(2)(q34q37),inv(3)(p25q25),der(4)t(4;13) (q35;q11),del(7)(q21q21),i(8)(q10),+10,del(12)(p12),-13,add(13) (q32),t(14;14)(q11.2;q32.1),add(17)(q11.2),i(17)(q10),+22[cp6]/ 46~47,idem,-der(1),+der(1)add(1)(p13)add(1)(q21),der(11) add(11)(p14)inv(11)(p14q12)[cp4]/46,XY[10]

aberrations including rearrangements of 1p, 1q, 13q, and 17q, deletions of 2q, 7q, and 12p, a pericentric inversion of chromosome 3, an unbalanced translocation between 4q and 13q, isochromosomes 8q and 17q leading to a net gain of 8q and 17q and a net loss of 8p and 17p (*TP53*), respectively. Clone 1 also exhibited a gain of chromosomes 10 and 22, monosomy 13, and a balanced t(14;14)(q11.2;q32.1)/*TRA-D::TCL1A* (Fig. 21.2.1). Clone 2 with four cells had similar aberrations as seen in clone 1 except for a minor variation of the aberration involving chromosome 1 and a new complex rearrangement involving chromosome 11 (Fig. 21.2.2).

Results with interpretations

The chromosome findings support the diagnosis of Mycosis fungoides/Sezary syndrome (MF/SS) in the context of the flow cytometry report. Although the t(14;14)/*TRA-D::TCL1A* is frequently seen in T-cell prolymphocytic leukemia, it has occasionally been reported in MF/SS. Furthermore, the complexity of the karyotype together with rearrangements of the

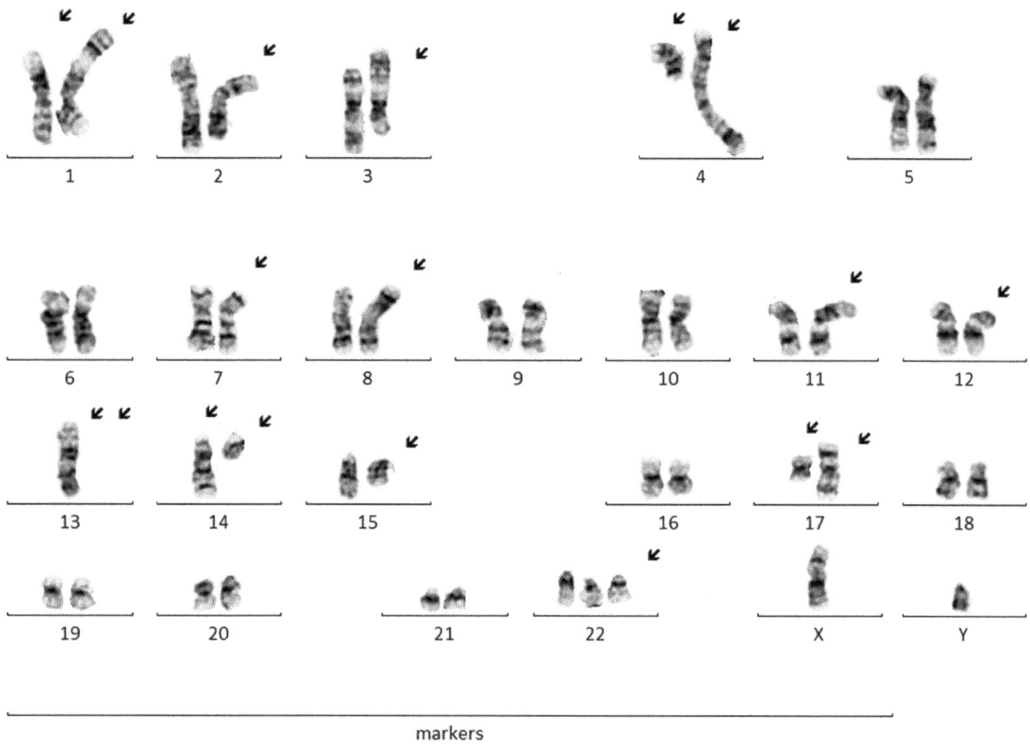

FIG. 21.2.2 Chromosome analysis of the bone marrow showed clone 2 from a complex karyotype with t(14;14)(q11.2;q32.1)/*TRA-D::TCL1A* and many other chromosome abnormalities that were related to clone 1.

short arm of chromosome 1 involving the critical regions 1p32~p36, and abnormalities involving chromosomes 3, 10, 11, 12, and 17p/*TP53* as seen here, has been reported in MF/SS.

Future testing and recommendations

Regular follow up and clinicopathologic correlation are recommended. Chromosome analysis can be ordered for monitoring disease progression and treatment efficacy in the future.

Case 21.3 T-cell leukemia/lymphoma with *TRB* rearrangement

Clinical indication

A 68-year-old male with a clinical history of metastatic renal cell carcinoma and T-cell lymphoma. A complete blood count revealed leukocytosis (white blood cells $19.4 \times 10e3/\mu L$). Previous flow cytometric findings revealed an aberrant mature

T lineage process in peripheral blood. Bone marrow aspirate was sent for the following laboratory tests.

Test ordered

- Immunophenotypic analysis of the bone marrow aspirate by flow cytometry
- T-cell beta (*TRB*) gene rearrangement PCR Analysis
- T-cell gamma (*TRG*) gene rearrangement PCR Analysis
- FISH: T-cell lymphoma panel
- Chromosome analysis of the bone marrow

Laboratory test performed

Flow cytometry analysis on bone marrow aspirate was performed at first. T-cell gene rearrangement assays by PCR were performed for the T-cell receptor beta and gamma chain loci. Each T-cell has a single productive T-cell receptor gene rearrangement that is unique in both length and sequence. Polymerase chain reaction (PCR) assays with specific primers for the T-cell receptor beta and the T-cell receptor gamma were used. DNA from a normal or polyclonal T-cell population produces a bell-shaped curve of amplicon products (Gaussian distribution). Clonal rearrangements are identified as prominent, single-sized products by capillary electrophoresis and GeneScanning. FISH study with T-cell lymphoma panel including probes for the following loci: *ALK* at 2p23, *MYB* at 6q23, *TRB* (T-cell receptor beta locus) at 7q34, and *TRA/TRD* (T-cell receptor alpha locus including nested T cell receptor delta locus) at 14q11.2. Chromosome analysis on bone marrow aspirate was performed using the culture condition optimized for T cell growth.

Test results

The flow cytometric analysis revealed an aberrant T-cell population at 32% of nonerythroid cells. Monoclonal amplicons were detected for both the T-cell receptor beta chain locus and the T-cell receptor gamma chain locus (Fig. 21.3.1A and B). The FISH analysis detected rearrangements of *TRB* at 7q34 but not *TRA/TRD* at 14q11.2 (Fig. 21.3.1C). Metaphase cells were obtained from the culture optimized for T-cell growth. The cytogenetic chromosome analysis on the metaphase cells revealed a complex abnormal karyotype including a pericentric inversion of chromosome 7 [inv(7)(p15q34)] along with a secondary translocation involving 7q and 10q, a complex rearrangement involving chromosome 3, and loss of the Y chromosome (Fig. 21.3.1D).

Results with interpretations

The flow cytometric findings are consistent with T-cell leukemia/lymphoma, but additional information is needed for classification.

Rearrangements between 7p15 and 7q34 are rare but have been reported in the medical literature, primarily in cases of T-cell lymphoblastic leukemias (T-ALL) [11]. The

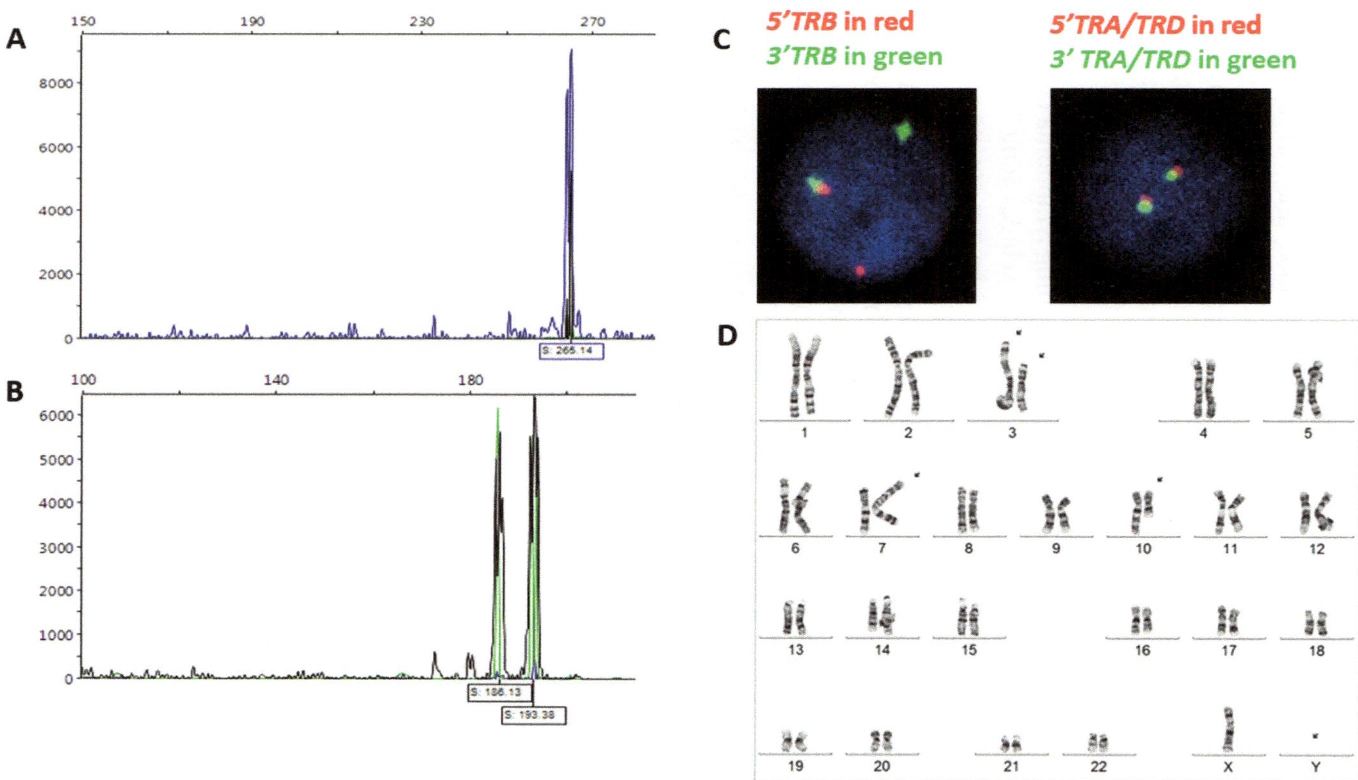

FIG. 21.3.1 Distinct monoclonal amplicon for the T-cell receptor beta chain locus (A). Distinct monoclonal amplicons for the T-cell receptor gamma chain locus (B). FISH images demonstrated an abnormal *TRB* rearrangement pattern with *one red*, *one green*, and one fusion signal (left) and a normal signal pattern with two fusion signals for *TRA/TRD* locus (C). Abnormal karyogram demonstrated the rearrangement involving chromosome band 7q34 where *TRB* localizes. ISCN: 45,X,-Y,der(3)ins(3;?3)(q13;p21p13),der(3)ins(3;?3)add(3)(q25),der(7)inv(7)(p15q34)t(7;10)(q34;q21),der(10)t(7;10)(q34;q21)

neoplastic T cells in the current specimen had the rearrangement and expressed dim CD10 (which may be present in T-ALL), but the CD45 intensity is higher than usually seen in immature T cells (T lymphoblasts). The neoplastic T cells lacked the traditional markers of immaturity (CD1a, CD34, and TdT) in the prior analysis of blood, and TdT should be expressed by more than 90% of T-ALL [12]. Given this information, the possibility of a rare variant of T-ALL seems unlikely but is difficult to exclude completely based on this data. The obvious question is whether the neoplastic cells look like small mature lymphocytes or lymphoblasts by light microscopy. Also, patients with T-ALL may have a mediastinal mass.

For mature T-cell leukemias and lymphomas, the differential diagnosis includes T-cell prolymphocytic leukemia (T-PLL), mycosis fungoides/Sezary syndrome, large granular lymphocytic leukemia, hepatosplenic T-cell lymphoma, circulating lymphoma cells of a mature T-cell lymphoma, etc. [13]. The lack of CD25 essentially excludes adult T-cell leukemia/lymphoma. Mycosis fungoides/Sezary syndrome would have a cutaneous rash, and the neoplastic cells would have the morphologic features of Sezary cells. The neoplastic cells in large granular lymphocytic leukemia would express CD57 (which was dimly expressed in the prior analysis of blood) and would have cytoplasmic granules by light microscopy. Hepatosplenic T-cell lymphoma is associated with prominent splenomegaly although that finding is not specific. Lymphocytosis from circulating lymphoma cells would be unusual in mature T-cell lymphomas although CD10 is often expressed by the lymphoma cells of angioimmunoblastic T-cell lymphoma. Among the mature T-cell leukemias/lymphomas, those most often associated with complex karyotypes include T-PLL, mycosis fungoides/Sezary syndrome, and mature T-cell lymphomas [13]. This may represent T-PLL, but this case has a few features that would be unusual for that diagnosis: the white blood cell count is lower than usually seen in T-PLL, and 90% of T-PLL have a rearrangement of *TRA/TRD*, which was absent in this case. Correlation with clinical and morphologic findings is required for final classification.

Future testing and recommendations

The monoclonality profiles by PCR assays for the *TRB* and *TRG* as well as FISH testing using the *TRB* break-apart probe can be used for future comparison and confirmation purposes. Flow cytometric immunophenotype and monoclonal profile analysis for T-cell receptor loci can be used to define the monoclonal T-cell population. FISH test and cytogenetic karyotype can detect recurrent chromosomal rearrangement involving T-cell receptor regions, which may support the establishment of an association with certain T-cell leukemia or lymphoma.

Case 21.4 Peripheral T-cell lymphoma with *TRA/TRD* rearrangement

Clinical indication

A 76-year-old female presented with a differential diagnosis including peripheral T-cell lymphoma. A complete blood count revealed leukocytosis (white blood cells

150.7 × 10e3/µL) and lymphocytosis (130.9 × 10e3/µL). Bone marrow aspirate was sent for the following laboratory tests.

Test ordered

- T-cell gamma (*TRG*) gene rearrangement by PCR Analysis
- FISH: Chronic Lymphocytic Leukemia (CLL) panel
- Chromosome analysis of the bone marrow

Laboratory test performed

T-cell gene rearrangement assays by PCR were performed for the T-cell receptor beta and gamma chain loci (see the Method in Case 21.3). The FISH study was performed with a CLL panel including the probe sets for 13q (13q14.3/13q34), *IGH/CCND1* (14q32/11q13) rearrangement, *MYB* (6q23), *ATM/TP53* (11q22.3/17p13.1), and the probe for the centromere of chromosome 12. Chromosome analysis was performed on the metaphase cells harvested from a culture with conditions optimized for T-cell growth. FISH study with the *TRA/TRD* break-apart probe was performed as a reflex assay based on the finding of chromosomal rearrangements at 14q11.2.

Test results

Distinct monoclonal amplicons were detected for the *TRG* locus by PCR analysis (Fig. 21.4.1A). The FISH study revealed a mono-allelic 13q14 deletion, loss of 5′ *MYB*, and gain of *IGH* (14q32) as demonstrated by distinct FISH signal patterns: one red and two green signals for the 13q probe set, one green and two red signals for the *MYB* probe set, and two red and three green signals for the *IGH::CCND1* probe set (Fig. 21.4.1B). Cytogenetic analysis revealed a complex abnormal female karyotype including a rearrangement involving Xq28 and 14q11.2, an isochromosome i(7)(q10), trisomy 8, del(13q), t(6;13) and an unbalanced translocation involving 6q and 14q in 13 out of 20 cells examined (Fig. 21.4.1C). The remaining seven cells were chromosomally normal. A reflex FISH test confirmed the rearrangement of *TRA/TRD* involving the translocation t(X;14) (Fig. 21.4.1D).

Results with interpretations

Detection of monoclonal amplicons for the *TRG* locus is consistent with the presence of a monoclonal T-cell population. The cytogenetic findings are consistent with T-cell lymphoma. It was noticed that all the abnormal cells were found in a culture under conditions optimized for T-cell growth, but not in another culture without stimulation for T-cell growth. The FISH findings are concordant with the complex karyotype. Reflex FISH further verified rearrangement of *TRA/TRD* resulted from the chromosome translocation t(X;14)(q28;q11.2), which has been mostly reported in T-prolymphocytic leukemia [14–16]. The presence of i(7)(q10) and trisomy 8 has been reported in hepatosplenic T-cell lymphoma [17–19]. Clinical and hematopathology correlation is required for a definitive diagnosis.

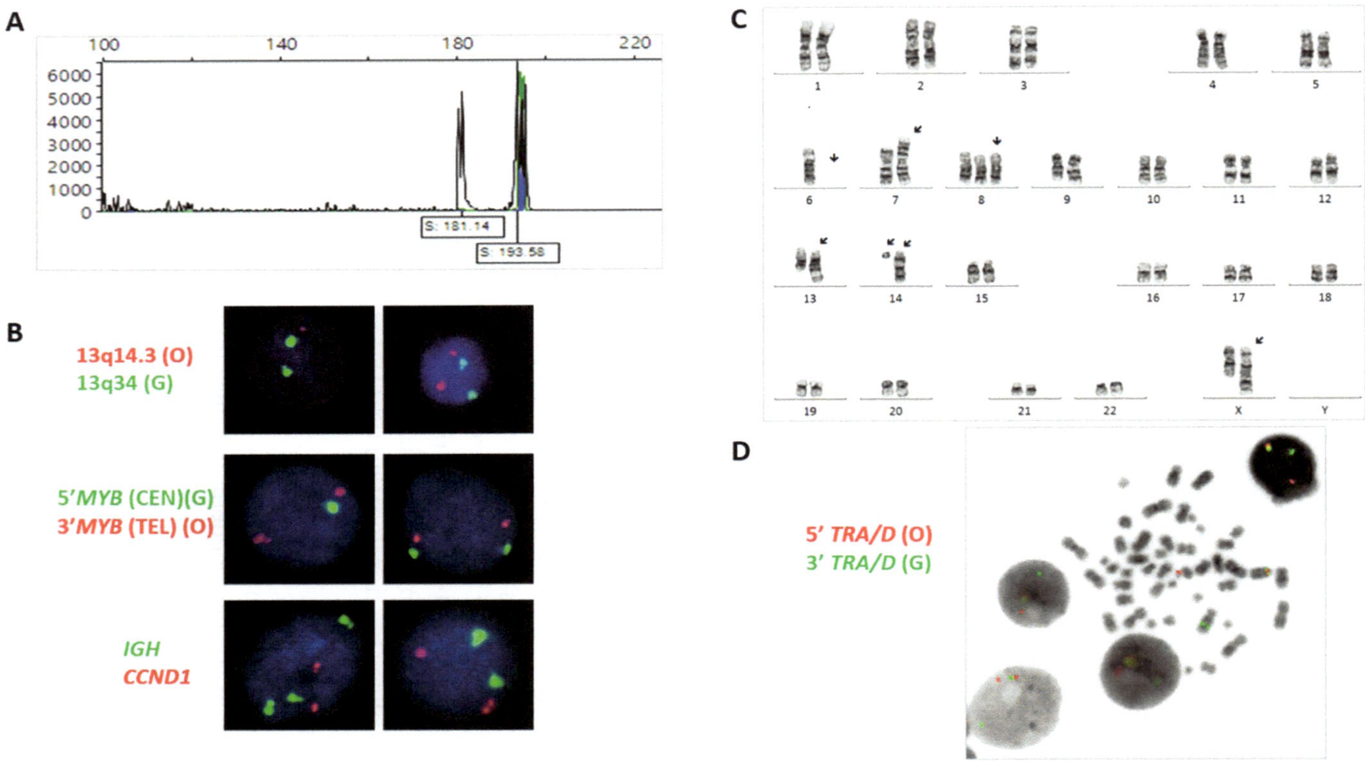

FIG. 21.4.1 Distinct monoclonal amplicons for the T-cell receptor gamma chain locus (A). FISH images demonstrated a mono-allelic 13q14 deletion, loss of pericentromeric region of chromosome 6, and gain of *IGH*. 13q14.3 in *red*, 13q34 in *green*, 3′*MYB* in *red*, 5′MYB in *green*, *CCND1* in *red*, and *IGH* in *green*. A pair of abnormal signal patterns (left) and normal signal patterns (right) was displayed for each probe set (B). An abnormal karyogram with rearrangement involving 14q11.2 was shown (C). ISCN: 46,X,t(X;14)(q28;q11.2),-6,+7,i(7)(q10),+8,der(13)del(13)(q12q14)t(6;13)(p12;q34), der(14)?t(6;14)(q23;q32) FISH results showed abnormal signal patterns for the *TRA/D* dual-color break-apart probe (5′ in *red*, 3′ in *green*). Interphase nuclei presented a typical positive break-apart signal pattern with one fusion, *one red*, and *one green signal*. Localization of these signals demonstrated that *TRA/TRD* rearrangement was associated with the chromosome translocation t(X;14) (D).

Future testing and recommendations

The monoclonal profile for the *TRG* locus can be used for future comparison and confirmation purposes. FISH testing using the *TRA/TRD* break-apart probe can be used for future monitoring as well.

PCR analysis can be used to assess the monoclonal profile of T-cell receptor loci, which can define the presence of monoclonal T-cell populations. The cytogenetic analysis can assess recurrent chromosomal abnormalities associated with certain T-cell leukemia/lymphoma. Chromosome aberrations involving the chromosome regions such as 7q34 and 14q11.2 should raise the need for a follow-up with further FISH studies to assess T-cell receptor gene rearrangements. The setup of multiple in vitro cultures with distinct conditions optimized to stimulate specific cell lineage proliferation is a common practice in the cytogenetics study process. The distribution of abnormal cell clones among different cultures can provide supportive information to define the cell lineage of abnormal cytogenetic clones. Correlation to flow-cytometric immunophenotypes, histology findings, and other histopathological findings is required for comprehensive interpretation.

Summary of key learning points

- Mature T-cell neoplasms have many different subtypes of diseases, and most have a poor prognosis compared with their B-cell counterparts.
- Classification of the disorder is complex. Both genetic testing and pathology evaluation from tissue, bone marrow, and blood are important.
- FISH, karyotyping, and NGS can identify mutations, gain/loss of chromosomes, and gene fusions that will help physicians to obtain valuable information on diagnosis and prognosis.

References

[1] R. Alaggio, et al., The 5th edition of the World Health Organization classification of haematolymphoid tumours: lymphoid neoplasms, Leukemia 36 (7) (2022) 1720–1748.

[2] E. Sabattini, et al., WHO classification of tumours of haematopoietic and lymphoid tissues in 2008: an overview, Pathologica 102 (3) (2010) 83–87.

[3] E.S. Jaffe, A. Nicolae, S. Pittaluga, Peripheral T-cell and NK-cell lymphomas in the WHO classification: pearls and pitfalls, Mod. Pathol. 26 (Suppl. 1) (2013) S71–S87.

[4] E. Matutes, The 2017 WHO update on mature T- and natural killer (NK) cell neoplasms, Int. J. Lab. Hematol. 40 (Suppl. 1) (2018) 97–103.

[5] V. Hathuc, F. Kreisel, Genetic landscape of peripheral T-cell lymphoma, Life (Basel) 12 (3) (2022).

[6] A. Stengel, et al., Genetic characterization of T-PLL reveals two major biologic subgroups and JAK3 mutations as prognostic marker, Genes Chromosomes Cancer 55 (1) (2016) 82–94.

[7] S. Gesk, et al., Molecular cytogenetic detection of chromosomal breakpoints in T-cell receptor gene loci, Leukemia 17 (4) (2003) 738–745.

[8] A.J. Ferreri, et al., Anaplastic large cell lymphoma, ALK-negative, Crit. Rev. Oncol. Hematol. 85 (2) (2013) 206–215.

[9] A.J. Ferreri, et al., Anaplastic large cell lymphoma, ALK-positive, Crit. Rev. Oncol. Hematol. 83 (2) (2012) 293–302.

[10] F.M. de Oliveira, et al., Translocations t(X;14)(q28;q11) and t(Y;14)(q12;q11) in T-cell prolymphocytic leukemia, Int. J. Lab. Hematol. 31 (4) (2009) 453–456.

[11] B. Cauwelier, et al., Clinical, cytogenetic and molecular characteristics of 14 T-ALL patients carrying the TCRbeta-HOXA rearrangement: a study of the Groupe Francophone de Cytogenetique Hematologique, Leukemia 21 (1) (2007) 121–128.

[12] A.S. Duffield, F. Racke, M. Borowitz, Precursor B- and T-cell neoplasms, in: E.S. Jaffe, D.A. Arber, E. Campo, N.L. Harris, L. Quintanilla-Martinez (Eds.), Hematopathology, Elsevier, Philadelphia, 2017.

[13] A. Porwit, M. Djokic, T-cell prolymphocytic leukemia, in: E.S. Jaffe, D.A. Arber, E. Campo, N.L. Harris, L. Quintanilla-Martinez (Eds.), Hematopathology, Elsevier, Philadelphia, 2017.

[14] Z. Hu, et al., Prognostic significance of cytogenetic abnormalities in T-cell prolymphocytic leukemia, Am. J. Hematol. 92 (5) (2017) 441–447.

[15] P.B. Staber, et al., Consensus criteria for diagnosis, staging, and treatment response assessment of T-cell prolymphocytic leukemia, Blood 134 (14) (2019) 1132–1143.

[16] T. Braun, et al., Advanced pathogenetic concepts in T-cell prolymphocytic leukemia and their translational impact, Front. Oncol. 11 (2021) 775363.

[17] E.L. Alonsozana, et al., Isochromosome 7q: the primary cytogenetic abnormality in hepatosplenic gammadelta T cell lymphoma, Leukemia 11 (8) (1997) 1367–1372.

[18] S. Coventry, et al., Consistency of isochromosome 7q and trisomy 8 in hepatosplenic gammadelta T-cell lymphoma: detection by fluorescence in situ hybridization of a splenic touch-preparation from a pediatric patient, Pediatr. Dev. Pathol. 2 (5) (1999) 478–483.

[19] Y. Shi, E. Wang, Hepatosplenic T-cell lymphoma: a Clinicopathologic review with an emphasis on diagnostic differentiation from other T-cell/natural killer-cell neoplasms, Arch. Pathol. Lab. Med. 139 (9) (2015) 1173–1180.

PART III

Solid tumors

22

Lung cancer

Xia Li and Guang Liu
SONORA QUEST LABORATORIES, PHOENIX, AZ, UNITED STATES

Background

Lung cancer is one of the most diagnosed cancers and the leading cause of cancer-related death worldwide. It is estimated that there are 2.2 million new cases and 1.79 million deaths each year [1] (http://www.cancer.org). With the advanced molecular and sequencing technologies, the adoption of clinical research, improvement of biomarker testing, and refinements of treatment have led to remarkable progress to bring a better outcome for lung cancer patients [2].

Lung cancer can occur in both smokers and nonsmokers. Public health efforts to reduce smoking rates have contributed to the reduction of the incidence of lung cancer in many countries, especially high-income countries [3–5]. The most effective ways to improve the outcome for lung cancer patients include screening, diagnosis, and treatment of non small cell lung cancer (NSCLC) and small cell lung cancer (SCLC).

Lung cancer is a heterogeneous disease with broad-ranging clinicopathologic features [6,7]. Among lung cancers, about 85% are NSCLC, and 15% are SCLC. The main subtypes of NSCLC are adenocarcinoma, squamous cell carcinoma, and large cell carcinoma. These subtypes, which start from different types of lung cells, are grouped as NSCLC because their treatments and prognoses (outlook) are often similar. In a person with small cell cancer, the cancerous cells appear small and round under a microscope. The cells of NSCLC are larger. Smoking is a major risk factor for both types. Of those who receive a diagnosis of SCLC, 95% have a history of smoking.

In this chapter, we would like to focus on the diagnosis of lung cancer, which will guide targeted therapy. So far, many biomarkers are used to guide targeted therapy and immunotherapy. In NSCLC, the biomarkers being observed include but are not limited to *NTRK1/2/3* (<1%), *ROS1* (1%), *RET*(~2%), *ALK* (~3%), *ERBB* (~4%), *MET* (~4%), *BRAF* (5%), *EGFR* (17%), *KRAS G12C* (12%), and *KRAS* Non-G12C (17%) [1]. For molecular testing, guidelines including the National Comprehensive Cancer Network (NCCN), College of American Pathologists (CAP), the International Association for the Study of Lung Cancer (IASLC), and the Association for Molecular Pathology (AMP) have recommended that NGS is the preferred method to test the driver genes including *EGFR, ALK, KRAS, ROS1, BRAF, ERBB2, NTRK1/2/3, MET* exon 14 skipping, and *RET* for guiding targeted therapy. For patients with resistance to *EFGR* inhibitors, *EGFR* T790M mutation analysis is indicated [8–10]. PDL1 expression is another important biomarker being assessed along with

Cases in Laboratory Genetics and Genomics (LGG) Practice. https://doi.org/10.1016/B978-0-323-99622-8.00017-4

the genes listed above for all patients with newly diagnosed advanced NSCLC because it informs the use of immune checkpoint inhibitors (ICIs) and identifies patients eligible for chemoimmunotherapy. Tumor mutational burden (TMB) is used to predict the response to ICIs, although prospective validation is still insufficient [1]. Tumor samples are usually obtained and tested upon diagnosis. During the disease progression and monitoring, liquid biopsy can be collected to see if new mutations are present in cell-free circulating tumor DNA (ctDNA) in plasma [11–13]. *EGFR* mutation-positive tumors respond well to tyrosine kinase inhibitors (TKIs). Most patients eventually develop resistance, 50%–60% due to secondary *EGFR* mutation T790M. Other mutations and transformations account for another 20%–25%; 15%–20% are unexplained [14]. More mutations including C797S for resistance to Osimertinib were reported [15]. In this chapter, six cases of lung cancer will be illustrated to show how biomarker testing by NGS will guide targeted therapy.

Case 22.1 Pulmonary adenocarcinoma with *ALK::EML4* fusion

Clinical indication

A 73-year-old female was diagnosed with grade II pulmonary adenocarcinoma. Initially, she was treated by lung lobectomy. She had a recurrence in the next 2 years and was treated with carboplatin and pemetrexed. Molecular testing from other labs showed negative results for *EGFR*, *ALK::EML4* fusion, and *KRAS* mutations. She was in remission until the second relapse 5 years later.

Test ordered

- NGS Lung Molecular Profile

  Hotspot: *AKT1, ALK, BRAF, CDK4, DDR2, EGFR, ERBB2, FGFR1, FGFR2, FGFR3, HRAS, JAK2, KRAS, MAP2K1, MET, MYC, NRAS, PDGFRA, RET, ROS1*
  Copy Number Variation (CNV): *CCND1, FGFR1, MET*
  Fusion: *ALK, MET, NTRK1, NTRK2, NTRK3, RET, ROS1*

Laboratory test performed

NGS Lung Molecular Profile assay was performed. This assay is designed to detect genetic alterations present in tumor tissue specimens. The test is designed to detect single-nucleotide variants (SNVs) and small insertions/deletions (In/Dels), copy number variations, and fusions using CHEF/S5 GeneStudio sequencing platforms. The 24 genes were selected based on the actionability of mutations identified in those genes using currently available evidence from national and international guidelines and literature. Actionability is defined as information a clinician might find useful to aid in diagnosis, prognosis, and/or treatment strategy for a patient. The results of the test should be correlated with clinical findings. This test was designed for the detection and annotation of somatic variants and is not intended to be a germline test. When the allele frequency is approximately 50%, the possibility of these being germline mutations cannot be excluded.

Test results

NGS Lung Molecular Profile was performed and detected *ALK::EML4* translocation (Fig. 22.1.1).

Results with interpretations

NGS detected *ALK::EML4* fusion from this patient. There are several FDA-recommended medications for targeted therapy for lung cancer. In addition, Immunohistochemistry showed that she had low PD-L1 (1%–4%). Therefore, she was treated with alectinib 600 mg twice daily (b.i.d.).

Alectinib (ALECENSA) is a kinase inhibitor indicated for the treatment of patients with anaplastic lymphoma kinase (*ALK*)-positive, metastatic NSCLC who have progressed on or are intolerant to crizotinib. This indication is approved under accelerated approval based on tumor response rate and duration of response. Continued approval for this indication may be contingent upon verification and description of clinical benefit in confirmatory trials (https://www.accessdata.fda.gov/drugsatfda_docs/label/2016/208434s001lbl.pdf).

NCCN Guidelines recommended that sensitizing *EGFR* TKI therapy is not effective in patients with *KRAS* mutations, *ALK* gene rearrangements, or *ROS1* rearrangements. Primary resistance to TKI therapy is associated with *KRAS* mutations and *ALK* or *ROS1* gene

Variant Summary — Indicated / Contraindicated

Gene Variant	Relevant Therapies (In this cancer type)	Clinical Trials
EML4-ALK fusion	brigatinib[1] crizotinib[1] ceritinib[1] alectinib[1] EGFR tyrosine kinase inhibitor	49

Sources included in relevant therapies: FDA[1], NCCN

Variant Details

Gene Fusions (RNA)

Genes	Variant ID	Locus
EML4-ALK	EML4-ALK.E13A20.AB462411	chr2:42522656 - chr2:29446394

FIG. 22.1.1 NGS detected *ALK::EML4* fusion. The FDA-approved drugs are marked in *green*. The contraindicated drugs, e.g., *EGFR* tyrosine kinase inhibitor are marked in *red*.

rearrangements [8]. Therefore, *EGFR* TKI should not be used for the treatment of patients with *ALK::EML4* rearrangement. She was checked by a CT thorax and had stable disease afterward.

Future testing and recommendations

Regular checkups with an oncologist were recommended. If any clinical exam suggests relapse in disease, molecular testing is indicated to see if any mutations/fusions have emerged.

Case 22.2 Pulmonary adenocarcinoma with *EGFR* p.(L858R) mutation

Clinical indication

A 65-year-old male presented with shortness of breath. A right pleural effusion and right middle mass were discovered, and both were positive for adenocarcinoma. The biopsy tissue was submitted for molecular analysis. He was found to have *EGFR* p.(L858R) mutation. He was treated and had no new disease for 3 years. In the 4th year, bilateral lung nodules were discovered, the largest nodule was 1.5 cm in the right lower lobe. Biopsy and repeat molecular analysis (NGS Lung Molecular Profile) were performed.

Test ordered

- NGS Lung Molecular Profile

Laboratory test performed

NGS Lung Molecular Profile assay was performed as described (see Case 22.1).

Test results

NGS Lung Molecular Profile detected two *EGFR* mutations: *EGFR* p.(L858R) c.2573T>G and *EGFR* p.(T790M) c.2369C>T (Fig. 22.2.1).

Results with interpretations

At initial diagnosis, the patient had *EGFR* L858R exon 21 mutations identified from the lung biopsy. At that moment, *ALK* rearrangement was negative. Following the results, he began single-agent adjuvant therapy with erlotinib (Tarceva). When the bilateral lung nodules were discovered, a biopsy was collected, and repeat NGS testing was performed. The results showed *EGFR* L858R and T790M mutations. *EGFR* T790M is conferring resistance to erlotinib, afatinib, or gefitinib therapy. Options for additional treatment are osimertinib (third-generation *EGFR* TKI) or afatinib plus cetuximab. He was in remission after the new treatment.

Relevant Biomarkers

Tier	Genomic Alteration	Relevant Therapies (In this cancer type)	Clinical Trials
IA	*EGFR p.(L858R) c.2573T>G*	**bevacizumab* + erlotinib** [2] **erlotinib + ramucirumab** [1] **gefitinib*** [2] **osimertinib** [1, 2] atezolizumab + bevacizumab + chemotherapy bevacizumab + gefitinib gefitinib + chemotherapy osimertinib + chemotherapy	46
IA	*EGFR p.(T790M) c.2369C>T*	**osimertinib** [1, 2]	23

Public data sources included in relevant therapies: FDA1, NCCN, **EMA**2, ESMO
Tier Reference: Li et al. *Standards and Guidelines for the Interpretation and Reporting of Sequence Variants in Cancer: A Joint Consensus Recommendation of the Association for Molecular Pathology, American Society of Clinical Oncology, and College of American Pathologists.* J Mol Diagn. 2017 Jan;19(1):4-23.
*** Includes biosimilars**

Alerts informed by public data sources: ⊘ Contraindicated, ⛉ Resistance

EGFR p.(T790M) c.2369C>T	⊘ **gefitinib*** [2] ⛉ afatinib, dacomitinib, erlotinib, gefitinib

Public data sources included in alerts: FDA1, NCCN, **EMA**2, ESMO

Variant Details

DNA Sequence Variants

Gene	Amino Acid Change	Coding	Locus	Allele Frequency	Variant Effect
EGFR	p.(T790M)	c.2369C>T	chr7:55249071	17.13%	missense
EGFR	p.(L858R)	c.2573T>G	chr7:55259515	22.36%	missense

FIG. 22.2.1 NGS detected *EGFR* p.(L858R) and p.(T790M) mutations. The FDA-approved drugs are indicated. The drugs that are contraindicated and resistant are also shown.

Future testing and recommendations

Regular checkups with an oncologist were recommended. If any clinical exam suggests relapse in disease, molecular testing is indicated to see if any mutations/fusions have emerged.

Case 22.3 Squamous cell carcinoma with *MET* exon 14 skipping mutation

Clinical indication

A 70-year-old female was admitted with a persistent cough for more than 6 months. CT chest revealed a right upper lobe (RUL) cavitary lesion with encasement of the right main-stem bronchus and thoracic lymphadenopathy.

Test ordered

- NGS Lung Molecular Profile

Laboratory test performed

NGS Lung Molecular Profile assay was performed as described (see Case 22.1).

Test results

NGS Lung Molecular Profile detected *MET* exon 14 skipping mutation (Fig. 22.3.1).

Results with interpretations

NGS identified *MET* exon 14 skipping and a splice variant p.(D1028H). According to FDA, NCCN, and European Society for Medical Oncology (ESMO) recommendations, targeted therapies for this genetic alteration include capmatinib, tepotinib, or crizotinib.

Future testing and recommendations

Regular checkups with an oncologist were recommended. If any clinical exam suggests relapse in disease, molecular testing is indicated to see if any mutations/fusions have emerged.

Relevant Biomarkers

Genomic Alteration	Relevant Therapies (In this cancer type)	Clinical Trials
MET exon 14 skipping	**capmatinib** [1] **tepotinib** [1] crizotinib	23

Public data sources included in relevant therapies: FDA1, NCCN, **EMA**2, ESMO

Variant Details

DNA Sequence Variants

Gene	Amino Acid Change	Coding	Locus	Allele Frequency	Variant Effect
MET	p.(D1028H)	c.3082G>C	chr7:116412043	41.16%	missense

Gene Fusions (RNA)

Genes	Variant ID	Locus
MET-MET	MET-MET.M13M15	chr7:116411708 - chr7:116414935

FIG. 22.3.1 NGS detected *MET* exon 14 skipping mutation. The FDA-approved drugs are shown. The splice site variant *MET* p.(D1028H) was also identified, which confirmed the finding of *MET* exon 14 skipping.

Case 22.4 Pulmonary adenocarcinoma with *KRAS* p.(G12C) mutation

Clinical indication

A 78-year-old male presented with shortness of breath and cough. The diagnostic imaging of the lung showed abnormal findings, and metastatic lung adenocarcinoma was confirmed by the pathology report.

Test ordered

- NGS Lung Molecular Profile

Laboratory test performed

NGS Lung Molecular Profile assay was performed as described (see Case 22.1).

Test results

NGS Lung Molecular Profile detected *KRAS* p.(G12C) c.34G>T mutation (Fig. 22.4.1).

Results with interpretations

NGS identified *KRAS* p.(G12C) c.34G>T mutation. The recommended FDA drug is LUMAKRAS (sotorasib). This is an inhibitor of the *RAS* GTPase family indicated for the treatment of adult patients with *KRAS* G12C-mutated locally advanced or metastatic non small cell lung cancer (NSCLC), as determined by an FDA-approved test, for those who have received at least one prior systemic therapy.

Relevant Biomarkers

Genomic Alteration	Relevant Therapies (In this cancer type)	Clinical Trials
KRAS p.(G12C) c.34G>T **Prognostic significance:** None **Diagnostic significance:** None	**sotorasib** [1]	62

Public data sources included in relevant therapies: FDA1, NCCN, **EMA**2, ESMO
Public data sources included in prognostic and diagnostic significance: NCCN, ESMO

Variant Details

DNA Sequence Variants

Gene	Amino Acid Change	Coding	Locus	Allele Frequency	Variant Effect
KRAS	p.(G12C)	c.34G>T	chr12:25398285	5.47%	missense

FIG. 22.4.1 NGS detected *KRAS* p.(G12C) c.34G>T mutation. The FDA-approved drug is shown.

Future testing and recommendations

Regular checkups with an oncologist were recommended. If any clinical exam suggests relapse in disease, molecular testing is indicated to see if any mutations/fusions have emerged.

Case 22.5 Pulmonary adenocarcinoma with *ERBB2* exon 20 insertion

Clinical indication

A 72-year-old female who is a never-smoker presented with a lung mass. The diagnostic imaging of the lung showed abnormal findings with enlarged mediastinal nodes. Non small cell carcinoma was diagnosed by a pathology report.

Test ordered

- NGS Lung Molecular Profile

Laboratory test performed

NGS Lung Molecular Profile assay was performed as described (see Case 22.1).

Test results

NGS Lung Molecular Profile detected an *ERBB2(HER2)* exon 20 insertion (Fig. 22.5.1).

Relevant Biomarkers

Genomic Alteration	Relevant Therapies (In this cancer type)	Clinical Trials
ERBB2 exon 20 insertion	ado-trastuzumab emtansine trastuzumab deruxtecan	17

Public data sources included in relevant therapies: FDA[1], NCCN, **EMA**[2], ESMO

Variant Details

DNA Sequence Variants

Gene	Amino Acid Change	Coding	Locus	Allele Frequency	Variant Effect
ERBB2	p.(G778_P780dup)	c.2331_2339dup	chr17:37881001	19.92%	nonframeshift Insertion

FIG. 22.5.1 NGS detected *ERBB2* p.(G778_P780dup), c.2331_2339dup mutation. The NCCN recommendation for therapy is shown.

Results with interpretations

NGS Lung Molecular Profile detected an *ERBB2 (HER2)* exon 20 insertion. NCCN-recommended therapy includes ado-trastuzumab emtansine and trastuzumab deruxtecan. Both are used for the treatment of metastatic NSCLC. The FDA has granted breakthrough designation for the *HER2*-directed antibody-drug conjugate, Enhertu (trastuzumab deruxtecan), for the treatment of *HER2* mutated metastatic NSCLC with disease progression on or after platinum-based therapy (https://www.astrazeneca.com/media-centre/press-releases/2020/enhertu-granted-breakthrough-therapy-designation-in-the-us-for-her2-mutant-metastatic-non-small-cell-lung-cancer.html#). The FDA has also granted fast track designation to BDTX-189 for solid tumors harboring a *HER2* mutation or an *EGFR* or *HER2* exon 20 insertion after progression on prior therapy (https://investors.blackdiamondtherapeutics.com/news-releases/news-release-details/black-diamond-therapeutics-granted-fasttrack-designation-fda).

Future testing and recommendations

Regular checkups with an oncologist were recommended. If any clinical exam suggests relapse in disease, molecular testing is indicated to see if any mutations/fusions have emerged.

Case 22.6 Pulmonary adenocarcinoma with *TPM3::NTRK1* fusion

Clinical indication

A 62-year-old male smoker presented with chest pain and shortness of breath. He had a fast-growing nodule in the left lower lobe that was negative for the Coccidiodes serology test. The fine needle aspirate report from the left lower lobe nodule of the lung revealed rare tumor cells, consistent with adenocarcinoma that was compatible with lung primary. PD-L1 (22C3) test showed a tumor proportion score (TPS) with 100% (High PD-L1 expression).

Test ordered

- NGS Lung Molecular Profile

Laboratory test performed

NGS Lung Molecular Profile assay was performed as described (see Case 22.1).

Test results

NGS Lung Molecular Profile detected *TPM3::NTRK1* fusion (Fig. 22.6.1).

Relevant Biomarkers

Genomic Alteration	Relevant Therapies (In this cancer type)	Clinical Trials
TPM3-NTRK1 fusion	**entrectinib** [1, 2] **larotrectinib** [1]	11

Public data sources included in relevant therapies: FDA1, NCCN, **EMA**2, ESMO

Variant Details

Gene Fusions (RNA)

Genes	Variant ID	Locus
TPM3-NTRK1	TPM3-NTRK1.T7N12	chr1:154142876 - chr1:156845312

FIG. 22.6.1 NGS detected a *TPM3::NTRK1* fusion.

Results with interpretations

The *NTRK* gene fusion *TPM3::NTRK1* was initially discovered in colorectal cancer in 1986 [16]. The *NTRK* gene fusions were discovered as oncogenic drivers from various adult and pediatric tumors. Up until now, multiple NTRK fusion partners have been reported in lung cancer. The most common fusions include *MPRIP::NTRK1* and *CD74::NTRK1*, which result in constitutive TRKA kinase activity and are oncogenic [17]. *TPM3* was the most common *NTRK1* fusion partner, and *TPM3::NTRK1* was reported as a resistance mechanism to both first-generation and third-generation *EGFR*-TKIs in NSCLC patients [18,19]. Based on FDA and NCCN recommendations, lung cancer patients with *NTRK* fusion can be treated with Entrectinib and Larotrectinib (https://www.accessdata.fda.gov/drugsatfda_docs/label/2022/212725s006lbl.pdf; https://www.accessdata.fda.gov/drugsatfda_docs/label/2021/210861s006lbl.pdf; NCCN Guidelines—NCCN-Non-Small Cell Lung Cancer [Version 3.2022]).

Future testing and recommendations

Regular checkups with an oncologist were recommended. If any clinical exam suggests relapse in disease, molecular testing is indicated to see if any mutations/fusions have emerged.

Summary of key learning points

- Lung cancer is the leading cause of cancer-related death worldwide.
- NSCLC consists of 85% of lung cancer.
- Biomarkers for diagnosis and targeted therapy of lung cancer are introduced.
- NGS is preferred over individual gene testing for lung cancer patients.
- Resistance mutations should be tested for patients not responding to *EGFR* TKIs.

References

[1] A.A. Thai, et al., Lung cancer, Lancet 398 (10299) (2021) 535–554.

[2] N. Howlader, et al., The effect of advances in lung-cancer treatment on population mortality, N. Engl. J. Med. 383 (7) (2020) 640–649.

[3] A. Jemal, et al., Global patterns of cancer incidence and mortality rates and trends, Cancer Epidemiol. Biomark. Prev. 19 (8) (2010) 1893–1907.

[4] M. Ng, et al., Smoking prevalence and cigarette consumption in 187 countries, 1980-2012, JAMA 311 (2) (2014) 183–192.

[5] R.L. Siegel, K.D. Miller, A. Jemal, Cancer statistics, 2020, CA Cancer J. Clin. 70 (1) (2020) 7–30.

[6] W.D. Travis, et al., Introduction to the 2015 World Health Organization classification of tumors of the lung, pleura, thymus, and heart, J. Thorac. Oncol. 10 (9) (2015) 1240–1242.

[7] W.D. Travis, et al., The 2015 World Health Organization classification of lung tumors: impact of genetic, clinical and radiologic advances since the 2004 classification, J. Thorac. Oncol. 10 (9) (2015) 1243–1260.

[8] D.S. Ettinger, et al., Non-small cell lung cancer, version 3.2022, NCCN clinical practice guidelines in oncology, J. Natl. Compr. Cancer Netw. 20 (5) (2022) 497–530.

[9] D.E. Wood, et al., NCCN guidelines(R) insights: lung cancer screening, version 1.2022, J. Natl. Compr. Cancer Netw. 20 (7) (2022) 754–764.

[10] N.I. Lindeman, et al., Updated molecular testing guideline for the selection of lung cancer patients for treatment with targeted tyrosine kinase inhibitors: guideline from the College of American Pathologists, the International Association for the Study of Lung Cancer, and the Association for Molecular Pathology, J. Mol. Diagn. 20 (2) (2018) 129–159.

[11] J. Zugazagoitia, et al., Clinical utility of plasma-based digital next-generation sequencing in oncogene-driven non-small-cell lung cancer patients with tyrosine kinase inhibitor resistance, Lung Cancer 134 (2019) 72–78.

[12] J. Zugazagoitia, et al., Clinical utility of plasma-based digital next-generation sequencing in patients with advance-stage lung adenocarcinomas with insufficient tumor samples for tissue genotyping, Ann. Oncol. 30 (2) (2019) 290–296.

[13] L.A. Diaz Jr., A. Bardelli, Liquid biopsies: genotyping circulating tumor DNA, J. Clin. Oncol. 32 (6) (2014) 579–586.

[14] H.L. Tumbrink, A. Heimsoeth, M.L. Sos, The next tier of EGFR resistance mutations in lung cancer, Oncogene 40 (1) (2021) 1–11.

[15] A. Leonetti, et al., Resistance mechanisms to osimertinib in EGFR-mutated non-small cell lung cancer, Br. J. Cancer 121 (9) (2019) 725–737.

[16] D. Martin-Zanca, S.H. Hughes, M. Barbacid, A human oncogene formed by the fusion of truncated tropomyosin and protein tyrosine kinase sequences, Nature 319 (6056) (1986) 743–748.

[17] A. Vaishnavi, et al., Oncogenic and drug-sensitive NTRK1 rearrangements in lung cancer, Nat. Med. 19 (11) (2013) 1469–1472.

[18] H. Xia, et al., Evidence of NTRK1 fusion as resistance mechanism to EGFR TKI in EGFR+ NSCLC: results from a large-scale survey of NTRK1 fusions in Chinese patients with lung cancer, Clin. Lung Cancer 21 (3) (2020) 247–254.

[19] F. Liu, et al., NTRK fusion in non-small cell lung cancer: diagnosis, therapy, and TRK inhibitor resistance, Front. Oncol. 12 (2022) 864666.

23

Colorectal cancer

Xia Li and Guang Liu
SONORA QUEST LABORATORIES, PHOENIX, AZ, UNITED STATES

Background

Colorectal cancer (CRC) is the third most common cancer diagnosed in the United States (behind breast first and lung cancers second). The American Cancer Society estimated that new cases of colon cancer in 2022 at 106,180, and cases of rectal cancer 44,850 (http://www.cancer.org). The death rate from colorectal cancer has been dropping in the last several decades, with one of the most likely causes being colon screening tests. These tests can identify polyps and have them removed before the polyps develop into cancer; the second reason is that the treatment for colorectal cancer has improved significantly over the last few decades.

It is estimated that 30% of CRC cases are hereditary diseases with germline mutations identified. These diseases include Lynch syndrome, familial adenomatous polyposis, *MUTYH*-associated polyposis, and certain hamartomatous polyposis conditions. The genes involved in hereditary CRC include but are not limited to *APC, MLH1, PMS2, GTBP, MSH6, LKB1, STK11, PTEN, DPC4, MBPR1a, MYH*, etc. [1–4]. The remaining cases are sporadic with either somatic mutations identified or unknown etiology. Sporadic CRC is a somatic genetic disease in which pathogenesis is impacted by the local colonic environment and the patient's genetic makeup. Next-generation sequencing (NGS) has allowed the identification of genetic alterations in CRC and provides treatment guidance. The most common findings of CRC testing include the hypermutated group with defective DNA mismatch repair for microsatellite instability (MSI), *POLE* mutations, *BRAF, KRAS*, and *PIK3CA* mutations, and loss of heterozygosity of tumor suppressor genes such as *APC* and *TP53*. These results are used as diagnostic, prognostic, and treatment biomarkers [5–7].

For genetic testing of colon cancer, NCCN has recommended testing be performed on formalin-fixed paraffin-embedded tissue (preferred) or blood-based assay. All patients with metastatic colorectal cancer should have their tumor genotyped for *KRAS, NRAS*, and *BRAF* mutations individually or as part of an NGS panel. Universal DNA mismatch repair (MMR) or MSI should be tested for all newly diagnosed patients with colon cancer [8]. Patients with any known *KRAS* mutation (exon 2, 3, 4) or *NRAS* mutation (exon 2, 3, 4) should not be treated with either cetuximab or panitumumab [9–12]. A *BRAF* V600E

Cases in Laboratory Genetics and Genomics (LGG) Practice. https://doi.org/10.1016/B978-0-323-99622-8.00021-6

mutation makes a response to panitumumab or cetuximab highly unlikely unless given with a *BRAF* inhibitor [13–15].

On August 8, 2020, the FDA approved that BRAFTOVI (encorafenib) is indicated in combination with cetuximab, for the treatment of adult patients with metastatic colorectal cancer (CRC) with a *BRAF* V600E mutation, as detected by an FDA-approved test, after prior therapy.

The clinical utility of NGS has been demonstrated for the identification of the mutations associated with CRC. In this chapter, we will illustrate several cases in which mutations were detected by NGS to facilitate tailored adjuvant chemotherapy to mitigate the risk of recurrence.

Case 23.1 Metastatic colon cancer with *BRAF* p.(V600E) mutation

Clinical indication

A 75-year-old male presented with a history of colon cancer with liver metastasis. The pathology report showed metastatic adenocarcinoma, consistent with a colonic origin in a colon biopsy.

Test ordered

- NGS Colon Molecular Profile
- Hotspot: *AKT1, BRAF, ERBB2, EGFR, KRAS, PIK3CA, NRAS*

Laboratory test performed

NGS Colon Molecular Profile assay was performed. This assay is designed to detect genetic alterations present in tumor tissue specimens. The test detects single-nucleotide variants (SNVs) and small insertions/deletions (In/Dels) and copy number variations using CHEF/S5 GeneStudio sequencing platforms. Seven genes were selected based on the actionability of mutations identified in those genes using currently available evidence from national and international guidelines and literature. Actionability is defined as information a clinician might find useful to aid in the diagnosis, prognosis, and/or treatment strategy for a patient. The results of the test should be correlated with clinical findings. This test was designed for the detection and annotation of somatic variants and is not intended to be a germline test. When the allele frequency is approximately 50%, the possibility of these being germline mutations cannot be excluded.

Test results

NGS Colon Molecular Profile was performed on a biopsy from the right colon mass. The results showed *BRAF* p.(V600E) c.1799T>A mutation (Fig. 23.1.1).

Relevant Biomarkers

Tier	Genomic Alteration	Relevant Therapies (In this cancer type)	Clinical Trials
IA	*BRAF p.(V600E) c.1799T>A*	**cetuximab + encorafenib**[1,2] encorafenib + panitumumab	35

Public data sources included in relevant therapies: **FDA**1, NCCN, **EMA**2, ESMO

Tier Reference: Li et al. *Standards and Guidelines for the Interpretation and Reporting of Sequence Variants in Cancer: A Joint Consensus Recommendation of the Association for Molecular Pathology, American Society of Clinical Oncology, and College of American Pathologists.* J Mol Diagn. 2017 Jan;19(1):4-23.

Variant Details

DNA Sequence Variants

Gene	Amino Acid Change	Coding	Locus	Allele Frequency	Variant Effect
BRAF	p.(V600E)	c.1799T>A	chr7:140453136	45.01%	missense

FIG. 23.1.1 NGS detected *BRAF* p.(V600E)) c.1799T>A mutation. The FDA-approved drugs are indicated.

Results with interpretations

NGS detected a *BRAF* V600E mutation in this patient. There are several FDA-recommended medications for targeted therapy for colon cancer. BRAFTOVI (encorafenib) is a kinase inhibitor indicated in combination with cetuximab, for the treatment of adult patients with metastatic colorectal cancer (CRC) with a *BRAF* V600E mutation, as detected by an FDA-approved test, after prior therapy.

Future testing and recommendations

Regular checkups with an oncologist are recommended. If any clinical exam suggests relapse in disease, molecular testing is indicated to see if any mutations/fusions have emerged.

Case 23.2 Metastatic colon cancer with *KRAS* p.(G12D) mutation

Clinical indication

A 57-year-old female presented with metastatic sigmoid colon adenocarcinoma with liver metastasis. The pathology report showed metastatic adenocarcinoma, consistent with a colonic origin in a colon biopsy.

Test ordered

- NGS Colon Molecular Profile

Laboratory test performed

The NGS Colon Molecular Profile assay method was described in this chapter, Case 23.1.

Test results

NGS Colon Molecular Profile was performed on a biopsy of the liver. The results showed *KRAS* p.(G12D) c.35G>A mutation (Fig. 23.2.1).

Results with interpretations

NGS detected *KRAS* exon 2 p.(G12D) mutation from this patient. There are no FDA-recommended treatments for this mutation, but an alert is shown. NCCN guidelines stated that patients with any known *KRAS* mutation (exon 2, 3, 4) or *NRAS* mutation (exon 2, 3, 4) should not be treated with either cetuximab or panitumumab medications for targeted therapy for colon cancer. However, the FDA has granted fast-track designation to the Polo-like Kinase 1 (*PLK1*) inhibitor, onvansertib, in combination with FOLFIRI and bevacizumab, for *KRAS* mutations in metastatic colorectal cancer in the second line (https://cardiffoncology.investorroom.com/2020-05-28-Cardiff-Oncology-Announces-Fast-Track-Designation-Granted-by-the-FDAto-Onvansertib-for-Second-Line-Treatment-of-KRAS-Mutated-Colorectal-Cancer).

Future testing and recommendations

Regular checkups with an oncologist are recommended. If any clinical exam suggests relapse in disease, molecular testing is indicated to see if any mutations/fusions have emerged.

Relevant Biomarkers

Tier	Genomic Alteration	Relevant Therapies (In this cancer type)	Clinical Trials
IA	*KRAS p.(G12D) c.35G>A*	None	43

Public data sources included in relevant therapies: FDA1, NCCN, **EMA**2, ESMO

Tier Reference: Li et al. *Standards and Guidelines for the Interpretation and Reporting of Sequence Variants in Cancer: A Joint Consensus Recommendation of the Association for Molecular Pathology, American Society of Clinical Oncology, and College of American Pathologists.* J Mol Diagn. 2017 Jan;19(1):4-23.

Alerts informed by public data sources: Contraindicated, Resistance

KRAS p.(G12D) c.35G>A	**cetuximab** [1,2], **cetuximab + chemotherapy** [2], **panitumumab** [1], **panitumumab + chemotherapy** [2]

Public data sources included in alerts: FDA1, NCCN, **EMA**2, ESMO

Variant Details

DNA Sequence Variants

Gene	Amino Acid Change	Coding	Locus	Allele Frequency	Variant Effect
KRAS	p.(G12D)	c.35G>A	chr12:25398284	56.41%	missense

FIG. 23.2.1 NGS detected *KRAS* p.(G12D) c.35G>A mutation. The contraindicated drugs are indicated.

Summary of key learning points

- Colorectal cancer is the third most common cancer in the United States.
- About 30% of colorectal cancers are hereditary, and the remaining 70% are sporadic or with unknown etiology.
- There are several biomarkers used in the diagnosis and targeted therapy for colon cancer.
- NCCN guideline recommends that NGS is preferred over individual gene testing for colon cancer patients.

References

[1] P. Lichtenstein, et al., Environmental and heritable factors in the causation of cancer—analyses of cohorts of twins from Sweden, Denmark, and Finland, N. Engl. J. Med. 343 (2) (2000) 78–85.

[2] E.R. Fearon, Molecular genetics of colorectal cancer, Annu. Rev. Pathol. 6 (2011) 479–507.

[3] K.W. Jasperson, et al., Hereditary and familial colon cancer, Gastroenterology 138 (6) (2010) 2044–2058.

[4] K.W. Jasperson, et al., Evaluating Lynch syndrome in very early onset colorectal cancer probands without apparent polyposis, Familial Cancer 9 (2) (2010) 99–107.

[5] J.M. Carethers, B.H. Jung, Genetics and genetic biomarkers in sporadic colorectal cancer, Gastroenterology 149 (5) (2015) 1177–1190. e3.

[6] S.H. Zaidi, et al., Landscape of somatic single nucleotide variants and indels in colorectal cancer and impact on survival, Nat. Commun. 11 (1) (2020) 3644.

[7] W.M. Grady, Genetic testing for high-risk colon cancer patients, Gastroenterology 124 (6) (2003) 1574–1594.

[8] A.B. Benson, et al., Colon cancer, version 2.2021, NCCN clinical practice guidelines in oncology, J. Natl. Compr. Cancer Netw. 19 (3) (2021) 329–359.

[9] C.M. Johnson, et al., Meta-analyses of colorectal cancer risk factors, Cancer Causes Control 24 (6) (2013) 1207–1222.

[10] E.K. Johnson, S.R. Steele, Evidence-based management of colorectal trauma, J. Gastrointest. Surg. 17 (9) (2013) 1712–1719.

[11] M.W. Lutgens, et al., Declining risk of colorectal cancer in inflammatory bowel disease: an updated meta-analysis of population-based cohort studies, Inflamm. Bowel Dis. 19 (4) (2013) 789–799.

[12] J. Cheng, et al., Meta-analysis of prospective cohort studies of cigarette smoking and the incidence of colon and rectal cancers, Eur. J. Cancer Prev. 24 (1) (2015) 6–15.

[13] K.M. De Bruijn, et al., Systematic review and meta-analysis of the association between diabetes mellitus and incidence and mortality in breast and colorectal cancer, Br. J. Surg. 100 (11) (2013) 1421–1429.

[14] K. Esposito, et al., Colorectal cancer association with metabolic syndrome and its components: a systematic review with meta-analysis, Endocrine 44 (3) (2013) 634–647.

[15] C.M. Kitahara, et al., Prospective investigation of body mass index, colorectal adenoma, and colorectal cancer in the prostate, lung, colorectal, and ovarian cancer screening trial, J. Clin. Oncol. 31 (19) (2013) 2450–2459.

24

Melanoma

Xia Li and Guang Liu
SONORA QUEST LABORATORIES, PHOENIX, AZ, UNITED STATES

Background

Melanomas are malignant tumors that develop from pigmented cells called melanocytes, the cells that are found in the skin, the eye, the mucosal epithelia, and many other locations of the body. Melanoma accounts for only about 1% of skin cancers but causes a large majority of skin cancer deaths. In 2022, about 99,780 new melanomas were diagnosed, and 7650 people are expected to die of melanoma (http://www.cancer.org).

There are many types of melanomas depending on where the tumors originated from. Ocular melanoma is a rare but potentially devastating malignancy arising from the melanocytes of the uveal tract, conjunctiva, or orbit; it represents less than 5% of all melanoma cases in the United States [1]. Cutaneous melanoma is a cancer that starts in the pigment-making cells of the skin and is the most common type of melanoma worldwide with 68,000 new cases diagnosed each year in the US population. Cutaneous melanoma is usually caused by too much exposure to ultraviolet rays from the sun and indoor tanning, and it can spread to other sites on the skin, such as the lymph nodes, lungs, brain, and soft tissue. It is known that uveal melanoma spreads through the blood, while cutaneous melanoma can spread through both the blood and the lymphatic system. The most common site of metastasis of uveal melanoma with >90% of cases is the liver [2–4].

Cutaneous, uveal, and conjunctival melanomas have different molecular signatures [5–8]. Whereas *BRAF, NRAS, KIT,* and *TERT* promoter mutations are extremely rare in uveal melanoma, they are more common in conjunctival and cutaneous melanomas [6–9]. Molecular markers commonly seen in uveal melanomas with prognostic significance alone include chromosomal abnormalities (particularly chromosomes 3 and 8) [10–12] and mutations in *GNAQ* or *GNA11* (>80% of uveal cases) [13,14], *BAP1* [15], *SF3B1* [16], and *EIFAX* [16,17].

Multiple tumor molecular markers are associated with increased risk and/or shorter time to development of distant metastases. Chromosomal changes including monosomy 3 and gain of 8q were among the first molecular markers to be found that are associated with the risk of distant metastasis in patients with uveal melanoma. Other chromosome abnormalities found were del(8p), del(1p), del(16q), and del(6q). A gain of 6q may be protective against metastasis, at least in the context of monosomy 3, and a gain of 8q [18].

According to NCCN guidelines of 2022, for diagnosis and prognostication of cutaneous and uveal melanomas, *BRAF, NRAS, KIT* mutations, and *BRAF* fusions should be included

Cases in Laboratory Genetics and Genomics (LGG) Practice. https://doi.org/10.1016/B978-0-323-99622-8.00012-5

in the molecular testing in stage IV patients. These mutations (*BRAF, KIT*) may impact treatment options for patients with metastatic melanoma, and some patients with activating *BRAF* mutations are likely to respond to *BRAF* inhibitors. PD-L1 expression study can also be used to help identify candidates for anti-PD-1 therapy [19,20].

Comparative genomic hybridization (CGH) or FISH may be helpful in the detection of copy number changes in atypical Spitz tumors [21]. The gene expression profile has been developed as well to evaluate the risk stratification of the melanoma patient [22,23].

MEKTOVI (binimetinib) is a kinase inhibitor indicated, in combination with BRAFTOVI (encorafenib), for the treatment of patients with unresectable or metastatic melanoma with a *BRAF* V600E or V600K mutation, as detected by an FDA-approved test. There are also many other drugs including cobimetinib + vemurafenib, dabrafenib + trametinib, etc., that can be used for patients with *BRAF*-positive mutations. For patients with *NRAS* mutations, anti-CTLA-4 + anti-PD-1 are used for the treatment of cutaneous melanoma with stage III and IV unresectable tumors [24,25].

The clinical utility of NGS has been shown in identifying the mutations associated with melanoma. In this chapter, we will illustrate two cases in which somatic mutations were detected by NGS to facilitate tailored adjuvant chemotherapy to mitigate the risk of recurrence.

Case 24.1 Metastatic melanoma with *BRAF* p.(V600K) mutation

Clinical indication

A 49-year-old female presented with a painful lump on her head, concerning osteoma or tumor. She was previously treated with craniectomy and cranioplasty. After resection, the underlying pigmented lesion was growing rapidly.

Test ordered

- NGS Melanoma Molecular Profile
- Hotspot: *AKT1, BRAF, CCND1, CD4K, CTNNB1, ERBB4, GNA11, GNAQ, KIT, MAP2K, NRAS.*

Laboratory test performed

NGS Melanoma Molecular Profile assay was performed. This assay is designed to detect genetic alterations present in tumor tissue specimens. The test detects single-nucleotide variants (SNVs) and small insertions/deletions (In/Dels) and copy number variations using CHEF/S5 GeneStudio sequencing platforms. Eleven genes were selected based on the actionability of mutations identified in those genes using currently available evidence from national and international guidelines and literature. Actionability is defined as information a clinician might find useful to aid in the diagnosis, prognosis, and/or treatment strategy for a patient. The results of the test should be correlated with clinical findings. This test was designed for the detection and annotation of somatic variants and is not intended to be a germline test. When the allele frequency is approximately 50%, the possibility of these being germline mutations cannot be excluded.

Test results

NGS Melanoma Molecular Profile was performed on a biopsy from the lesion on the patient's head. The results showed *BRAF* p.(V600K) c.1798_1799delGTinsAA (Fig. 24.1.1).

Results with interpretations

NGS detected *BRAF* V600K mutation from this patient. There are several FDA-recommended medications for targeted therapy of melanoma. MEKTOVI is a kinase inhibitor indicated, in combination with encorafenib, for the treatment of patients with unresectable or metastatic melanoma with a *BRAF* V600E or V600K mutation, as detected by an FDA-approved test.

Relevant Biomarkers

Tier	Genomic Alteration	Relevant Therapies (In this cancer type)	Clinical Trials
IA	*BRAF p.(V600K) c.1798_1799delGTinsAA* **Prognostic significance:** None **Diagnostic significance:** None	**atezolizumab + cobimetinib + vemurafenib** [1] **binimetinib + encorafenib** [1, 2] **cobimetinib + vemurafenib** [1, 2] **dabrafenib** [2] **dabrafenib + trametinib** [1, 2] **trametinib** [1, 2] BRAF inhibitor + MEK inhibitor encorafenib ipilimumab + nivolumab vemurafenib	28

Public data sources included in relevant therapies: FDA1, NCCN, **EMA**2, ESMO
Public data sources included in prognostic and diagnostic significance: NCCN, ESMO
Tier Reference: Li et al. *Standards and Guidelines for the Interpretation and Reporting of Sequence Variants in Cancer: A Joint Consensus Recommendation of the Association for Molecular Pathology, American Society of Clinical Oncology, and College of American Pathologists.* J Mol Diagn. 2017 Jan;19(1):4-23.

Variant Details

DNA Sequence Variants

Gene	Amino Acid Change	Coding	Variant ID	Locus	Allele Frequency	Transcript	Variant Effect
BRAF	p.(V600K)	c.1798_1799delGTins AA	COSM473	chr7:140453136	22.06%	NM_004333.4	missense

FIG. 24.1.1 NGS detected *BRAF* V600K mutation. The FDA-approved drugs are indicated.

Future testing and recommendations

Regular checkups with an oncologist are recommended. If any clinical exam suggests relapse in disease, molecular testing is indicated to see if any mutations/fusions have emerged.

Case 24.2 Metastatic melanoma with *NRAS* p.(Q61L) mutation

Clinical indication

A 74-year-old male presented with a left axillary lymph node, concerning metastatic melanoma with recurrence. He had a history of left scapula melanoma excised in 2018.

Test ordered

- NGS Melanoma Molecular Profile

Laboratory test performed

The NGS Melanoma Molecular Profile assay method was described in this chapter, Case 24.1.

Test results

NGS Melanoma Molecular Profile was performed on a lymph node biopsy from this patient. The results showed *NRAS* p.(Q61L) c.182A>T mutation (Fig. 24.2.1).

Relevant Biomarkers

Tier	Genomic Alteration	Relevant Therapies (In this cancer type)	Clinical Trials
IA	*NRAS p.(Q61L) c.182A>T*	anti-CTLA-4 + anti-PD-1 anti-PD-1 binimetinib	16

Public data sources included in relevant therapies: FDA1, NCCN, **EMA**2, ESMO
Tier Reference: Li et al. *Standards and Guidelines for the Interpretation and Reporting of Sequence Variants in Cancer: A Joint Consensus Recommendation of the Association for Molecular Pathology, American Society of Clinical Oncology, and College of American Pathologists.* J Mol Diagn. 2017 Jan;19(1):4-23.

Variant Details

DNA Sequence Variants

Gene	Amino Acid Change	Coding	Locus	Allele Frequency	Variant Effect
NRAS	p.(Q61L)	c.182A>T	chr1:115256529	52.03%	missense

FIG. 24.2.1 NGS detected *NRAS* Q61K mutation. The FDA-approved drugs are indicated.

Results with interpretations

NGS detected *NRAS* Q61K mutation from this patient. There are FDA-recommended medications for targeted therapy of melanoma. Anti-CTLA-4 + anti-PD-1 is used for the first-line therapy to treat cutaneous melanoma with stage III or IV, unresectable disease. The NCCN has recommended binimetinib for the treatment of cutaneous, metastatic, unresectable melanoma as a second-line therapy.

Future testing and recommendations

Regular checkups with an oncologist are recommended. If any clinical exam suggests relapse in disease, molecular testing is indicated to see if any mutations/fusions have emerged.

Summary of key learning points

- Melanoma causes a large majority of skin cancer deaths in the United States.
- There are many types of melanomas, of which cutaneous melanoma is the most common.
- There are several biomarkers used in the diagnosis and targeted therapy for melanomas.
- NCCN guidelines recommend that NGS is preferred over individual gene testing for melanoma patients.

References

[1] E.S. Blum, et al., Clinical management of uveal and conjunctival melanoma, Oncology (Williston Park) 30 (1) (2016) 29–32. 34-43, 48.

[2] W.J. Kheir, J.S. Kim, M.A. Materin, Multiple uveal melanoma, Ocul. Oncol. Pathol. 6 (5) (2020) 368–375.

[3] M.A. Materin, M. Faries, H.M. Kluger, Molecular alternations in uveal melanoma, Curr. Probl. Cancer 35 (4) (2011) 211–224.

[4] E.A. Hurst, J.W. Harbour, L.A. Cornelius, Ocular melanoma: a review and the relationship to cutaneous melanoma, Arch. Dermatol. 139 (8) (2003) 1067–1073.

[5] M.J. de Lange, et al., Distribution of GNAQ and GNA11 mutation signatures in uveal melanoma points to a light dependent mutation mechanism, PLoS One 10 (9) (2015), e0138002.

[6] W. Zuidervaart, et al., Activation of the MAPK pathway is a common event in uveal melanomas although it rarely occurs through mutation of BRAF or RAS, Br. J. Cancer 92 (11) (2005) 2032–2038.

[7] H. Gear, et al., BRAF mutations in conjunctival melanoma, Invest. Ophthalmol. Vis. Sci. 45 (8) (2004) 2484–2488.

[8] C. Beadling, et al., KIT gene mutations and copy number in melanoma subtypes, Clin. Cancer Res. 14 (21) (2008) 6821–6828.

[9] S.C. Edmunds, et al., Absence of BRAF gene mutations in uveal melanomas in contrast to cutaneous melanomas, Br. J. Cancer 88 (9) (2003) 1403–1405.

[10] B. Damato, J.A. Dopierala, S.E. Coupland, Genotypic profiling of 452 choroidal melanomas with multiplex ligation-dependent probe amplification, Clin. Cancer Res. 16 (24) (2010) 6083–6092.

[11] N. Cassoux, et al., Genome-wide profiling is a clinically relevant and affordable prognostic test in posterior uveal melanoma, Br. J. Ophthalmol. 98 (6) (2014) 769–774.

[12] R. Caines, et al., Cluster analysis of multiplex ligation-dependent probe amplification data in choroidal melanoma, Mol. Vis. 21 (2015) 1–11.

[13] C.D. Van Raamsdonk, et al., Mutations in GNA11 in uveal melanoma, N. Engl. J. Med. 363 (23) (2010) 2191–2199.

[14] A.E. Koopmans, et al., Patient survival in uveal melanoma is not affected by oncogenic mutations in GNAQ and GNA11, Br. J. Cancer 109 (2) (2013) 493–496.

[15] J.W. Harbour, et al., Frequent mutation of BAP1 in metastasizing uveal melanomas, Science 330 (6009) (2010) 1410–1413.

[16] J.W. Harbour, et al., Recurrent mutations at codon 625 of the splicing factor SF3B1 in uveal melanoma, Nat. Genet. 45 (2) (2013) 133–135.

[17] M. Martin, et al., Exome sequencing identifies recurrent somatic mutations in EIF1AX and SF3B1 in uveal melanoma with disomy 3, Nat. Genet. 45 (8) (2013) 933–936.

[18] C.L. Shields, et al., Personalized prognosis of uveal melanoma based on cytogenetic profile in 1059 patients over an 8-year period: the 2017 Harry S. Gradle lecture, Ophthalmology 124 (10) (2017) 1523–1531.

[19] P.K. Rao, et al., NCCN guidelines insights: uveal melanoma, version 1.2019, J. Natl. Compr. Cancer Netw. 18 (2) (2020) 120–131.

[20] S.M. Swetter, et al., NCCN guidelines(R) insights: melanoma: cutaneous, version 2.2021, J. Natl. Compr. Cancer Netw. 19 (4) (2021) 364–376.

[21] L. Raskin, et al., Copy number variations and clinical outcome in atypical spitz tumors, Am. J. Surg. Pathol. 35 (2) (2011) 243–252.

[22] P. Gerami, et al., Gene expression profiling for molecular staging of cutaneous melanoma in patients undergoing sentinel lymph node biopsy, J. Am. Acad. Dermatol. 72 (5) (2015) 780–5 e3.

[23] P. Gerami, et al., Development of a prognostic genetic signature to predict the metastatic risk associated with cutaneous melanoma, Clin. Cancer Res. 21 (1) (2015) 175–183.

[24] O. Michielin, et al., Cutaneous melanoma: ESMO clinical practice guidelines for diagnosis, treatment and follow-updagger, Ann. Oncol. 30 (12) (2019) 1884–1901.

[25] A.A. van de Loosdrecht, I. Mandac Smoljanovic, EHA endorsement of ESMO clinical practice guidelines for diagnosis, treatment, and follow-up for myelodysplastic syndromes, Hemasphere 6 (3) (2022), e695.

25

Breast cancer

Guang Liu and Xia Li
SONORA QUEST LABORATORIES, PHOENIX, AZ, UNITED STATES

Background

Breast cancer is the most common form of tumor and is the leading cause of cancer-related deaths in the female population worldwide [1]. Breast cancer is diagnosed in approximately 12% of women in the United States throughout their lifetimes [2]. On the molecular level, breast cancer is a heterogeneous disease; molecular features include activation of human epidermal growth factor receptor 2 (*HER2*, encoded by *ERBB2*), activation of hormone receptors (HR, estrogen receptor/ER, and progesterone receptor/PR), mutations of *BRCA*, *PIK3CA*, and some other genes [3]. The genes that are commonly altered in hereditary breast and ovarian cancer are *BRCA* genes (*BRCA1* and *BRCA2*). About 3% of breast cancers and 10% of ovarian cancers result from inherited mutations in the *BRCA1* and *BRCA2* genes [4]. Treatment strategies for breast cancer differ according to molecular subtype. Management of breast cancer is multidisciplinary and includes locoregional (surgery and radiation therapies) and systemic therapy approaches. Systemic therapies include endocrine therapy for hormone receptor-positive disease, chemotherapy, trastuzumab-based treatments for *HER2*-positive disease, poly(ADP-ribose) polymerase inhibitors for *BRCA* mutation carriers, immunotherapy, and PI3K Inhibitors for *PIK3CA* mutations in HR-positive/*HER2* negative advanced breast cancer [1,5].

HER2 is overexpressed and/or amplified in approximately 20% of breast cancers. This somatically acquired genetic alteration is associated with shorter disease-free survival and poorer overall survival in patients without the *HER2*-targeted therapy. Because *HER2*-targeted therapies have significantly improved outcomes for patients with this alteration, accurate assessment of this alteration is considered critically important [6]. To standardize evaluations of *HER2* status, the American Society of Clinical Oncology (ASCO) and the College of American Pathologists (CAP) have convened committees to establish guidelines for the assessment of *HER2* status. These guidelines were published in 2007, 2013, and 2018. The interpretative approaches to the assessment of *HER2* gene amplification by FISH recommended by the ASCO/CAP guidelines have changed over the years, with current guidelines designating five different groups according to *HER2* FISH ratio and average *HER2* gene copy number per tumor cell (Table 25.1) [7–9].

HER2-low is a new classification of the *HER2* subtype. It describes a new subtype of breast cancer that has some HER2 proteins on the cell surface, but not enough to be classified as HER2-positive. On Aug. 5, 2022, FDA approved Enhertu (fam-trastuzumab-

Cases in Laboratory Genetics and Genomics (LGG) Practice. https://doi.org/10.1016/B978-0-323-99622-8.00024-1

Table 25.1 Algorithm for evaluation of *HER2* gene amplification by an ISH assay using the dual color probes.

ASCO/CAP result category	HER2:D17Z1 ratio Average HER2 copies per cell	Reporting approach per 2018 ASCO/CAP guidelines
Group 1	HER2:D17Z1 ≥2.0 HER2/cell ≥4.0	Positive
Group 2	HER2:D17Z1 ≥2.0 HER2/cell <4.0	Reflex IHC; FISH reanalysis if 2+
Group 3	HER2:D17Z1 <2.0 HER2/cell ≥6.0	Reflex IHC; FISH reanalysis if 2+
Group 4	HER2:D17Z1 <2.0 HER2/cell ≥4.0 and <6.0	Reflex IHC; FISH reanalysis if 2+
Group 5	HER2:D17Z1 <2.0 HER2/cell <4.0	Negative

deruxtecan-nxki), an IV infusion for the treatment of patients with unresectable (unable to be removed) or metastatic (spread to other parts of the body) *HER2*-low breast cancer (http://www.FDA.gov). In a clinical trial involving patients with *HER2*-low metastatic breast cancer, trastuzumab deruxtecan resulted in significantly longer progression-free survival and overall survival than the physician's choice of chemotherapy [10].

PIK3CA is the most frequently mutated gene in HR-positive (HR+)/*HER2* negative (*HER2*−) breast cancer; approximately 40% of patients living with HR+/*HER2*- breast cancer have this mutation. *PIK3CA* mutations are associated with tumor growth, resistance to endocrine treatment, and a poor overall prognosis [11]. Piqray (alpelisib) was the first ever treatment approved by the FDA specifically for HR+/*HER2*− advanced breast cancer with a *PIK3CA* mutation [5]. *PIK3CA* somatic mutations in breast cancer are highly heterogeneous. Compared with the PCR method, NGS allows coverage of more different variants simultaneously, even from low input of DNA. It also provides information on the fraction of alleles carrying the mutation [12]. This chapter uses a few clinical cases as examples to demonstrate how the identification of *HER2* amplification and *PIK3CA* mutation can guide targeted therapy.

Case 25.1 Invasive ductal carcinoma of breast origin with *HER2* amplification

Clinical indication

A 55-year-old female presented to her primary care physician with a lump in her left breast. She had a 4.6 cm left breast mass and 2 palpable axillary lymph nodes. Her ultrasound and mammogram confirmed these physical findings. She was referred to an oncologist and had a core needle biopsy, which returned a diagnosis of invasive ductal carcinoma. She then underwent a mastectomy and clearance of axillary nodes on the left side. She received six cycles of docetaxel, trastuzumab, and pertuzumab, followed by trastuzumab and pertuzumab for another 12 months. She did not have a family history of breast cancer.

Test ordered

- FISH: *HER2* gene amplification
- Hormone receptors (PR/ER) by immunohistochemistry

Laboratory test performed

The *HER2* DNA probe (Dual-color *HER2/CEP17* DNA probe kit, Abbott Molecular, Inc.) is designed to detect the amplification of the *HER2*/neu gene in interphase nuclei from formalin-fixed, paraffin-embedded breast tissue. The formalin fixation time was between 6 and 72 hours. The *HER2*/neu gene located on chromosome 17 and in normal tissue is expressed as 2 copies/cell. Evaluation of patient amplification status was achieved by determining *HER2*/neu copy number relative to the number of centromeric chromosome 17 (*CEP 17*) derived signals present in each cell nucleus. Additionally, the number of *HER2*/neu copy numbers per cell is calculated. A minimum of 40 interphase tumor cell nuclei were studied. A replicate histology preparation of this tumor was reviewed by a consulting pathologist, and regions for the study were identified.

The status of hormone receptors (PR/ER) was evaluated via immunohistochemistry in the pathology lab.

Test results

HER2 (*ERBB2*) is positive for amplification (Group 1) by FISH analysis (Fig. 25.1.1)

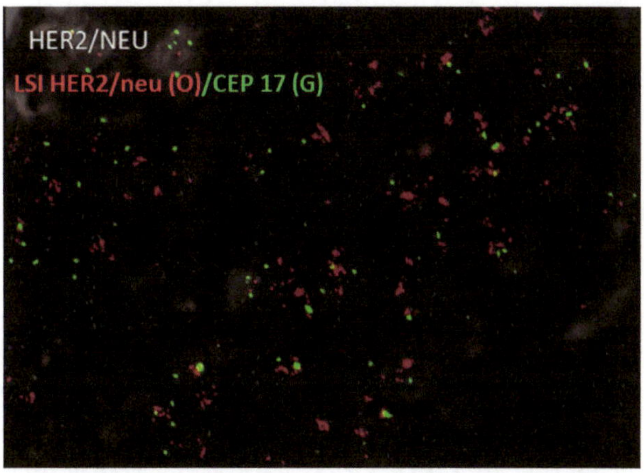

FIG. 25.1.1 FISH was positive for *HER2* amplification. ISCN: nuc ish(D17Z1x1~3,ERBB2x5~20)[40]

Ratio of HER2 to D17Z1: 4.1
Average HER2 copies/cell: 9.8
Average centromere 17 copies/cell: 2.3
Number of cells scored: 40

The status of hormone receptors (PR/ER) was evaluated, and the results showed negative.

Results with interpretations

FISH was positive for *HER2* amplification (Group 1). Both ER and PR were negative by Immunohistochemistry. Therefore, the anti-HER2 antibody, trastuzumab, approved by the FDA can be used for this patient.

Future testing and recommendations

Regular history/physical examinations are recommended every 4–6 months for the first 5 years after primary therapy and then annually thereafter. Mammogram should be performed annually. If any clinical exam suggests relapse in disease, molecular testing is indicated to see if any mutations/fusions have emerged.

Case 25.2 Adenocarcinoma of breast origin with *PIK3CA* p.(E545K) mutation

Clinical indication

A 67-year-old female was diagnosed with right invasive ductal carcinoma (IDC) and initially treated by lumpectomy, then radiation brachytherapy. Estrogen receptor (ER) and progesterone receptor (PR) showed positive results, and *HER2* was negative. She declined chemotherapy and hormonal therapy (which she took for 1 week and discontinued due to side effects). After 10 years, she had pain in her ribs. A bone scan showed bilateral multifocal osseous metastatic disease. A subcarinal lymph node biopsy was performed. She did not have a family history of breast cancer.

Test ordered

- FISH: HER2 gene amplification
- Hormone receptor (PR/ER) by immunohistochemistry
- NGS Solid Tumor Panel
- Hotspot: *AKT1, ALK, AR, BRAF, CDK4, CTNNB1, DDR2, EGFR, ERBB2, ERBB3, ERBB4, ESR1, FGFR2, FGFR3, GNA11, GNAQ, HRAS, IDH1, IDH2, JAK1, JAK2, JAK3, KIT, KRAS, MAP2K1, MAP2K2, MET, MTOR, NRAS, PDGFRA, PIK3CA, RAF1, RET, ROS1, SMO*
- Copy number variation (CNV): *ALK, AR, BRAF, CCND1, CDK4, CDK6, EGFR, ERBB2, FGFR1, FGFR2, FGFR3, FGFR4, KIT, KRAS, MET, MYC, MYCN, PDGFRA, PIK3CA*
- Fusion: *ABL1, AKT3, ALK, AXL, BRAF, EGFR, ERBB2, ERG, ETV1, ETV4, ETV5, FGFR1, FGFR2, FGFR3, MET, NTRK1, NTRK2, NTRK3, PDGFRA, PPARG, RAF1, RET, ROS1*

Laboratory test performed

The *HER2* FISH method was described in this chapter, Case 25.1.

The status of hormone receptors (PR/ER) was evaluated via immunohistochemistry in the pathology lab.

NGS Solid Tumor Molecular Profile assay was performed. This assay is designed to detect genetic alterations present in tumor tissue specimens. The test detects single-nucleotide variants (SNVs) and small insertions/deletions (In/Dels), copy number variations, and fusions using CHEF/S5 GeneStudio sequencing platforms. The 52 genes were selected based on the actionability of mutations identified in those genes using currently available evidence from national and international guidelines and literature. Actionability is defined as information a clinician might find useful to aid in the diagnosis, prognosis, and/or treatment strategy for a patient. The results of the test should be correlated with clinical findings. This test was designed for the detection and annotation of somatic variants and is not intended to be a germline test. When the allele frequency is approximately 50%, the possibility of these being germline mutations cannot be excluded.

Test results

HER2 (*ERBB2*) was negative for amplification (Group 5) by FISH analysis (Fig. 25.2.1)

Ratio of HER2 to D17Z1: 1.0

Average HER2 copies/cell: 2.0

Average centromere 17 copies/cell: 1.9

Number of cells scored: 40

The status of hormone receptors (PR/ER) was evaluated in the pathology lab and the results showed positive.

NGS Solid Tumor panel was conducted and detected *PIK3CA* p.(E545K) c.1633G>A mutation (Fig. 25.2.2).

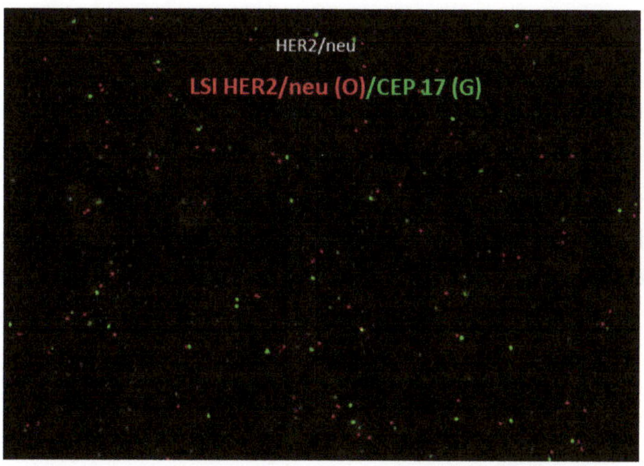

FIG. 25.2.1 FISH was negative for *HER2* amplification. ISCN: nuc ish(D17Z1,ERBB2)x2[40]

Relevant Biomarkers

Tier	Genomic Alteration	Relevant Therapies (In this cancer type)	Clinical Trials
IA	*PIK3CA p.(E545K) c.1633G>A*	alpelisib + hormone therapy [1, 2]	24

Public data sources included in relevant therapies: FDA1, NCCN, EMA2, ESMO
Tier Reference: Li et al. *Standards and Guidelines for the Interpretation and Reporting of Sequence Variants in Cancer: A Joint Consensus Recommendation of the Association for Molecular Pathology, American Society of Clinical Oncology, and College of American Pathologists.* J Mol Diagn. 2017 Jan;19(1):4-23.

Variant Details

DNA Sequence Variants

Gene	Amino Acid Change	Coding	Locus	Allele Frequency	Variant Effect
PIK3CA	p.(E545K)	c.1633G>A	chr3:178936091	80.83%	missense

FIG. 25.2.2 NGS detected *PIK3CA* p.(E545K) mutation. FDA-approved drugs and clinical trials are available.

Results with interpretations

FISH was negative for *HER2* amplification (Group 5). Both ER and PR were positive by Immunohistochemistry. NGS detected *PIK3CA* p.(E545K) mutation; therefore, FDA-approved orally available α-selective *PIK3CA* inhibitor, alpelisib, along with a hormone drug can be used for this patient (https://www.fda.gov/news-events/press-announcements/fda-approves-first-pi3k-inhibitor-breast-cancer).

Future testing and recommendations

Regular history/physical examinations are recommended every 4–6 months for the first 5 years after primary therapy and then annually thereafter. Mammogram should be performed annually. If any clinical exam suggests relapse in disease, molecular testing is indicated to see if any mutations/fusions have emerged.

Case 25.3 Adenocarcinoma of breast origin with *PIK3CA* p.(E545K) mutation and *FGFR1* amplification

Clinical indication

An 87-year-old female with a history of metastatic cancer to the bone presented with generalized pain, weakness, and decreased appetite since her last radiation treatment. The diagnosis from pathology examination was metastatic poorly differentiated adenocarcinoma, of breast origin. The tissue from the right acetabulum bone was sent for genetic testing.

Test ordered

- NGS Solid Tumor Panel (see Case 25.2)

Relevant Biomarkers

Tier	Genomic Alteration	Relevant Therapies (In this cancer type)	Clinical Trials
IIC	*PIK3CA p.(E545K) c.1633G>A*	None	13
IIC	*FGFR1 amplification*	None	4

Public data sources included in relevant therapies: FDA1, NCCN, **EMA**2, ESMO
Tier Reference: Li et al. *Standards and Guidelines for the Interpretation and Reporting of Sequence Variants in Cancer: A Joint Consensus Recommendation of the Association for Molecular Pathology, American Society of Clinical Oncology, and College of American Pathologists.* J Mol Diagn. 2017 Jan;19(1):4-23.

Variant Details

DNA Sequence Variants

Gene	Amino Acid Change	Coding	Locus	Allele Frequency	Variant Effect
PIK3CA	p.(E545K)	c.1633G>A	chr3:178936091	38.50%	missense

Copy Number Variations

Gene	Locus	Copy Number
FGFR1	chr8:38271445	19.48

FIG. 25.3.1 NGS detected *PIK3CA* p.(E545K) mutation. Fast-track treatment and clinical trials are available.

Laboratory test performed

The NGS Solid Tumor assay method was described in this chapter, Case 25.2.

Test results

NGS Solid Tumor panel was performed and detected *PIK3CA* p.(E545K) c.1633G>A, and *FGFR1* amplification (Fig. 25.3.1).

Results with interpretations

NGS detected *PIK3CA* p.(E545K) mutation. Since hormone receptors (PR/ER) status was not known, therefore, no FDA-approved orally available α-selective *PIK3CA* inhibitor, alpelisib, can be used for this patient (https://www.fda.gov/news-events/press-announcements/fda-approves-first-pi3k-inhibitor-breast-cancer).

In addition, the FDA has granted fast-track designation to the *FGFR 1-3* inhibitor debio 1347 for *FGFR1/2/3* alterations in unresectable or metastatic solid tumors (https://www.debiopharm.com/drug-development/press-releases/fda-grants-fast-track-designation-to-debiopharm-internationalsdebio-1347-for-the-treatment-of-patients-with-unresectable-or-metastatic-tumors-with-a-specific-fgfr-gene-alteration/).

Future testing and recommendations

Regular history/physical examinations are recommended every 4–6 months for the first 5 years after primary therapy and then annually thereafter. Mammogram should be performed annually. If any clinical exam suggests relapse in disease, molecular testing is indicated to see if any mutations/fusions have emerged.

Summary of key learning points

- Breast cancer is the most common form of tumor and is the leading cause of cancer-related deaths in the female population worldwide.
- On the molecular level, breast cancer is a heterogeneous disease; molecular features include activation of *HER2*, activation of hormone receptors, and mutations of *BRCA*, *PIK3CA*, and some other genes.
- *HER2* gene is amplified in approximately 20% of breast cancer and is associated with shorter disease-free survival and poorer overall survival in breast cancers. *HER2*-targeted therapies have significantly improved outcomes. Therefore, an accurate assessment of *HER2* status is critical.
- *PIK3CA* mutations are highly represented in HR+/*HER2*− breast cancer and are of clinical interest due to the availability of targeted therapy.

References

[1] N. Harbeck, et al., Breast cancer, Nat. Rev. Dis. Primers 5 (1) (2019) 66.

[2] A.G. Waks, E.P. Winer, Breast cancer treatment: a review, JAMA 321 (3) (2019) 288–300.

[3] S. Loibl, et al., Breast cancer, Lancet 397 (10286) (2021) 1750–1769.

[4] R.L. Nussbaum, R.R. McInnes, H.F. Willard, Thompson & Thompson Genetics in Medicine, eighth ed., Elsevier, Philadelphia, PA, 2016. xi, 546 pages.

[5] O. Martinez-Saez, et al., Frequency and spectrum of PIK3CA somatic mutations in breast cancer, Breast Cancer Res. 22 (1) (2020) 45.

[6] O. Martinez-Saez, A. Prat, Current and future management of HER2-positive metastatic breast cancer, JCO Oncol. Pract. 17 (10) (2021) 594–604.

[7] A.C. Wolff, et al., Human epidermal growth factor receptor 2 testing in breast cancer: American Society of Clinical Oncology/College of American Pathologists clinical practice guideline focused update, J. Clin. Oncol. 36 (20) (2018) 2105–2122.

[8] A.M. Gordian-Arroyo, D.L. Zynger, G.H. Tozbikian, Impact of the 2018 ASCO/CAP HER2 guideline focused update, Am. J. Clin. Pathol. 152 (1) (2019) 17–26.

[9] W.J. Gradishar, et al., Breast cancer, version 3.2022, NCCN clinical practice guidelines in oncology, J. Natl. Compr. Cancer Netw. 20 (6) (2022) 691–722.

[10] S. Modi, et al., Trastuzumab deruxtecan in previously treated HER2-low advanced breast cancer, N. Engl. J. Med. 387 (1) (2022) 9–20.

[11] T.W. Miller, et al., Mutations in the phosphatidylinositol 3-kinase pathway: role in tumor progression and therapeutic implications in breast cancer, Breast Cancer Res. 13 (6) (2011) 224.

[12] N. Fusco, et al., PIK3CA mutations as a molecular target for hormone receptor-positive, HER2-negative metastatic breast cancer, Front. Oncol. 11 (2021), 644737.

26

Thyroid cancer

Xia Li and Guang Liu
SONORA QUEST LABORATORIES, PHOENIX, AZ, UNITED STATES

Background

Thyroid cancer occurs in the cells of the thyroid and can be classified into four types according to the origin of cells and the rate of cancer cell division: papillary thyroid carcinoma (PTC), follicular thyroid cancer (FTC), medullary thyroid cancer (MTC), and anaplastic thyroid cancer (ATC).

PTC is the most common type of thyroid cancer and consists of 70%–80% of thyroid cancer. FTC accounts for less than 15% of thyroid cancers. PTC and FTC, as well as the less common Hürthle cell carcinoma (a variant of FTC), are classified as differentiated thyroid carcinoma (DTC) [1,2], which originated from follicular epithelial thyroid cells. Both PTC and FTC are slow to progress and usually have a good prognosis, especially if diagnosed early. MTC accounts for about 3% of all thyroid cancers [3] and is developed by C-cells or parafollicular cells that produce calcitonin. When calcitonin level is elevated, it indicates cancer. MTC is more likely to run in the family [4]. ATC is rare thyroid cancer, which accounts for less than 2% of all thyroid cancers. ATC originates from follicular cells but does not have its original biological characteristics [5]. It is the most invasive type among all thyroid cancers [6] and is not sensitive to conventional treatment [7]. The prognosis is also the worst, with a 5-year survival rate of 5% [8]. Overall, thyroid cancers including PTC, FTC, and ATC are more common in women than men except for MTC (in which women and men are equally affected).

Most thyroid cancers are sporadic, and only 5% are familial. About 25% of MTC is inherited as an autosomal trait [9]. Certain genetic syndromes increase the risk of thyroid cancer, such as familial MTC and multiple endocrine neoplasms. Multiple endocrine neoplasia type 1 (MEN1) is caused by mutations in the *MEN1* gene and transmitted as an autosomal dominant trait. MEN1 may cause tumors in the parathyroid gland, pituitary gland, and pancreas. Multiple endocrine neoplasia type 2 (MEN2) includes three types (MEN2A, MEN2B, and FMTC) and is caused by *RET* gene mutations. MEN2 may develop tumors such as MTC, pheochromocytomas, hyperparathyroidism, and other conditions [9].

Many genetic alterations play an important role in thyroid cancer development. Chromosomal translocations were found in thyroid cancer, such as peroxisome proliferation-activated receptor (PPAR gamma) translocations in about 30% of follicular thyroid cancer cases [10]. The most common gene mutations reported are those from *BRAF* [11] and the *RAS* family [3].

Cases in Laboratory Genetics and Genomics (LGG) Practice. https://doi.org/10.1016/B978-0-323-99622-8.00001-0

According to NCCN guidelines of 2022, for diagnosis and prognostication of patients with thyroid cancer, molecular testing should include *BRAF, NTRK, ALK, RET,* MSI and dMMR, etc. These mutations (e.g., *BRAF*) may impact treatment options for patients with metastatic cancer, and some patients with activating *BRAF* mutations are likely to respond to BRAF inhibitors. PD-L1 expression study can also be used to help identify candidates for anti-PD-1 therapy [12].

The clinical utility of NGS has been shown for identifying the mutations associated with thyroid cancer. In this chapter, we will illustrate two cases of thyroid cancer in which somatic mutations were detected by NGS to facilitate tailored adjuvant chemotherapy to mitigate the risk of recurrence.

Case 26.1 Anaplastic thyroid carcinoma with *BRAF* p.(V600E) mutation

Clinical indication

A 70-year-old female presented with neck pain, fever, chills, and headache. Neck CT showed thickened epiglottis. The left thyroid mass was biopsied. The surgical pathology report stated that anaplastic thyroid carcinoma was the final diagnosis.

Test ordered

- NGS solid tumor molecular profile

Laboratory test performed

NGS solid tumor molecular profile assay method was described previously in Chapter 25, Case 25.1.

Test results

NGS solid tumor molecular profile was performed on a biopsy from the left thyroid. The results showed *BRAF* p.(V600E) c.1799 T>A mutation (Fig. 26.1.1).

Results with interpretations

Immunohistochemical stains were performed, and the results showed strong cytoplasmic expression of BRAF protein in most tumor cells, suggesting a *BRAF* mutation (results not shown). Concordantly, NGS detected a *BRAF* V600E mutation from this patient. There are several FDA-recommended medications for targeted therapy of thyroid cancer. MEKINIST (trametinib) is indicated, in combination with dabrafenib, for the treatment of patients with locally advanced or metastatic anaplastic thyroid cancer (ATC) with *BRAF* V600E mutation and with no satisfactory locoregional treatment options.

Relevant Biomarkers

Tier	Genomic Alteration	Relevant Therapies (In this cancer type)	Clinical Trials
IA	*BRAF p.(V600E) c.1799T>A* **Prognostic significance:** ATA-DTC Risk: Low to Intermediate **Diagnostic significance:** None	**dabrafenib + trametinib** [1] dabrafenib vemurafenib	17

Public data sources included in relevant therapies: FDA1, NCCN, **EMA**2, ESMO
Public data sources included in prognostic and diagnostic significance: NCCN, ESMO
Tier Reference: Li et al. *Standards and Guidelines for the Interpretation and Reporting of Sequence Variants in Cancer: A Joint Consensus Recommendation of the Association for Molecular Pathology, American Society of Clinical Oncology, and College of American Pathologists.* J Mol Diagn. 2017 Jan;19(1):4-23.

Variant Details

DNA Sequence Variants

Gene	Amino Acid Change	Coding	Locus	Allele Frequency	Variant Effect
BRAF	p.(V600E)	c.1799T>A	chr7:140453136	19.89%	missense

FIG. 26.1.1 NGS detected *BRAF* V600E mutation. The FDA-approved drugs are indicated.

Future testing and recommendations

Regular checkups with an oncologist are recommended. If any clinical exam suggests relapse in disease, molecular testing is indicated to see if any mutations/fusions have emerged.

Case 26.2 Malignant thyroid cancer with *KRAS* p.(G12V) mutation

Clinical indication

A 71-year-old female presented with a thyroid mass. She complained of dysphagia and shortness of breath. The surgical pathology report showed high-grade malignant neoplasm. Thyroid cancer was suspected.

Test ordered

- NGS solid tumor molecular profile

Laboratory test performed

NGS solid tumor molecular profile assay method was described previously in Chapter 25, Case 25.1.

Test results

NGS solid tumor molecular profile was performed on a biopsy from the thyroid isthmus. The results showed *KRAS* p.(G12V) c.35G>T mutation (Fig. 26.2.1).

Results with interpretations

NGS detected *KRAS* G12V mutation from this patient. Although ESMO clinical practice guidelines have recommended treating thyroid gland medullary carcinoma with cabozantinib for metastatic thyroid cancer patients, the diagnosis of this patient is undifferentiated (anaplastic) thyroid carcinoma [13]. She had a tracheostomy and was followed up with radiation therapy.

Future testing and recommendations

Regular checkups with an oncologist are recommended. If any clinical exam suggests relapse in disease, molecular testing is indicated to see if any mutations/fusions have emerged.

Summary of key learning points

- Thyroid cancers have four different types.
- PTC and FTC are slow to progress and have a good prognosis; ATC is the most aggressive cancer; MTC mostly runs in families.
- There are several biomarkers used in the diagnosis and targeted therapy for thyroid cancer.
- NCCN guidelines recommend that NGS is preferred over individual gene testing for thyroid cancer patients.

Relevant Biomarkers

Tier	Genomic Alteration	Relevant Therapies (In this cancer type)	Clinical Trials
IA	*KRAS p.(G12V) c.35G>T* **Prognostic significance:** None **Diagnostic significance:** None	cabozantinib	23

Public data sources included in relevant therapies: **FDA**1, NCCN, **EMA**2, ESMO
Public data sources included in prognostic and diagnostic significance: NCCN, ESMO
Tier Reference: Li et al. *Standards and Guidelines for the Interpretation and Reporting of Sequence Variants in Cancer: A Joint Consensus Recommendation of the Association for Molecular Pathology, American Society of Clinical Oncology, and College of American Pathologists.* J Mol Diagn. 2017 Jan;19(1):4-23.

Variant Details

DNA Sequence Variants

Gene	Amino Acid Change	Coding	Locus	Allele Frequency	Variant Effect
KRAS	p.(G12V)	c.35G>T	chr12:25398284	59.70%	missense

FIG. 26.2.1 NGS detected *KRAS* G12V mutation. The FDA-approved drug is indicated.

References

[1] J. Arribas, et al., Expression of YY1 in differentiated thyroid cancer, Endocr. Pathol. 26 (2) (2015) 111–118.

[2] R. Nagy, M.D. Ringel, Genetic predisposition for nonmedullary thyroid cancer, Horm. Cancer 6 (1) (2015) 13–20.

[3] Y.E. Nikiforov, M.N. Nikiforova, Molecular genetics and diagnosis of thyroid cancer, Nat. Rev. Endocrinol. 7 (10) (2011) 569–580.

[4] R.M. Carneiro, et al., Targeted therapies in advanced differentiated thyroid cancer, Cancer Treat. Rev. 41 (8) (2015) 690–698.

[5] S. Chiacchio, et al., Anaplastic thyroid cancer: prevalence, diagnosis and treatment, Minerva Endocrinol. 33 (4) (2008) 341–357.

[6] B. Xu, R. Ghossein, Genomic landscape of poorly differentiated and anaplastic thyroid carcinoma, Endocr. Pathol. 27 (3) (2016) 205–212.

[7] G. Chiappetta, et al., PATZ1 acts as a tumor suppressor in thyroid cancer via targeting p53-dependent genes involved in EMT and cell migration, Oncotarget 6 (7) (2015) 5310–5323.

[8] J.K. Hoang, X.V. Nguyen, L. Davies, Overdiagnosis of thyroid cancer: answers to five key questions, Acad. Radiol. 22 (8) (2015) 1024–1029.

[9] M.B. Lodish, C.A. Stratakis, RET oncogene in MEN2, MEN2B, MTC and other forms of thyroid cancer, Expert. Rev. Anticancer. Ther. 8 (4) (2008) 625–632.

[10] P. Raman, R.J. Koenig, Pax-8-PPAR-gamma fusion protein in thyroid carcinoma, Nat. Rev. Endocrinol. 10 (10) (2014) 616–623.

[11] Cancer Genome Atlas Research, N, Integrated genomic characterization of papillary thyroid carcinoma, Cell 159 (3) (2014) 676–690.

[12] S. Ahn, et al., Comprehensive screening for PD-L1 expression in thyroid cancer, Endocr. Relat. Cancer 24 (2) (2017) 97–106.

[13] S. Filetti, et al., Thyroid cancer: ESMO clinical practice guidelines for diagnosis, treatment and follow-updagger, Ann. Oncol. 30 (12) (2019) 1856–1883.

27

Pediatric solid tumors

Mikako Warren
CHILDREN'S HOSPITAL LOS ANGELES, LOS ANGELES, CA, UNITED STATES

Background

Pediatric cancers

Among children (0–14 years) and adolescents (0–19 years) in the United States, cancer is the second most common cause of death, following injury-related deaths, and the leading cause of death by disease [1–3]. The case distributions by International Classification of Childhood Cancer (ICCC) cancer types are leukemias (28% and 13% in children and adolescents, respectively), followed by central nervous system neoplasms (27%, 21%), lymphomas (12%, 19%), neuroblastoma and other peripheral nervous cell tumors (6%, <1%), nephroblastoma and other nonepithelial renal tumors (5%, <1%), malignant bone tumors (4%, 5%), rhabdomyosarcoma (3%), germ cell and gonadal tumors (3%), hepatic tumors (2%, <1%), retinoblastoma (2%, <1%), thyroid carcinoma (2%, 11%), and malignant melanoma (1%, 3%) [1–3]. The overall cancer incidence in children and adolescents has increased by 0.6%–0.7% per year. However, the mortality rates have decreased from 6.3 and 7.1/100,000 children and adolescents in 1970 to 2.0 and 2.9/100,000 in 2018, respectively, which are 68% and 59% reductions [1–3]. The survival rates have improved in many cancers. However, it remains low for some tumor types (e.g., diffuse intrinsic pontine glioma), for some age groups (e.g., Wilms' tumors in older patients), tumors in higher stages (metastasis), and tumors with unfavorable histology (e.g., neuroblastoma, Wilms tumor).

Pediatric extracranial solid tumors

Pediatric cancers are divided into three broad groups: hematologic, central nervous system (CNS) tumors, and extracranial solid tumors (EST). Extracranial solid tumors account for approximately 40% of all pediatric cancers [1,2]. Pediatric EST is roughly subdivided into two groups: (1) mesenchymal-derived tumors (sarcomas) in bone and soft tissue and less commonly in other organs, and (2) embryonal to predominantly epithelial-type tumors with morphologic features following the embryonic differentiation specific to the organs; many of these tumors are called "-blastomas." Although there are challenging cases, the typical histologic features of "-blastomas" are helpful for correct diagnosis. In contrast, pediatric sarcomas often show monotonous "small round cell tumor" or "spindle cell tumor" morphology resembling each other. The diagnosis, therefore, is very

Cases in Laboratory Genetics and Genomics (LGG) Practice. https://doi.org/10.1016/B978-0-323-99622-8.00029-0

challenging solely from a histology point of view and usually requires ancillary testing, such as immunohistochemical staining and molecular testing.

Roles of molecular testing in pediatric oncologic patient care

In recent decades, the roles of molecular testing in pediatric oncologic patient care have become increasingly important. In the pediatric setting, molecular testing has been used: (1) as it provides gene alterations specific for cancers (**diagnostic utility**); (2) for detecting molecular markers that are targetable by chemotherapeutic drugs (**therapeutic utility**); (3) for risk stratification by providing prognostic information associated with specific gene alterations (**prognostic utility**); (4) for identifying gene alterations (**germline mutations**) associated with cancer predisposition syndromes (Table 27.1).

In addition to FISH and chromosomal microarray (CMA), we designed the OncoKids next-generation sequencing (NGS) testing to detect these gene alterations specific to many pediatric tumors (Table 27.2).

Diagnostic utility (Cases 27.1, 27.2, 27.3, 27.5, 27.6, 27.7, and *27.8)*

Cancer driver genes in pediatric cancers are unique. Only 45% of pediatric cancer driver genes are shared with adult cancers [3–6]; therefore, unique therapeutic agents are required for pediatric cancer. In adults, most solid tumors are carcinomas of epithelial cells. The development of carcinoma typically follows the carcinogenesis steps: low-grade dysplasia, high-grade dysplasia, and malignancy. Multiple harmful environmental events and gene mutations ("hits") accelerate the process. In contrast, carcinoma is rare in the pediatric population, and pediatric tumors tend to have fewer molecular alterations. Many "small round blue cell" and "spindle cell" tumors have one of the driver gene fusions or mutations that are diagnostic for the tumors.

Therapeutic utility (Cases 27.1, 27.4, and 27.9)

The development of NGS has played an essential role in detecting targetable molecular markers in the era of personalized medicine (Table 27.3) [3–6].

Prognostic utility (Cases 27.3 and 27.9)

Oncologists establish a treatment regimen for each patient based on the risk group specific to the tumor type. In recent decades, molecular and cytogenetic markers have been incorporated into the risk stratifications [4–8].

Wilms' tumor (WT) is the most common renal malignancy in children, with approximately 500 new cases diagnosed in the United States each year. Two-thirds of patients with WT are diagnosed before 5 years of age. WT is likely caused by abnormal renal development, resulting in the proliferation of the metanephric blastema without normal tubular and glomerular differentiation. WT is associated with loss-of-function mutations of tumor suppressor and transcription genes (e.g., *WT1, CTNNB1, AMER1 (WTX) p53, FWT1,* and *FWT2* and genes at the 11p15.5 locus). Several prognostic factors are associated with an

Table 27.1 Roles of molecular testing in pediatric oncologic patient care.

	Diagnostic	Therapeutic	Prognostic	Germline
Definition	Gene alterations specific for cancers	Molecular makers that are targetable by chemotherapeutic drugs	Prognostic information associated with specific gene alterations	Gene alterations (germline mutations) associated with cancer predisposition syndromes
Examples	**Cancers diagnosed by NGS at CHLA (2017–20)** • Rhabdomyosarcoma (alveolar): *PAX3/PAX7::FOXO1* • Ewing sarcoma: *EWSR1::FLI1/ERG* • Synovial cell sarcoma: *SYT::SSX1/SSX2/SSX4* • Infantile fibrosarcoma: *ETV6::NTRK3* • Desmoplastic small round cell tumor: *EWSR1::WT1* • Alveolar soft part sarcoma: *TFE3::ASPSCR1(ASPS)* • Clear cell sarcoma: *EWSR1::ATF1, EWSR1::CREB1* • Inflammatory myofibroblastic tumor: *ALK::TPM3/4, CLTC, ATIC, TFG::ROS1, FN1::ROS1* • Low-grade fibromyxoid sarcoma: *FUC::CREBB3L2/CREBB3L1* • Dermatofibrosarcoma protuberans: *COL1A::PDGFB* • Angiomatoid fibrous histiocytoma: *EWSR1::ATF1, EWSRI::CREB1* • Epithelioid hemangioendothelioma: *WWTR1::CAMTA1* • Mesenchymal chondrosarcoma: *HEY1::NCOA2* • Ewing-like sarcomas: *BCOR::CCNB3, CIC::DUX4* • Midline carcinoma: *NUT::BRD4* • Low-grade fibromyxoid sarcoma: *FUS::CREB1L1* and *FUS::CREB1L3* • Lipoblastoma: *PLAGI* fusion • Aneurysmal bone cyst, nodular fasciitis, myositis ossificans: *USP6* fusion • Mucoepidermoid carcinoma: *CRTC1:MAML2* • Adenoid cystic carcinoma: *MYB::NFIB* • Rhabdoid tumor: *SMARCB1/SMARCA4* mutations • DICER1- associated sarcoma: *DICER1* mutations • NTRK-associated mesenchymal tumor: *NTRK1::LMNA, NTRK1::TPR*	**Targetable gene alterations identified by NGS at CHLA (2017–20)** • *ALK* mutation/fusion • *MYCN* amplification • *BRAF* mutation/fusion • *FGFR1/2/3* mutation/fusion/ amplification • *N/KRAS* mutation • *BRACA1/2* mutation • *EWSR1::FLI1* fusion • *PTEN* loss of function • *PLK3CA* mutation • *VEGF* mutation • (see Table 27.2)	**Prognostic factors used for risk stratification and selection of therapy** • Neuroblastoma Age at diagnosis, tumor stage, tumor histology, DNA ploidy, *MYCN* amplification status, chromosomal aberrations (e.g., LOH of 1p, 14q, and 11q and gain of 17q) • Wilms tumor Age at diagnosis, tumor stage, tumor histology, molecular and genetic alterations (e.g., loss of heterozygosity at chromosome 16q, 1p, and 11p15 and 1q gain) • Rhabdomyosarcoma Tumor stage, tumor histology (embryonal, alveolar, vs spindle cell/ sclerosing subtypes), surgery results	**Germline mutations identified by NGS at CHLA (2017–20)** • *PTEN*: *PTEN* hamartoma tumor syndrome (Bannayan-Riley-Ruvalcaba syndrome, Cowden syndrome, multiple hamartoma syndrome, proteus-like syndrome) • *NF1*: Neurofibromatosis type 1 • *APC*: Familial polyposis syndrome • *TP53*: Li Fraumeni syndrome • *DICER1*: DICER1 syndrome • *IDH1*: Ollier syndrome • *RB1*: Hereditary retinoblastoma • *RET*: MENIIb • *PIK3CA*: *PIK3CA*-related segmental overgrowth syndrome • *WT1*: Denys-Drash syndrome

Table 27.2 Gene list included in Oncokids targeted next generation sequencing panel.

Mutation hotspots (86)			Full-length coding (44)		High-level amplification (28)	OncoKids fusion list driver genes (122)			
ABL1	FGFR3	PDGFRA	APC	SOCS2	ABL2	ABL1	FGFR1	MYBL1	RECK
ABL2	FLT3	PDGFRB	ARID1A	SUFU	ALK	ABL2	FGFR2	MYH11	RELA
ACVR1	GATA2	PIK3CA	ARID1B	SUZ12	BRAF	AFF3	FGFR3	MYH9	RET
AKT1	GNA11	PIK3R1	ATRX	TCF3	CCND1	ALK	FLNA	NCOA2	ROS1
ALK	GNAQ	PPM1D	CDKN2A	TET2	CDK4	ASXL1	FLT3	NCOR1	RPL22
ASXL1	H3F3A	PTPN11	CDKN2B	TP53	CDK6	ATF7IP	FOSB	NOTCH1	RUNX1
ASXL2	HDAC9	RAF1	CEBPA	TSC1	EGFR	BCL11B	FOXO1	NOTCH2	SETBP1
BRAF	HIST1H3B	RET	CHD7	TSC2	ERBB2	BCL2	FUS	NOTCH4	SS18
CALR	HRAS	RHOA	CRLF1	WHSC1	ERBB3	BCL2L11	GLI1	NPM1	SSBP2
CBL	IDH1	SETBP1	DDX3X	WT1	FGFR1	BCOR	GLIS2	NR4A3	STAG2
CCND1	IDH2	SETD2	DICER1	XIAP	FGFR2	BCR	GPS2	NSD1	STAT6
CCND3	IL7R	SH2B3	EBF1		FGFR3	BRAF	GTF2I	NSD2 (WHSC1)	TAF15
CCR5	JAK1	SH2D1A	EED		FGFR4	CAMTA1	HMGA2	NTRK1	TAL1
CDK4	JAK2	SMO	FAS		GLI1	CBFB	JAK2	NTRK2	TBL1XR1
CIC	JAK3	STAT3	GATA1		GLI2	CCND1	KAT6A	NTRK3	TCF12
CREBBP	KDM4C	STAT5B	GATA3		IGF1R	CIC	KMT2A	NUP214	TCF3
CRLF2	KDR	TERT	GNA13		JAK1	COL1A1	KMT2B	NUP98	TFE3
CSF1R	KIT	TPMT	ID3		JAK2	CREBBP	KMT2C	NUTM1	THRAP3
CSF3R	KRAS	USP7	IKZF1		JAK3	CRLF2	KMT2D	NUTM2B	TP53
CTNNB1	MAP2K1	ZMYM3	KDM6A		KIT	CSF1R	LMO2	PAX3	TP63
DAXX	MAP2K2		KMT2D		KRAS	CTNNB1	LZTR1	PAX5	TSLP
DNMT3A	MET		MYOD1		MDM2	CUL1	MACF1	PAX7	TSPAN4
EGFR	MPL		NF1		MDM4	DACH1	MAML2	PDGFB	UBTF
EP300	MSH6		NF2		MET	DAZAP1	MAN2B1	PDGFRA	USP6
ERBB2	MTOR		PHF6		MYC	DDX3X	MECOM	PDGFRB	WT1
ERBB3	MYC		PRPS1		MYCN	DUSP22	MEF2D	PLAG1	YAP1
ERBB4	MYCN		PSMB5		PDGFRA	EGFR	MET	PRKAR1A	ZMYM2
ESR1	NCOR2		PTCH1		PIK3CA	EP300	MKL1	PSIP1	ZMYND11
EZH2	NOTCH1		PTEN			ETV6	MLLT10	RAF1	ZNF384
FASLG	NPM1		RBI			EWSR1	MN1	RANBP17	
FBXW7	NRAS		RUNX1			FAT1	MYB	RARA	
FGFR1	NT5C2		SMARCA4						
FGFR2	PAX5		SMARCB1						

increased risk of tumor recurrence or death, including unfavorable tumor histology, higher tumor stage, molecular and genetic alterations (e.g., loss of heterozygosity at chromosome 16q, 1p, and 11p15 and 1q gain), and age >2 years. Therefore, these prognostic factors are considered when determining the initial treatment regimen (Children's Oncology Group).

Rhabdomyosarcoma (RMS, Case 27.3) is the most common soft tissue sarcoma in childhood. It is important to identify the alveolar RMS subtype for therapeutic management because it has a worse prognosis. Sclerosing and spindle cell RMS recently emerged

Table 27.3 Selected targeted treatment for pediatric solid tumors.

Genomic alteration	Target structure	Medication	Example
ALK mutation/fusion	*ALK*	Crizotinib, certinib, entrectinib	Neuroblastoma, inflammatory myofibroblastic tumor, anaplastic large cell lymphoma
MYCN amplification	AURKA	Alisertib	Neuroblastoma
BRAF mutation/fusion	*BRAF*	Dabrafenib, vemurafenib	LCH, melanoma
FGFR1/2/3 mutation/ fusion/amplification	*FGFR*	Dovitinib, pinatinib	Rhabdomyosarcoma, Ewing sarcoma
N/KRAS mutation	MEK	Trametinib	Melanoma
BRCA1/2 mutation	PARP1	Olaparib	Osteosarcoma
EWSR1::FLI1 fusion		Rucaparib	Ewing sarcoma
PTEN loss	P13K/mTOR	Everolimus	Sarcomas
PIK3CA mutation	P13k/mTOR	Sirolimus/tensirolimus	Vascular malformation
VEGF	Multikinases	Pazopanib	Sarcomas
NTRK	*NTRK*	Larotrectinib, entrectinib	Multiple, primarily sarcomas, gliomas
ROS1	*ROS1*	Crizotinib, enterctinib	Inflammatory myofibroblastic tumor, meningioma, gliomas

as a separate pathologic entity, in which two subtypes are molecularly defined, an infantile subset with *VGLL2, TEAD1,* and *SRF* fusion and a subset with an *MYOD1* gene mutation. The presence of *MYOD1* alteration is associated with poor outcomes.

For example, neuroblastoma (Case 27.9) is risk-stratified by most significant prognostic factors, including the child's age at diagnosis, tumor stage, tumor histology, DNA ploidy, *MYCN* amplification status, and the presence of specific chromosomal aberrations, such as loss of heterozygosity (LOH) of 1p, 14q, and 11q and gain of 17q.

Identification of cancer predisposition syndrome (Cases 27.7 and 27.8)

Cancer predisposition syndrome (CPS, also called hereditary cancer syndrome, familial cancer syndrome, etc.) terms a disorder with a germline genetic alteration that increases the risk of developing specific cancer(s), which can be multiple in sites and types. Up to 10% of all cancers show one of the germline mutations that are diagnostic for CPS. More than 100 genes have been reported to cause CPS (Table 27.1) [9,10].

The onsets of hereditary cancers are typically earlier than the same tumors without germline mutation. Many occur in childhood, but carcinomas are rare in children, even if they are associated with CPS. For example, carcinomas with germline *BRCA1/BRACA2* mutations in hereditary breast-ovarian cancer syndrome and DNA mismatch repair genes (*MLH1, MSH2, MSH6, PMS1,* and *PMS2*) in hereditary nonpolyposis colorectal cancer syndrome (Lynch syndrome) commonly occur in adulthood.

The characteristics of CPS currently used for its screening are (1) Family history with (a) $\geq$2 malignancies that occurred in family members before 12 years; (b) parent or sibling with a history of cancer before 45 years of age; (c) $\geq$2 first or second-degree relatives in the same parental lineage with cancer before the age of 45 years; (d) the parents of the child with cancer who are consanguineous; (2) cancer type and features known to be associated with CPS; (3) a child with ≥2 malignancies (secondary, bilateral, multifocal, and/or metachronous); (4) a child with cancer and congenital or other anomalies; (5) a child with excessive treatment toxicity; (6) multiple types of cancer occurring in a single person; (7) genetic tumor analysis that reveals defects suggesting a germline predisposition.

Identification of CPS is crucial because it can change the approach to treating the patient. For example, intensive radiologic and chemotherapeutic treatments should not be used for patients with Li-Fraumeni syndrome because these treatments may increase the risk of developing secondary cancers. All patients with CPS and their family members should be monitored closely for bilateral, multifocal, and metachronous malignancies. NGS, which can detect numerous gene alterations simultaneously, has dramatically improved the efficacy of screening children for CPS.

Case 27.1 Ewing sarcoma (ES)

Clinical indication

The patient was a 10-year-old male who presented with right knee pain and a soft tissue lesion of the right distal femur. It was initially thought to be osteomyelitis with soft tissue abscess and treated with antibiotics.

Radiology

There was a heterogeneous T2 hyper signal intensity area involving the distal femur, with rim enhancement. This was associated with subperiosteal soft tissue enhancement (0.6 × 4.2 cm). Adjacent to this area, there was also a heterogeneous collection with septation and thick wall rim enhancement measuring at least 4.5 × 2.0 × 3.1 cm. The findings mimicked intraosseous abscess and osteomyelitis involving the distal femur with soft tissue and subperiosteal abscesses (Fig. 27.1.1A).

Histology

The hematoxylin and eosin (H&E)-stained histologic slides demonstrated fragments of trabecular bone and bone marrow contents. In the bone marrow space, there were multiple, small cohesive clusters of "small round blue cells," characterized by monotonous hyperchromatic nuclei with inconspicuous nucleoli and scant cytoplasm (Fig. 27.1.1B).

Immunohistochemical staining

The immunohistochemical (IHC) staining demonstrated diffuse strong membranous staining for CD99 (Fig. 27.1.1C1) and diffuse strong nuclear staining for NKX2.2 (Fig. 27.1.1C2), and negative staining for muscle markers (myogenin, desmin, and MyoD1)

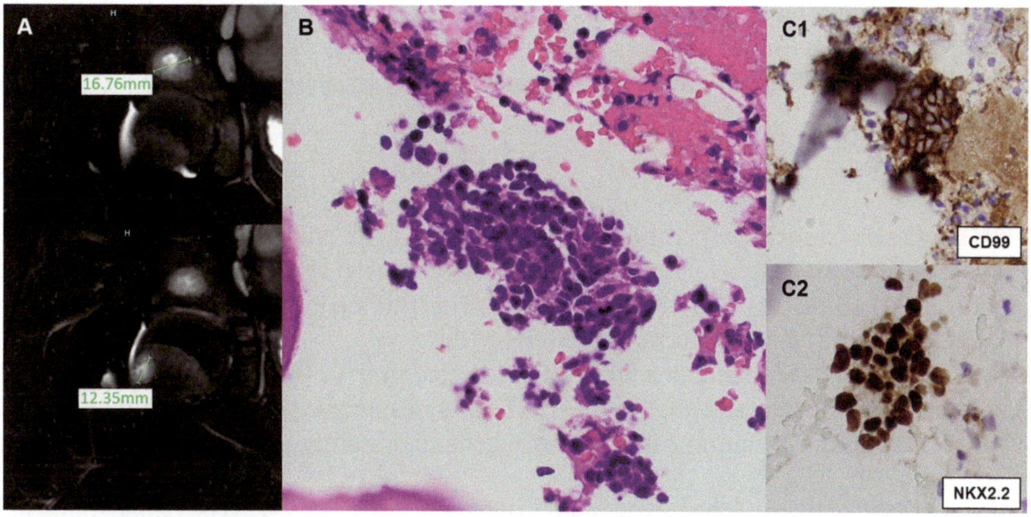

FIG. 27.1.1 MRI and microphotograph of the patient. (A) MRI of right femur. A heterogeneous T2 hyper signal intensity area involving the right distal femur mimicked intraosseous abscess and osteomyelitis. (B) Microphotograph of the right femur lesion (H&E, 400X). In the bone marrow space, there were multiple, small cohesive clusters of "small round blue cells," characterized by monotonous hyperchromatic nuclei with inconspicuous nucleoli and scant cytoplasm. (C) Microphotograph of the right femur lesion (immunohistochemical staining (IHC), C1: CD99, 400X, C2: NKX2.2, 400X. The IHC staining demonstrated diffuse strong membranous staining for CD99 (C1) and diffuse strong nuclear staining for NKX2.2 (C2), and negative staining for muscle markers (myogenin, desmin, and MyoD1) and hematopoietic cell markers (CD3, CD20, CD34, CD43, and TdT). The staining pattern was highly suggestive of Ewing sarcoma.

and hematopoietic cell markers (CD3, CD20, CD34, CD43, and TdT. The staining pattern was highly suggestive of Ewing sarcoma.

Test ordered

- FISH: *EWSR1* break-apart probe
- Chromosome microarray (CMA)
- Targeted NGS

Laboratory test performed

FISH with a break-apart probe for *EWSR1* was used for testing. When red and green signals are separated, it indicates a rearrangement of *EWSR1*. Chromosome microarray methods were described previously in Chapters 1 and 12.

The OncoKids panel is an amplification-based NGS assay designed to detect various molecular aberrations across the spectrum of pediatric malignancies, including leukemias, brain tumors, and extracranial solid tumors. Targeted aberrations include single-nucleotide variants (SNV), multinucleotide variants (MNV), insertions/deletions (InDels), amplification ($\geq$8 estimated copies), and RNA-based gene fusions. This panel uses low-input amounts of DNA (20 ng) and RNA (20 ng) and is compatible with formalin-fixed, paraffin-embedded (FFPE) and frozen tissue, bone marrow, and peripheral blood. The DNA assay of this panel covers the full coding regions of 44 cancer predisposition loci, tumor suppressor genes, and oncogenes; hotspots for mutations in 82 genes; and amplification events in 24 genes. The RNA assay includes 1421 targeted gene fusions (Table 27.2).

Test results

FISH studies on a touch-preparation slide, using the *EWSR1* dual-color break-apart probe, showed an abnormal signal pattern consisting of one fused signal and one split (red and green) signal in 97% [194/200] of analyzed cells. This finding indicates a rearrangement of the *EWSR1* (22q12) locus in tumor cells (Fig. 27.1.2).

Using CMA analysis, focal deletions were detected in the 11q24.3 and 22q12.2 regions, encompassing portions of the *FLI1* and *EWSR1* genes, respectively. These deletions occurred at the breakpoints of a chromosomal rearrangement resulting in an *EWSR1::FLI1* fusion (Fig. 27.1.3).

An *EWSR1::FLI1* gene fusion was identified by the targeted NGS panel (data not shown).

Results with interpretations

The diagnosis of this case is Ewing sarcoma (ES). ES has similar histologic to IHC features and shares specific, nonrandom translocations. ES can arise in any bone or soft tissue site, but it is mainly seen in the pelvis, axial skeleton, and femur.

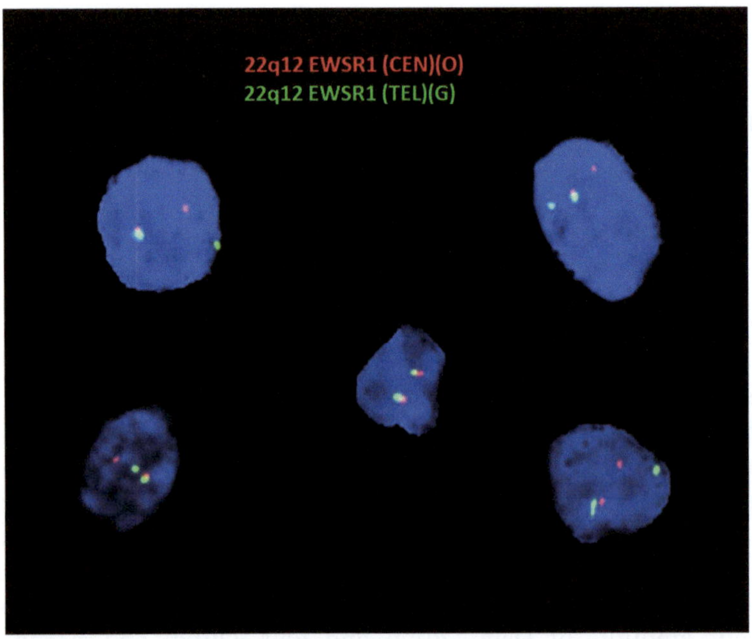

FIG. 27.1.2 *EWSR1* dual-color break-apart FISH. It indicated a rearrangement of the *EWSR1* (22q12) locus in tumor cells.

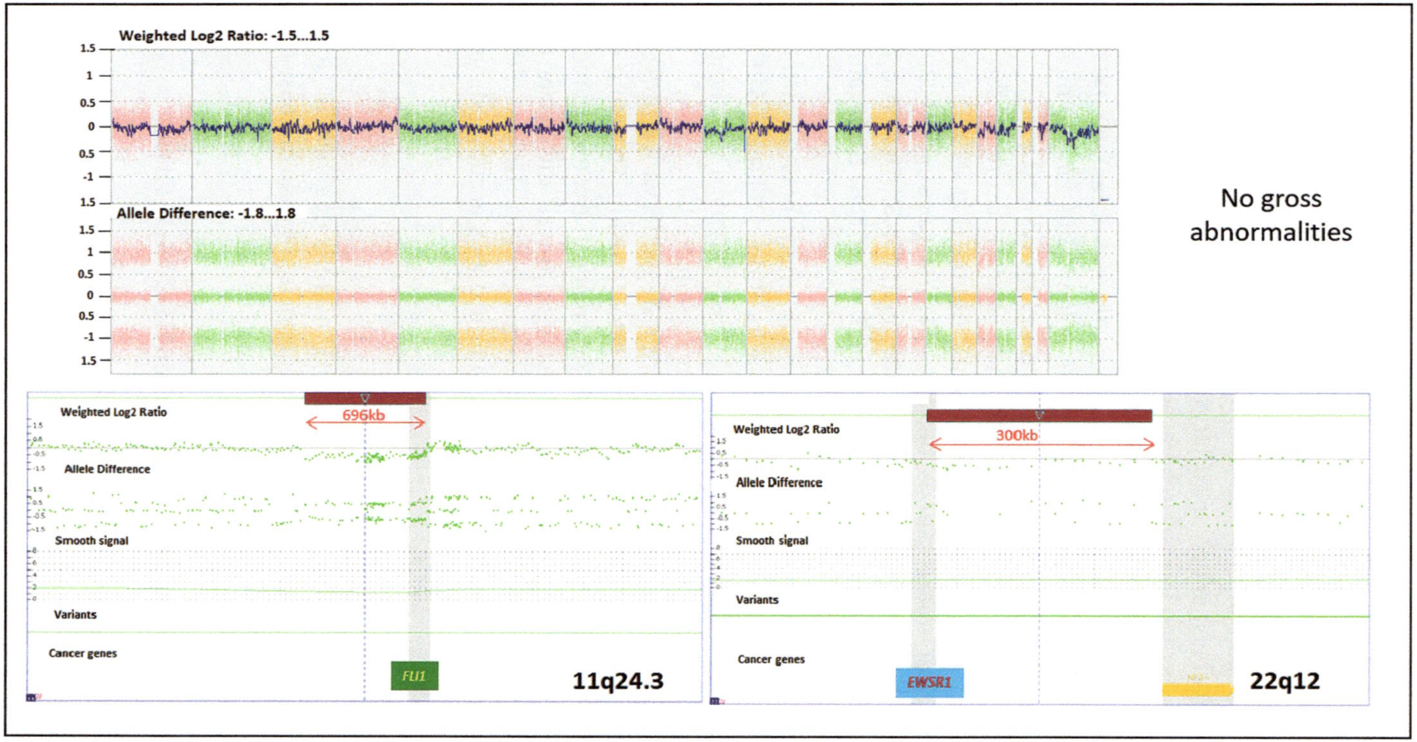

FIG. 27.1.3 Chromosomal microarray (CMA). Focal deletions were detected in the 11q24.3 and 22q12.2 regions, encompassing portions of the *FLI1* and *EWSR1* genes, respectively. These deletions occurred at the breakpoints of a chromosomal rearrangement resulting in a *FLI1-EWSR1* fusion.

On H&E-stained sections, "typical" ES are sheets of "small round blue cells (SRBC)" with hyperchromatic nuclei and scant clear cytoplasm containing abundant glycogen. Nuclear pleomorphism (difference in size and shape) is typically minimal. "Atypical ES" represents a histologic variant of ES and ES-like sarcoma, demonstrating SRBC often with additional "atypical" histologic features (e.g., larger-sized cells, a greater degree of cellular pleomorphism, a higher mitotic rate, and unusual organoid, lobular, and alveolar architectural patterns).

IHC staining has been widely used for distinguishing ES from other undifferentiated SRBC tumors, such as rhabdomyosarcoma, neuroblastoma, a small-cell variant of osteosarcoma, mesenchymal chondrosarcoma, and hematopoietic malignancies. Most ES expresses a cell surface glycoprotein, CD99/MIC2. Therefore, the membranous CD99 expression can be a sensitive marker for ES, but its specificity is low. Thus far, the most sensitive and specific IHC marker is *NKX2.2*, the protein product of the *NKX2.2* gene.

EWSR1, along with *FOXO1* (13q14) and *SS18* (18q11), is one of the most common translocation gene partners identified in pediatric tumors (Fig. 27.1.4). Most ES expresses one of the reciprocal translocations between *EWSR1* (located at 22q12) and either *FLI1* (11q24) or *ERG* (21q22) as a partner gene (Fig. 27.1.5). However, *EWSR1* may be replaced by *FUS* (16p11) in a small population of ES, and *EWSR1* FISH cannot detect these fusions.

Recently, *EWSR1*-negative SRBC tumors, which often show "atypical ES" histology, have been further characterized by novel translocation, including SRBC tumors with *CIC::DUX4* fusion and *BCOR* fusions. Therefore, these "atypical" ES should be identified as Ewing-like sarcoma, an entity different from ES (Fig. 27.1.4).

EWSR1 FISH has been widely used but may not detect ES with a *FUS* gene fusion and Ewing-like sarcoma that are not associated with *EWSR1* fusion. Therefore, targeted NGS is the most specific form of testing in this case. Advances in treatment, including surgery, radiotherapy, and multidrug chemotherapy, have improved the 5-year overall survival (OS) rate of patients with localized ES to approximately 75%. However, the 5-year OS of metastasis patients remains low (20%–45%). Prompt and accurate diagnosis is critical [11–14].

Future testing and recommendations

The current NGS panel testing included a limited number of genes. Testing with broader gene coverage (e.g., exome sequencing, RNA sequencing) would increase testing sensitivity and breadth of treatment options.

Case 27.2 *CIC-DUX* fusion-associated sarcoma

Clinical indication

The patient was a 5-year-old male who presented with a subcutaneous mass on the abdominal wall, measuring $1.8 \times 1.2 \times 1.0$ cm. He underwent a resection of the mass.

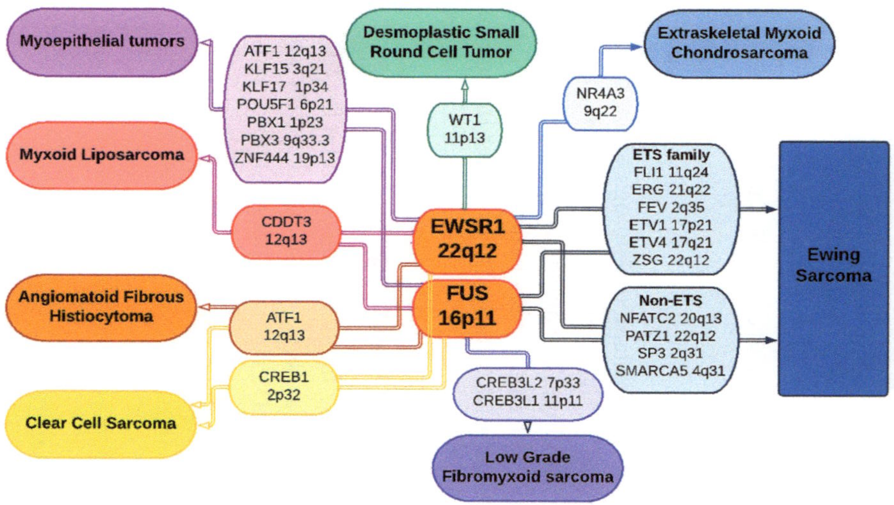

Examples of non-EWSR1/FUS fusion tumors	
Rhabdomyosarcoma (alveolar)	*PAX3/PAX7::FOXO1*
Synovial cell sarcoma	*SYT::SSX1/SSX2/SSX4*
Infantile fibrosarcoma	*ETV6::NTRK3*
Alveolar soft part sarcoma	*TFE3::ASPSCR1 (ASPS)*
Inflammatory myofibroblastic tumor	*ALK::TPM3/4, CLTC, ATIC, TFG::ROS1, FN1::ROS1*
Low grade fibromyxoid sarcoma	*FUS::CREBB3L2/CREBB3L1*
Dermatofibrosarcoma protuberans	*COL1A::PDGFB*
Epithelioid hemangioendothelioma	*WWTR1::CAMTA1*
Mesenchymal chondrosarcoma	*HEY1::NCOA2*
Ewing like sarcomas	*BCOR::CCNB3, CIC::DUX4*
Midline carcinoma	*NUT::BRD4*
Lipoblastoma	*PLAG1 fusion*
Mucoepidermoid carcinoma	*CRTC1::MAML2*
Adenoid cystic carcinoma	*MYB::NFIB*
Rhabdoid tumor	*SMARCB1/SMARCA4* mutations
DICER1-associated sarcoma	*DICER1* mutations
NTRK-associated mesenchymal tumor	*NTRK1::LMNA, NTRK1::TPR*
Aneurysmal bone cyst, nodular fasciitis, myositis ossificans	*USP6* fusion

FIG. 27.1.4 Main non-*EWSR1/FUS* fusion tumors in pediatrics.

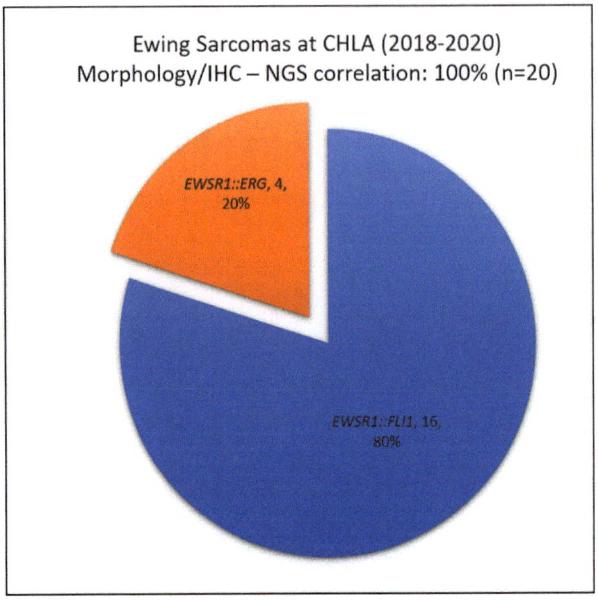

FIG. 27.1.5 Ewing sarcomas at Children's Hospital Los Angeles (2018–20).

Histology

The hematoxylin and eosin (H&E)-stained histologic slides demonstrated sheets of "small round blue cell tumor (SRBCT)" with an infiltrating border (Fig. 27.2.1A). The tumor cells showed an "atypical" ES histology. The nuclei were slightly larger and more vesicular than those seen in "typical ES." The tumor cells had occasional 1–2 conspicuous nucleoli and scant cytoplasm. There were brisk mitoses, up to 40 per 10 high-power fields, and areas of tumor necrosis (Fig. 27.2.1B).

Immunohistochemical (IHC) staining

The tumor demonstrated patchy membranous staining for CD99 (Fig. 27.2.2A) and negative staining for Phox2B and synaptophysin (markers for neuroblastoma), myogenin, desmin (Fig. 27.2.2B), and MyoD1 (muscle markers), and pan-cytokeratin.

Test ordered

- FISH: *EWSR1* break-apart probe
- Targeted NGS

Laboratory test performed

FISH and targeted NGS methods were described previously (see Case 27.1).

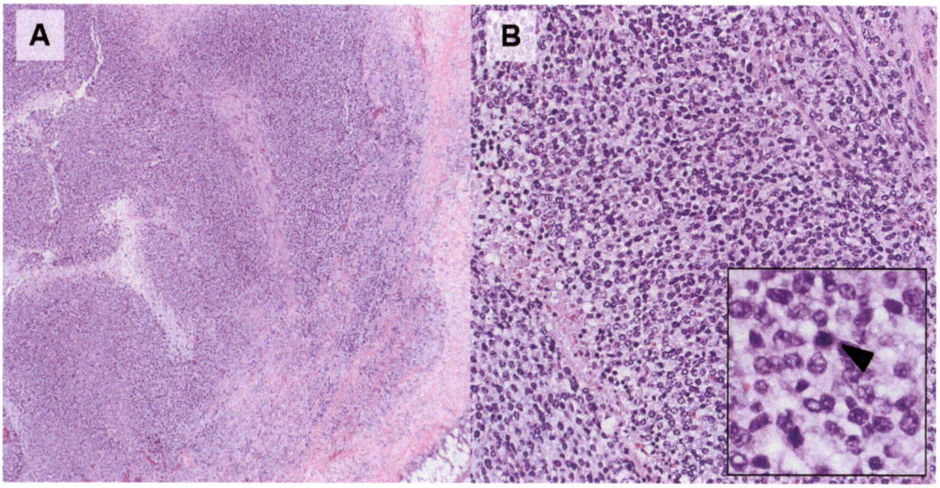

FIG. 27.2.1 Microphotograph of the abdominal soft tissue lesion (H&E, A: 40X; B: 400X). The tumor demonstrated sheets of "SRBCT" with an infiltrating border (A). The tumor cells show an "atypical" ES histology. The nuclei were slightly larger and more vesicular than those seen in "typical ES" (B).

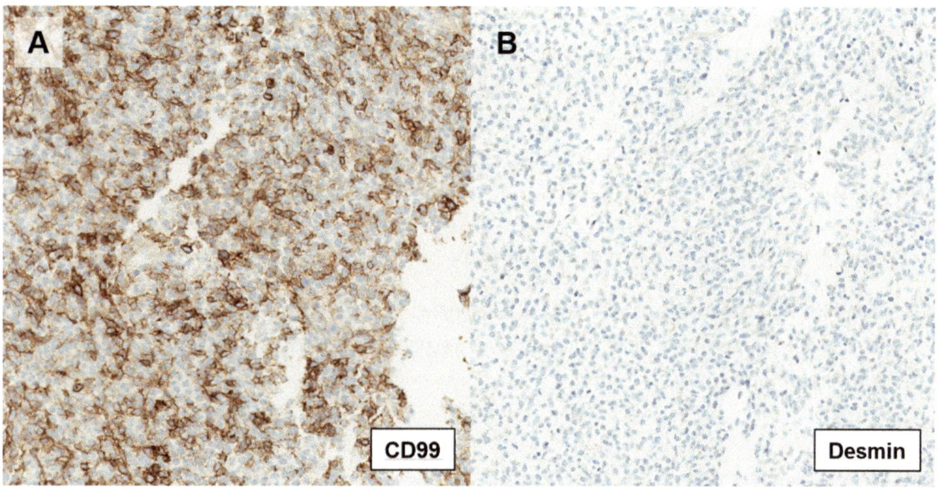

FIG. 27.2.2 Microphotograph of the abdominal soft tissue lesion (immunohistochemical staining (IHC). A: CD99, 400X; B: desmin, 400X. The tumor demonstrated patchy membranous staining for CD99 (A) and negative staining for Phox2B and synaptophysin (a marker for neuroblastoma), myogenin, desmin (B), and MyoD1 (muscle markers), and pan-cytokeratin.

Test results

FISH is negative for *EWSR1* rearrangement. A *CIC::DUX4* gene fusion was identified by the targeted NGS.

Results with interpretations

The diagnosis is *CIC::DUX4*-associated sarcoma. *CIC::DUX4*-associated sarcoma is the most common *EWSR1*-negative small round blue cell tumor (SRBCT). A study including 115 patients with *CIC::DUX4* fusion-associated sarcoma showed a mean age of 32 years and a slight male predominance, and that most tumors arose in the soft tissue (86%), predominantly in the trunk and extremity, followed by visceral locations (12%), and rarely in the bone (3%). The study also demonstrated that the 5-year survival rate of *CIC::DUX4*-associated sarcoma was significantly lower than that of ES [15].

On H&E-stained sections, *CIC::DUX4* fusion-associated sarcoma shows SRBCT with an "atypical" ES histology. The nuclei are slightly larger and more vesicular (with a more open chromatin pattern) than those in "typical" ES. Mitoses and necrosis are commonly present (Fig. 27.2.1A and B). IHC staining is often nonspecific. CD99 shows variable cytoplasmic and membranous expression, often in a patchy distribution (Fig. 27.2.2A). Although staining was not performed on this case, most cases were reported to be diffusely nuclear positive for C- and N-terminus *WT1* and *ETV4* [16,17].

CIC::DUX4 fusion involves *CIC*, a transcriptional repressor (located at 19q13.1), and *DUX4*, a double homeobox transcription factor (either chromosome 4q35 or 10q26.3). Some tumors harbor *FOXO4* (Xq13.1), *LEUTX* (19q13.2), *NUTM1* (15q14), or *NUTM2A* (10q23) genes instead of *DUX4* as a partner gene [18–21]. In this case, *EWSR1* FISH was negative. *CIC* FISH is not available at many laboratories. Thus far, targeted NGS is the most specific testing for diagnosing SRBCT.

Future testing and recommendations

The current NGS panel testing included a limited number of genes. Testing with broader gene coverage (e.g., exome sequencing, RNA sequencing) would increase testing sensitivity and treatment options.

Case 27.3 Rhabdomyosarcoma

Clinical indication

The patient was a 9-year-old female who presented with abdominal pain and hematemesis.

Radiology

There was a thoracoabdominal mass involving the middle/posterior mediastinum to the superior retroperitoneum, measuring $10.4 \times 9.4 \times 6.0$ cm (red circle), and left pleural effusion (red arrow, Fig. 27.3.1).

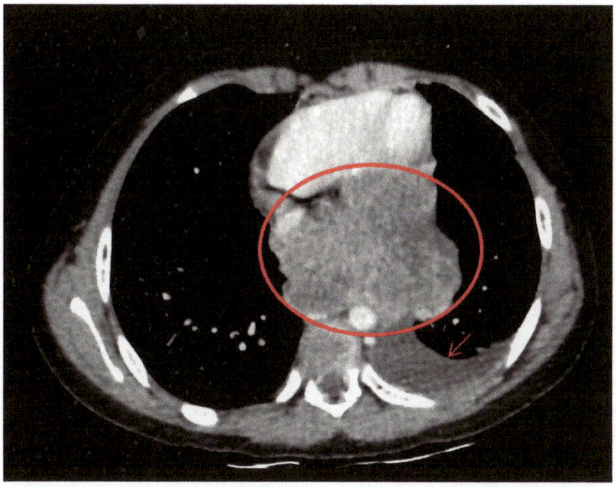

FIG. 27.3.1 MRI of the mediastinal mass lesion. There was a thoracoabdominal mass involving the middle/posterior mediastinum to the superior retroperitoneum, measuring 10.4 × 9.4 × 6.0 cm *(red circle)* and left pleural effusion *(red circle)*.

Histology

The H&E slides demonstrated sheets of SRBCT infiltrating into fibro adipose and skeletal muscle tissue. The tumor cells showed monotonous, oval to slightly angulated nuclei with finely granular chromatin, and occasional prominent nucleoli and scant cytoplasm. There were frequent mitoses (20–30 per 10 high-power fields) and karyorrhexis debris. Areas of tumor necrosis were present (Fig. 27.3.2A).

Immunohistochemical (IHC) staining

The IHC staining demonstrated diffuse strong cytoplasmic, nuclear, and nuclear staining for desmin (Fig. 27.3.2B1), MyoD1, and myogenin (Fig. 27.3.2B2), respectively. In addition, the tumor cells were negative for HMGA2, CD99, pan-cytokeratin, WT1, Phox2b, and hematopoietic cell markers (CD45 and TdT).

Test ordered

- FISH: *FOXO1* and *SS18* break-apart probes
- Chromosome microarray (CMA)
- Targeted NGS

Laboratory test performed

FISH, CMA, and NGS methods were described previously (see Case 27.1).

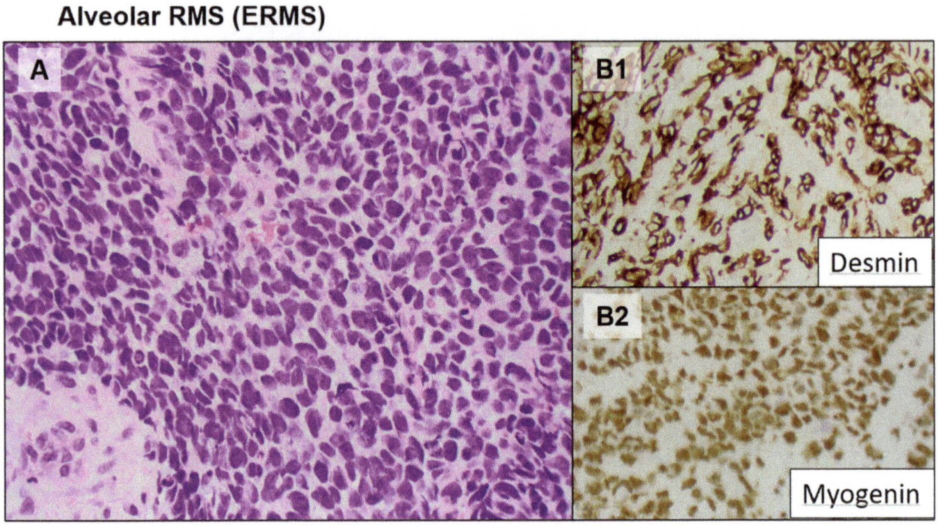

FIG. 27.3.2 Histology of alveolar rhabdomyosarcoma. (A) Microphotograph of the mediastinal mass lesion (H&E, 600X). The H&E slides demonstrated sheets of SRBCT. The tumor cells showed monotonous, oval to slightly angulated nuclei with finely granular chromatin, and occasional prominent nucleoli and scant cytoplasm. (B) Microphotograph of the mediastinal mass lesion [immunohistochemical staining (IHC), B1: desmin, 400X; B2: myogenin, 400X]. The IHC staining demonstrated diffuse strong cytoplasmic, nuclear, and nuclear staining for Desmin (B1), MyoD1, and myogenin (B2), respectively.

Test results

Fluorescence in situ hybridization (FISH)

FISH studies on interphase tissue cells using the *FOXO1* (13q14) dual-color break-apart probe showed an abnormal signal pattern consisting of two copies of the *FOXO1* fused signal, two copies of the telomeric *FOXO1* signal, and two copies of the centromeric *FOXO1* signal (signal pattern: 2F2R2G) in 78/200 (39%) of the cells examined. The observed signal pattern indicates the presence of an abnormal clone with a rearrangement involving the *FOXO1* gene in tetraploid (4N) cells (Fig. 27.3.3A). FISH studies on interphase tissue cells using the *SS18* (18q11.2) dual-color break-apart probe showed four copies of the *SS18* signal (signal pattern: 4F) in 174/200 (87%) of the cells examined; this pattern indicates the presence of tetraploid (4N) cells, with no evidence of a rearrangement involving the *SS18* gene (Fig. 27.3.3B).

Chromosomal microarray (CMA)

This tumor demonstrated amplification of the *CDK4* and *GLI1* genes in chromosome 12, a common alteration in malignant sarcomas, including ARMS. There was no evidence of a deletion involving *TP53*. The t(2;13) that results in a *PAX3::FOXO1* fusion is not detected by this assay in the balanced state (Fig. 27.3.3C).

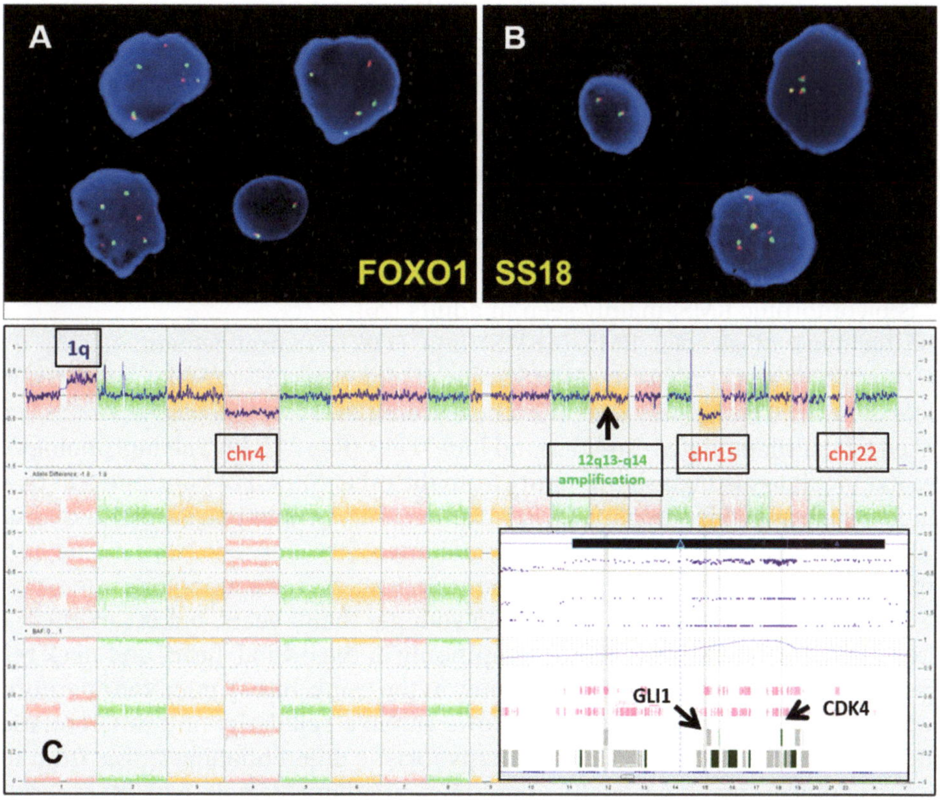

FIG. 27.3.3 FISH and CMA results from the patient. (A) *FOXO1* dual-color break-apart FISH. *FOXO1* (13q14) dual-color break-apart probe showed an abnormal signal pattern consisting of two copies of the *FOXO1* fused signal, two copies of the telomeric *FOXO1* signal, and two copies of the centromeric *FOXO1* signal. The observed signal pattern indicates a rearrangement involving the *FOXO1* gene in tetraploid (4N) cells. *FOXO1* 3′FISH probe is in *green*; *FOXO1* 5′FISH probe is in *orange*. (B) *SS18* dual-color break-apart FISH. FISH studies on interphase tissue cells using the *SS18* (18q11.2) dual-color break-apart probe showed four copies of the *SS18* signal (signal pattern: 4F); this pattern indicates the presence of tetraploid (4N) cells, with no evidence of a rearrangement involving the *SS18* gene (B). *SS18* 3′FISH probe is in *green*; *SS18* 5′FISH probe is in *orange*. (C) Chromosomal microarray (CMA). This tumor demonstrated amplification of the *CDK4* and *GLI1* genes in chromosome 12, a common alteration in malignant sarcomas, including ARMS. There was no evidence of a deletion involving *TP53*. The t(2;13) that results in a *PAX3::FOXO1* fusion is not detected by this assay in the balanced state.

Targeted next-generation sequencing (NGS)

The *PAX3::FOXO1* gene fusion and amplification of *GLI1* and *CDK4* were identified by NGS, concordant with the findings from FISH and CMA (data not shown). *GLI1* and *CDK4* amplification have been reported in ARMS. Approximately 25% of *PAX3::FOXO1*(+) alveolar rhabdomyosarcomas demonstrate *CDK4* amplification. *CDK4* amplification may be associated with a more aggressive disease course in alveolar rhabdomyosarcoma [22–25].

Results with interpretations

The diagnosis of this case is rhabdomyosarcoma (RMS), alveolar type. RMS is the most common pediatric soft tissue sarcoma, accounting for 3% of childhood tumors and 50% of pediatric sarcoma with slight male predominance [23]. At diagnosis, RMS is usually larger than 5 cm and shows poorly circumscribed, infiltrative masses.

Pediatric RMS is divided into three histologic subtypes with necessary treatment and prognostic implications, embryonal, alveolar, and spindle cell/sclerosing RMS. Another subtype listed in the WHO Classification of Tumors (5th Ed.) along with these three subtypes is pleomorphic RMS, mainly seen in adults [26].

The hallmark of alveolar RMS (ARMS) is a *FOXO1* rearrangement, t(1:13) (*PAX3::FOXO1*) or t(2:13) (*PAX7::FOXO1*). ARMS accounts for 25%–40% of RMS. It is often found in adolescents and commonly occurs in the extremities. Morphologically, ARMS is characterized by undifferentiated "small round blue" cells (Fig. 27.3.2). The morphology often makes it difficult to distinguish this entity from the other SRBCT of bone and soft tissue in childhood, such as Ewing sarcoma, Ewing-like sarcoma, lymphoma, a small cell variant of osteosarcoma, and mesenchymal chondrosarcoma.

Embryonal RMS (ERMS) is the most common RMS that accounts for up to 60% of all RMS. One-third of ERMS occurs in children younger than 5 years, but it can affect older children and adults. ERMS arises most commonly in the orbital, head, and neck regions and genitourinary system, and less commonly in the trunk, biliary tract, retroperitoneum, and abdomen. Morphologically, ERMS shows variable cellularity and differentiation of rhabdomyoblastic tumor cells. The rhabdomyoblastic differentiation ranges from undifferentiated "small round blue" to differentiating/differentiated cells with an expanded or elongated eosinophilic cytoplasm, sometimes with cross-striations (e.g., "strap cells," "tadpole cells," Fig. 27.3.4A). Scattered anaplastic cells (with markedly enlarged nuclei and atypical mitotic figures) can be seen. ERMS is negative for *FOXO1* rearrangement [26].

Spindle cell/sclerosing RMS (SC/SRMS) is characterized by "spindle cell" morphology with elongated nuclei and long cellular fascicles with or without intervening stroma with hyalinization/sclerosis (Fig. 27.3.4B). There are two distinct groups with recurrent molecular genetic abnormalities in pediatric SC/SRMS. The presence of an *MYOD1* mutation characterizes the first group. It is the most common SC/SRMS that occurs mainly in adolescents and young adults. The second group is SC/SRMS associated with gene fusions, including the *VGLL2, TEAD1, SRF,* and *NCOA2* genes, such as *TEAD1-NCOA2, VGLL2-NCOA2,* and *VGLL2-CITED2*. SC/SRMS in this group occurs almost exclusively in young children <2 years [26–29].

IHC staining is a useful tool to distinguish RMS from other sarcomas. RMS cells are typically positive for muscle markers, such as myogenin, desmin, and MyoD1; however, the intensity and distribution of the staining vary among the histologic subtypes. ARMS usually shows diffuse, strong positivity for myogenin, desmin, and MyoD1. ERMS is positive for muscle markers, but the staining intensity is often variable, and the distribution is patchy or focal. An IHC pattern with diffuse strong MyoD1 positivity and negative myogenin and desmin is a good clue for *MYOD1* mutated SC/SRMS (Fig. 27.3.4B). Recently

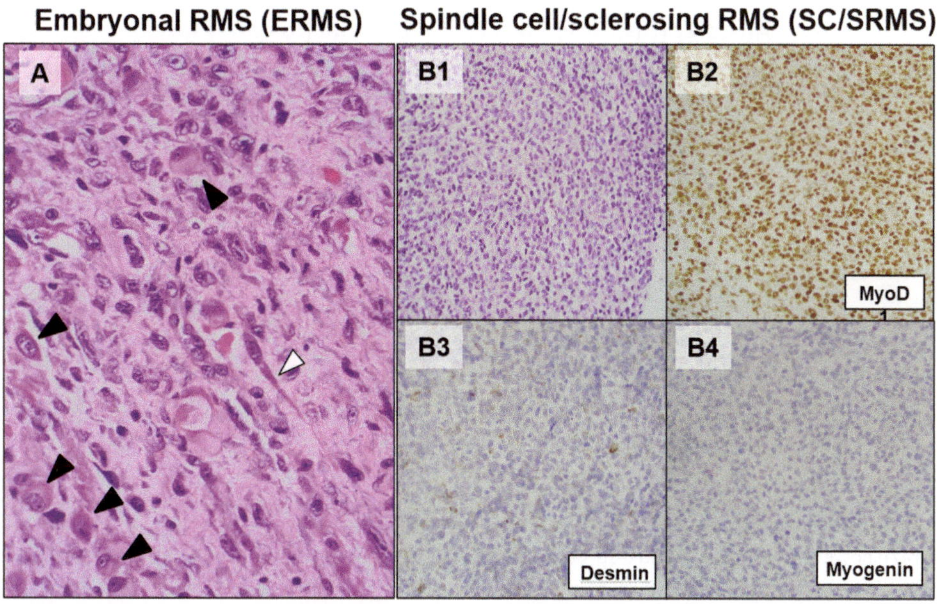

FIG. 27.3.4 Histology of embryonal (A) and spindle cell/sclerosing (B) rhabdomyosarcoma (RMS). (A) Microphotograph of an embryonal RMS (H&E, 600X). Morphologically, ERMS shows variable cellularity and rhabdomyoblastic differentiation. The rhabdomyoblastic differentiation ranges from undifferentiated "small round blue" to differentiating/differentiated cells with an expanded or elongated eosinophilic cytoplasm, sometimes with cross-striations (e.g., "strap cells," "tadpole cells," *black* and *white* arrowheads, respectively). Scattered anaplastic cells (with markedly enlarged nuclei and atypical mitotic figures) can be seen. (B) Microphotograph of a spindle cell/sclerosing RMS [H&E, 400X; immunohistochemical staining (IHC), B1: MyoD1, 200X; B2: desmin, 200X; B3: myogenin, 200X]. Spindle cell/sclerosing RMS (SC/SRMS) is characterized by "spindle cell" morphology with elongated nuclei and long cellular fascicles with or without intervening stroma with hyalinization/sclerosis (B1). An IHC pattern with diffuse strong MyoD1 positivity and negative myogenin and desmin is typical for *MYOD1* mutated SC/SRMS (B2–B4).

useful markers distinguishing fusion-negative RMS (ERMS) and fusion-negative RMS (ARMS) have been introduced (Fig. 27.3.5) [30].

All children with RMS are treated with surgery, chemotherapy, and/or radiation. The intensity and duration of the chemotherapy depend on the Risk Group Assignment. Each patient is categorized into either the low-, intermediate-, or high-risk group, which is decided by the stage (according to the initial clinicopathologic review of tumor; site, size, degree of tumor extension, and lymph node and distant metastatic status), the group (depending on completeness or extent of initial surgical resection after pathologic review of the tumor specimens), and the histology subtype (ERMS vs AEMS). Therefore, the presence or absence of *FOXO1* gene fusion is essential to determine the treatment option [31,32].

Future testing and recommendations

The current NGS panel testing included a limited number of genes. Testing with broader gene coverage (e.g., exome sequencing, RNA sequencing) would increase testing sensitivity and treatment options.

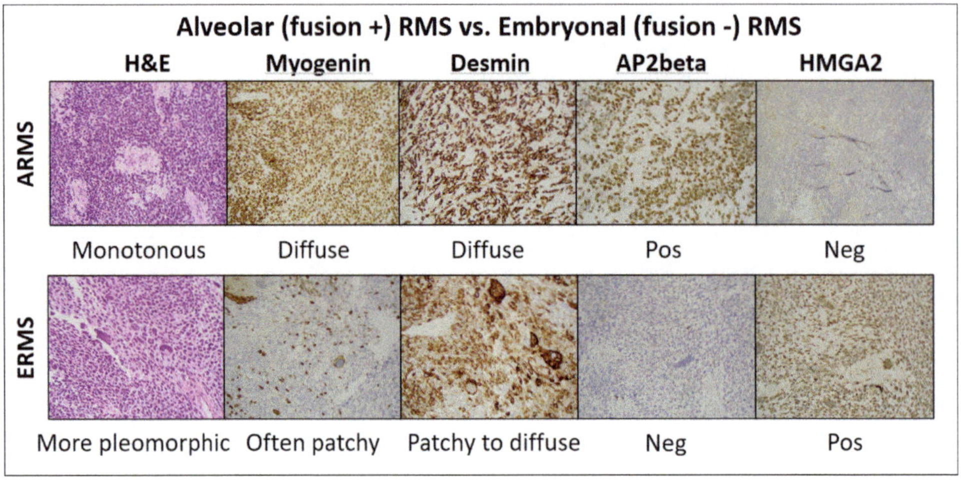

FIG. 27.3.5 Immunohistochemical staining (IHC) pattern of alveolar and embryonal RMS. IHC is a useful tool to distinguish RMS from other sarcomas. RMS cells are typically positive for muscle markers, such as myogenin, desmin, and MyoD1; however, the intensity and distribution of the staining vary among the histologic subtypes. Alveolar RMS (ARMS) usually shows diffuse, strong positivity for myogenin, desmin, and MyoD1. Embryonal RMS (ERMS) is positive for muscle markers, but the staining intensity is often variable, and the distribution is patchy or focal. APbeta and HMGA2 IHC staining patterns are typically "positive and negative" and "negative and positive" in ARMS and ERMS, respectively.

Case 27.4 *NTRK1*-associated sarcoma

Clinical indication

The patient was an 11-year-old male who presented with a persistent cough.

Radiology

The chest X-ray demonstrated a round, well-circumscribed mass measuring approximately 8.0 cm in diameter at the posterior medial left lower lung (Fig. 27.4.1A). The chest CT also detected a 5.9 × 7.4 × 6.9 cm left lower lobe mass posteromedially abutting the left posterior hilum (Fig. 27.4.1B).

Histology

The H&E slides of the core biopsy demonstrated sheets of monotonous spindle cells. The tumor cells showed spindle cells with monotonous spindle nuclei with a fine chromatin pattern and inconspicuous nucleoli and cytoplasm with elongated eosinophilic fascicles. Only mild pleomorphism (size difference) was seen in the nuclei. There were occasional mitoses (up to 3 per 10 high-power fields). Small areas of tumor necrosis were present (Fig. 27.4.2A1 and A2).

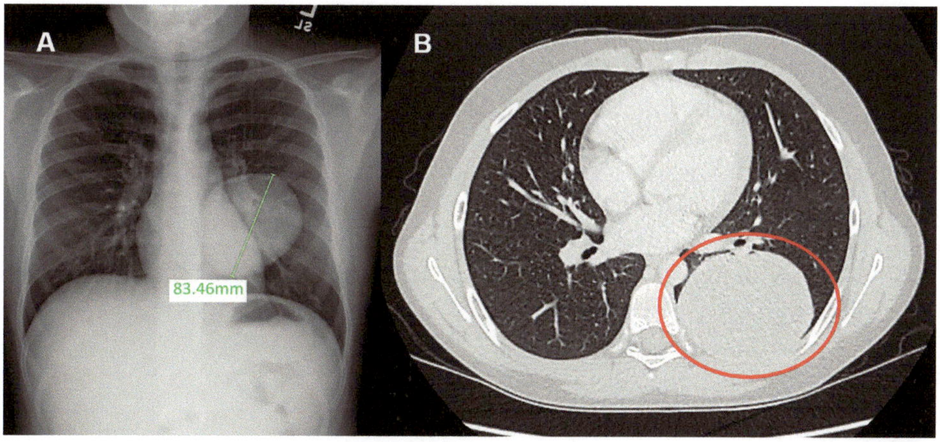

FIG. 27.4.1 Chest X-ray results. Chest-X ray (A) and CT (B) of the chest wall mass lesion. The chest X-ray demonstrated a round, well-circumscribed mass measuring approximately 8.0 cm in diameter at the posterior medial left lower lung (A). The chest CT also detected a $5.9 \times 7.4 \times 6.9$ cm left lower lobe mass posteromedially abutting the left posterior hilum (B).

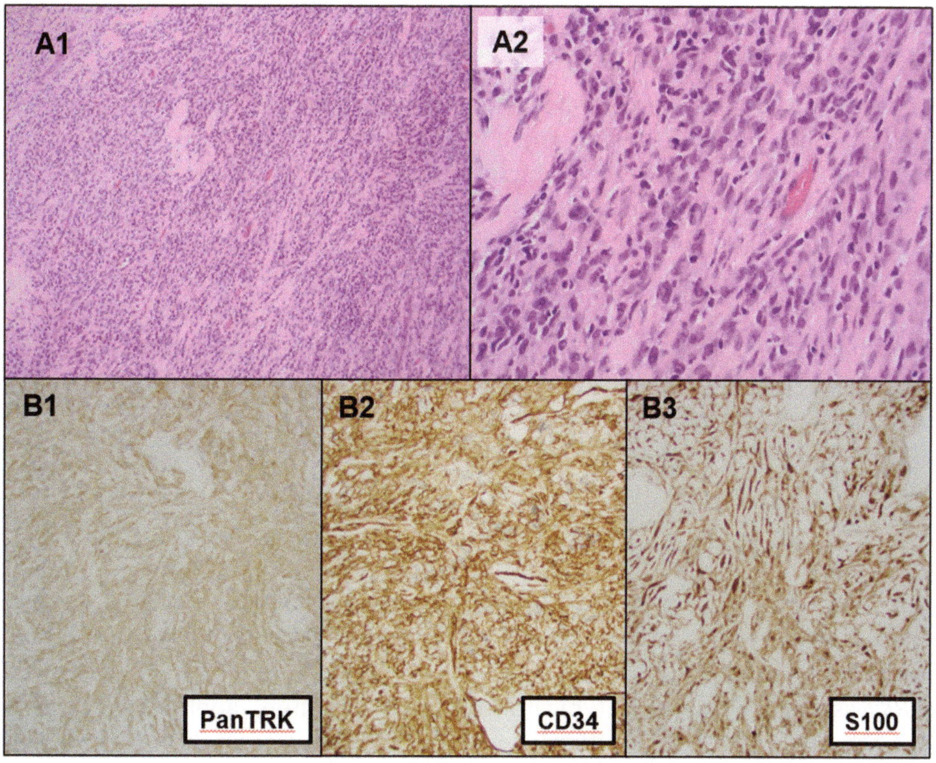

FIG. 27.4.2 Microphotographic results. (A) Microphotograph of the posterior chest wall mass lesion (H&E, A: 200X; B: 400X). The H&E slides of the core biopsy demonstrated sheets of monotonous spindle cells (A1). The tumor cells showed spindle cells with relatively monotonous spindle nuclei with a fine chromatin pattern and inconspicuous nucleoli and elongated eosinophilic fascicles (A2). Only mild pleomorphism (size difference) was seen in the nuclei. There were occasional mitoses (up to 3 per 10 high-power fields). Small areas of tumor necrosis were present. (B) Microphotograph of the posterior chest wall mass lesion [immunohistochemical staining (IHC), B1: panTRK, 400X; B2: CD34, 400X; B3: S100, 400X]. The IHC staining demonstrated diffuse strong cytoplasmic staining for panTRK (A), CD34 (B), and S100 (C). Desmin and SMA showed focal, weak cytoplasmic positivity. Beta-catenin was negative for nuclear staining. Pan-cytokeratin, desmin, myogenin, SMA, TLE, and hematopoietic markers (CD30, CD21, CD23, CD43, and myeloperoxidase) were negative.

Immunohistochemical (IHC) staining

The IHC staining demonstrated diffuse strong cytoplasmic staining for NTRK1, CD34, S100 (Fig. 27.4.2B1–B3), and PanTRK. Desmin and SMA showed focal, weak cytoplasmic positivity. Beta-catenin was negative for nuclear staining. Pan-cytokeratin, desmin, myogenin, SMA, TLE, and hematopoietic markers (CD30, CD21, CD23, CD43, and myeloperoxidase) were negative.

Test ordered

- Chromosome microarray (CMA)
- Targeted next NGS

Laboratory test performed

CMA and NGS methods were described previously (see Case 27.1).

Test results

Chromosomal microarray results showed a 302 kb deletion in the 1q23.1 region with a breakpoint within the *NTRK1* gene and a 3 Mb deletion in the 1q25.3q31.1 region with a breakpoint within the *TPR* gene, which were consistent with the *TPR::NTRK1* fusion (Fig. 27.4.3A). Overlapping deletions observed in the p21.3 region of chromosome 9 resulted in a loss of both copies (bi-allelic loss) of the *CDKN2A* and *CDKN2B* tumor suppressor genes. *CDKN2A/2B* abnormalities are frequently observed in different types of sarcomas (Fig. 27.4.3B) [33]. The 2.6 Mb copy number gain in the p13.33p13.32 region of chromosome 12 involves the *CCND2* gene. *CCND2*, encoding the cyclin D2 protein that regulates cyclin-dependent kinases in the cell cycle, functions as an oncogene in a variety of tumor types, including sarcomas (Fig. 27.4.3C).

Targeted NGS detected a *TPR::NTRK1* gene fusion of strong clinical significance (data not shown).

Results with interpretations

The diagnosis of this case is *NTRK1*-associated sarcoma. *NTRK1*-associated sarcoma is a low-grade, locally aggressive neoplasm that commonly arises in the trunk and proximal extremities.

NTRK1-associated sarcoma typically occurs in children, adolescents, and young adults, with rare reports in the third-fourth decade and no gender preference. These tumors may recur locally and have the metastatic potential [33–38]. *NTRK1* encodes a tyrosine kinase receptor. *NTRK1* fusions lead to ligand-independent activation of tyrosine kinase and downstream signaling pathways, which promote tumorigenesis. *NTRK1* fusions are observed in many other tumor types.

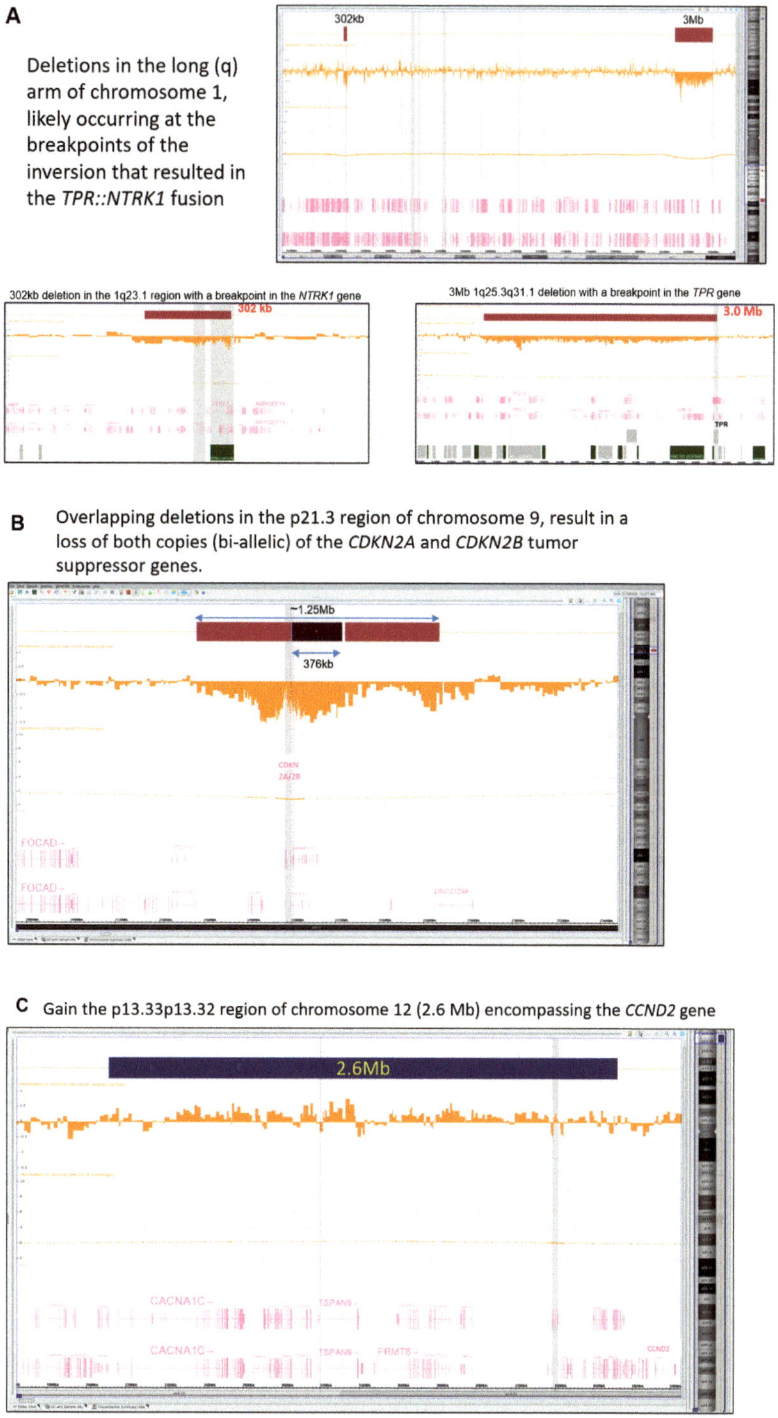

FIG. 27.4.3 Chromosomal microarray (CMA). (A) The 302 kb deletion in the 1q23.1 region with a breakpoint within the *NTRK1* gene and the 3 Mb deletion in the 1q25.3q31.1 region with a breakpoint within the *TPR* gene were consistent with the *TPR*::*NTRK1* fusion. (B) Overlapping deletions observed in the p21.3 region of chromosome 9 result in a loss of both copies (bi-allelic loss) of the *CDKN2A* and *CDKN2B* tumor suppressor genes. (C) The 2.6 Mb copy number gain in the p13.33p13.32 region of chromosome 12 involves the *CCND2* gene.

Histologically, *NTRK1*-associated sarcoma is characterized by spindle cells with long fascicles in a myxoid to the fibrous background. It often shows brisk mitoses (3–4 high power fields, with some cases up to 10–15 per 10 high power fields) and CD34/S100 positivity by immunohistochemical staining (IHC). In addition to S100 positivity, the tumors can have areas with admixed mature adipose tissue; therefore, they were initially called lipofibromatosis-like neural tumors [33,35]. *NTRK*1-associated sarcoma may demonstrate overlapping clinical and histologic features with other spindle cell sarcomas, such as infantile fibrosarcoma (IFS) or dermatofibrosarcoma protuberans (DFSP) and which manifests a diagnostic difficulty. IFS is the most common soft-tissue sarcoma in infants, characterized by hypercellular spindle cell lesions with an infiltrative border, harboring an *ETV6::NTRK3* fusion in approximately 85% of tumors. DFSP is also a hypercellular spindle cell neoplasm with an infiltrative border caused by *PDGFB* rearrangements. In such diagnostically challenging cases, screening with panTRK (or *NTRK1/2*) IHC may guide subsequent molecular testing to identify *TRK* fusions [33].

NGS is a critical diagnostic adjunct in which a multiplex sequencing panel can detect *TRK* fusions and other potentially targetable molecular alterations. Current first-line therapy for pediatric patients with advanced solid tumors includes cytotoxic chemotherapy, radiation therapy, and sometimes disfiguring surgeries. Larotrectinib is a TRK inhibitor with durable response rates and a favorable toxicity profile [34,38–41].

Future testing and recommendations

The current NGS panel testing included a limited number of genes. Testing with broader gene coverage (e.g., exome sequencing, RNA sequencing) would increase testing sensitivity and treatment options.

Case 27.5 Aneurysmal bone cyst

Clinical indication

Patient 1 was an 18-year-old female who presented with back pain. She was found to have a sacral expansile, lytic lesion, which locally recurred 5 months following the initial procedure.

Patient 2 was a 13-year-old male who presented with bone pain in the right knee and was found to have an expansile lytic lesion in the posterior mid-shaft of the right femur.

Radiology

Patient 1: The MRI demonstrated an expansile lytic lesion in the posterior left neural arch at the level of S3, measuring 5.7×3.4 cm, containing multiple fluid levels, and extending into the sacral spinal canal with compression of S3 nerve roots (Fig. 27.5.1A).

Patient 2: Radiographs demonstrated an eccentric (cortical based) lytic lesion extending in the proximal to mid femoral diaphysis with soft tissue involvement. There was increased periosteal bone formation with Codman triangles (Fig. 27.5.1B).

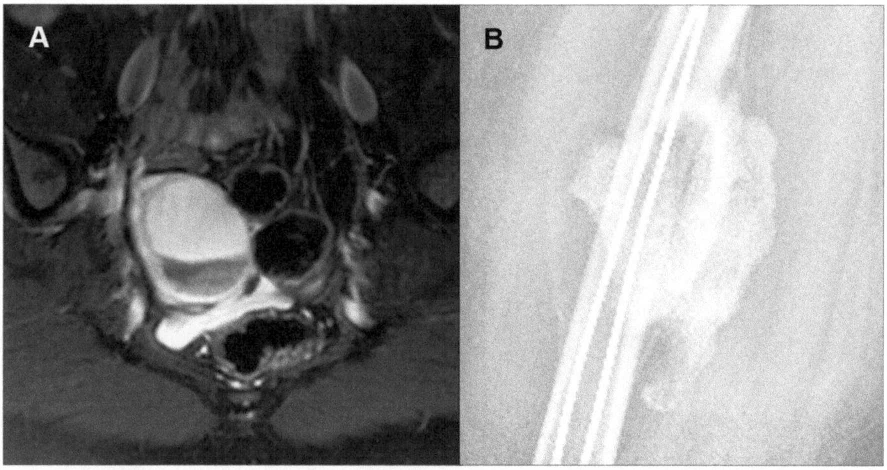

FIG. 27.5.1 MRI and radiographs of patients 1 and 2. (A) Patient 1: The MRI demonstrated an expansile lytic lesion in the posterior left neural arch at the level of S3, measuring 5.7 × 3.4 cm, containing multiple fluid levels, and extending into the sacral spinal canal with compression of S3 nerve roots. (B) Patient 2: Radiographs demonstrated an eccentric (cortical based) lytic lesion extending in the proximal to mid femoral diaphysis with soft tissue involvement. There was increased periosteal bone formation with Codman triangles.

Histology

Patient 1: H&E-stained sections demonstrated characteristic histologic features of ABC with thin fibrous septa of the multilocular cysts (Fig. 27.5.2A). The septa consisted of fibrous stroma with spindled cells, numerous osteoclastic giant cells, lymphohistiocytic inflammatory cells, extravasated red blood cells, and scattered hemosiderin deposits. Reactive bone formation and dystrophic calcification were occasionally seen.

Patient 2: H&E-stained sections from the bone and soft tissue components showed cystic areas with fibrous septa, as seen in patient 1 (Fig. 27.5.2B1), admixed with focal solid expansion (Fig. 27.5.2B2). The solid areas contained a mixture of spindle cells, osteoclastic giant cells, and inflammatory cells (Fig. 27.5.2B3). No pleomorphic atypical cells, osteoid formation, or necrosis was identified. Occasional reactive bone formations were seen.

Immunohistochemical (IHC) staining

IHC staining was not relevant.

Test ordered

- Targeted NGS (Patient 1 and 2)
- Sanger Sequencing (Patient 1)
- RT-PCR (Patient 1)
- FISH: *RUNX2::USP6* dual-color dual-fusion (Patient 2)

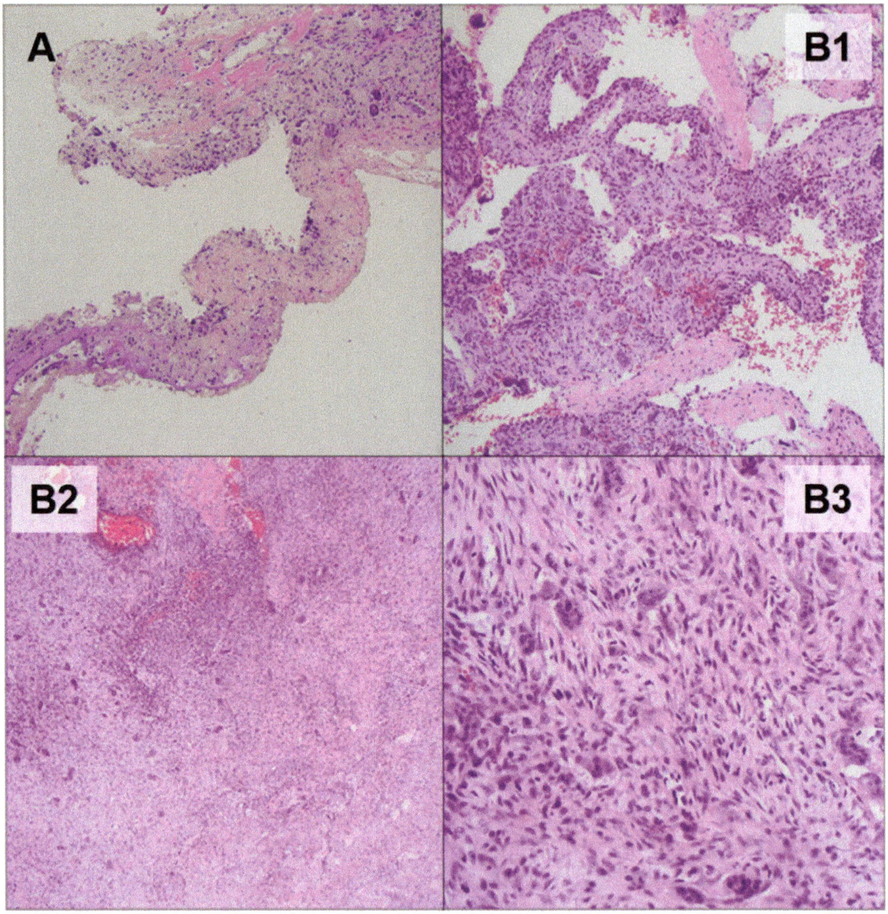

FIG. 27.5.2 Microphotograph of the bone lesions (H&E, A, B1, and B2: 200X; B3: 400X). (A) Patient 1: The sections demonstrated characteristic histologic features of ABC with thin fibrous septa of the multilocular cysts. The septa consisted of a fibrous stroma with spindled cells, numerous osteoclastic giant cells, lymphohistiocytic inflammatory cells, extravasated red blood cells, and scattered hemosiderin deposits. Reactive bone formation and dystrophic calcification were occasionally seen. (B) Patient 2: The sections from the bone and soft tissue components showed cystic areas with fibrous septa-like patient 1 (B1), admixed with focal, solid expansion (B2). The solid areas contained spindle cells, osteoclastic giant cells, and inflammatory cells (B3). No atypical pleomorphic cells, osteoid formation, or necrosis were identified. Occasional reactive bone formation was seen.

Laboratory test performed

NGS, RT-PCR, FISH, and CMA methods were described previously in this chapter, Case 27.1. Sanger sequencing methods were described in Chapter 5, Case 5.4.

Test results

Targeted next-generation sequencing (NGS)

NGS identified novel *USP6* fusions, *PAFAH1B1::USP6* from patient 1 (Fig. 27.5.3A), and *RUNX2::USP6* fusion from patient 2 (Fig. 27.5.4A).

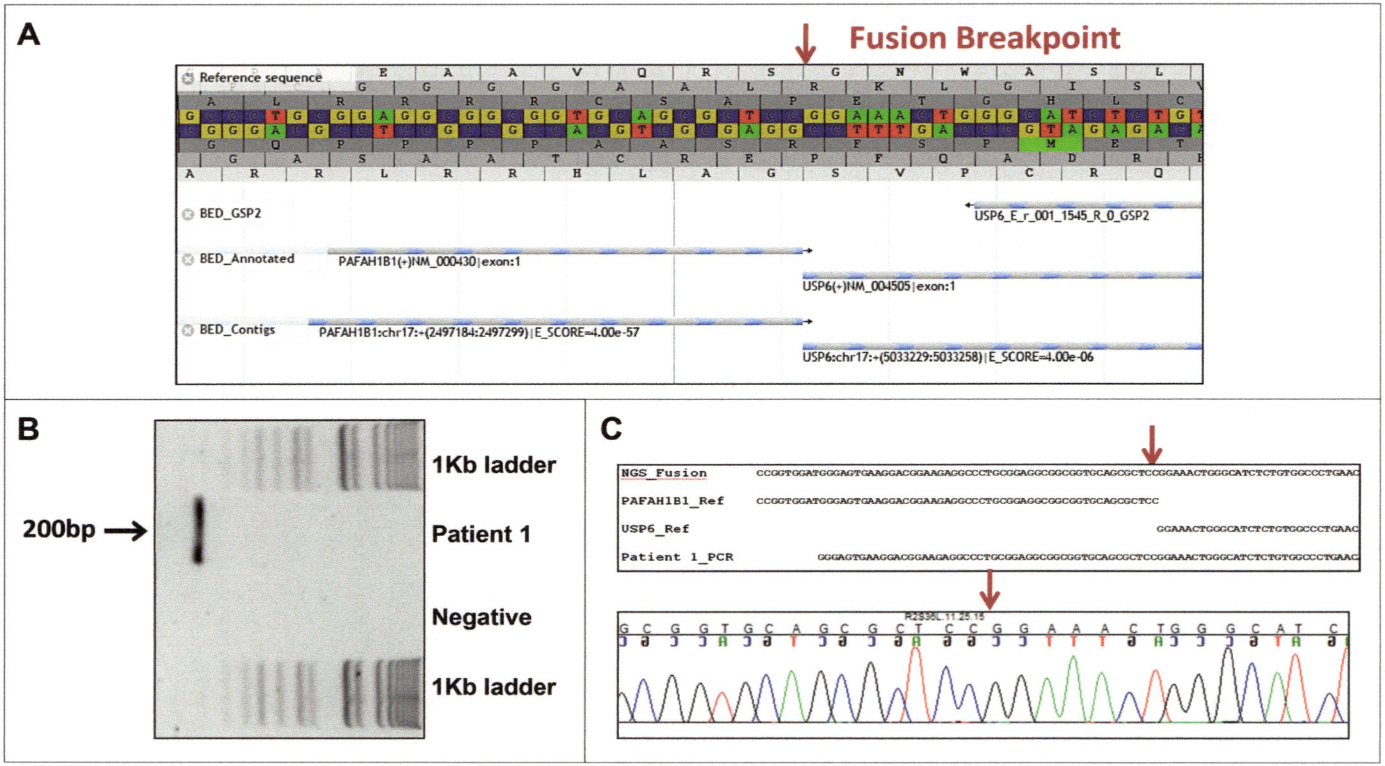

FIG. 27.5.3 Molecular testing results of patient 1. (A) NGS showed novel *USP6* fusions, *PAFAH1B1*::*USP6*. (B) RT-PCR using the forward primer from *PAFAH1B1* and the reverse primer from *USP6* (2.5 Mb apart). RT-PCR amplified a 148 bp single band of fusion transcript. (C) The PCR product was purified and sequenced via the Sanger method.

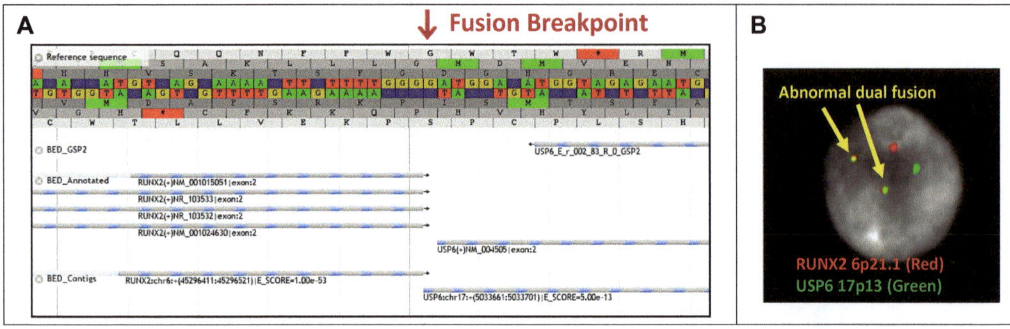

FIG. 27.5.4 Molecular testing results of patient 2. (A) NGS showed a novel *USP6* fusion, *RUNX2::USP6* fusion. (B) FISH using dual-color, dual-fusion probes (RP3-427I13 for RUNX2 on 6p21.1, and RP11-46I8 for *USP6* on 17p13.2). The interphase FISH results showed two *yellow* fusion signals, one *red*, and one *green*, which refer to one normal *RUNX2* and one normal *USP6* signal.

Confirmation of the novel fusions

Patient 1: To confirm the *PAFAH1B1::USP6* fusion, we performed RT-PCR by using the forward primer from *PAFAH1B1* and the reverse primer from *USP6*, because both genes are located on the short arm of chromosome 17 and are 2.5 Mb apart. A 148 bp single band of fusion transcript was amplified by RT-PCR (Fig. 27.5.3B), and the PCR product was purified and sequenced by the Sanger method (Fig. 27.5.3C).

Patient 2: To confirm the *RUNX2::USP6* fusion, FISH with dual-color, dual-fusion probes (RP3-427I13 for *RUNX2* on 6p21.1, and RP11-46I8 for *USP6* on 17p13.2) was performed. The interphase FISH results showed two yellow fusion signals, one red, and one green, which refer to one normal *RUNX2* signal and one normal *USP6* signal (Fig. 27.5.4B).

Results with interpretations

The diagnosis is aneurysmal bone cyst (ABC). A spectrum of *USP6*-rearranged lesions includes ABC, nodular fasciitis (NF), myositis ossificans (MO), fibro-osseous pseudotumor of digits (FOPD), and fibroma of tendon sheath. The molecular-genetic hallmark of these lesions is the presence of *USP6* (Ubiquitin specific peptidase 6, also known as *TRE2*, located at 17p13) fusion with one of the various partner genes [42–45].

ABC presents with a benign bone tumor with multiloculated cysts often containing bloody fluid. Histologically, ABC shows a well-circumscribed cystic mass separated by fibrous septae with myofibroblastic proliferation admixed with osteoclast-like multinucleated giant cells. *USP6::CDH11* is the most common gene fusion reported in approximately 30% of *USP6* fusion-positive ABC [44]. Other known *USP6* fusion partners include *TRAP150, ZNF9, OMD, COL1A1, SEC31A, EIF1, FOSL2, RUNX2, STAT3, PAFAH1B1, CTNNB1, USP9X, ANGPTL2, ASAP1, FAT1, SAR1A,* and *TNC* [42].

NF presents as a rapidly growing, benign myofibroblastic proliferation that can be self-limiting. It often occurs in the subcutis of the upper extremities, trunk, head and neck, and rarely in the deep sites, such as intramuscular, intraarticular, and periosteal regions.

USP6-MYH9 is the most detected gene fusion. *CDH11* and *COL1A*, which are frequent gene partners in ABC, are rarely seen in NF. A growing number of novel fusion partners have been detected, including *RRBP1, CALU, CTNNB1, MIR22HG, SPARC, THBS2, COL6A2, SERPINH1, COL3A1*, and *EIF5A*. An aggressive NF with a *PPP6R3-USP6* fusion and amplification was also described [45].

MO and FOPD are now thought to be the same entity. They also present with a rapidly growing, benign myofibroblastic proliferation in the subcutis and skeletal muscle, which can be self-limiting. Woven-bone formation rimmed by bland osteoblasts is the feature that distinguishes MO/FP from NF. *USP6* rearrangement, often with *COL1A1::USP6* fusion, has been detected in these tumors [43].

Future testing and recommendations

The current NGS panel testing included a limited number of genes. Testing with broader gene coverage (e.g., exome sequencing, RNA sequencing) would increase testing sensitivity and treatment options.

Case 27.6 Lipoblastoma

Clinical indication

A previously healthy 3-month-old female with no abnormal prenatal/birth history presented with a 2-week history of a firm, rapidly growing, well-defined left neck mass.

Radiology

An MRI demonstrated a homogeneous T2 hyperintense and T1 hypointense supraclavicular mass measuring $6 \times 5.5 \times 2$ cm (Fig. 27.6.1A).

Histology

An incisional biopsy demonstrated a homogeneous, cellular proliferation of oval to spindled mesenchymal cells in a prominent myxoid background with a delicate vascular network. The tumor cells have bland cytology with open chromatin and small indistinct nucleoli with no significant atypia. No necrosis was identified. Adipose differentiation and lipoblasts were absent (Fig. 27.6.1B1 and B2). A complete resection of the lesion was performed 3 weeks after the initial biopsy. The tumor demonstrated the histology as seen in the initial biopsy except for focal, small areas with adipose differentiation demonstrating lobules of primitive adipose tissue (Fig. 27.6.1B3, black arrowheads) and lipoblasts (Fig. 27.6.1B4, white arrowheads).

Immunohistochemical (IHC) staining

The IHC stain demonstrated diffuse strong nuclear PLAG1 expression (Fig. 27.6.1C) and diffuse strong vimentin cytoplasmic positivity. CD34 accentuated the background

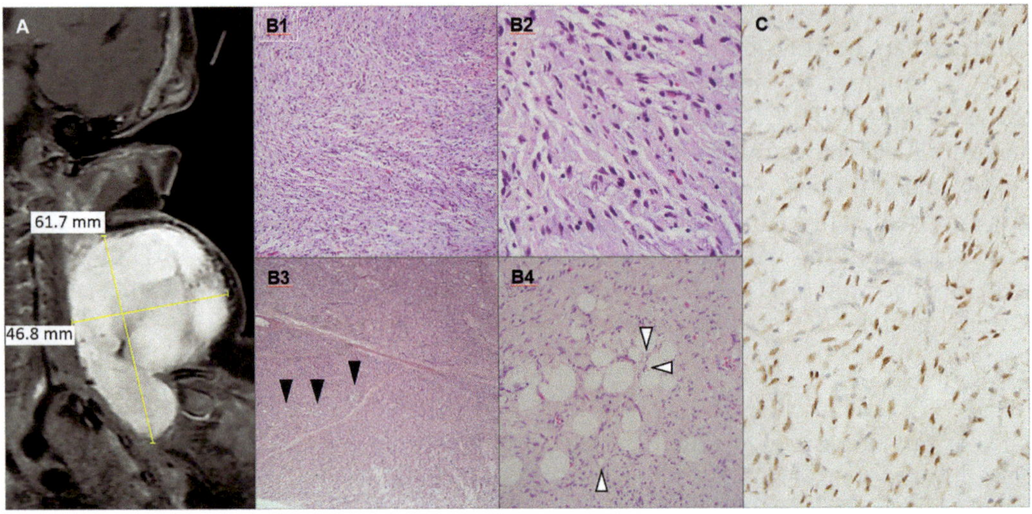

FIG. 27.6.1 MRI and microphotograph of the patient. (A) An MRI demonstrated a homogeneous T2 hyperintense and T1 hypointense supraclavicular mass measuring 6 × 5.5 × 2 cm. (B) Microphotograph of the neck soft tissue mass lesions [H&E, B1 and B3: 40X; B4: 200X; B4: 400X; immunohistochemical staining (IHC): C: PLAG1, 400X]. An initial incisional biopsy demonstrated a homogeneous, cellular proliferation of oval to spindled mesenchymal cells in a prominent myxoid background with a delicate vascular network. The tumor cells have bland cytology with open chromatin and small indistinct nucleoli with no significant atypia. No necrosis was identified. Adipose differentiation and lipoblasts are absent (B1 and B2). The tumor resection specimen demonstrated similar histology to that seen in the initial biopsy except for focal, small areas with adipose differentiation showing lobules of primitive adipose tissue (B3, *black* arrowheads) and lipoblasts (B4, *white* arrowheads). (C) The tumor cells demonstrated diffuse strong nuclear PLAG1 expression.

vasculatures and showed diffuse weak positivity in the tumor cells. SMA, S100, myogenin, CD99, and CD117 were negative.

Test ordered

- FISH: *FUS, ETV6, EWSR1* and *PLAG1* break-apart probes
- Conventional chromosome analysis
- Chromosome microarray (CMA)
- Targeted NGS

Laboratory test performed

FISH, karyotyping, CMA, and NGS methods were described previously in this chapter, Case 27.1.

Test results

FISH analyses using *FUS, ETV6*, and *EWSR1* break-apart probes showed normal patterns, while FISH analysis using a *PLAG1* break-apart probe on metaphase spreads demonstrated a rearrangement of *PLAG1*: the 3′end of *PLAG1* (green) localized normally to 8q, and the 5′end of *PLAG1* (orange) abnormally localized to 11q (Fig. 27.6.2A1). Additional FISH analyses using *PLAG1* probe (Red) and *HAS2* probe (Green) demonstrated a *PLAG1::HAS2* fusion signal localizing to the derivative chromosome 8 created by the insertion of 8q23q12 within the derivative chromosome 11 (Fig. 27.6.2A2).

Conventional chromosome analysis using G-banding was performed on tumor cells and the results showed 46,XX,ins(11,8)(q23;q23q12) (Fig. 27.6.2B).

Chromosome microarray analysis (Illumina, CytoSNP 850K BeadChip) revealed a 1.9 Mb deletion of 8q24.1 (123,954,718-125,898,882), possibly associated with the insertion of 8q23q12 within 11q23 (Fig. 27.6.3).

Finally, the fusion of exon 3 of *PLAG1* with exon 1 of *HAS2* was identified by NGS (Fig. 27.6.4).

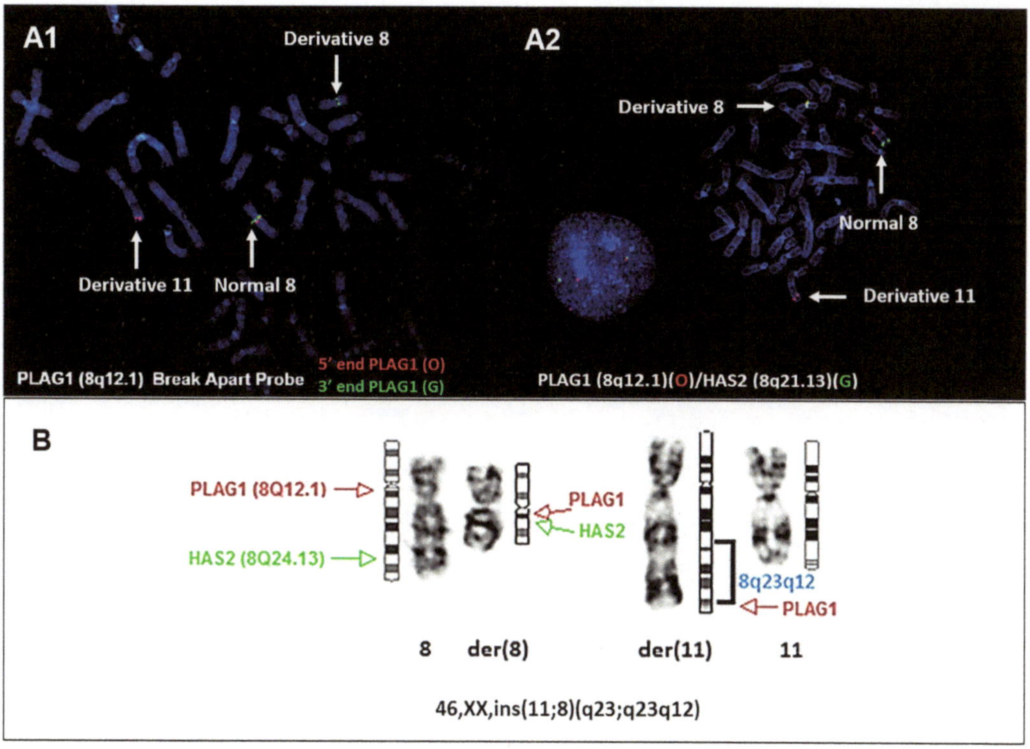

FIG. 27.6.2 FISH and karyotype results of the patient. (A) FISH analyses using *FUS, ETV6,* and *EWSR1* break-apart probes showed normal patterns, while FISH analysis using a *PLAG1* break-apart probe on metaphase spreads demonstrated a rearrangement of *PLAG1*: the 3′end of *PLAG1 (green)* localized normally to 8q, and the 5′end of *PLAG1 (orange)* abnormally localized to 11q (A1). Additional FISH analyses using the *PLAG1* probe *(red)* and *HAS2* probe *(green)* demonstrated a *PLAG1::HAS2* fusion signal localizing to the derivative chromosome 8 created by the insertion of 8q23q12 within the derivative chromosome 11 (A2). (B) Conventional chromosome analysis using G-banding was performed on tumor cells, and the results showed 46,XX,ins(11;8)(q23;q23q12).

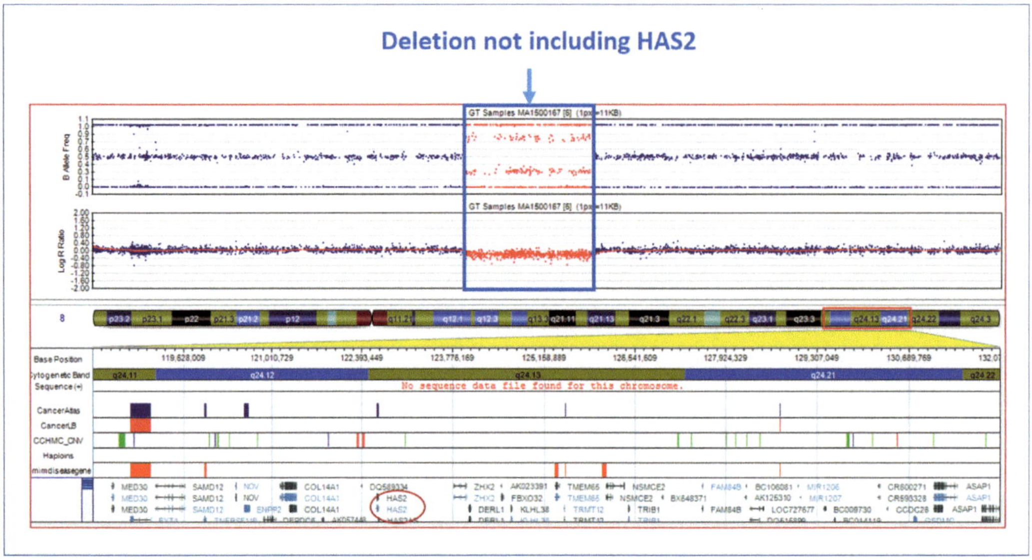

FIG. 27.6.3 Chromosomal microarray (CMA) results from the patient. DNA SNP microarray analysis (Illumina, CytoSNP 850K BeadChip) revealed a 1.9 Mb deletion of 8q24.1 (123,954,718-125,898,882), possibly associated with the insertion of 8q23q12 within 11q23.

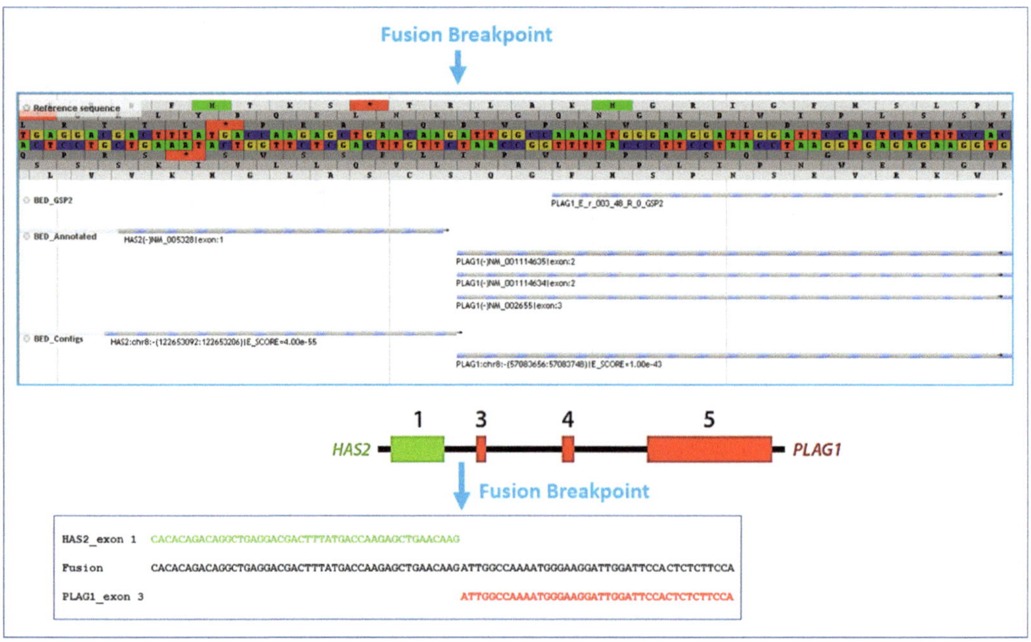

FIG. 27.6.4 Targeted NGS showed that the exon 3 of *PLAG1* fused with the exon 1 of *HAS2*.

Results with interpretations

The diagnosis for this case is lipoblastoma. Most lipoblastomas are detected in young children, with 40% of cases diagnosed during the first year and over 80% occurring before 3 years of age. There is a male predominance of approximately 2:1. Lipoblastoma arises predominantly in the trunk and extremities but may involve the mediastinum, retroperitoneum, and viscera.

Lipoblastomas are benign, with an excellent prognosis, and treated only by surgical excision. However, local recurrences are common, especially when the tumors are excised incompletely [46,47].

Many lipoblastomas are well-circumscribed, but a subset has an infiltrative border and is referred to as "diffuse lipoblastoma" or "lipoblastomatosis." Histologically, lipoblastoma is characterized by the presence of immature adipocytes in various stages of development, ranging from primitive mesenchymal cells (Fig. 27.6.1B1 and B2) and lipoblasts (Fig. 27.6.1B4, white arrowhead) to mature adipocytes, with variable amounts of a myxoid matrix in the background. Diagnosing lipoblastomas with inconspicuous lipoblasts can be challenging. At the most immature end, lipoblastoma shows an exclusively myxoid spindle cell pattern that mimics primitive myxoid mesenchymal tumor of infancy (PMMTI) and the most matured lipoblastoma can be indistinguishable from lipoma (lipoma-like lipoblastoma) [48].

PLAG1 immunohistochemical staining (IHC) has a high sensitivity for the diagnosis of lipoblastoma, including histologically unusual lipoblastomas (myxoid and lipoma-like lipoblastomas). As IHC is rapid and inexpensive as contrasted with molecular genetic analysis, it can be considered a first-line diagnostic method.

The genetic hallmark of lipoblastoma is a fusion involving *PLAG1* and one of the many partner genes, including: *HAS2* (located at 8q24.13) [49], *COL1A2* (7q21.3) [50], *COL3A1* (2q32.2) [50], *RAB2A* (8q12.1-q12.2) [50], *RAD51L1* (14q24.1) [51], *BOC* (3q13.2) [52], *ZEB2* (2q22.3) [53], *CHCHD7* (8q12.1), *CTDSP2* (12q14.1), *HNRNPC* (14q11.2), *SRSF3* (6p21.3-p21.2), *PCMTD1* (8p11.23), *YWHAZ* (8q22.3), *PPP2R2A* (8p21.2), *DDX6* (11q23.3), *KLF10* (8q22.3), and *KANSL1L* (2q34) [54,55]. In addition, *HMG2* rearrangements instead of the *PLAG1* fusion have been identified by FISH [56] and NGS [54] in a small subset of lipoblastomas. So far, two fusion partner genes with *HMG2, FGD6,* and *EP400,* have been reported [54]. FISH did not detect many of these rearrangements. *CHCHD7* and *PLAG1* are located only a few hundred base pairs apart at 8q12.1, and the fusion in general is cryptic by conventional karyotyping and FISH. *CTDSP2::PLAG1* and *PPP2R2A::PLAG1* were also cryptic and usually could not be identified by cytogenetic methods [55]. *HAS2::PLAG1, DDX6::PLAG1, KLF10::PLAG1, KANSL1L::PLAG1,* and *COL3A1::PLAG1* could not be detected by *PLAG1* FISH [54]. These studies suggest that cryptic chromosomal rearrangements are common in lipoblastoma. NGS is an excellent method to detect *PLAG1* fusions; however, commercially available targeted RNA sequencing panels may not cover newly emerged fusion partners.

Future testing and recommendations

The current NGS panel testing included a limited number of genes. Testing with broader gene coverage (e.g., exome sequencing, RNA sequencing) would increase testing sensitivity and treatment options.

Case 27.7 Pleuropulmonary blastoma-*DICER1*-associated tumors

Clinical indication

A previously healthy 3-year-old male presented with wheezing and shortness of breath.

Radiology

An MRI demonstrated a large multiloculated, pleural-based lung mass with a predominantly fluid intensity at the base of the left lower lobe (Fig. 27.7.1A).

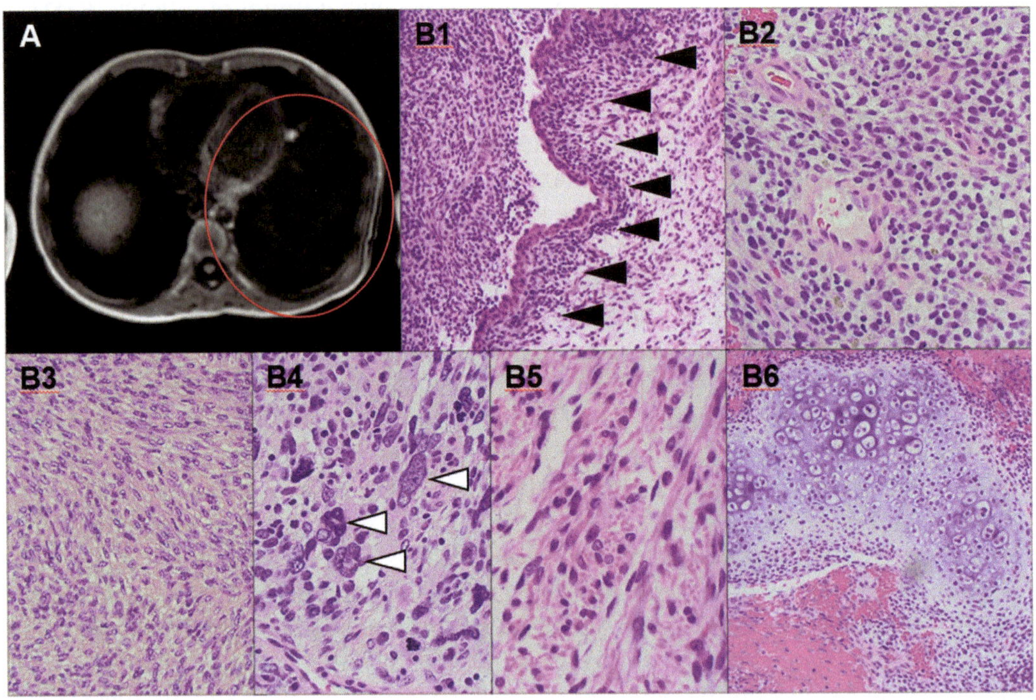

FIG. 27.7.1 MRI and microphotographic results from the patient. (A) An MRI demonstrated a large multiloculated, pleural-based lung mass with a predominantly fluid intensity at the base of the left lower lobe. (B) Microphotographs of the lung mass lesion (H&E, B1 and B6: 100X, B2–B5: 400X). H&E histologic sections of the tumor demonstrated heterogeneous features, including undifferentiated small round *blue* cells (B1 and B2), poorly differentiated spindle cells (B3), and large bizarre pleomorphic cells (anaplasia, B4 *white* arrowheads), often with rhabdomyoblastic (B1 and B5) and/or chondroid differentiation (B6). The rhabdomyoblastic cells aggregated beneath the epithelial lining ("cambium layer," B1, *black* arrowheads).

Histology

H&E histologic sections of the tumor demonstrated heterogeneous features, including undifferentiated small round blue cells (Fig. 27.7.1B1 and B2), poorly differentiated spindle cells (Fig. 27.7.1B3), and large bizarre pleomorphic cells (anaplasia, Fig. 27.7.1B4, white arrowheads), often with rhabdomyoblastic (Fig. 27.7.1B1 and B5) and/or chondroid differentiation (Fig. 27.7.1B6). The rhabdomyoblastic cells aggregated beneath the epithelial lining ("cambium layer," Fig. 27.7.1B1, black arrowheads).

Test ordered

- Chromosome microarray (CMA)
- Targeted NGS

Laboratory test performed

CMA and NGS methods were described previously in this chapter, Case 27.1.

Test results

CMA demonstrated evidence for a cytogenetically evolving tumor, as copy number alterations were present in small subpopulations of cells. The alterations in the significant subclone included a gain in distal 1q, a loss of the distal 7p, a gain of chromosome 8, and a deletion of a region in 17p13 (including the *TP53* locus). There were no alterations noted in chromosome 14 (Fig. 27.7.2A and B). Although not specific to this entity, the results are consistent with the histologic diagnosis.

Two *DICER1* mutations were identified in this study by Targeted NGS: a nonsense mutation (p.Tyr1205Ter) and a hotspot missense mutation (p.Glu1813Gln) (Fig. 27.7.3).

Results with interpretations

The diagnosis is pleuropulmonary blastoma or *DICER1* syndrome. The *DICER1* gene encodes an endoribonuclease required for processing microRNA (miRNA), which regulates the translation and degradation of messenger RNA. Dysregulation of miRNA by *DICER1* mutations causes activation of oncogenes [57]. Patients with *DICER1* syndrome commonly have a heterozygous germline loss-of-function (LOF) variant leading to truncation and loss of function of the protein [57–61]. Characteristic "hotspot" somatic mutations (E1705, E1813, D1709, D1810, or G1809) are located at the four metal-binding sites of the RNase IIIb domain that affect the catalytic activity of *DICER1* (Online Mendelian Inheritance in Man number 601200) [62].

DICER1 syndrome is associated with numerous benign and malignant conditions/neoplasms, including lung cysts, pleuropulmonary blastoma (PPB), cystic nephroma, Wilms' tumor, multinodular goiter, thyroid adenoma and carcinoma, juvenile-type intestinal polyps, nasal chondromesenchymal hamartoma, ovarian tumors (mainly Sertoli-Leydig cell tumors) [62–67], and sarcomas (e.g., genitourinary embryonal rhabdomyosarcoma

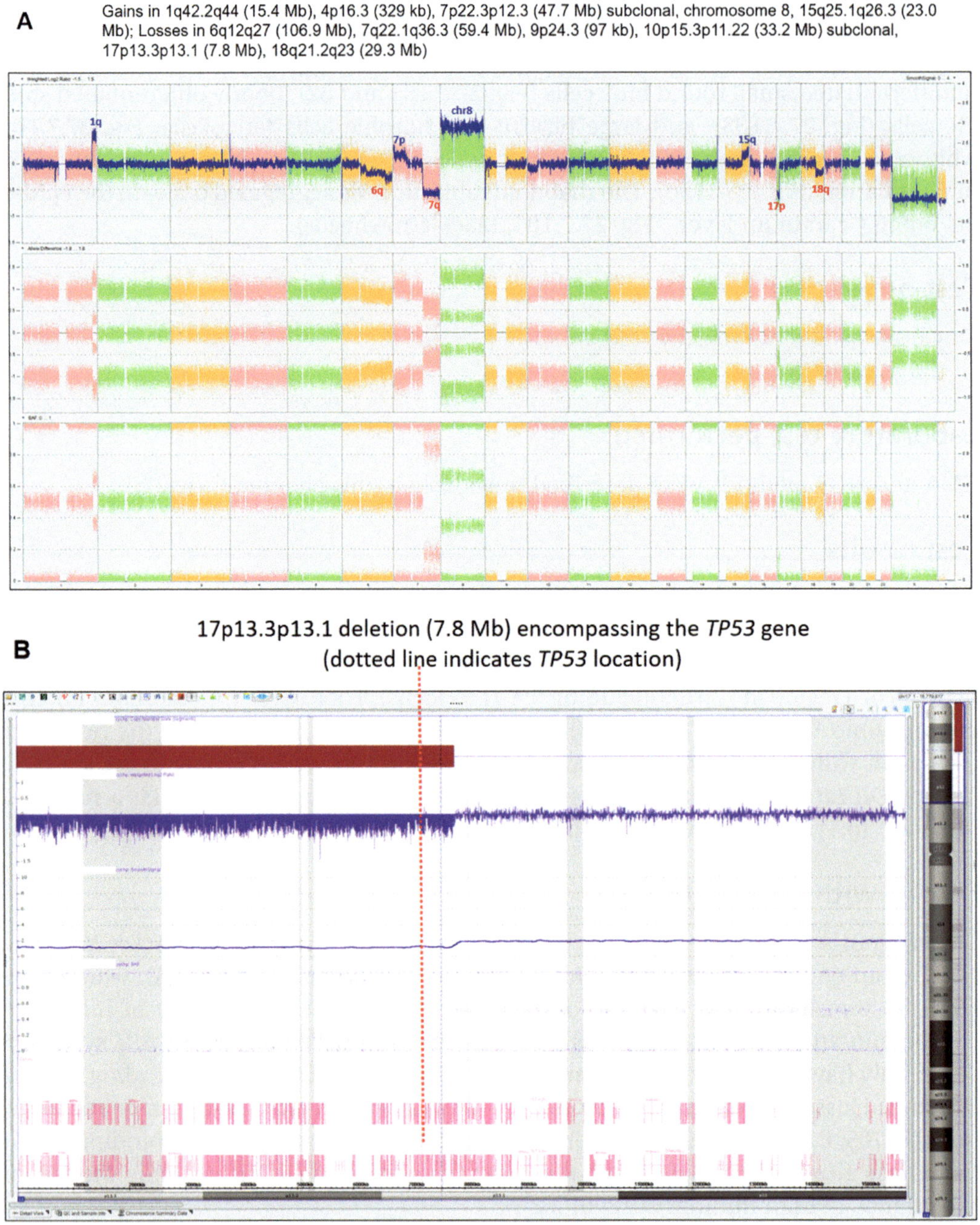

FIG. 27.7.2 Chromosomal microarray (CMA) results from the patient. CMA showed copy number alterations in small subpopulations of cells. The alterations in the significant subclone included a gain in distal 1q, a loss of the distal 7p, a gain of chromosome 8 (A), and a deletion of a region in 17p13 (including the *TP53* locus. B). No alterations were noted on chromosome 14.

DNA

Classification	Gene Name	Exon	Genomic Position (hg19)	DNA Change	Protein Change	Variant Allele Frequency
Strong Clinical Significance (Tier I)	DICER1	23	Chr14: 95,570,118	c.3615G>T (NM_030621.4)	p.Tyr1205Ter	47%
Strong Clinical Significance (Tier I)	DICER1	27	Chr14: 95,557,630	c.5437C>G (NM_030621.4)	p.Glu1813Gln	42%

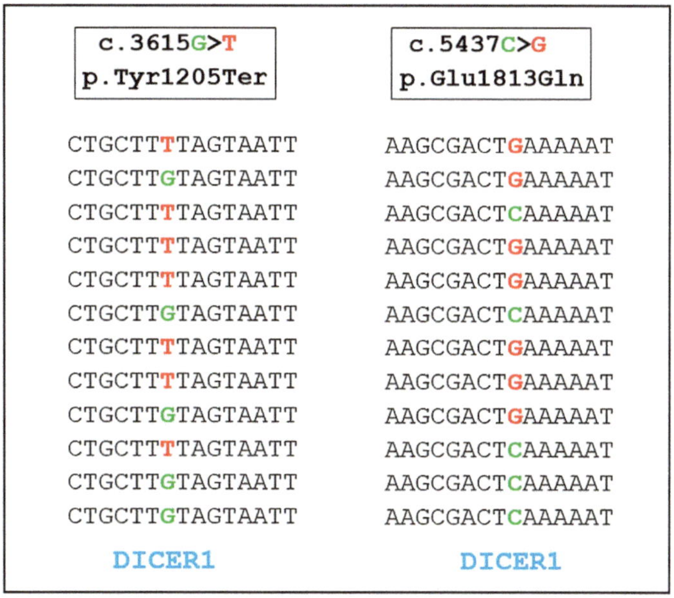

FIG. 27.7.3 Two *DICER1* mutations were identified in this study: a nonsense mutation (p.Tyr1205Ter) and a hotspot missense mutation (p.Glu1813Gln).

of the bladder or uterine cervix and anaplastic sarcoma of the kidney) [68]. PPB, rhabdomyosarcoma, and other soft tissue sarcomas can metastasize to the central nervous system. Primary CNS manifestations include pituitary blastoma, pineoblastoma, ciliary body medulloepithelioma, and macrocephaly [62]. Recently, primary *DICER1*-associated CNS sarcoma and embryonal tumors with multilayered rosettes-like infantile cerebellar tumors have also been reported [69].

A biallelic pathogenic variant in *DICER1* is detected in these neoplasms: usually a germline LOF pathogenic variant in one allele and a somatic hotspot variant in the other allele. Identifying pathogenic *DICER1* variants would facilitate optimized genetic counseling, individual and caregiver education, and appropriate imaging-based surveillance.

Future testing and recommendations

The current NGS panel testing included a limited number of genes. Testing with broader gene coverage (e.g., exome sequencing, RNA sequencing) would increase testing sensitivity and treatment options.

Case 27.8 Malignant peripheral nerve sheath tumor (MPNST)-Neurofibromatosis type 1

Clinical indication

A previously healthy 11-year-old female presented with a left thigh mass with throbbing pain.

Radiology

An MRI demonstrated a large multilobulated heterogeneously enhancing mass with cystic and nodular areas centered in the left adductor muscles, measuring $12.2 \times 8.3 \times 9.1$ cm (Fig. 27.8.1A). The gross photo of the tumor is also shown (Fig. 27.8.1B).

Histology

H&E histologic sections of the diagnostic core biopsy demonstrated tumor tissue and broad areas of necrosis (Fig. 27.8.1C1). The tumor comprises poorly differentiated spindle cells with occasional pleomorphic cells and frequent mitotic figures (Fig. 27.8.1C2).

Immunohistochemical (IHC) staining

IHC staining is nonspecific for any entities of differentiation. H3K27me3 expression was lost in a subset of tumor cells (Fig. 27.8.1C3).

Test ordered

- Chromosome microarray (CMA)
- Targeted NGS

Laboratory test performed

CMA and NGS methods were described previously in this chapter, Case 27.1.

Test results

Chromosomal microarray (CMA)

CMA demonstrated a complex pattern of segmental gains and losses affecting almost every chromosome. MPNST typically shows a similar complex CMA pattern. In addition,

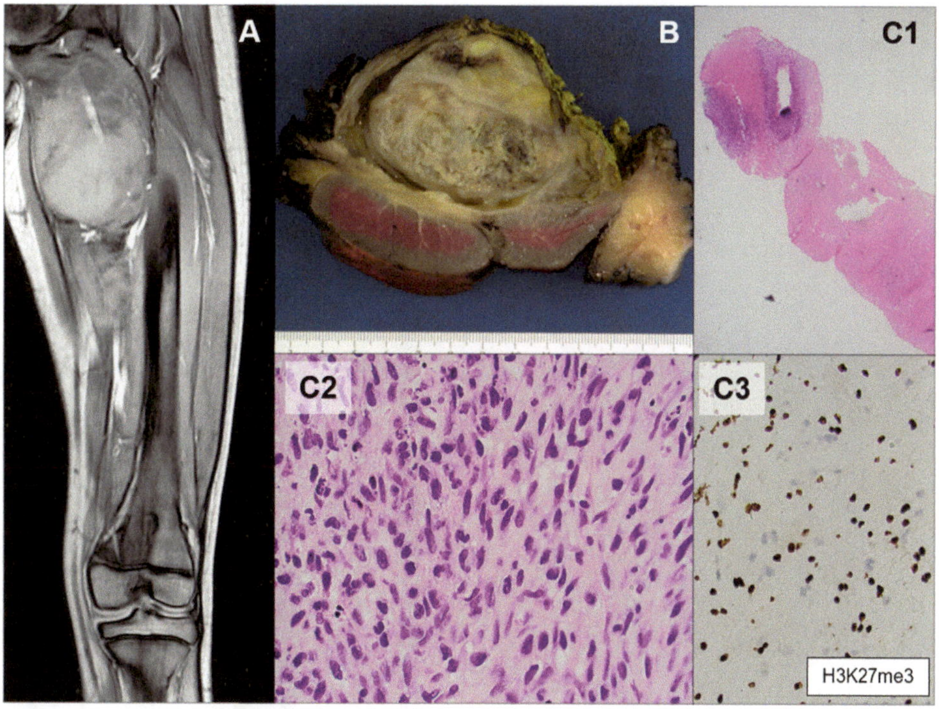

FIG. 27.8.1 MRI and microphotographic results from the patient. (A) An MRI demonstrated a large multilobulated heterogeneously enhancing mass with cystic and nodular areas centered in the left adductor muscles, measuring 12.2 × 8.3 × 9.1 cm. (B) The gross photo of the tumor. (C) Microphotographs of the left thigh soft tissue tumor [H&E, C1: 40X, C2: 600X; immunohistochemical staining (IHC), C3: H3K27me3, 400X]. H&E histologic sections of the diagnostic core biopsy demonstrated tumor tissue broad areas of necrosis (C1). The tumor comprises poorly differentiated spindle cells with occasional pleomorphic cells and frequent mitotic figures (C2). H3K27me3 IHC expression was lost in a subset of tumor cells (C3).

a copy number loss (deletion) is noted in a locus encompassing the *NF1* gene in 17q11.2; together with the NGS finding resulting in bi-allelic *NF1* alterations (Fig. 27.8.2).

Targeted next generation sequencing (NGS)

Variants of strong clinical significance were identified in *NF1* (p.Tyr489Cys) and *TP53* (p. Arg248Gln) (data not shown).

Results with interpretations

The diagnosis for this case is malignant peripheral nerve sheath tumor (MPNST), also called neurofibromatosis type 1. Neurofibromatosis type 1 (NF1) is caused by mutations in the *NF1* gene (17q11.2) that encodes neurofibromin (NF1), a tumor suppressor protein that downregulates the RAS/MAPK pathway. NF1 is an autosomal dominant disorder with an incidence of approximately 1:2600 to 1:3000 individuals. About half of the NF1 cases present as familial (inherited) forms [26,70,71].

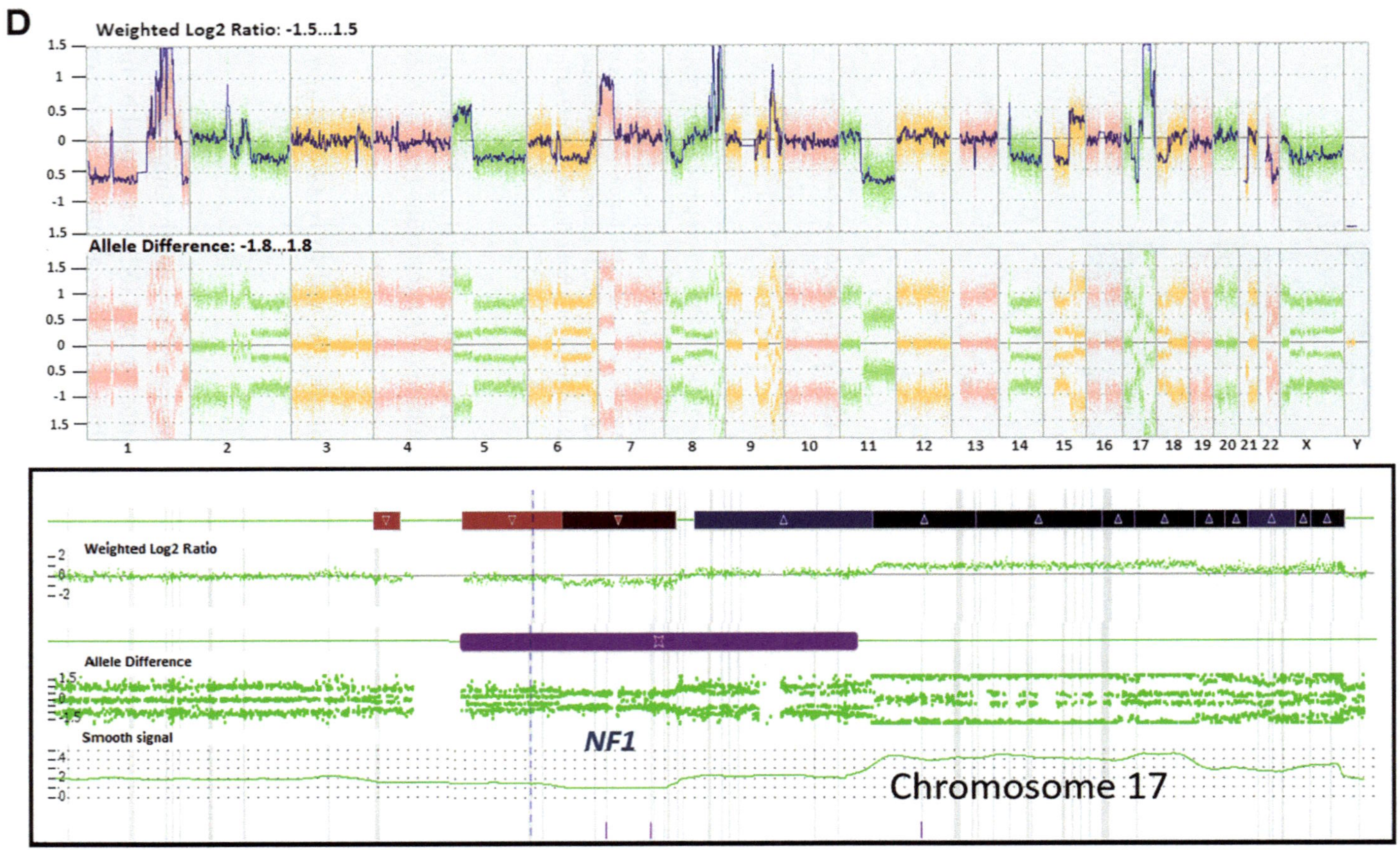

FIG. 27.8.2 CMA results from the patient. CMA demonstrated a complex pattern of segmental gains and losses affecting almost every chromosome. In addition, a copy number loss (deletion) was noted in a locus encompassing the *NF1* gene in 17q11.2; together with the NGS finding resulting in bi-allelic *NF1* alterations.

NF1 regulates neural crest-derived cells, such as Schwann cells, melanocytes, and glial cells. Therefore, the common clinical manifestations are café-au-lait spots, axillary and inguinal freckling, Lisch nodules of the iris, multiple neurofibromas (NF), and increased risk for various malignancies in this lineage [70]. Children with NF1 have an increased risk of developing peripheral nervous system (PNS) tumors, ranging from benign cutaneous NF, plexiform NF, to MPNST. Plexiform NF arises in 50% of individuals with NF1 (childhood-onset in 20%). Plexiform NF is a benign tumor but may transform to MPNST [26,71,72]. MPNST often arises in adolescents and young adults in the pediatric setting, usually with neurofibromatosis type 1 (NF1). MPNST has metastatic potential and usually presents with aggressive features, such as rapid growth, pain, and local neurologic dysfunction. The most common locations of MPNST are the trunk and extremities, followed by the head and neck [26]. Optic pathway glioma (OPG) is the most common *NF1*-associated CNS neoplasm detected in 15%–20% of children with NF1. OPG is usually low-grade but can lead to vision loss and premature or delayed puberty. Other *NF1*-associated neoplasms include rhabdomyosarcoma, gastrointestinal stromal tumor, glomus tumors, juvenile myelomonocytic leukemia, and pheochromocytoma [71].

Histologically, MPNST comprises hypercellular monotonous spindle tumor cells with hyperchromatic nuclei and pale to eosinophilic wavy cytoplasm. The tumor demonstrates variable nuclear pleomorphism, increased mitotic figures, and geographical necrosis. In addition, the tumor cells may show heterogeneous differentiations such as epithelioid, skeletal muscle (malignant Triton tumor), and glandular differentiations. In the NF1 setting, MPNST often arises in a preexisting neurofibroma. Immunohistochemically, MPNST may be positive for S100 (<50% of tumors) and SOX10 (<70%), typically with a focal or patchy distribution. In addition, MPNST often lacks H3K27me3, a marker for epigenetic modification of histone H3; high-grade MPNST shows more frequent loss than low-grade MPNST.

Patients with *NF1*-associated tumors usually have a germline *NF1* mutation and a second-hit somatic *NF1* alteration. Identifying pathogenic *NF1* variants would require optimized genetic counseling, individual and caregiver education, and appropriate imaging-based surveillance and oncologic management [73]. For example, radiation should be used with caution since it increases the risk of developing MPNST in NF1 patients.

Future testing and recommendations

The current NGS panel testing included a limited number of genes. Testing with broader gene coverage (e.g., exome sequencing, RNA sequencing) would increase testing sensitivity and treatment options.

Case 27.9 Neuroblastoma

Clinical indication

A previously healthy 3-year-old female presented with intermittent and sharp abdominal pain.

Radiology

CT chest demonstrated a large central and upper retroperitoneal mass, measuring 12.9 × 11.1 × 8.2 cm, arising in the right adrenal gland and extending into the hilum of the right kidney (Fig. 27.9.1A). MIBG scan showed MIBG accumulation at the site of the patient's right-sided intra-abdominal mass (Fig. 27.9.1B).

Histology

H&E histologic sections of the initial core biopsy demonstrated sheets of undifferentiated small round blue cells with no significant neuropil production. There were brisk mitotic figures (Fig. 27.9.2A, yellow circles) and high karyorrhexis (high MKI: >400/5,000 cells, >4%) (Fig. 27.9.2A, red circles). Foci of necrosis were present. No Schwannian stroma or calcification was identified.

Immunohistochemical (IHC) staining

IHC demonstrated diffuse, strong PHOX2B nuclear staining in the tumor cells.

Test ordered

- FISH: *MYCN* amplification
- Chromosome microarray (CMA)
- Targeted NGS

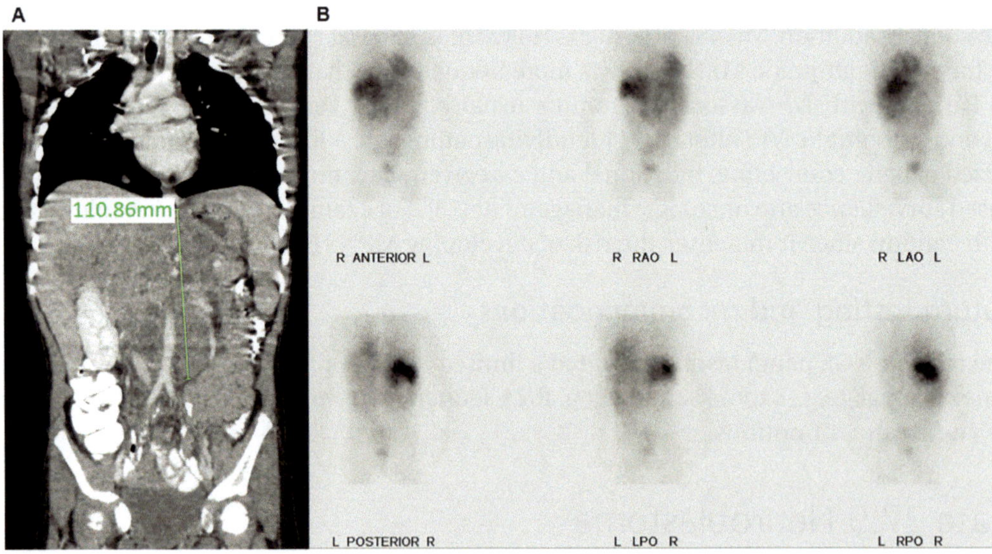

FIG. 27.9.1 Chest CT and MIBG results from the patient. (A) Chest CT demonstrated a large central and upper retroperitoneal mass, measuring 12.9 × 11.1 × 8.2 cm, arising in the right adrenal gland, and extending into the hilum of the right kidney. (B) MIBG scan showed MIBG accumulation at the site of the patient's right-sided intra-abdominal mass.

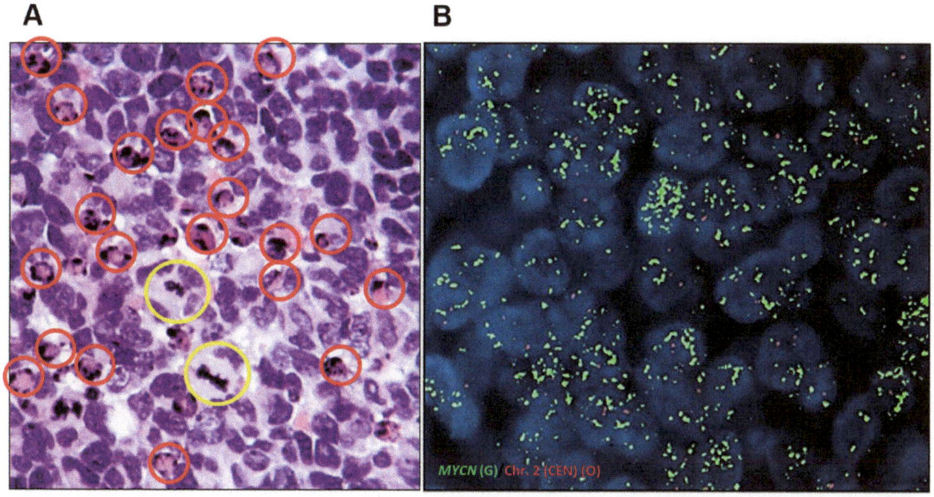

FIG. 27.9.2 Microphotograph and FISH results from the patient. (A) Microphotograph of the upper retroperitoneal mass lesion (H&E, 600X). H&E histologic sections of the initial core biopsy demonstrated sheets of undifferentiated small round blue cells with no significant neuropil production. There were brisk mitotic figures (*yellow* circles) and high karyorrhexis (high MKI: >400/5,000 cells, >4%), *red* circles). No Schwannian stroma was identified. (B) FISH analysis with probes for *MYCN (green)* and chromosome 2 centromere *(orange)* revealed amplification of *MYCN*.

Laboratory test performed

FISH, CMA, and NGS methods were described previously in this chapter, Case 27.1.

Test results

FISH analysis with probes for *MYCN* (green) and chromosome 2 centromere (orange) revealed amplification of *MYCN* (Fig. 27.9.2B).

CMA analysis revealed the following abnormalities:

> CN-LOH in 1p and a relative copy number gain in the long arm of chromosome 17. The analysis showed a polyploid tumor with amplification involving the *MYCN* locus in 2p24.3p24.2 (Fig. 27.9.3A).
>
> *MYCN* gene amplification was identified by Targeted NGS (Fig. 27.9.3B). *MYCN* gene amplification is a recurrent driver alteration in neuroblastomas and is associated with a poor prognosis.

Results with interpretations

The diagnosis for this case is neuroblastoma (Schwannian stroma-poor), undifferentiated subtype, high MKI, >18 months of age with unfavorable histology.

Neuroblastoma (NB) is the third most common pediatric malignancy after leukemia and brain tumors. It is the most common solid extracranial tumor in children, with more than 600 new cases diagnosed in the United States each year [11].

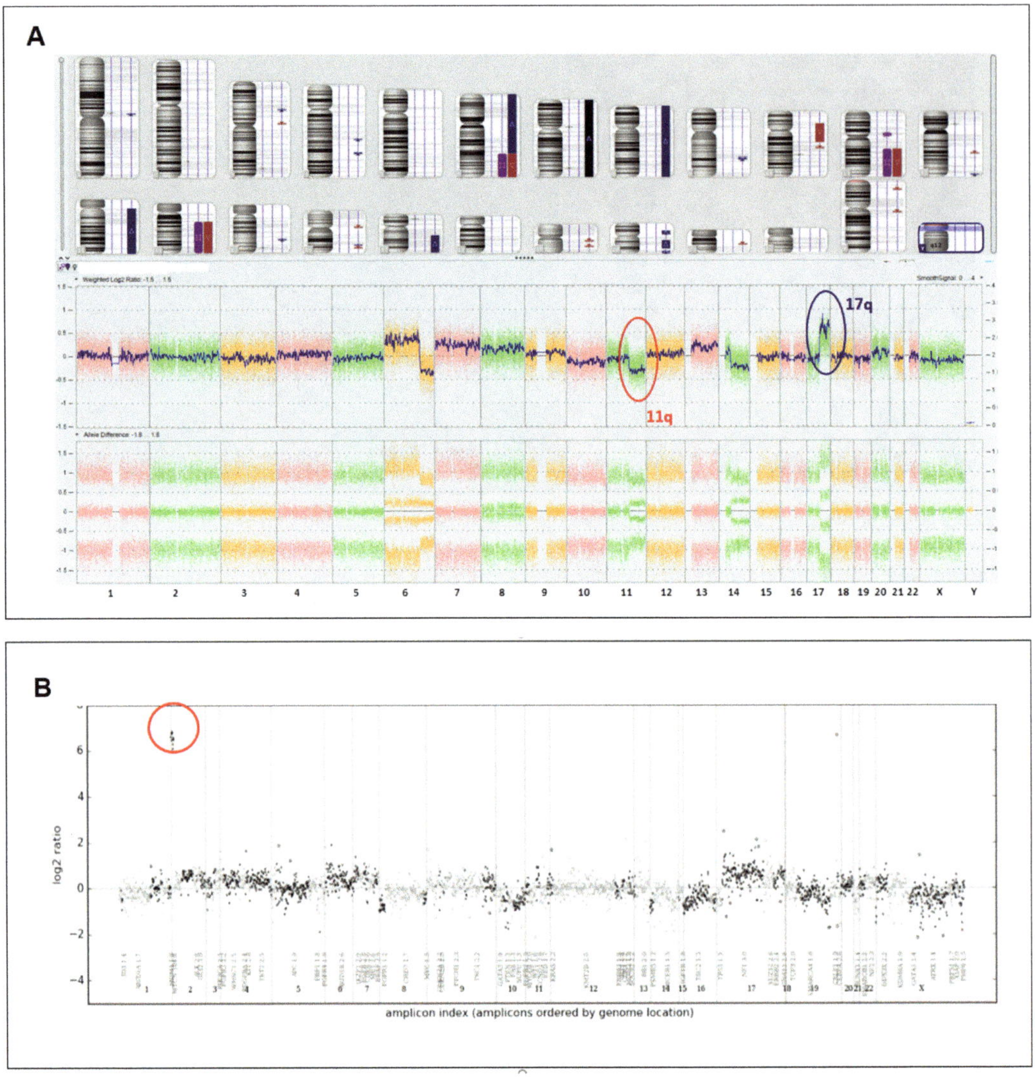

FIG. 27.9.3 Chromosomal microarray (CMA) results from the patient. (A) Chromosomal microarray (CMA) analysis revealed CN-LOH in 1p and a relative copy number gain in the long arm of chromosome 17. (B) The analysis also showed a polyploid tumor with amplification involving the *MYCN* locus in 2p24.3p24.2.

NB refers to a spectrum of peripheral neuroblastic tumors (PNT) that arises from the cell group differentiating to the autonomic nervous system in the neural crest. Therefore, the PNT primarily occurs in the adrenal medulla and sympathetic chains in the neck and thoracic and abdominal cavities and secretes catecholamine.

NB was once called an "enigmatic disease" because its clinical presentation is highly variable. It ranges from indolent disease, with spontaneous regression and maturation to a benign ganglioneuroma, to a highly aggressive disease with extensive local invasion, disseminated metastasis, and death. Clinical heterogeneity is associated with numerous clinical and biological prognostic factors. The prognostic factors are used for risk stratification, which is essential for determining the initial treatment regimen [74–76].

1. Age:
   a. Younger patients (under 12–18 months) have better outcomes.
2. Tumor stage:
   a. The stage represents the degree of local extension and metastasis of the tumor. Currently, there are two systems are used for NB risk stratification as follows:
      i. INRGSS: the International Neuroblastoma Risk Group Staging System (INRGSS) incorporates Image-defined risk factors (IDRF) based on the preoperative imaging results (e.g., CT, MRI, and MIBG scans).
      ii. INSS: the International Neuroblastoma Staging System (INSS) considers surgery results (complete vs incomplete resection).
3. Tumor histology based on "International Neuroblastoma Pathology Classification (INPC)" (Fig. 27.9.4).
   a. PNT comprises two histologic cell types as follows:
      i. Neuroblastic tumor cells.
      ii. Schwann cells and Schwannian stroma induced by the neuroblastic cells during differentiation.

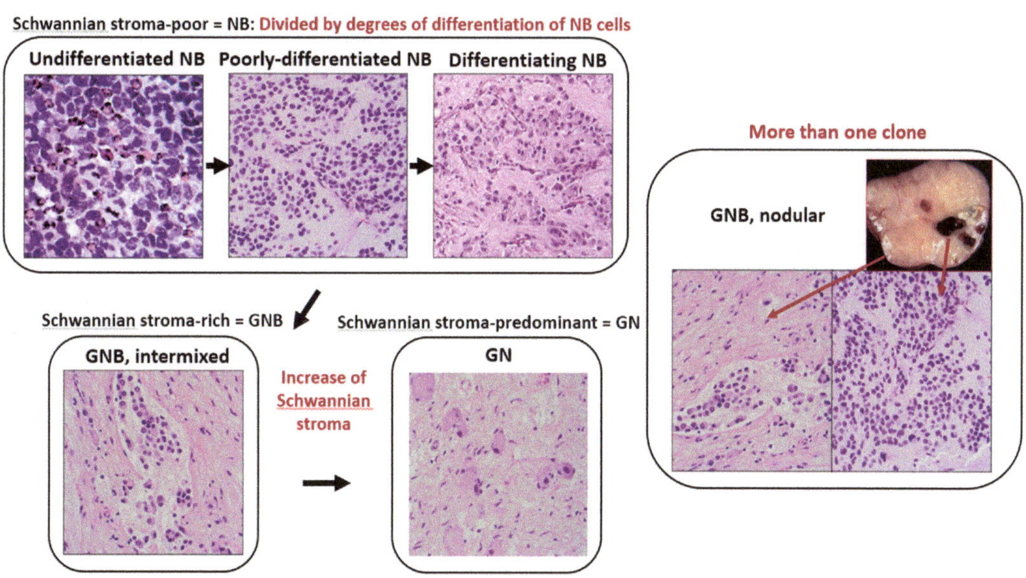

FIG. 27.9.4 Histology images for PNT. Peripheral neuroblastic tumor (PNT) histology based on "International Neuroblastoma Pathology Classification (INPC)." PNT is divided into histologic subtypes depending on its degrees of differentiation and maturation of neuroblastic tumor cells and Schwann cells/Schwannian stroma.

b. PNT is divided into histologic subtypes depending on its degrees of differentiation and maturation of neuroblastic tumor cells and Schwann cells/Schwannian stroma.
   i. **Neuroblastoma (NB, Schwannian stroma-poor forms, <50% of the tumor volume)**
      (1) NB is further divided into three groups based on the degrees of neuroblastic cell differentiation.
         (a) **Undifferentiated NB (UD-NB)**: undifferentiated small round blue neuroblastic cells with no neuropil production
         (b) **Poorly differentiated NB (PD-NB)**: mildly differentiated neuroblastic cells (<5% of tumor cells) with neuropil production
         (c) **Differentiating NB (D-NB)**: ≥5% differentiated neuroblastic cells with neuropil production
   ii. **Ganglioneuroblastoma, intermixed (GNB, Schwannian stroma-rich form, ≥50%)**
      (1) **Nodular GNB** is a form of GNB. It is a composite tumor (containing more than one clone), with a Schwannian stroma-poor (NB) component and rich/predominant (GNB/GN) component.
   iii. **Ganglioneuroma (GN, Schwannian stroma-dominant form):**
      (1) This is the most mature subtype with maturing/matured neuroblastic cells (ganglion cells); neuropil production is no longer identified.

4. According to INPC, PNT is divided into "**favorable histology (FH)**," which is associated with a better outcome, and "**unfavorable histology (UH)**," which tends to have a worse prognosis.
   a. To be classified as FH, the tumor should demonstrate **differentiation and maturation of the neuroblastic, and Schwann cells appropriate for the patient's age**.
   b. **The hallmark of UH is the "arrest of age-appropriate differentiation."**
      i. For example, PD-NB is classified as FH in patients younger than 18 months (unless it shows a high mitotic-karyorrhexis index (MKI), a surrogate marker of "MYCN amplification"). However, PD-NB is UH if the patient is ≥18 months of age.
      ii. Studies support the "arrest of age-appropriate differentiation" of UH is caused by an additional molecular, cytogenetic, and epigenetic alteration(s).
   c. The most sensitive and specific IHC marker for PNT is PHOX2B[79].
5. DNA ploidy (DNA index)
6. Hyperploidy in NB is associated with a better prognosis, particularly in children younger than two years of age. However, DNA ploidy is not a good predictive factor for older patients.
7. *MYCN* gene amplification:
   a. The presence of *MYCN* amplification (located 2p24, >10 copies) is a powerful adverse prognostic factor found in up to 20% of NB.

8. Other gene alterations identified in NB are as follows:
   a. Chromosomal alterations:
      i. Segmental chromosomal aberrations (SCA), such as loss of heterozygosity (LOH) of 1p, 14q, and 11q and gain of 17q are adverse prognostic factors.
   b. *ALK*:
      i. Activation mutations of *ALK* are implicated as an oncogenic driver of NB.
      ii. Mutations are found in the majority of familial NB (<1% of total NB) and in 6%–10% of sporadic NB [77].
      iii. *ALK* inhibitors are included in clinical trials for ALK-mutated NB.
   c. *PHOX2B*:
      i. *PHOX2B* and *PHOX2A* are primary drivers of ANS differentiation of neural crest cells.
      ii. Germline *PHOX2B* mutations cause congenital central hypoventilation syndrome, also known as "Ondine's curse," a rare neurological disorder causing ANS dysregulation (e.g., hypoventilation), Hirschsprung disease, and NB [77].
      iii. Germline *PHOX2B* mutations are found in a subset of familial NB and up to 4% of sporadic NB [82].
   d. *ATRX*:
      i. *ATRX* mutated NB is a distinct subtype of NB in older children and adolescents.

Future testing and recommendations

The current NGS panel testing included a limited number of genes. Testing with broader gene coverage (e.g., exome sequencing, RNA sequencing) would increase testing sensitivity and treatment options.

Summary of key learning points

- Pediatric extracranial solid tumors (EST) are roughly subdivided into two groups: (1) mesenchymal-derived tumors (sarcomas), and (2) embryonal to predominantly epithelial-type tumors often suffixed "-blastomas."
- Molecular testing has been used: (1) to detect gene alterations specific to cancers (diagnostic utility), (2) to detect molecular markers that are targetable by chemotherapeutic drugs (therapeutic utility), (3) for risk stratification by providing prognostic information associated with specific gene alterations (prognostic utility), and (4) to identify gene alterations (germline mutations) associated with cancer predisposition syndromes.

References

[1] W. Allen-Rhoades, S.B. Whittle, N. Rainusso, Pediatric solid tumors in children and adolescents: an overview, Pediatr. Rev. 39 (9) (2018) 444–453.

[2] W. Allen-Rhoades, S.B. Whittle, N. Rainusso, Pediatric solid tumors of infancy: an overview, Pediatr. Rev. 39 (2) (2018) 57–67.

[3] E. Butler, et al., Recent progress in the treatment of cancer in children, CA Cancer J. Clin. 71 (4) (2021) 315–332.

[4] E.A. Sweet-Cordero, J.A. Biegel, The genomic landscape of pediatric cancers: implications for diagnosis and treatment, Science 363 (6432) (2019) 1170–1175.

[5] X. Ma, et al., Pan-cancer genome and transcriptome analyses of 1,699 paediatric leukaemias and solid tumours, Nature 555 (7696) (2018) 371–376.

[6] J. Trubicka, W. Grajkowska, B. Dembowska-Baginska, Molecular markers of pediatric solid tumors-diagnosis, optimizing treatments, and determining susceptibility: current state and future directions, Cells 11 (7) (2022).

[7] P.E. Grundy, et al., Loss of heterozygosity for chromosomes 1p and 16q is an adverse prognostic factor in favorable-histology Wilms tumor: a report from the National Wilms Tumor Study Group, J. Clin. Oncol. 23 (29) (2005) 7312–7321.

[8] N. Bown, et al., Cytogenetic abnormalities and clinical outcome in Wilms tumor: a study by the U.K. cancer cytogenetics group and the U.K. Children's Cancer Study Group, Med. Pediatr. Oncol. 38 (1) (2002) 11–21.

[9] F. Saletta, L. Dalla Pozza, J.A. Byrne, Genetic causes of cancer predisposition in children and adolescents, Transl. Pediatr. 4 (2) (2015) 67–75.

[10] T. Ripperger, et al., Childhood cancer predisposition syndromes-a concise review and recommendations by the Cancer Predisposition Working Group of the Society for Pediatric Oncology and Hematology, Am. J. Med. Genet. A 173 (4) (2017) 1017–1037.

[11] A. Bleyer, A. Viny, R. Barr, Cancer in 15- to 29-year-olds by primary site, Oncologist 11 (6) (2006) 590–601.

[12] H. Zhan, et al., A SEER-based nomogram accurately predicts prognosis in Ewing's sarcoma, Sci. Rep. 11 (1) (2021) 22723.

[13] N. Gaspar, et al., Ewing sarcoma: current management and future approaches through collaboration, J. Clin. Oncol. 33 (27) (2015) 3036–3046.

[14] M. Stahl, et al., Risk of recurrence and survival after relapse in patients with Ewing sarcoma, Pediatr. Blood Cancer 57 (4) (2011) 549–553.

[15] C.R. Antonescu, et al., Sarcomas with CIC-rearrangements are a distinct pathologic entity with aggressive outcome: a clinicopathologic and molecular study of 115 cases, Am. J. Surg. Pathol. 41 (7) (2017) 941–949.

[16] S. Le Guellec, et al., ETV4 is a useful marker for the diagnosis of CIC-rearranged undifferentiated round-cell sarcomas: a study of 127 cases including mimicking lesions, Mod. Pathol. 29 (12) (2016) 1523–1531.

[17] Y.P. Hung, C.D. Fletcher, J.L. Hornick, Evaluation of ETV4 and WT1 expression in CIC-rearranged sarcomas and histologic mimics, Mod. Pathol. 29 (11) (2016) 1324–1334.

[18] S. Sugita, et al., NUTM2A-CIC fusion small round cell sarcoma: a genetically distinct variant of CIC-rearranged sarcoma, Hum. Pathol. 65 (2017) 225–230.

[19] S. Sugita, et al., A novel CIC-FOXO4 gene fusion in undifferentiated small round cell sarcoma: a genetically distinct variant of Ewing-like sarcoma, Am. J. Surg. Pathol. 38 (11) (2014) 1571–1576.

[20] S.C. Huang, et al., Recurrent CIC gene abnormalities in angiosarcomas: a molecular study of 120 cases with concurrent investigation of PLCG1, KDR, MYC, and FLT4 gene alterations, Am. J. Surg. Pathol. 40 (5) (2016) 645–655.

[21] F. Le Loarer, et al., Clinicopathologic features of CIC-NUTM1 Sarcomas, a new molecular variant of the family of CIC-fused sarcomas, Am. J. Surg. Pathol. 43 (2) (2019) 268–276.

[22] S. Park, et al., Aberrant CDK4 amplification in refractory rhabdomyosarcoma as identified by genomic profiling, Sci. Rep. 4 (2014) 3623.

[23] F.G. Barr, et al., Genomic and clinical analyses of 2p24 and 12q13-q14 amplification in alveolar rhabdomyosarcoma: a report from the Children's Oncology Group, Genes Chromosom. Cancer 48 (8) (2009) 661–672.

[24] P. Ragazzini, et al., Amplification of CDK4, MDM2, SAS and GLI genes in leiomyosarcoma, alveolar and embryonal rhabdomyosarcoma, Histol. Histopathol. 19 (2) (2004) 401–411.

[25] M.E. Olanich, et al., CDK4 amplification reduces sensitivity to CDK4/6 Inhibition in fusion-positive rhabdomyosarcoma, Clin. Cancer Res. 21 (21) (2015) 4947–4959.

[26] J.H. Choi, J.Y. Ro, The 2020 WHO classification of tumors of soft tissue: selected changes and new entities, Adv. Anat. Pathol. 28 (1) (2021) 44–58.

[27] J.W. Tsai, et al., The expanding morphological and genetic spectrum of MYOD1-mutant spindle cell/sclerosing rhabdomyosarcomas: a clinicopathological and molecular comparison of mutated and non-mutated cases, Histopathology 74 (6) (2019) 933–943.

[28] R. Alaggio, et al., A molecular study of pediatric spindle and sclerosing rhabdomyosarcoma: identification of novel and recurrent VGLL2-related fusions in infantile cases, Am. J. Surg. Pathol. 40 (2) (2016) 224–235.

[29] N.P. Agaram, et al., MYOD1-mutant spindle cell and sclerosing rhabdomyosarcoma: an aggressive subtype irrespective of age. A reappraisal for molecular classification and risk stratification, Mod. Pathol. 32 (1) (2019) 27–36.

[30] E.R. Rudzinski, et al., Myogenin, AP2beta, NOS-1, and HMGA2 are surrogate markers of fusion status in rhabdomyosarcoma: a report from the soft tissue sarcoma committee of the children's oncology group, Am. J. Surg. Pathol. 38 (5) (2014) 654–659.

[31] S. Hettmer, et al., Rhabdomyosarcoma: current challenges and their implications for developing therapies, Cold Spring Harb. Perspect Med. 4 (11) (2014), a025650.

[32] H. Kaseb, J. Kuhn, H.M. Babiker, Rhabdomyosarcoma, StatPearls, Treasure Island, FL, 2022.

[33] J.L. Davis, et al., Infantile NTRK-associated mesenchymal tumors, Pediatr. Dev. Pathol. 21 (1) (2018) 68–78.

[34] R.C. Doebele, et al., An oncogenic NTRK fusion in a patient with soft-tissue sarcoma with response to the tropomyosin-related kinase inhibitor LOXO-101, Cancer Discov. 5 (10) (2015) 1049–1057.

[35] D. Pavlick, et al., Identification of NTRK fusions in pediatric mesenchymal tumors, Pediatr. Blood Cancer 64 (8) (2017).

[36] V. Wong, et al., Evaluation of a congenital infantile fibrosarcoma by comprehensive genomic profiling reveals an LMNA-NTRK1 gene fusion responsive to crizotinib, J. Natl. Cancer Inst. 108 (1) (2016).

[37] N. Zhou, et al., A primary undifferentiated pleomorphic sarcoma of the lumbosacral region harboring a LMNA-NTRK1 gene fusion with durable clinical response to crizotinib: a case report, BMC Cancer 18 (1) (2018) 842.

[38] A. Amatu, A. Sartore-Bianchi, S. Siena, NTRK gene fusions as novel targets of cancer therapy across multiple tumour types, ESMO Open 1 (2) (2016), e000023.

[39] A. Drilon, et al., Efficacy of larotrectinib in TRK fusion-positive cancers in adults and children, N. Engl. J. Med. 378 (8) (2018) 731–739.

[40] M.C. Hiemenz, et al., OncoKids: a comprehensive next-generation sequencing panel for pediatric malignancies, J. Mol. Diagn. 20 (6) (2018) 765–776.

[41] N.Q. Bui, et al., A clinico-genomic analysis of soft tissue sarcoma patients reveals CDKN2A deletion as a biomarker for poor prognosis, Clin. Sarcoma Res. 9 (2019) 12.

[42] L.S. Hiemcke-Jiwa, et al., USP6-associated neoplasms: a rapidly expanding family of lesions, Int. J. Surg. Pathol. 28 (8) (2020) 816–825.

[43] M. Svajdler, et al., Fibro-osseous pseudotumor of digits and myositis ossificans show consistent COL1A1-USP6 rearrangement: a clinicopathological and genetic study of 27 cases, Hum. Pathol. 88 (2019) 39–47.

[44] A.M. Oliveira, et al., USP6 and CDH11 oncogenes identify the neoplastic cell in primary aneurysmal bone cysts and are absent in so-called secondary aneurysmal bone cysts, Am. J. Pathol. 165 (5) (2004) 1773–1780.

[45] M.F. Amary, et al., Detection of USP6 gene rearrangement in nodular fasciitis: an important diagnostic tool, Virchows Arch. 463 (1) (2013) 97–98.

[46] A.V. Dilley, et al., Lipoblastoma: pathophysiology and surgical management, J. Pediatr. Surg. 36 (1) (2001) 229–231.

[47] K.Y. Kok, P.U. Telisinghe, Lipoblastoma: clinical features, treatment, and outcome, World J. Surg. 34 (7) (2010) 1517–1522.

[48] M. Warren, et al., Undifferentiated myxoid lipoblastoma with PLAG1-HAS2 fusion in an infant; morphologically mimicking primitive myxoid mesenchymal tumor of infancy (PMMTI)—diagnostic importance of cytogenetic and molecular testing and literature review, Cancer Genet. 209 (1-2) (2016) 21–29.

[49] D. Gisselsson, et al., PLAG1 alterations in lipoblastoma: involvement in varied mesenchymal cell types and evidence for alternative oncogenic mechanisms, Am. J. Pathol. 159 (3) (2001) 955–962.

[50] H. Yoshida, et al., Identification of COL3A1 and RAB2A as novel translocation partner genes of PLAG1 in lipoblastoma, Genes Chromosom. Cancer 53 (7) (2014) 606–611.

[51] M. Deen, et al., A novel PLAG1-RAD51L1 gene fusion resulting from a t(8,14)(q12;q24) in a case of lipoblastoma, Cancer Genet. 206 (6) (2013) 233–237.

[52] Y. Nitta, et al., Identification of a novel BOC-PLAG1 fusion gene in a case of lipoblastoma, Biochem. Biophys. Res. Commun. 512 (1) (2019) 49–52.

[53] L. Krskova, et al., Novel ZEB2-PLAG1 fusion gene identified by RNA sequencing in a case of lipoblastoma, Pediatr. Blood Cancer 68 (3) (2021), e28691.

[54] K. Fritchie, et al., Lipoblastomas presenting in older children and adults: analysis of 22 cases with identification of novel PLAG1 fusion partners, Mod. Pathol. 34 (3) (2021) 584–591.

[55] O. Lopez-Nunez, et al., New molecular insights into the pathogenesis of lipoblastomas: clinicopathologic, immunohistochemical, and molecular analysis in pediatric cases, Hum. Pathol. 104 (2020) 30–41.

[56] F. Pedeutour, et al., Rearrangement of HMGA2 in a case of infantile lipoblastoma without Plag1 alteration, Pediatr. Blood Cancer 58 (5) (2012) 798–800.

[57] W.D. Foulkes, J.R. Priest, T.F. Duchaine, DICER1: mutations, microRNAs and mechanisms, Nat. Rev. Cancer 14 (10) (2014) 662–672.

[58] T. Rio Frio, et al., DICER1 mutations in familial multinodular goiter with and without ovarian Sertoli-Leydig cell tumors, JAMA 305 (1) (2011) 68–77.

[59] J.R. Priest, et al., Ciliary body medulloepithelioma: four cases associated with pleuropulmonary blastoma—a report from the International Pleuropulmonary Blastoma Registry, Br. J. Ophthalmol. 95 (7) (2011) 1001–1005.

[60] D.R. Stewart, et al., Nasal chondromesenchymal hamartomas arise secondary to germline and somatic mutations of DICER1 in the pleuropulmonary blastoma tumor predisposition disorder, Hum. Genet. 133 (11) (2014) 1443–1450.

[61] M.K. Wu, et al., Biallelic DICER1 mutations occur in Wilms tumours, J. Pathol. 230 (2) (2013) 154–164.

[62] K.A.P. Schultz, et al., DICER1 and associated conditions: identification of at-risk individuals and recommended surveillance strategies, Clin. Cancer Res. 24 (10) (2018) 2251–2261.

[63] D.A. Hill, et al., DICER1 mutations in familial pleuropulmonary blastoma, Science 325 (5943) (2009) 965.

[64] A. Bahubeshi, et al., Germline DICER1 mutations and familial cystic nephroma, J. Med. Genet. 47 (12) (2010) 863–866.

[65] K.A. Schultz, et al., Ovarian sex cord-stromal tumors, pleuropulmonary blastoma and DICER1 mutations: a report from the International Pleuropulmonary Blastoma Registry, Gynecol. Oncol. 122 (2) (2011) 246–250.

[66] L. de Kock, et al., Germ-line and somatic DICER1 mutations in pineoblastoma, Acta Neuropathol. 128 (4) (2014) 583–595.

[67] L. de Kock, et al., Pituitary blastoma: a pathognomonic feature of germ-line DICER1 mutations, Acta Neuropathol. 128 (1) (2014) 111–122.

[68] M. Warren, et al., Expanding the spectrum of dicer1-associated sarcomas, Mod. Pathol. 33 (1) (2020) 164–174.

[69] L. de Kock, et al., An update on the central nervous system manifestations of DICER1 syndrome, Acta Neuropathol. 139 (4) (2020) 689–701.

[70] D.G. Evans, et al., Birth incidence and prevalence of tumor-prone syndromes: estimates from a UK family genetic register service, Am. J. Med. Genet. A 152A (2) (2010) 327–332.

[71] S.R. Plotkin, et al., Quantitative assessment of whole-body tumor burden in adult patients with neurofibromatosis, PLoS One 7 (4) (2012), e35711.

[72] N. Ortonne, et al., Cutaneous neurofibromas: current clinical and pathologic issues, Neurology 91 (2 Suppl 1) (2018) S5–S13.

[73] D.T. Miller, et al., Health supervision for children with neurofibromatosis type 1, Pediatrics 143 (5) (2019).

[74] W.H. Liang, et al., Tailoring therapy for children with neuroblastoma on the basis of risk group classification: past, present, and future, JCO Clin. Cancer Inform. 4 (2020) 895–905.

[75] J.M. Maris, Recent advances in neuroblastoma, N. Engl. J. Med. 362 (23) (2010) 2202–2211.

[76] C. Chung, et al., Neuroblastoma, Pediatr. Blood Cancer 68 (Suppl 2) (2021), e28473.

[77] M. Warren, et al., Utility of Phox2b immunohistochemical stain in neural crest tumours and non-neural crest tumours in paediatric patients, Histopathology 72 (4) (2018) 685–696.

Index

Note: Page numbers followed by *f* indicate figures.

C

D

E

N

Case Index